DIGITAL COMMUNICATIONS

Simon Haykin
McMaster University

WILEY

JOHN WILEY & SONS
New York   Chichester   Brisbane   Toronto   Singapore

Library of Congress Cataloging in Publication Data:

Haykin, Simon S., 1931–
    Digital communications / Simon Haykin.
        p. cm.
    Includes index.
    ISBN 0-471-62947-2
    1. Digital communications.   I. Title.
TK5103.7.H39 1988
621.38'041—dc19   87-28573
                                                                                        CIP

Printed and bound in the United States of America

10

ABOUT THE AUTHOR

Simon Haykin received his B.Sc. (First Class Honors), Ph.D., and D.Sc., all in electrical engineering, from the University of Birmingham, England. He is a Fellow of the Institute of Electrical and Electronics Engineers (IEEE), and a Fellow of the Royal Society of Canada. He is a recipient of the McNaughton Gold Medal, awarded by the IEEE, Region 7.

He is Professor of Electrical and Computer Engineering, and founding Director of the Communications Research Laboratory, McMaster University, Hamilton, Ontario, Canada.

Professor Haykin's research interests include detection and estimation, spectrum analysis, adaptive systems, and radar systems. He has taught undergraduate courses on Communication Systems and Digital Communications, and graduate courses on Detection and Estimation, Adaptive Systems, and Digital Signal Processing. He is the author of *Communication Systems*, Second Edition, (Wiley, 1983), *An Introduction to Analog and Digital Communications* (Wiley, 1988), *Adaptive Filter Theory* (Prentice-Hall, 1986), and the editor of *Array Signal Processing* (Prentice-Hall, 1985), and *Nonlinear Methods of Spectral Analysis*, Second Edition (Springer-Verlag, 1983).

DIGITAL
COMMUNICATIONS

ACKNOWLEDGMENTS

I thank the following reviewers: R. W. Rochelle, the University of Tennessee; P. S. Passupathy, the University of Toronto; and J. Delansky, Pennsylvania State University, who have contributed many suggestions for improvements. I am also grateful to N. S. Jayant and A. Wyner, both of AT & T Bell Laboratories, and V. J. Bhargava, the University of Victoria, for reading selected chapters of the book and for their constructive inputs.

I am indebted to my colleagues at the Communications Research Laboratory, McMaster University: R. de Buda, T. Todd, N. Secord, G. Pottie, P. Weber, K. M. Wong and D. P. Taylor for their helpful inputs.

I am grateful to the following two publishers, company, and institution for permission to reproduce selected figures in the book:

1. John Wiley for permission to reproduce the following two figures:
    (a) Figure 4.10 from the book entitled *Introduction to Communication Engineering* by R. M. Gagliardi, 1978.
    (b) Figure 7.23 from the book entitled *Introduction to Statistical Communication Theory* by J. B. Thomas, 1969.
2. Prentice-Hall and AT & T Bell Laboratories, for permission to reproduce Fig. 5.14 from the book entitled *Digital Coding of Waveforms* by N. S. Jayant and P. Noll, 1984.
3. The Institute of Electrical and Electronic Engineers for permission to reproduce the following two figures:
    (a) Figure 3.28 from J. L. Flanagan et al., "Speech Coding," *IEEE Transactions on Communications*, vol. COM-27, pp. 710–37, April 1979.
    (b) Figure 10.6 from M. Gerla, "Controlling Routes, Traffic, and Buffer Allocation in Packet Networks," *IEEE Communications Magazine*, vol. 22, pp. 11–23, November 1984.

The guidance and encouragement from my Editor, Christina Mediate, is appreciated. I am also grateful to Ann Meader, Copy Editor, and Frank Doonan, Production Supervisor, and other staff members of John Wiley for work done on the book.

Finally, I wish to thank my secretary, Lola Brooks, for her patience and extra care in typing several versions of the manuscript for the book.

Simon Haykin

PREFACE

This book is devoted to the study of principles of digital communications. The focus is on basic issues, relating theory to practice wherever possible. An effort has been made to develop the material not only in a logical and interesting manner but also to help the reader develop insight into the many facets of digital communications. Numerous examples, worked out in detail, have been included to help the reader develop an intuitive grasp of the theory under discussion. Each chapter begins with introductory remarks and (except for Chapter 1) concludes with a summary and discussion of the main points covered in the chapter. Most chapters include practical applications. Also, except for Chapter 1, each chapter includes a set of problems intended not only to help readers test their understanding of the material covered in the chapter, but also to challenge them to extend this material. For the interested reader, the text includes footnotes that provide suggestions for further reading.

The book consists of 10 chapters and 8 appendices. Many of the topics considered in the book are mature enough to justify the writing of books devoted to them individually. Indeed, this is already the case, as evidenced by the list of references and bibliography included at the end of the book. In the present book, an attempt has been made to integrate the various topics under one cover.

In the text, each reference is identified by the name(s) of the author(s) and the year of the publication. To help the reader, a list of abbreviations is also included at the end of the book.

The book is written at a level suitable for a one-semester course in digital communications, which may be taught to electrical engineering students at the undergraduate or graduate level. It is expected that the reader has a working knowledge of Fourier techniques and probability. These topics are covered in undergraduate courses on Signals and Systems, and Probability and Random Processes, which usually precede a course on communications. For the benefit of those readers who may not have such a background, reviews of these topics are included in appendices at the end of the book.

The book covers a broad range of topics in digital communications. It should therefore satisfy a variety of backgrounds and interests, thereby allowing considerable flexibility in making up the course material. For example, material for an undergraduate course on digital communications may be made up as follows:

1. Introduction to digital communications, including some historical notes: Chapter 1.
2. Geometric representation of signals and the matched filter: Sections 3.1 to 3.9.
3. The sampling process: Chapter 4.
4. Pulse-code modulation, differential pulse-code modulation, delta modulation, and applications: Sections 5.1 to 5.6, and 5.8.
5. Baseband shaping for data transmission: Chapter 6.
6. Digital modulation techniques: Chapter 7.
7. Other topics such as information theoretic concepts, coding, and data networks, selected from Chapters 2, 8, and 10.

Material for a graduate course on digital communications may be developed as follows:

1. Fundamental limits on the performance of sources and channels for digital communications: Chapter 2.
2. Statistical theory of detection and estimation, including adaptive filters: Chapter 3.
3. Robust quantization for pulse-code modulation, and coding speech at low bit rates; Sections 5.4 and 5.7.
4. Digital modulation techniques, with emphasis on quadrature modulation and M-ary modulation techniques: Chapter 7.
5. Error-control coding: Chapter 8.
6. Spread-spectrum modulation: Chapter 9.

The foregoing two lists of topics for possible courses on digital communications are offered as possible models, which would naturally be tuned or modified by the course teachers to reflect their own special interests.

As aids to the teachers of the course, the following material is available from the publisher:

1. Solutions Manual that presents detailed solutions to all of the problems in the book.
2. Master transparencies for all the figures and tables used in the book.
3. Laboratory Manual containing 6 computer-oriented experiments.

It is hoped that the book will also be found useful by practicing communications engineers.

Simon Haykin

CONTENTS

DIGITAL
COMMUNICATIONS

CHAPTER ONE

INTRODUCTION

The purpose of a *communication system* is to transport an *information-bearing signal* from a source to a user destination via a communication channel. Basically, a communication system is of an analog or digital type. In an *analog* communication system, the information-bearing signal is continuously varying in both amplitude and time, and it is used directly to modify some characteristic of a sinusoidal carrier wave, such as amplitude, phase, or frequency. In a *digital* communication system, on the other hand, the information bearing signal is processed so that it can be represented by a sequence of *discrete messages*. In this book, we are concerned with the study of digital communications.

The discipline of digital communications has experienced a phenomenal growth in both scope and application. The growth of digital communications is largely due to the following factors:

1. The impact of the *computer,* not only as a source of data but also as a tool for communications, and the demands of other digital services such as telex and facsimile.
2. The use of digital communications offers *flexibility* and *compatibility* in that the adoption of a common digital format makes it possible for a transmission system to sustain many different sources of information in a flexible manner.
3. The improved *reliability* made possible by use of digital communications.
4. The availability of *wide-band channels* provided by geostationary satellites, optical fibers, and coaxial cables.
5. The ever-increasing availability of integrated *solid-state electronic* technology, which has made it possible to increase system complexity by orders of magnitude in a cost-effective manner.

Indeed, the trend toward digital communications (and away from analog communications) will continue, so much so that the second half of the twentieth century will be recorded in history as the era of digital communications.

In this introductory chapter, we present an overview of digital communications and some historical notes on their evolution. We begin the chapter by describing types of sources and signals encountered in communications.

1

1.1 SOURCES AND SIGNALS

A *source of information* generates a *message,* examples of which include human voice, television picture, teletype data, atmospheric temperature and pressure. In these examples, the message is not electrical in nature, and so a *transducer* is used to convert it into an electrical waveform called the *message signal*. The waveform is also referred to as a *baseband signal;* the term "baseband" is used to designate the band of frequencies representing the message signal generated at the source.

The message signal can be of an *analog* or *digital* type. An analog signal is one in which both amplitudes and time vary *continuously* over their respective intervals. A speech signal, a television signal, and a signal representing atmospheric temperature or pressure at some location are examples of analog signals. In a digital signal, on the other hand, both amplitude and time take on *discrete values.* Computer data and telegraph signals are examples of digital signals.

An analog signal can always be converted into digital form by combining three basic operations: *sampling, quantizing,* and *encoding,* as shown in the block diagram of Fig. 1.1. In the sampling operation, only *sample values* of the analog signal at uniformly spaced discrete instants of time are retained. In the quantizing operation, each sample value is approximated by the nearest level in a *finite set of discrete levels.* In the encoding operation, the selected level is represented by a *code word* that consists of a prescribed number of *code elements.* The analog-to-digital conversion process so described is illustrated in Fig. 1.2. Part (*a*) of the figure shows a segment of an analog waveform. Part (*b*) shows the corresponding digital waveform, based on the use of a *binary code.* In this example, symbols 0 and 1 of the binary code are represented by zero and 1 volt, respectively. The code word consists of four *binary digits* (bits), with the last bit assigned the role of a *sign bit* that signifies whether the sample value in question is positive or negative. The remaining three bits are chosen to provide a numerical representation for the absolute value of a sample in accordance with Table 1.1.

As a result of sampling and quantizing operations, *errors* are introduced into the digital signal. These errors are *nonreversible* in that it is not possible to produce an exact replica of the original analog signal from its digital representation. However, the errors are under a designer's control. Indeed, by proper selections of the sampling rate and code-word length (i.e., number of quantizing levels), the errors due to sampling and quantizing can be made so small that the difference between the analog signal and its digital reconstruction is not discernible by a human observer.

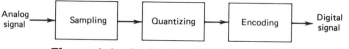

Figure 1.1 Analog-to-digital conversion.

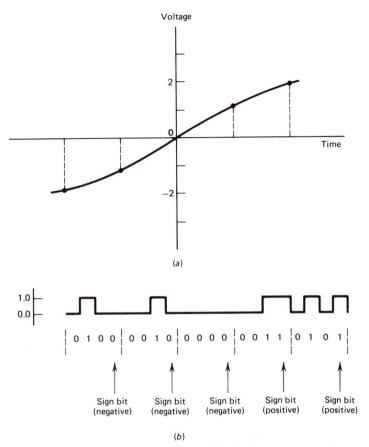

(a)

(b)

Figure 1.2   (a) Analog waveform. (b) Digital representation.

Table 1.1   Binary Representation of Quantized Levels

Ordinal Number of Quantized Level	Level Number Expressed as Sum of Powers of 2	Binary Number
0		000
1	2^0	001
2	2^1	010
3	$2^1 + 2^0$	011
4	2^2	100
5	$2^2 \quad + 2^0$	101
6	$2^2 + 2^1$	110
7	$2^2 + 2^1 + 2^0$	111

1.2 BASIC SIGNAL PROCESSING OPERATIONS IN DIGITAL COMMUNICATIONS

Figure 1.3 shows the block diagram of a digital communication system. In this diagram, three basic signal-processing operations are identified: *source coding, channel coding,* and *modulation.* It is assumed that the source of information is digital by nature or converted into it by design.

In source coding, the *encoder* maps the digital signal generated at the source output into another signal in digital form. The mapping is one-to-one, and the objective is to eliminate or reduce redundancy so as to provide an *efficient representation* of the source output. Since the source encoder mapping is one-to-one, the *source decoder* simply performs the inverse mapping and thereby delivers to the user destination a reproduction of the original digital source output. The primary benefit thus gained from the application of source coding is a reduced bandwidth requirement.

In channel coding, the objective is for the *encoder* to map the incoming digital signal into a channel input and for the *decoder* to map the channel output into an output digital signal in such a way that the effect of channel noise is minimized. That is, the combined role of the channel encoder and decoder is to provide for *reliable communication* over a noisy channel. This provision is satisfied by introducing *redundancy* in a prescribed fashion in the channel encoder and exploiting it in the decoder to reconstruct the original encoder input as accurately as possible. Thus, in source coding, we remove redundancy, whereas in channel coding, we introduce controlled redundancy.

Clearly, we may perform source coding alone, channel coding alone, or the two together. In the latter case, naturally, the source encoding is performed first, followed by channel encoding in the transmitter as illustrated in Fig. 1.3. In the receiver, we proceed in the reverse order; channel decoding is performed first, followed by source decoding. Whichever combination is used, the result-

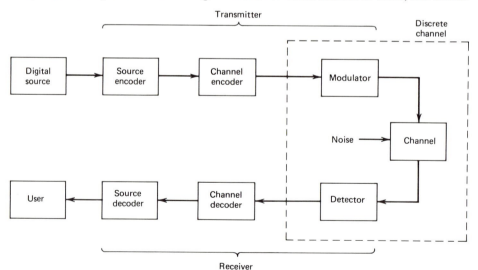

Figure 1.3 Block diagram of digital communication system.

ing improvement in system performance is achieved at the cost of increased circuit complexity.

As for modulation, it is performed with the purpose of providing for the *efficient transmission* of the signal over the channel. In particular, the *modulator* (constituting the last stage of the transmitter in Fig. 1.3) operates by keying shifts in the *amplitude, frequency,* or *phase* of a sinusoidal carrier wave to the channel encoder output. The digital modulation technique for so doing is referred to as *amplitude-shift keying, frequency-shift keying,* or *phase-shift keying,* respectively. The detector (constituting the first stage of the receiver in Fig. 1.3) performs demodulation (the inverse of modulation), thereby producing a signal that follows the time variations in the channel encoder output (except for the effects of noise).

The combination of modulator, channel, and detector, enclosed inside the dashed rectangle shown in Fig. 1.3, is called a *discrete channel*. It is so called since both its input and output signals are in discrete form.

Traditionally, coding and modulation are performed as separate operations, and the introduction of redundant symbols by the channel encoder appears to imply increased transmission bandwidth. In some applications, however, these two operations are performed as one function in such a way that the transmission bandwidth need not be increased. In situations of this kind, we define the joint function of the channel encoder and modulator as the *imposition of distinct patterns on the transmitted signal,* which are discernible by the combined action of the channel decoder and detector in the receiver.

1.3   CHANNELS FOR DIGITAL COMMUNICATIONS

The details of modulation and coding used in a digital communication system depend on the characteristics of the channel and the application of interest. The two channel characteristics, *bandwidth* and *power,* constitute the *primary communication resources* available to the designer. Other channel characteristics of particular concern are the degree to which the amplitude and phase responses of the channel are determined, whether the channel is *linear* or *nonlinear,* and how free the channel is from external *interference.* In the sequel, we discuss these issues in the context of five specific channels: telephone channels, coaxial cables, optical fibers, microwave radio, and satellite channels.*

A *telephone channel* is designed to provide voice-grade communication. During the past 100 years, it has evolved into a worldwide network that encompasses a variety of transmission media (open-wire lines, cables, optical fibers, microwave radio, and satellites) and a complex of switching systems. This makes the telephone channel an excellent candidate for data communication over long distances. The channel has a band-pass characteristic occupying the frequency range 300 to 3400 Hz, a high signal-to-noise ratio of about 30 dB, and approximately linear response. The fine tuning of the channel to accommodate

---

* For a discussion of transmission media for digital communications, see Chapter 1 entitled "Transmission Media" by D. N. Hatfield in the book edited by Bartee (1985).

the transmission of voice signals results in a flat amplitude response over the pass-band of the channel. However, no particular attention is given to the phase response since the ear is relatively insensitive to phase-delay variations. On the other hand, data and image transmissions are strongly influenced by phase-delay variations. Accordingly, the efficient transmission of such signals over a telephone channel requires the use of an *equalizer* designed to maintain a flat amplitude response and a linear phase response over the frequency band of interest. Transmission rates up to 16.8 kilobits per second (kb/s) have been achieved over telephone lines by combining the use of sophisticated modulation techniques with *adaptive* equalization. The term "adaptive" refers to the fact that the coefficients of the equalizer vary in accordance with the operating conditions, such that efficient transmission is maintained.

A *coaxial cable* consists of a single-wire conductor centered inside an outer conductor, which are insulated from each other by means of a dielectric material. The two primary advantages of coaxial cables as a transmission medium are a relatively wide bandwidth and freedom from external interference. However, their operational characteristics require the use of closely spaced *repeaters*. Indeed, efficient digital transmission systems using coaxial cables have been built to operate at a *data rate* of 274 megabits per second (Mb/s), with repeaters spaced at 1 km intervals.

An *optical fiber* consists of a very fine inner *core* made of silica glass, surrounded by a concentric layer called *cladding* that is also made of glass. The glass in the core has a refractive index (or optical density) slightly higher than that of the glass in the cladding. A basic property of light is that when a ray of light passes from a medium of high refractive index to another medium of low refractive index, the ray is bent back toward the medium with the higher refractive index. Accordingly, if a ray of light is launched into an optical fiber at the right oblique acceptance angle, it is continually refracted into the core by the cladding. That is, the difference between the refractive indices of the core and the cladding helps guide the propagation of the ray of light inside the core of the fiber from one end to the other. Compared to coaxial cables, optical fibers are smaller and they offer higher transmission bandwidths and longer repeater separations.

A *microwave radio,* operating on a line-of-sight link, consists basically of a transmitter and a receiver that are equipped with antennas of their own. The antennas are placed on towers at sufficient height to have the transmitter and receiver in *line-of-sight* of each other. The operating frequencies range from about 1 to 30 GHz. For a large fraction of the time in most locations, the propagation conditions do not vary significantly from a direct (line-of-sight) path between the transmitter and receiver. Under such conditions, the radio channel operates as a nondispersive transmission medium, capable of highly reliable, high-speed digital transmission. At other times, however, anomalous propagation conditions develop in the channel due to meteorological variations, causing severe degradation in the radio system's performance. These conditions manifest themselves in a phenomenon known as *multipath fading*. The term "multipath" refers to the fact that propagation between the transmitter and receiver takes place along several paths of different electrical lengths.

The receiver thus sees a weighted sum of delayed replicas of the transmitted signal from these multiple paths, interfering with each other constructively or destructively. Consequently, the received signal experiences "fading" in that its amplitude varies with time. When the replicas arrive at the receiver *in-phase*, they reinforce each other; however, when they arrive in *anti-phase*, they cancel each other. In order to design a digital radio system to work in such an environment, it is therefore necessary that provisions be made to overcome the effects of multipath fading.

A *satellite channel* consists of a satellite in geostationary orbit, an uplink from a ground station, and a downlink to another ground station. Typically, the uplink and the downlink operate at microwave frequencies, with the uplink frequency higher than the downlink frequency. On board the satellite there is a low-power amplifier, which is usually operated in its nonlinear mode for high efficiency. Thus, the satellite channel may be viewed as a repeater in the sky, permitting communication (from one ground station to another) over long distances at high bandwidths and relatively low cost. The nonlinear nature of the channel restricts its use to constant envelope modulation techniques (i.e., phase modulation, frequency modulation).

The five channels described illustrate the diverse nature of physical media that support the transmission of digital data. The list of channels for data transmission include many others that offer unique features of their own.

1.4 SOME HISTORICAL NOTES

In this section, we present some historical notes on communications, with emphasis on digital communications and related issues. The material is organized under separate categories.

(1) Binary Code

The origins of the binary code, basic to the operation of digital communications, may be traced back to the early work of Francis Bacon at the beginning of the seventeenth century. Bacon used five-letter combinations of only two distinct letters (symbols) to represent 24 letters of the alphabet as a means of encoding secret messages. Bacon's two-letter alphabet was published in 1605. Later on, in 1641, John Wilkins published a book in which Bacon's two-letter alphabet was not only explained but also expanded by employing three- and five-letter alphabets. It was thus demonstrated for the first time that more code elements are needed for the average code-word length to become shorter.

In 1703, Gottfried Wilhelm Leibnitz gave a lecture to the Royal Academy of Sciences in Paris, entitled "Explication de l'arithmetique binaire." The text of his lecture was published in the proceedings of the Academy in 1705. Leibnitz used the numbers 0 and 1 for his binary code. It appears that Leibnitz's binary code was developed independently from Bacon and Wilkins.*

---

* For more details on the history of the binary code, see the following two references: Heath (1972), Aschoff (1983).

(2) Telegraphy

In 1837, the telegraph was perfected by Samuel Morse. With the words "What hath God wrought," transmitted by Morse's electric telegraph between Washington and Baltimore in 1844, a completely revolutionary means of real-time, long-distance communication was triggered. The telegraph is the forerunner of digital communications in that the *Morse code* is a *variable-length* binary code utilizing two symbols, a *dot* and a *dash,* which are represented by short and long electrical pulses, respectively. This type of signaling is ideal for manual keying. Subsequently, Emile Baudot developed a *fixed-length* binary code for telegraphy in 1875. In Baudot's telegraphic code, well-suited for use with teletypewriters, each code word consists of five equal-length code elements, and each element is assigned one of two possible states: a *mark* or a *space* (i.e., symbol 1 or 0 in today's terminology).

(3) Telephony

In 1874, the *telephone* was conceived by Alexander Graham Bell in Brantford, Ontario, and it was born in Boston, Massachusetts, in 1875. The telephone made real-time transmission of speech by electrical encoding and replication of sound a practical reality. The first version of the telephone was crude and weak, enabling people to talk over short distances only. The quality and range of the telephone was greatly enhanced by the invention and development of the carbon microphone and the induction coil during 1877–1890. The use of loading coils in transmission lines made telephony possible over distances extending up to 2000 miles. However, it was not until 1913 that transcontinental telephony came into being with the use of the then-new electronic amplification (using vacuum tubes).

When telephone service was only a few years old, interest developed in automating it. Notably, in 1897, Strowger devised the automatic *step-by-step switch* that bears his name. Of all the electromechanical switches devised over the years, the Strowger switch was the most popular and widely used.

The invention of the transistor in 1948 spurred the application of electronics to switching. The motivation was to improve reliability, increase capacity, and reduce cost. The first call through a stored-program system was placed in March 1958 at Bell Laboratories. The first commercial telephone service with digital switching began in Morris, Illinois, in June 1960.

In 1937, Alec Reeves invented *pulse-code modulation* for the digital encoding of speech signals. The technique was developed during World War II to enable the encryption of speech signals. Indeed, a full-scale, 24-channel system was used in the field by the United States military at the end of the war. In 1945, DeLoraine invented *time-division multiplexing,* a natural for digital carrier systems. Here too, the first application was in integrated military tactical telephone networks.

The concept of integrated digital transmission and switching was proposed in the late 1950s, about the same time as the deployment of the first digital switching system. The move toward an *integrated digital network* (IDN) was motivated by expectations of continued reductions in component and interface costs. The first step to extend the digital network to the telephones was made in 1974.*

The challenge for the future is to meet the needs for a broad range of services (data, voice, and video) that are being proposed. It is assumed that it is economical to convert the signals generated by these services into digital form for digital transmission. The part that new switching approaches will play in an *integrated services digital network* (ISDN) will be determined by the desired degree of integration.

(4)   Radio

In 1864, James Clerk Maxwell formulated the electromagnetic theory of light and predicted the existence of radio waves. The existence of radio waves was established by Heinrich Hertz in 1887. Then, on December 12, 1901, Guglialmo Marconi received a radio signal at Signal Hall in Newfoundland; the radio signal had originated in Cornwall, England, 1700 miles away. The way was thereby opened toward a tremendous broadening of the scope of communications.

It appears that digital modulation techniques were first employed for microwave radio transmission in France in the 1930s. Then, after a long pause, *digital radio* (i.e., digital communications by radio) experienced a renaissance in the early 1970s. Renewed interest in microwave radio as a transmission medium for digital communications was largely due to the introduction of digital switching at that time.

(5)   Satellite Communications

In 1945, Arthur C. Clarke proposed the idea of using an *earth-orbiting* satellite as a relay point for communication between two earth stations. In 1957, the Soviet Union launched Sputnik I, which transmitted telemetry signals for 21 days. This was followed shortly by the launching of Explorer I by the United States in 1958, which transmitted telemetry signals for about five months. A major experimental step in communications satellite technology was taken with the launching of Telstar I from Cape Canaveral on July 10, 1962. The Telstar satellite was built by the Bell Telephone Laboratories, which had acquired considerable knowledge from pioneering work by John R. Pierce. The satellite was capable of relaying TV programs across the Atlantic; this was made possible only through the use of maser receivers and large antennas. In July 1964, INTELSAT, a multinational organization, was formed. The purpose of INTELSAT was to design, develop, construct, establish, and maintain the operation of the space segment of a global commercial communications satellite system. Early Bird (INTELSAT I), a geostationary communications satellite, was launched in April 1965. In a period of seven years, four generations of

---

* This historical account of telecommunication switching is based on Joel (1984).

satellites (INTELSAT I through IV) were launched and placed in commercial operation.* Capacity increased from 240 telephone circuits and 1 television channel in INTELSAT I to 6000 telephone circuits and 12 TV channels in INTELSAT IV. This spectacular growth was made possible by increased spacecraft size, weight, and power, increased antenna gain, and increased transponder bandwidth. INTELSAT V provided the first international use of *time-division multiple-access* (TDMA), a digital technique for accessing a geostationary satellite.

(6) Optical Communications

The use of optical means (e.g., smoke and fire signals) for the transmission of information dates back to prehistoric times. However, no major breakthrough in optical communications was made until 1966, when Kao and Hockham proposed the use of a clad glass fiber as a dielectric waveguide. The *laser* (acronym for *l*ight *a*mplification by *s*timulated *e*mission of *r*adiation) had been invented and developed in 1959 and 1960. Kao and Hockham pointed out that (1) the attenuation in an optical fiber was due to impurities in the glass, and (2) the intrinsic loss, determined by Rayleigh scattering, is very low. Indeed, they predicted that a loss of 20 dB/km should be attainable. This was a remarkable prediction, made at a time when the power loss in a glass fiber was about 1000 dB/km. In 1970, Kapron, Keck, and Maurer of Corning Glass Works achieved the goal of 20 dB/km by fabricating a silica-doped clad fiber.†

(7) Computer Communications

Computers and terminals started communicating with each other over long distances in the early 1950s. The links used were initially voice-grade telephone channels operating at low speeds (300 to 1200 b/s). Today, telephone channels are routinely used to support data transmission at rates of 9.6 kb/s or even as high as 16.8 kb/s. Various factors have contributed to this dramatic increase in data transmission rates. Notable among them are the idea of *adaptive equalization*, pioneered by Lucky in 1965, and efficient modulation techniques.

Another idea widely employed in computer communications is that of *automatic request for retransmission* (ARQ). The ARQ method was originally devised by van Duuren‡ during World War II, and published in 1946. It was used to improve radio-telephony for telex transmission over long distances. van Duuren's alphabet had 35 letters; all combinations of four marks and three spaces were interpreted by the receiver as legal, and all other combinations led to a repeat request.

During 1950–1970, various studies were made on *computer networks*. How-

---

* For a historical account of satellite communications, see Pritchard (1984).

† For an assessment of optical communications from conception to prominence in 20 years, see Schwartz (1984).

‡ van Duuren's contribution is described in Stumpers (1984).

ever, the most significant of them in terms of impact on computer communications was the *Advanced Research Project Agency Net*work (ARPANET), first put into service in 1971. The development of ARPANET was sponsored by the Advanced Research Projects Agency of the U.S. Department of Defense. The pioneering work on *packet switching* was done on ARPANET; packet switching provides a method for efficient utilization of the capacity of a data communication network.*

The ARPANET is an example of *point-to-point store-and-forward* network. Another type of packet transmission network is the (single-hop) *multiple-access network,* which is exemplified by the *ALOHA network* invented by Abramson and Kuo in the early 1970s. Yet a third type of packet transmission network is the (multiple-hop) *store-and-forward multiple-access network* that combines the features of the first two types. An example of this third type is the *packet radio network* (PRNET), sponsored by the Advanced Research Projects Agency.

(8) Theory

In 1928, Harry Nyquist published a classic paper on the theory of signal transmission in telegraphy.† In particular, Nyquist developed criteria for the correct reception of telegraph signals transmitted over dispersive channels in the absence of noise. Much of Nyquist's early work was applied later to the transmission of digital data.

In 1943, North devised the *matched filter* for the optimum detection of a known signal in additive white noise.‡ A similar result was obtained in 1946 independently by Van Vleck and Middleton, who coined the term "matched filter."£

The development of *detection theory* relies heavily on the use of decision-theoretic concepts. *Statistical decision theory* is a major area of statistical inference, dating back to the classic work of Thomas Bayes in the middle of the eighteenth century.§ Significant contributions to statistical decision theory were also made by Neyman and Pearson in the early 1930s.**

Another major area of statistical inference, which is important to the study of statistical communication theory, is *estimation theory*. Significant contributions in this area were made by Fisher in the mid 1920s; in particular, he originated a widely used method known as *maximum likelihood estimation*.††

---

* For historical accounts of computer communications, see the following references: Moreau (1984), Green (1984).

† Nyquist (1928).

‡ North's derivation of the matched filter was first described in a classified report (RAC Laboratories Report PTR-6C, June 1943). The report was published 20 years later in the open literature (North, 1963).

£ Van Vleck and Middleton (1946).

§ See Bayes (1764).

** See Neyman and Pearson (1933).

†† See Fisher (1925).

Then, during the late 1930s and early 1940s, the first studies of *minimum mean-square estimation* in stochastic processes were made independently by Norbert Wiener in the United States and Kolmogorov in the Soviet Union.* Wiener formulated the continuous-time linear prediction problem and derived an explicit formula for the optimum prediction in 1942. Wiener also considered the "filtering" problem of estimating a process corrupted by an additive "noise" process. The explicit formula for the optimum estimate required the solution of an integral equation known as the Wiener-Hopf equation. Kolmogorov, on the other hand, developed a comprehensive treatment of the linear prediction problem for discrete-time stochastic processes in 1939.

In 1947, the geometric representation of signals was developed by Kotel'nikov in a doctoral dissertation presented before the Academic Council of the Molotov Energy Institute in Moscow.† This method was subsequently brought to full fruition by Wozencraft and Jacobs in a landmark textbook published in 1965.‡

In 1948, the theoretical foundations of digital communications were laid by Claude Shannon in a paper entitled "A Mathematical Theory of Communication." Shannon's paper£ was received with immediate and enthusiastic acclaim. It was perhaps this response that emboldened Shannon to amend the title of his paper to "The Mathematical Theory of Communication," when it was reprinted a year later in book form.§

Digital communications has a rich history. The material presented in this section provides some highlights of this history, hoping that it serves as a source of inspiration and motivation for the reader.

---

* For a historical review of minimum mean-square estimation theory and related issues, see Kailath (1974).

† A translation of the original doctoral dissertation of Kotel'nikov was published by Dover in 1960 (Kotel'nikov, 1960).

‡ Wozencraft and Jacobs (1965), pp. 211–284.

£ Shannon's paper was originally published in two parts in the *Bell System Technical Journal* (Shannon, 1948). The paper, with the correction of minor errata and the inclusion of some additional references, was reproduced in the form of a book (Shannon and Weaver, 1949).

§ For an assessment of Shannon's 1948 paper, see (Massey, 1984).

CHAPTER TWO

FUNDAMENTAL LIMITS ON PERFORMANCE

A key issue in evaluating the performance of a digital communication system concerns the efficiency with which information from a given source can be represented. Another key issue pertains to the rate at which information can be transmitted reliably over a noisy channel. The *fundamental limits* on these key aspects of system performance have their roots in *information theory,* which has developed into a mature discipline in its own right. Information theory, originally known as the *mathematical theory of communication,* deals only with mathematical modeling and analysis of a communication system rather than with physical sources and physical channels. Specifically, given an *information source* and a *noisy channel,* information theory provides limits on (1) the *minimum number of bits per symbol* required to fully *represent* the source, and (2) the *maximum rate* at which *reliable* communication can take place over the channel.

In this chapter, we discuss fundamental concepts and limits in information theory* in the context of a digital communication system. It is logical that we begin the discussion with a measure of information, which we do in the next section.

2.1 UNCERTAINTY, INFORMATION, AND ENTROPY

Suppose that a *probabilistic experiment* involves the observation of the output emitted by a discrete source every unit of time (signaling interval). The source output is modeled as a discrete random variable, S, which takes on symbols from a fixed finite *alphabet*

$$\mathscr{S} = \{s_0, s_1, \ldots, s_{K-1}\} \tag{2.1}$$

---

* For introductory treatment of information theory, see Wyner (1981), Chapter 15 of Blake (1979), and the books of Hamming (1980) and Abramson (1963). For advanced treatment of the subject, see Chapter 1 of Viterbi and Omura (1979), and the books of McEliece (1977), Gallager (1968), and Ash (1965). For a collection of papers on the development of information theory (including the 1948 classical paper by Shannon), see Slepian (1973).

with probabilities

$$P(S = s_k) = p_k \qquad k = 0, 1, \ldots, K - 1 \tag{2.2}$$

Of course, this set of probabilities must satisfy the condition

$$\sum_{k=0}^{K-1} p_k = 1 \tag{2.3}$$

We assume that the symbols emitted by the source during successive signaling intervals are statistically independent. A source having the properties just described is called a *discrete memoryless source,** memoryless in the sense that the symbol emitted at any time is independent of previous choices.

Can we find a measure of how much "information" is produced by such a source? The idea of information is closely related to that of "uncertainty" or "surprise," as described next.

Consider the event $S = s_k$, describing the emission of symbol s_k by the source with probability p_k, as defined in Eq. 2.2. Clearly, if the probability $p_k = 1$ and $p_i = 0$ for all $i \neq k$, then there is no "surprise" and therefore no "information" when symbol s_k is emitted, since we know what the message from the source must be. If, on the other hand, the source symbols occur with different probabilities, and the probability p_k is low, then there is more "surprise" and therefore "information" when symbol s_k is emitted by the source than when symbol s_i, $i \neq k$, with higher probability is emitted. Thus, the words "uncertainty," "surprise," and "information" are all related. Before the event $S = s_k$ occurs, there is the amount of uncertainty. When the event $S = s_k$ occurs there is the amount of surprise. After the occurrence of the event $S = s_k$, there is gain in the amount of information. All three amounts are obviously the same. Moreover, the amount of information is related to the *inverse* of the probability of occurrence.

We define the amount of information gained after observing the event $S = s_k$, which occurs with probability p_k, as the *logarithmic* function†

$$I(s_k) = \log\left(\frac{1}{p_k}\right) \tag{2.4}$$

This definition exhibits the following important properties that are intuitively satisfying:

1.
$$I(s_k) = 0 \quad \text{for } p_k = 1 \tag{2.5}$$

Obviously, if we are absolutely *certain* of the outcome of an event, even before it occurs, there is *no* information gained.

---

* The discrete memoryless source is rather restrictive for some applications. A more general type of information source is one in which the occurrence of a source symbol s_k may depend upon m of the preceding symbols. Such a source is called a *Markov source of order m*. For a discussion of Markov information sources, see Abramson (1963, pp. 22–33).

† The use of a logarithmic measure of information was first suggested by Hartley (1928); however, Hartley used logarithms to base 10.

2.
$$I(s_k) \geq 0 \quad \text{for } 0 \leq p_k \leq 1 \tag{2.6}$$

That is to say, the occurrence of an event $S = s_k$ either provides some or no information but never brings about a *loss* of information.

3.
$$I(s_k) > I(s_i) \quad \text{for } p_k < p_i \tag{2.7}$$

That is, the less probable an event is, the more information we gain when it occurs.

4. $I(s_k s_l) = I(s_k) + I(s_l)$, if s_k and s_l are statistically independent.

The base of the logarithm in Eq. 2.4 is quite arbitrary. Nevertheless, it is the standard practice today to use a logarithm to base 2. The resulting unit of information is called the *bit* (a contraction of *binary unit*). We thus write

$$I(s_k) = \log_2\left(\frac{1}{p_k}\right) \qquad k = 0, 1, \ldots, K - 1 \tag{2.8}$$

When $p_k = 1/2$, we have $I(s_k) = 1$ bit. Hence, *one bit is the amount of information that we gain when one of two possible and equally likely (i.e., equiprobable) events occurs.* Note, however, that a bit also refers to a binary digit. In this book, we will use the term "bit" as a unit of information when dealing with the information content of a source or channel output, and the term "bit" as an acronym for *binary digit* when dealing with a sequence of 1s and 0s.

The amount of information, $I(s_k)$, produced by the source during an arbitrary signaling interval depends on the symbol s_k emitted by the source at that time. Indeed, $I(s_k)$ is a discrete random variable that takes on the values $I(s_0)$, $I(s_1)$, $\ldots$, $I(s_{K-1})$ with probabilities p_0, p_1, $\ldots$, p_{K-1}, respectively. The mean value of $I(s_k)$ over the source alphabet $\mathcal{S}$ is given by

$$H(\mathcal{S}) = E[I(s_k)]$$

$$= \sum_{k=0}^{K-1} p_k I(s_k)$$

$$= \sum_{k=0}^{K-1} p_k \log_2\left(\frac{1}{p_k}\right) \tag{2.9}$$

The important quantity, $H(\mathcal{S})$, is called the *entropy** of a discrete memoryless source with source alphabet $\mathcal{S}$. It is a measure of the *average information content per source symbol*. Note that the entropy $H(\mathcal{S})$ depends only on the probabilities of the symbols in the alphabet $\mathcal{S}$ of the source. Thus, $\mathcal{S}$ in $H(\mathcal{S})$ is not an argument of a function but rather a label for a source.

---

* In *statistical physics*, the *entropy* of a physical system is defined by (Reif, 1965, p. 147)

$$\mathcal{S} = k \ln \Omega$$

where k is Boltzmann's constant, Ω is the number of states accessible to the system, and ln denotes the natural logarithm. This entropy has the dimensions of energy because its definition involves the constant k. In particular, it provides a *quantitative measure of the degree of randomness of the system.*

Comparing the entropy of statistical physics with that of information theory, we see that they have a similar form. For a detailed discussion of the relation between them, see Pierce (1961, pp. 184–207).

(1)   Some Properties of Entropy

Consider a discrete memoryless source whose mathematical model is defined by Eqs. 2.1–2.2. The entropy $H(\mathcal{S})$ of such a source is bounded as follows

$$0 \leqslant H(\mathcal{S}) \leqslant \log_2 K \tag{2.10}$$

where K is the *radix* (number of symbols) of the alphabet $\mathcal{S}$ of the source. Furthermore, we may state that

1. $H(\mathcal{S}) = 0$, if and only if the probability $p_k = 1$ for some k, and the remaining probabilities in the set are all zero. This lower bound on entropy corresponds to *no uncertainty*.
2. $H(\mathcal{S}) = \log_2 K$, if and only if $p_k = 1/K$ for all k (i.e., all the symbols in the alphabet $\mathcal{S}$ are *equiprobable*). This upper bound on entropy corresponds to *maximum uncertainty*.

To prove these properties of $H(\mathcal{S})$, we proceed as follows. First, since each probability p_k is less than or equal to unity, it follows that each term $p_k \log_2(1/p_k)$ in Eq. 2.9 is always nonnegative, and so $H(\mathcal{S}) \geqslant 0$. Next, we note that $p_k \log_2(1/p_k) = 0$ if, and only if, $p_k = 0$ or 1. We therefore deduce that $H(\mathcal{S}) = 0$ if, and only if, $p_k = 0$ or 1, that is, $p_k = 1$ for some k and all the rest are zero. This completes the proofs of the lower bound in Eq. 2.10 and point (1).

To prove the upper bound in Eq. 2.10 and point (2), we shall make use of a property of the logarithm:

$$\ln x \leqslant x - 1 \qquad x \geqslant 0 \tag{2.11}$$

This inequality can be readily verified by plotting the functions $\ln x$ and $(x - 1)$ versus x, as shown in Fig. 2.1. Here we see that the line $y = x - 1$ always lies above the curve $y = \ln x$. The equality holds *only* at the point $x = 1$, where the line is tangential to the curve.

To proceed with the proof, consider first any two probability distributions $\{p_0, p_1, \ldots, p_{K-1}\}$ and $\{q_0, q_1, \ldots, q_{K-1}\}$ on the alphabet $\mathcal{S} = \{s_1, s_2, \ldots, s_K\}$ of a discrete memoryless source. We may then write

$$\sum_{k=0}^{K-1} p_k \log_2\left(\frac{q_k}{p_k}\right) = \frac{1}{\log_2 e} \sum_{k=0}^{K-1} p_k \ln\left(\frac{q_k}{p_k}\right)$$

where e is the base of the natural algorithm. Hence, using the inequality of Eq. 2.11, we get

$$\sum_{k=0}^{K-1} p_k \log_2\left(\frac{q_k}{p_k}\right) \leqslant \frac{1}{\log_2 e} \sum_{k=0}^{K-1} p_k \left(\frac{q_k}{p_k} - 1\right)$$

$$\leqslant \frac{1}{\log_2 e} \sum_{k=0}^{K-1} (q_k - p_k)$$

$$\leqslant \frac{1}{\log_2 e} \left(\sum_{k=0}^{K-1} q_k - \sum_{k=0}^{K-1} p_k\right) = 0$$

We thus have the *fundamental inequality*

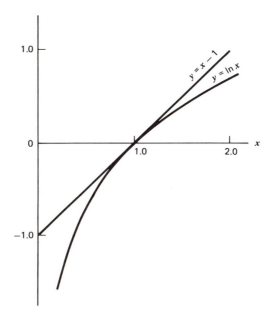

Figure 2.1 Plots of the functions $x - 1$ and $\ln x$.

$$\sum_{k=0}^{K-1} p_k \log_2\left(\frac{q_k}{p_k}\right) \leq 0 \qquad (2.12)$$

where the equality holds if $q_k = p_k$ for all k.

Suppose we next put

$$q_k = \frac{1}{K} \qquad k = 0, 1, \ldots, K - 1 \qquad (2.13)$$

which corresponds to an alphabet $\mathcal{S}$ with *equiprobable* symbols. The entropy of a discrete memoryless source with such a characterization equals

$$\sum_{k=0}^{K-1} q_k \log_2\left(\frac{1}{q_k}\right) = \log_2 K \qquad (2.14)$$

Also, the use of Eq. 2.13 in Eq. 2.12 yields

$$\sum_{k=0}^{K-1} p_k \log_2\left(\frac{1}{p_k}\right) \leq \log_2 K$$

Equivalently, the entropy of a discrete memoryless source with an arbitrary probability distribution for the symbols of its alphabet $\mathcal{S}$ is bounded as

$$H(\mathcal{S}) \leq \log_2 K$$

Thus, $H(\mathcal{S})$ is always less than or equal to $\log_2 K$. The equality holds only if the symbols in the alphabet $\mathcal{S}$ are equiprobable, as in Eq. 2.13. This completes the proof of Eq. 2.11 and points (1) and (2).

EXAMPLE 1 ENTROPY OF BINARY MEMORYLESS SOURCE

To illustrate the properties of $H(\mathcal{S})$, consider a binary source for which symbol 0 occurs with probability p_0 and symbol 1 with probability $p_1 = 1 - p_0$. We assume that the source is memoryless so that successive symbols emitted by the source are statistically independent.

The entropy of such a source equals

$$H(\mathcal{S}) = -p_0\log_2 p_0 - p_1\log_2 p_1$$
$$= -p_0\log_2 p_0 - (1 - p_0)\log_2(1 - p_0), \text{ bits} \qquad (2.15)$$

We note that

1. When $p_0 = 0$, the entropy $H(\mathcal{S}) = 0$. This follows from the fact that $x \log x \to 0$ as $x \to 0$.
2. When $p_0 = 1$, the entropy $H(\mathcal{S}) = 0$.
3. The entropy $H(\mathcal{S})$ attains its maximum value, $H_{max} = 1$ bit, when $p_1 = p_0 = 1/2$, that is, symbols 1 and 0 are equally probable.

The function of p_0 given in Eq. 2.15 is frequently encountered in information theory problems. It is therefore customary to assign a special symbol to this function. Specifically, we define

$$\mathcal{H}(p_0) = -p_0\log_2 p_0 - (1 - p_0)\log_2(1 - p_0) \qquad (2.16)$$

We refer to $\mathcal{H}(p_0)$ as the *entropy function*. The distinction between Eq. 2.15 and Eq. 2.16 should be carefully noted. The $H(\mathcal{S})$ of Eq. 2.15 gives the entropy of a discrete memoryless source with source alphabet $\mathcal{S}$. The $\mathcal{H}(p_0)$ of Eq. 2.16, on the other hand, is a function of the prior probability p_0 defined on the interval [0,1]. Accordingly, we may plot the entropy function $\mathcal{H}(p_0)$ versus p_0, defined on the interval [0,1], as in Fig. 2.2. Clearly, the curve in Fig. 2.2 highlights the observations made under points 1, 2, and 3.

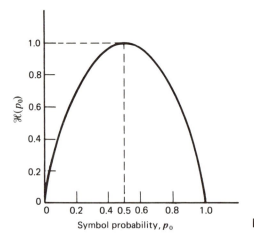

Figure 2.2 Entropy function $\mathcal{H}(p_0)$.

(2)   Extension of a Discrete Memoryless Source

In discussing information theoretic concepts, we often find it useful to consider *blocks* rather than individual symbols, with each block consisting of n successive source symbols. We may view each such block as being produced by an *extended source* with a source alphabet $\mathcal{S}^n$ that has K^n *distinct* blocks, where K is the number of distinct symbols in the source alphabet $\mathcal{S}$ of the original source. In the case of a discrete memoryless source, the source symbols are statistically independent. Hence, the probability of a source symbol in $\mathcal{S}^n$ is equal to the product of the probabilities of the n source symbols in $\mathcal{S}$ constituting the particular source symbol in $\mathcal{S}^n$. We may thus intuitively expect that $H(\mathcal{S}^n)$, the entropy of the extended source, is equal to n times $H(\mathcal{S})$, the entropy of the original source. That is, we may write

$$H(\mathcal{S}^n) = nH(\mathcal{S}) \tag{2.17}$$

EXAMPLE 2

Consider a discrete memoryless source with source alphabet $\mathcal{S} = \{s_0, s_1, s_2\}$ with probabilities

$$p(s_0) = p_0 = \frac{1}{4}$$

$$p(s_1) = p_1 = \frac{1}{4}$$

and

$$p(s_2) = p_2 = \frac{1}{2}$$

Hence, the use of Eq. 2.9 yields the entropy of the source as

$$\begin{aligned}
H(\mathcal{S}) &= p_0 \log_2\left(\frac{1}{p_0}\right) + p_1 \log_2\left(\frac{1}{p_1}\right) + p_2 \log_2\left(\frac{1}{p_2}\right) \\
&= \frac{1}{4} \log_2(4) + \frac{1}{4} \log_2(4) + \frac{1}{2} \log_2(2) \\
&= \frac{3}{2} \text{ bits}
\end{aligned}$$

Consider next the second-order extension of the source. With the source alphabet $\mathcal{S}$ consisting of three symbols, it follows that the source alphabet $\mathcal{S}^2$ of the extended source has nine symbols. The first row of Table 2.1 presents the nine symbols of $\mathcal{S}^2$, denoted as $\sigma_0, \sigma_1, \ldots, \sigma_8$. The second row of the table presents the compositions of these nine symbols in terms of the corresponding sequences of source symbols s_0, s_1, and s_2, taken two at a time. The probabilities of the nine source symbols of the extended source are presented in the last row of the table. Accordingly, the use of Eq. 2.9 yields the entropy of the extended source as

Table 2.1 Alphabet Particulars of Second-order Extension of a Discrete Memoryless Source

Symbols of $\mathscr{S}^2$	σ_0	σ_1	σ_2	σ_3	σ_4	σ_5	σ_6	σ_7	σ_8
Corresponding sequences of symbols of $\mathscr{S}$	$s_0 s_0$	$s_0 s_1$	$s_0 s_2$	$s_1 s_0$	$s_1 s_1$	$s_1 s_2$	$s_2 s_0$	$s_2 s_1$	$s_2 s_2$
Probability $p(\sigma_i)$, $i = 0, 1, \ldots, 8$	$\frac{1}{16}$	$\frac{1}{16}$	$\frac{1}{8}$	$\frac{1}{16}$	$\frac{1}{16}$	$\frac{1}{8}$	$\frac{1}{8}$	$\frac{1}{8}$	$\frac{1}{4}$

$$H(\mathscr{S}^2) = \sum_{i=0}^{8} p(\sigma_i) \log_2 \frac{1}{p(\sigma_i)}$$

$$= \frac{1}{16} \log_2(16) + \frac{1}{16} \log_2(16) + \frac{1}{8} \log_2(8) + \frac{1}{16} \log_2(16)$$

$$+ \frac{1}{16} \log_2(16) + \frac{1}{8} \log_2(8) + \frac{1}{8} \log_2(8) + \frac{1}{8} \log_2(8) + \frac{1}{4} \log_2(4)$$

$$= 3 \text{ bits}$$

We thus see that $H(\mathscr{S}^2) = 2H(\mathscr{S})$ in accordance with Eq. 2.17.

2.2 SOURCE CODING THEOREM

An important problem in communications is the *efficient* representation of data generated by a discrete source. The process by which this representation is accomplished is called *source encoding*. The device that performs the representation is called a *source encoder*. For the source encoder to be *efficient*, we require knowledge of statistics of the source. In particular, if some source symbols are known to be more probable than others, then we may exploit this feature in the generation of a *source code* by assigning *short* code-words to *frequent* source symbols, and *long* code-words to *rare* source symbols. We refer to such a source code as a *variable-length code*. The *Morse code* is an example of a variable-length code. In the Morse code, the letters of the alphabet and numerals are encoded into streams of *marks* and *spaces,* denoted as dots ".." and dashes "-", respectively. Since in the English language, the letter *E* occurs more frequently than the letter *Q*, for example, the Morse code encodes *E* into a single dot ".", the shortest code-word in the code, and it encodes *Q* into "--.-", the longest code-word in the code.

Our primary interest is in the development of an efficient source encoder that satisfies two functional requirements:

1. The code-words produced by the encoder are in *binary* form.
2. The source code is *uniquely decodable,* so that the original source sequence can be reconstructed perfectly from the encoded binary sequence.

Consider then the arrangement shown in Fig. 2.3 that depicts a discrete memoryless source whose output s_k is converted by the source encoder into a stream of 0s and 1s, denoted by b_k. We assume that the source has an alphabet with K different symbols, and that the kth symbol s_k occurs with probability p_k,

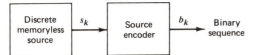

Figure 2.3 Source encoding.

$k = 0, 1, \ldots, K - 1$. Let the binary code-word assigned to symbol s_k by the encoder have length l_k, measured in bits. We define the average code-word length, $\bar{L}$, of the source encoder as

$$\bar{L} = \sum_{k=0}^{K-1} p_k l_k \qquad (2.18)$$

In physical terms, the parameter $\bar{L}$ represents the *average number of bits per source symbol* used in the source encoding process. Let $L_{\min}$ denote the *minimum* possible value of $\bar{L}$. We then define the *coding efficiency* of the source encoder as

$$\eta = \frac{L_{\min}}{\bar{L}} \qquad (2.19)$$

With $\bar{L} \geq L_{\min}$, we clearly have $\eta \leq 1$. The source encoder is said to be *efficient* when η approaches unity.

But how is the minimum value $L_{\min}$ determined? The answer to this fundamental question is embodied in Shannon's first theorem: the *source-coding theorem*, which may be stated as follows:*

Given a discrete memoryless source of entropy $H(\mathcal{S})$, the average code-word length $\bar{L}$ for any source encoding is bounded as

$$\bar{L} \geq H(\mathcal{S}) \qquad (2.20)$$

Accordingly, the entropy $H(\mathcal{S})$ represents a *fundamental limit* on the average number of bits per source symbol, $\bar{L}$, necessary to represent a discrete memoryless source in that $\bar{L}$ can be made as small as, but no smaller than, the entropy $H(\mathcal{S})$. Thus with $L_{\min} = H(\mathcal{S})$, we may rewrite the efficiency of a source encoder in terms of the entropy $H(\mathcal{S})$ as

$$\eta = \frac{H(\mathcal{S})}{\bar{L}} \qquad (2.21)$$

---

* For the original proof of the source-coding theorem, see Shannon (1948). A general proof of the source-coding theorem is also given in the following books: Viterbi and Omura (1979, pp. 13–19), McEliece (1977, Chapter 3), and Gallager (1968, pp. 38–55). The source-coding theorem is also referred to in the literature as the *noiseless coding theorem,* noiseless in the sense that it establishes the condition for error-free encoding to be possible.

A problem related to source coding is that of *data compression* where we use a "rate" R bits per source symbol that is less than the source entropy $H(\mathcal{S})$. Obviously, for such rates we cannot find error-free encoding, and we now study the problem of just how well we can do by minimizing a suitably defined *fidelity criterion.* This branch of information theory is called *rate distortion* theory. Detailed treatment of the subject is presented in Berger (1971).

In the sequel, we consider a type of source code known as a *prefix code,* which is not only decodable but also offers the possibility of realizing an average code-word length that can be made arbitrarily close to the source entropy.

(1) Prefix Coding

Consider a discrete memoryless source with source alphabet $\{s_0, s_1, \ldots, s_{K-1}\}$ and source statistics $\{p_0, p_1, \ldots, p_{K-1}\}$. For a source code representing the output of this source to be of practical use, the code has to be uniquely decodable. This restriction ensures that for each finite sequence of symbols emitted by the source, the corresponding sequence of code-words is different from the sequence of code-words corresponding to any other source sequence. We are specifically interested in a special class of codes satisfying a restriction known as the *prefix condition.* To define the prefix condition, let the kth code-word assigned to source symbol s_k be denoted by $(m_{k_1}, m_{k_2}, \ldots, m_{k_n})$, where the individual elements $m_{k_1}, \ldots, m_{k_n}$ are 0s and 1s, and n is the code-word length. The initial part of the code-word is represented by the elements $m_{k_1}, \ldots, m_{k_i}$ for some $i \leqslant n$. Any sequence made up of the initial part of the code-word is called a *prefix* of the code-word. A *prefix code* is defined as a code in which no code-word is the prefix of any other code-word.

To illustrate the meaning of a prefix code, consider the three source codes described in Table 2.2. Code I is not a prefix code since the bit 0, the code-word for s_0, is a prefix of 00, the code-word for s_2. Likewise, the bit 1, the code-word for s_1, is a prefix of 11, the code-word for s_3. Similarly, we may show that code III is not a prefix code, but code II is.

In order to decode a sequence of code-words generated from a prefix source code, the *source decoder* simply starts at the beginning of the sequence and decodes one code-word at a time. Specifically, it sets up what is equivalent to a *decision tree,* which is a graphical portrayal of the code-words in the particular source code. For example, Fig. 2.4 depicts the decision tree corresponding to code II in Table 2.2. The tree has an *initial state* and four *terminal states* corresponding to source symbols s_0, s_1, s_2, and s_3. The decoder always starts in the initial state. The first received bit moves the decoder to the terminal state s_0 if it is 0, or else to a second decision point if it is 1. In the latter case, the second bit moves the decoder one step further down the tree, either to terminal state s_1 if it is 0, or else to a third decision point if it is 1, and so on. Once each terminal state emits its symbol, the decoder is reset to its initial state. Note also that each bit in the received encoded sequence is examined only once. For example,

Table 2.2 Illustrating the Definition of a Prefix Code

Source Symbol	Probability of Occurrence	Code I	Code II	Code III
s_0	0.5	0	0	0
s_1	0.25	1	10	01
s_2	0.125	00	110	011
s_3	0.125	11	111	0111

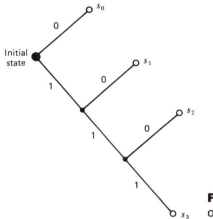

Figure 2.4   Decision tree for code II of Table 2.2.

the encoded sequence 1011111000 . . . is readily decoded as the source sequence $s_1s_3s_2s_0s_0$ . . . . The reader is invited to carry out this decoding.

A prefix code has the important property that it is *always* uniquely decodable. Indeed, if a prefix code has been constructed for a discrete memoryless source with source alphabet $\{s_0, s_1, \ldots, s_{K-1}\}$ and source statistics $\{p_0, p_1, \ldots, p_{K-1}\}$, and the code-word for symbol s_k has length l_k, $k = 0, 1, \ldots, K - 1$, then the code-word lengths of the code satisfy a certain inequality known as the *Kraft-McMillan inequality*. In mathematical terms, we may state that*

$$\sum_{k=0}^{K-1} 2^{-l_k} \leq 1 \tag{2.22}$$

where the factor 2 refers to the radix (number of symbols) in the binary alphabet. Conversely, we may state that if the code-word lengths of a code for a discrete memoryless source satisfy the Kraft-McMillan inequality, then a prefix code with these code-word lengths can be constructed.

Although all prefix codes are uniquely decodable, the converse is not true. For example, code III in Table 2.2 does not satisfy the prefix condition, and yet it is uniquely decodable since the bit 0 indicates the beginning of each code-word in the code.

Prefix codes distinguish themselves from other uniquely decodable codes by the fact that the end of a code-word is always recognizable. Hence, the decoding of a prefix code can be accomplished as soon as the binary sequence representing a source symbol is fully received. For this reason, prefix codes are also referred to as *instantaneous codes*.

Given a discrete memoryless source of entropy $H(\mathcal{G})$, the average code-word length of a prefix code is bounded as follows:†

$$H(\mathcal{G}) \leq \bar{L} < H(\mathcal{G}) + 1 \tag{2.23}$$

---

* For proof of the Kraft-McMillan inequality, see Appendix F.

† For a proof of Eq. 2.23, see McEliece (1977, pp. 241–242).

The equality (on the left side) holds under the condition that symbol s_k be emitted by the source with probability

$$p_k = 2^{-l_k} \tag{2.24}$$

where l_k is the length of the code word assigned to source symbol s_k. We then have

$$\sum_{k=0}^{K-1} 2^{-l_k} \leq \sum_{k=0}^{K-1} p_k = 1$$

Under this condition, the Kraft-McMillan inequality of Eq. 2.22 implies that we can construct a prefix code, such that the length of the code-word assigned to source symbol s_k is l_k. For such a code, the average code-word length is

$$\bar{L} = \sum_{k=0}^{K-1} \frac{l_k}{2^{l_k}}$$

and the corresponding entropy of the source is

$$H(\mathcal{S}) = \sum_{k=0}^{K-1} \left(\frac{1}{2^{l_k}}\right) \log_2(2^{l_k})$$

$$= \sum_{k=0}^{K-1} \frac{l_k}{2^{l_k}}$$

Hence, in this special (rather meretricious) case, the prefix code is *matched* to the source in that $\bar{L} = H(\mathcal{S})$.

But how do we match the prefix code to an arbitrary discrete memoryless source? The answer to this problem lies in the use of an *extended code*. For the nth extension of a code, a source encoder operates on blocks of n samples rather than individual samples. From our earlier discussion on the extensions of discrete memoryless sources, we recall that the entropy of an extended source with order n and source alphabet $\mathcal{S}^n$ is equal to n times the entropy of the original source with source alphabet $\mathcal{S}$, as shown by Eq. 2.17. Let $\bar{L}_n$ denote the average code-word length of the extended prefix code. For a uniquely decodable code, $\bar{L}_n$ is the smallest possible. From Eq. 2.23, we deduce that

$$H(\mathcal{S}^n) \leq \bar{L}_n < H(\mathcal{S}^n) + 1 \tag{2.25}$$

Substituting Eq. 2.17 in Eq. 2.25, we get

$$nH(\mathcal{S}) \leq \bar{L}_n < nH(\mathcal{S}) + 1$$

or, equivalently,

$$H(\mathcal{S}) \leq \frac{\bar{L}_n}{n} < H(\mathcal{S}) + \frac{1}{n} \tag{2.26}$$

In the limit, as n approaches infinity, the lower and upper bounds in Eq. 2.26 converge, as shown by

$$\lim_{n \to \infty} \frac{1}{n} \bar{L}_n = H(\mathcal{S}) \tag{2.27}$$

We may therefore state that by making the order n of an extended prefix source encoder large enough, we can make the code faithfully represent the discrete memoryless source $\mathscr{S}$ as closely as desired. In other words, the average code-word length of an extended prefix code can be made as small as the entropy of the source provided the extended code has a high enough order, in accordance with Shannon's source-coding theorem. However, the price we have to pay for decreasing the average code-word length is increased decoding complexity, which is brought about by the high order of the extended prefix code.

2.3 HUFFMAN CODING

The *Huffman code** is a source code whose average word length approaches the fundamental limit set by the entropy of a discrete memoryless source, namely, $H(\mathscr{S})$. The Huffman code is *optimum* in the sense that no other uniquely decodable set of code-words has a smaller average code-word length for a given discrete memoryless source. The essence of the *algorithm* used to synthesize the code is to replace the prescribed set of source statistics of a discrete memoryless source with a simpler one. This *reduction* process is continued in a step-by-step manner until we are left with a final set of source statistics (symbols) of only two, for which (0,1) is an optimal code. Starting from this trivial code, we then work backward and thereby construct an optimal code for the given source.

Specifically, the Huffman *encoding algorithm* proceeds as follows:

1. The source symbols are listed in order of decreasing probability. The two source symbols of lowest probability are assigned a 0 and a 1. (This part of the step is referred to as a *splitting* stage.)
2. These two source symbols are regarded as being *combined* into a new source symbol with probability equal to the sum of the two original probabilities. (The list of source symbols, and therefore source statistics, is thereby *reduced* in size by one.) The probability of the new symbol is placed in the list in accordance with its value.
3. The procedure is repeated until we are left with a final list of source statistics (symbols) of only two for which a 0 and a 1 are assigned.

The code for each (original) source symbol is found by working backward and tracing the sequence of 0s and 1s assigned to that symbol as well as its successors.

EXAMPLE 3

The five source symbols of the alphabet of a discrete memoryless source and their probabilities are shown in the left-most two columns of Fig. 2.5a. Following through the Huffman algorithm, we reach the end of the computation in four steps, as shown in Fig. 2.5a. The code-words of the Huffman code for the source are tabulated in Fig. 2.5b. The average code-word length is therefore

---

* The algorithm is named after its inventor, Huffman (1962).

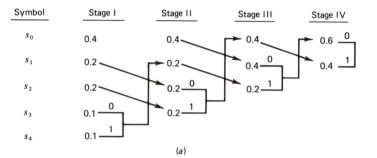

Symbol	Stage I	Stage II	Stage III	Stage IV

(a)

Symbol	Probability	Code word
s_0	0.4	00
s_1	0.2	10
s_2	0.2	11
s_3	0.1	010
s_4	0.1	011

(b)

Figure 2.5 (a) Example of the Huffman encoding algorithm. (b) Source code.

$$\bar{L} = 0.4(2) + 0.2(2) + 0.2(2) + 0.1(3) + 0.1(3)$$
$$= 2.2$$

The entropy of the specified discrete memoryless source is calculated from Eq. 2.9 as follows

$$H(\mathcal{S}) = 0.4 \log_2 \left(\frac{1}{0.4}\right) + 0.2 \log_2 \left(\frac{1}{0.2}\right) + 0.2 \log_2 \left(\frac{1}{0.2}\right)$$
$$+ 0.1 \log_2 \left(\frac{1}{0.1}\right) + 0.1 \log_2 \left(\frac{1}{0.1}\right)$$
$$= 0.52877 + 0.46439 + 0.46439 + 0.33219 + 0.33219$$
$$= 2.12193$$

For the example at hand, we may make two observations:

1. The average code-word length $\bar{L}$ exceeds the entropy $H(\mathcal{S})$ by only 3.67 percent.
2. The average code-word length $\bar{L}$ does indeed satisfy Eq. 2.23.

It is noteworthy that the Huffman encoding process is not unique. In particular, we may site two variations in the process that are responsible for the non-uniqueness of the Huffman code. First, at each splitting stage in the construction of a Huffman code, there is arbitrariness in the way a 0 and a 1 are assigned to the last two source symbols. Whichever way, however, the resulting differences are trivial. Second, ambiguity arises when the probability of a *combined* symbol (obtained by adding the last two probabilities pertinent to a particular step) is found to equal another probability in the list. We may proceed by placing the probability of the new symbol as *high* as possible, as in Example 3. Alternatively, we may place it as *low* as possible. (It is presumed that whichever way the placement is made, high or low, it is consistently adhered to throughout the encoding process.) But, this time, noticable differences arise in that the code words in the resulting source codes can have different lengths. Nevertheless, the average code-word length remains the same.

As a measure of the variability in code-word lengths of a source code, we define the *variance* of the average code-word length $\bar{L}$ over the ensemble of source symbols as

$$\sigma^2 = \sum_{k=0}^{K-1} p_k(l_k - \bar{L})^2 \qquad (2.28)$$

where $p_0, p_1, \ldots, p_{K-1}$ are the source statistics, and l_k is the length of the code-word assigned to source symbol s_k. It is usually found that when a combined symbol is moved as high as possible, the resulting Huffman code has a significantly smaller variance σ^2 than when it is moved as low as possible. On this basis, it is reasonable to choose the former Huffman code over the latter.

In Example 3, a combined symbol was moved as high as possible. In Example 4, presented next, a combined symbol is moved as low as possible. Thus, by comparing the results of these two examples, we are able to appreciate the subtle differences and similarities between the two Huffman codes.

EXAMPLE 4

Consider again the same discrete memoryless source described in Example 3. This time, however, we move the probability of a combined symbol as low as possible. The results obtained are shown in Fig. 2.6a. Working backward and tracing through the various steps, we find that the code words of this second Huffman code for the source are as tabulated in Fig. 2.6b. The average code-word length for the second Huffman code is therefore

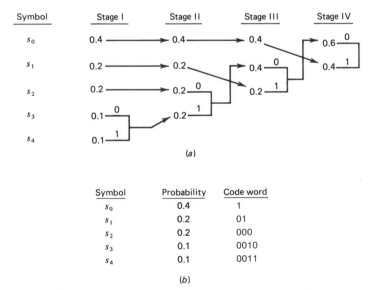

Figure 2.6 (*a*) Example illustrating nonuniqueness of the Huffman encoding algorithm. (*b*) Another source code.

$$\bar{L} = 0.4(1) + 0.2(2) + 0.2(3) + 0.1(4) + 0.1(4)$$
$$= 2.2$$

which is exactly the same as that for the first Huffman code of Example 3. However, as remarked earlier, the individual code words of the second Huffman code have different lengths, compared to the corresponding ones of the first Huffman code.

The use of Eq. 2.28 yields the variance of the first Huffman code obtained in Example 3 as

$$\sigma_1^2 = 0.4(2 - 2.2)^2 + 0.2(2 - 2.2)^2 + 0.2(2 - 2.2)^2$$
$$+ 0.1(3 - 2.2)^2 + 0.1(3 - 2.2)^2$$
$$= 0.16$$

On the other hand, for the second Huffman code obtained in this example, we have from Eq. 2.28:

$$\sigma_2^2 = 0.4(1 - 2.2)^2 + 0.2(2 - 2.2)^2 + 0.2(3 - 2.2)^2$$
$$+ 0.1(4 - 2.2)^2 + 0.1(4 - 2.2)^2$$
$$= 1.36$$

These results confirm that the minimum variance Huffman code is obtained by moving the probability of a combined symbol as high as possible.

2.4 DISCRETE MEMORYLESS CHANNELS

Up to this point in the chapter, we have been preoccupied with discrete memoryless sources. Indeed, much of the discussion was devoted to the use of entropy to define the efficiency of a discrete memoryless source and to the synthesis of Huffman codes containing as little redundancy as possible. In the remainder of the chapter, we shift the focus of our attention from information generation to information transmission, with particular emphasis on reliability. We start the shift in this section by considering a discrete memoryless channel, the counterpart of a discrete memoryless source.

A *discrete memoryless channel* is a statistical model with an input X and an output Y that is a *noisy* version of X; both X and Y are random variables. Every unit of time, the channel accepts an input symbol X selected from an alphabet $\mathcal{X}$ and, in response, it emits an output symbol Y from an alphabet $\mathcal{Y}$. The channel is said to be "discrete" when both of the alphabets $\mathcal{X}$ and $\mathcal{Y}$ have *finite* sizes. It is said to be "memoryless" when the current output symbol depends *only* on the current input symbol and *not* any of the previous ones.

Figure 2.7 depicts a view of a discrete memoryless channel. The channel is described in terms of an *input alphabet*

$$\mathcal{X} = \{x_0, x_1, \ldots, x_{J-1}\}, \tag{2.29}$$

an *output alphabet*,

$$\mathcal{Y} = \{y_0, y_1, \ldots, y_{K-1}\}, \tag{2.30}$$

and a set of *transition probabilities*

$$p(y_k|x_j) = P(Y = y_k|X = x_j) \qquad \text{for all } j \text{ and } k \tag{2.31}$$

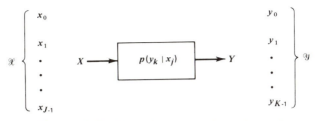

Figure 2.7   Discrete memoryless channel.

Naturally, we have

$$0 \leq p(y_k|x_j) \leq 1 \qquad \text{for all } j \text{ and } k \tag{2.32}$$

Also, the input alphabet $\mathcal{X}$ and output alphabet $\mathcal{Y}$ need not have the same size. For example, in channel coding, the size K of the output alphabet $\mathcal{Y}$ may be larger than the size J of the input alphabet $\mathcal{X}$; thus, $K \geq J$. On the other hand, we may have a situation in which the channel emits the same symbol when either one of two input symbols is sent; we then have $K \leq J$.

The transition probability $p(y_k|x_j)$ is just the conditional probability that the channel output $Y = y_k$, given that the channel input $X = x_j$. Owing to the physical limits on the fidelity with which information can be transmitted over the channel, there is the possibility of *errors* arising from the process of information transmission. Thus, when $k = j$, the transition probability $p(y_k|x_j)$ represents a *conditional probability of correct reception,* and when $k \neq j$, it represents a *conditional probability of error.* The subject of probability of error calculations is discussed in Chapters 5 and 7.

A convenient way of describing a discrete memoryless channel is to arrange the various transition probabilities of the channel in the form of a matrix as follows

$$\mathbf{P} = \begin{bmatrix} p(y_0|x_0) & p(y_1|x_0) & \cdots & p(y_{K-1}|x_0) \\ p(y_0|x_1) & p(y_1|x_1) & \cdots & p(y_{K-1}|x_1) \\ \vdots & \vdots & & \vdots \\ p(y_0|x_{J-1}) & p(y_1|x_{J-1}) & \cdots & p(y_{K-1}|x_{J-1}) \end{bmatrix} \tag{2.33}$$

The J-by-K matrix $\mathbf{P}$ is called the *channel matrix*. Note that each *row* of the channel matrix $\mathbf{P}$ corresponds to a *fixed channel input,* whereas each column of the matrix corresponds to a *fixed channel output*. Note also that a fundamental property of the channel matrix $\mathbf{P}$ is that the sum of the elements along any row of the matrix is always equal to one; that is to say

$$\sum_{k=0}^{K-1} p(y_k|x_j) = 1 \qquad \text{for all } j \tag{2.34}$$

Suppose now that the inputs to a discrete memoryless channel are selected according to the *probability distribution* $\{p(x_j), j = 0, 1, \ldots, J - 1\}$. In other words, the event, the channel input $X = x_j$, occurs with probability

$$p(x_j) = P(X = x_j) \qquad \text{for } j = 0, 1, \ldots, J - 1 \qquad (2.35)$$

Having specified the random variable X denoting the channel input, we may now specify the second random variable Y denoting the channel output. The *joint probability distribution* of the random variables X and Y is given by

$$
\begin{aligned}
p(x_j, y_k) &= P(X = x_j, Y = y_k) \\
&= P(Y = y_k | X = x_j) P(X = x_j) \\
&= p(y_k | x_j) p(x_j) \qquad\qquad\qquad\qquad (2.36)
\end{aligned}
$$

The *marginal probability distribution* of the output random variable Y is obtained by averaging out the dependence of $p(x_j, y_k)$ on x_j, as shown by

$$
\begin{aligned}
p(y_k) &= P(Y = y_k) \\
&= \sum_{j=0}^{J-1} P(Y = y_k | X = x_j) P(X = x_j) \\
&= \sum_{j=0}^{J-1} p(y_k | x_j) p(x_j) \qquad \text{for } k = 0, 1, \ldots, K - 1 \qquad (2.37)
\end{aligned}
$$

For $J = K$, the *average probability of symbol error*, P_e, is defined as the probability that the output random variable Y_k is different from the input random variable X_j, averaged over all $k \neq j$. We thus write

$$
\begin{aligned}
P_e &= \sum_{\substack{k=0 \\ k \neq j}}^{K-1} P(Y = y_k) \\
&= \sum_{j=0}^{J-1} \sum_{\substack{k=0 \\ k \neq j}}^{K-1} p(y_k | x_j) p(x_j) \qquad\qquad (2.38)
\end{aligned}
$$

The difference $1 - P_e$ is the *average probability of correct reception*.

The probabilities $p(x_j)$ for $j = 0, 1, \ldots, J - 1$, are known as the *a priori probabilities* of the various input symbols. Equation 2.37 states that if we are given the input a priori probabilities $p(x_j)$ and the channel matrix, the matrix of transition probabilities $p(y_k | x_j)$, then we may calculate the probabilities of the various output symbols, the $p(y_k)$. For the remainder of the chapter, we assume that we are given the $p(x_j)$ and the $p(y_k | x_j)$, in which case the $p(y_k)$ may be calculated by using Eq. 2.37.

EXAMPLE 5   BINARY SYMMETRIC CHANNEL

The *binary symmetric* channel is of great theoretical interest and practical importance. It is a special case of the discrete memoryless channel with $J = K = 2$. The channel has two input symbols ($x_0 = 0$, $x_1 = 1$) and two output symbols ($y_0 = 0$, $y_1 = 1$). The channel is symmetric because the probability of receiving a 1 if a 0 is sent is the same as the probability of receiving a 0 if a 1 is sent. This common transition probability or conditional probability of error is denoted by p. The *transition probability diagram* of a binary symmetric channel is thus as shown in Fig. 2.8.

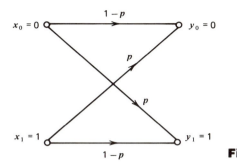

Figure 2.8 Binary symmetric channel.

2.5 MUTUAL INFORMATION

Given that we think of the channel output Y (selected from alphabet $\mathcal{Y}$) as a noisy version of the channel input X (selected from alphabet $\mathcal{X}$), and that the entropy $H(\mathcal{X})$ is a measure of the prior uncertainty about X, how can we measure the uncertainty about X after observing Y? In order to answer this question, we extend the ideas developed in Section 2.1 by defining the *conditional entropy* of X selected from alphabet $\mathcal{X}$, given that $Y = y_k$, as

$$H(\mathcal{X}|Y = y_k) = \sum_{j=0}^{J-1} p(x_j|y_k) \log_2\left[\frac{1}{p(x_j|y_k)}\right] \quad (2.39)$$

This quantity is itself a random variable that takes on the values $H(\mathcal{X}|Y = y_0)$, . . . , $H(\mathcal{X}|Y = y_{K-1})$ with probabilities $p(y_0)$, . . . , $p(y_{K-1})$, respectively. The mean value of $H(\mathcal{X}|Y = y_k)$ over the output alphabet $\mathcal{Y}$ is therefore given by

$$\begin{aligned}
H(\mathcal{X}|\mathcal{Y}) &= \sum_{k=0}^{K-1} H(\mathcal{X}|Y = y_k)p(y_k) \\
&= \sum_{k=0}^{K-1}\sum_{j=0}^{J-1} p(x_j|y_k)p(y_k) \log_2\left[\frac{1}{p(x_j|y_k)}\right] \\
&= \sum_{k=0}^{K-1}\sum_{j=0}^{J-1} p(x_j, y_k) \log_2\left[\frac{1}{p(x_j|y_k)}\right] \quad (2.40)
\end{aligned}$$

where in the last line, we have made use of Eq. 2.36 rewritten in the following form

$$p(x_j, y_k) = p(x_j|y_k)p(y_k)$$

The quantity $H(\mathcal{X}|\mathcal{Y})$ is called a *conditional entropy*. It represents *the amount of uncertainty remaining about the channel input after the channel output has been observed.*

Since the entropy $H(\mathcal{X})$ represents our uncertainty about the channel input *before* observing the channel output, and the conditional entropy $H(\mathcal{X}|\mathcal{Y})$ represents our uncertainty about the channel input *after* observing the channel output, it follows that the difference $H(\mathcal{X}) - H(\mathcal{X}|\mathcal{Y})$ must represent our uncertainty about the channel input that is *resolved* by observing the channel output.

This important quantity is called the *mutual information* of the channel. Denoting the mutual information by $I(\mathcal{X};\mathcal{Y})$, we may thus write

$$I(\mathcal{X};\mathcal{Y}) = H(\mathcal{X}) - H(\mathcal{X}|\mathcal{Y}) \tag{2.41}$$

(1) Properties of Mutual Information
The mutual information $I(\mathcal{X};\mathcal{Y})$ has the following important properties:

PROPERTY 1
The mutual information of a channel is symmetric; that is

$$I(\mathcal{X};\mathcal{Y}) = I(\mathcal{Y};\mathcal{X}) \tag{2.42}$$

where the mutual information $I(\mathcal{X};\mathcal{Y})$ is a measure of the uncertainty about the channel input that is resolved by *observing* the channel output, and the mutual information $I(\mathcal{Y};\mathcal{X})$ is a measure of the uncertainty about the channel output that is resolved by *sending* the channel input.

To prove this property, we first use the formula for entropy and then use Eqs. 2.34 and 2.36 to express $H(\mathcal{X})$ as

$$
\begin{aligned}
H(\mathcal{X}) &= \sum_{j=0}^{J-1} p(x_j) \log_2\left[\frac{1}{p(x_j)}\right] \\
&= \sum_{j=0}^{J-1} p(x_j) \log_2\left[\frac{1}{p(x_j)}\right] \sum_{k=0}^{K-1} p(y_k|x_j) \\
&= \sum_{j=0}^{J-1} \sum_{k=0}^{K-1} p(y_k|x_j)p(x_j) \log_2\left[\frac{1}{p(x_j)}\right] \\
&= \sum_{j=0}^{J-1} \sum_{k=0}^{K-1} p(x_j, y_k) \log_2\left[\frac{1}{p(x_j)}\right]
\end{aligned}
\tag{2.43}
$$

Hence, substituting Eqs. 2.40 and 2.43 into Eq. 2.41, and then combining terms we get

$$I(\mathcal{X};\mathcal{Y}) = \sum_{j=0}^{J-1} \sum_{k=0}^{K-1} p(x_j, y_k) \log_2\left[\frac{p(x_j|y_k)}{p(x_j)}\right] \tag{2.44}$$

From *Bayes' rule* for conditional probabilities, we have

$$\frac{p(x_j|y_k)}{p(x_j)} = \frac{p(y_k|x_j)}{p(y_k)} \tag{2.45}$$

Hence, substituting Eq. 2.45 into Eq. 2.44, and interchanging the order of summation, we may write

$$
\begin{aligned}
I(\mathcal{X};\mathcal{Y}) &= \sum_{k=0}^{K-1} \sum_{j=0}^{J-1} p(x_j, y_k) \log_2\left[\frac{p(y_k|x_j)}{p(y_k)}\right] \\
&= I(\mathcal{Y};\mathcal{X})
\end{aligned}
\tag{2.46}
$$

which is the desired result.

PROPERTY 2
The mutual information is always nonnegative; that is

$$I(\mathscr{X};\mathscr{Y}) \geq 0 \tag{2.47}$$

To prove this property, we first note that

$$p(x_j|y_k) = \frac{p(x_j, y_k)}{p(y_k)} \tag{2.48}$$

Hence, substituting Eq. 2.48 in Eq. 2.44, we may express the mutual information of the channel as

$$I(\mathscr{X};\mathscr{Y}) = \sum_{j=0}^{J-1} \sum_{k=0}^{K-1} p(x_j, y_k) \log_2\left(\frac{p(x_j, y_k)}{p(x_j)p(y_k)}\right) \tag{2.49}$$

Next, a direct application of the fundamental inequality (defined by Eq. 2.12) yields the desired result

$$I(\mathscr{X};\mathscr{Y}) \geq 0$$

with equality if, and only if,

$$p(x_j, y_k) = p(x_j)p(y_k) \qquad \text{for all } j \text{ and } k \tag{2.50}$$

Property 2 states that *we cannot lose information, on the average, by observing the output of a channel.* Moreover, the average information is zero if, and only if, the input and output symbols of the channel are statistically independent, as in Eq. 2.50.

PROPERTY 3
The mutual information of a channel may be expressed in terms of the entropy of the channel output as

$$I(\mathscr{X};\mathscr{Y}) = H(\mathscr{Y}) - H(\mathscr{Y}|\mathscr{X}) \tag{2.51}$$

where $H(\mathscr{Y}|\mathscr{X})$ is a conditional entropy.

This property follows directly from the defining Eq. 2.41 for mutual information of a channel, and Property 1.

PROPERTY 4
The mutual information of a channel is related to the joint entropy of the channel input and channel output by

$$I(\mathscr{X};\mathscr{Y}) = H(\mathscr{X}) + H(\mathscr{Y}) - H(\mathscr{X}, \mathscr{Y}) \tag{2.52}$$

where the joint entropy $H(\mathscr{X}, \mathscr{Y})$ is defined by

$$H(\mathscr{X}, \mathscr{Y}) = \sum_{j=0}^{J-1} \sum_{k=0}^{K-1} p(x_j, y_k) \log_2\left(\frac{1}{p(x_j, y_k)}\right) \tag{2.53}$$

To prove Eq. 2.52, we first rewrite the definition for joint entropy $H(\mathscr{X}, \mathscr{Y})$ as

$$H(\mathcal{X}, \mathcal{Y}) = \sum_{j=0}^{J-1} \sum_{k=0}^{K-1} p(x_j, y_k) \log_2 \left[\frac{p(x_j)p(y_k)}{p(x_j, y_k)} \right]$$

$$+ \sum_{j=0}^{J-1} \sum_{k=0}^{K-1} p(x_j, y_k) \log_2 \left[\frac{1}{p(x_j)p(y_k)} \right] \qquad (2.54)$$

The first double summation term on the right side of Eq. 2.54 is recognized as the negative of the mutual information of the channel, $I(\mathcal{X}; \mathcal{Y})$, previously given in Eq. 2.49. As for the second summation term, we manipulate it as follows

$$\sum_{j=0}^{J-1} \sum_{k=0}^{K-1} p(x_j, y_k) \log_2 \left[\frac{1}{p(x_j)p(x_k)} \right] = \sum_{j=0}^{J-1} \log_2 \left[\frac{1}{p(x_j)} \right] \sum_{k=0}^{K-1} p(x_j, y_k)$$

$$+ \sum_{k=0}^{K-1} \log_2 \left[\frac{1}{p(y_k)} \right] \sum_{j=0}^{J-1} p(x_j, y_k)$$

$$= \sum_{j=0}^{J-1} p(x_j) \log_2 \left[\frac{1}{p(x_j)} \right]$$

$$+ \sum_{k=0}^{K-1} p(y_k) \log_2 \left[\frac{1}{p(y_k)} \right]$$

$$= H(\mathcal{X}) + H(\mathcal{Y}) \qquad (2.55)$$

Accordingly, using Eqs. 2.49 and 2.55 in Eq. 2.54, we get the result

$$H(\mathcal{X}, \mathcal{Y}) = -I(\mathcal{X}; \mathcal{Y}) + H(\mathcal{X}) + H(\mathcal{Y}) \qquad (2.56)$$

Rearranging this equation, we readily get the result given in Eq. 2.52, thereby confirming Property 3.

We conclude our discussion of the mutual information of a channel by providing a diagramatic interpretation of Eqs. 2.41, 2.51, and 2.56. The interpretation is given in Fig. 2.9. The entropy of channel input X is represented by the circle on the left. The entropy of channel output Y is represented by the circle

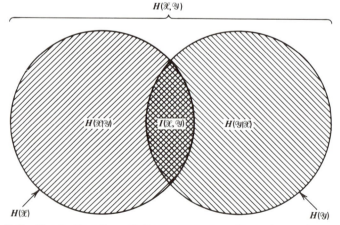

Figure 2.9 Illustrating the relations among various channel parameters.

on the right. The mutual information of the channel is represented by the overlap between these two circles.

2.6 CHANNEL CAPACITY

Consider a discrete memoryless channel with input alphabet $\mathcal{X}$, output alphabet $\mathcal{Y}$, and transition probabilities $p(y_k|x_j)$. The mutual information of the channel is defined by the first line of Eq. 2.46, which is reproduced here for convenience:

$$I(\mathcal{X};\mathcal{Y}) = \sum_{j=0}^{J-1} \sum_{k=0}^{K-1} p(x_j, y_k) \log_2\left[\frac{p(y_k|x_j)}{p(y_k)}\right] \tag{2.57}$$

Here we note that (see Eq. 2.36)

$$p(x_j, y_k) = p(y_k|x_j)p(x_j)$$

Also, from Eq. 2.37, we have

$$p(y_k) = \sum_{j=0}^{J-1} p(y_k|x_j)p(x_j)$$

From these three equations we see that it is necessary for us to know the input probability distribution $\{p(x_j), j = 0, 1, \ldots, J - 1\}$ so that we may calculate the mutual information $I(\mathcal{X};\mathcal{Y})$. The mutual information of a channel, therefore, depends not only on the channel, but also on the way in which the channel is used.

The input probability distribution $\{p(x_j)\}$ is obviously independent of the channel. We can then maximize the average mutual information $I(\mathcal{X};\mathcal{Y})$ of the channel with respect to $\{p(x_j)\}$. Hence, *we define the channel capacity of a discrete memoryless channel as the maximum average mutual information $I(\mathcal{X};\mathcal{Y})$ in any single use of the channel (i.e., signaling interval), where the maximization is over all possible input probability distributions $\{p(x_j)\}$ on $\mathcal{X}$.* The channel capacity is commonly denoted by C. We thus write

$$C = \max_{\{p(x_j)\}} I(\mathcal{X};\mathcal{Y}) \tag{2.58}$$

The channel capacity C is measured in *bits per channel use*.

Note that the channel capacity C is a function only of the transition probabilities $p(y_k|x_j)$, which define the channel. The calculation of C involves maximization of the average mutual information $I(\mathcal{X};\mathcal{Y})$ over J variables [i.e., the input probabilities $p(x_0), \ldots, p(x_{J-1})$] with both inequality constraints $p(x_j) \geq 0$, for all j, and an equality constraint $\sum_j p(x_j) = 1$. In general, the problem of finding the channel capacity C can be quite challenging.

EXAMPLE 6 BINARY SYMMETRIC CHANNEL (REVISITED)

Consider again the *binary symmetric channel*, which is described by the *transition probability diagram* of Fig. 2.8. This diagram is uniquely defined by the conditional probability of error, p.

By symmetry, the capacity for the binary symmetric channel is achieved with channel input probability $p(x_0) = p(x_1) = \frac{1}{2}$, where x_0 and x_1 are each 0 or 1. Hence,

$$C = I(\mathscr{X}; \mathscr{Y})_{|p(x_0)=p(x_1)=\frac{1}{2}} \tag{2.59}$$

From Fig. 2.8, we have

$$p(y_0|x_1) = p(y_1|x_0) = p$$

and

$$p(y_0|x_0) = p(y_1|x_1) = 1 - p$$

Therefore, substituting these channel transition probabilities into Eq. 2.57 with $J = K = 2$, and then setting the input probability $p(x_0) = p(x_1)$ in accordance with Eq. 2.16, we find that the capacity of the binary symmetric channel is

$$C = 1 + p\log_2 p + (1 - p)\log_2(1 - p) \tag{2.60}$$

Based on Eq. 2.16, we may define the *entropy function:*

$$\mathscr{H}(p) = p\log_2\left(\frac{1}{p}\right) + (1 - p)\log_2\left(\frac{1}{1 - p}\right) \tag{2.61}$$

Hence, we may rewrite Eq. 2.60 as

$$C = 1 - \mathscr{H}(p) \tag{2.62}$$

The channel capacity C varies with the probability of error p as shown in Fig. 2.10. Comparing the curve in this figure with that in Fig. 2.2, we may make the following observations:

1. When the channel is *noise-free,* permitting us to set $p = 0$, the channel capacity C attains its maximum value of one bit per channel use, which is exactly the information in each channel input. At this value of p, the entropy function $\mathscr{H}(p)$ attains its minimum value of zero.
2. When the channel is *noisy,* producing a conditional probability of error $p = 1/2$, the channel capacity C attains its minimum value of zero, whereas the entropy function $\mathscr{H}(p)$ attains its maximum value of unity.

2.7   CHANNEL CODING THEOREM

The inevitable presence of *noise* in a channel causes discrepancies (errors) between the output and input data sequences of a digital communication system. For a relatively noisy channel, the probability of error may have a value as high as 10^{-2}, which means that (on the average) 99 out of 100 transmitted bits are received correctly. For many applications, this *level of reliability* is found to be far from adequate. Indeed, a probability of error equal to 10^{-6} or even lower is often a necessary requirement. In order to achieve such a high level of performance, we may have to resort to the use of channel coding.

The design goal of channel coding is to increase the resistance of a digital communication system to channel noise. Specifically, *channel coding* consists of *mapping* the incoming data sequence into a channel input sequence, and

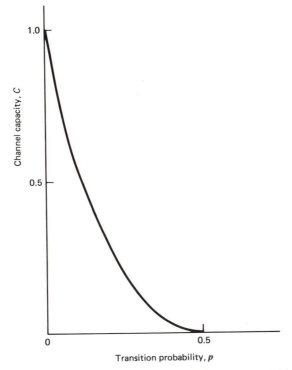

Figure 2.10   Variation of channel capacity of a binary symmetric channel with transition probability p.

inverse mapping the channel output sequence into an output data sequence in such a way that the overall effect of channel noise on the system is minimized. The first mapping operation is performed in the transmitter by means of an *encoder,* whereas the inverse mapping operation is performed in the receiver by means of a *decoder,* as shown in the block diagram of Fig. 2.11.

Both the encoder and the decoder in Fig. 2.11 are under the designer's control. The approach taken is to introduce *redundancy* in the encoder in a prescribed manner, and then exploit this redundancy in the decoder so as to reconstruct the original source sequence as accurately as possible. Thus, in a rather loose sense, we may view channel coding as the *dual* of source coding in that the former introduces controlled redundancy to improve reliability, whereas the latter reduces redundancy to improve efficiency.

The subject of channel coding is treated in detail in Chapter 8. For the purpose of our present discussion, it suffices to confine our attention to *block codes*. In this class of codes, the message sequence is subdivided into sequential blocks each k bits long, and each k-bit block is *mapped* into an n-bit block, where $n > k$. The number of redundant bits added by the encoder to each

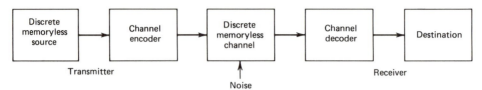

Figure 2.11 Block diagram of digital communication system.

transmitted block is $n - k$ bits. The ratio k/n is called the *code rate*. Using r to denote the code rate, we may thus write

$$r = \frac{k}{n}$$

where, of course, r is less than unity.

The accurate reconstruction of the original source sequence at the destination requires that the *average probability of symbol error* be arbitrarily low. This raises the following important question: *Does there exist a sophisticated coding scheme such that the probability that a message bit will be in error is less than any positive number ε (i.e., as small as we want it), and yet the coding scheme is efficient in that the code rate need not be too small?* The answer to this fundamental question is an emphatic "yes." Indeed, the answer to the question is provided by Shannon's second theorem in terms of the channel capacity C, as described in the sequel. Up until this point, *time* has not played an important role in our discussion of channel capacity. Suppose then the discrete memoryless source in Fig. 2.11 has the source alphabet $\mathcal{S}$ and entropy $H(\mathcal{S})$ bits per source symbol. We assume that the source emits symbols once every T_s seconds. Hence, the *average information rate* of the source is $H(\mathcal{S})/T_s$ bits per second. The decoder delivers decoded symbols to the destination from the source alphabet $\mathcal{S}$ and at the same source rate of one symbol every T_s seconds. The discrete memoryless channel has a channel capacity equal to C bits per use of the channel. We assume that the channel is capable of being used once every T_c seconds. Hence, the *channel capacity per unit time* is C/T_c bits per second, which represents the maximum rate of information transfer over the channel. We are now ready to state Shannon's second theorem, known as the channel coding theorem.

Specifically, the *channel coding theorem* for a discrete memoryless channel is stated in two parts as follows.*

(i) Let a discrete memoryless source with an alphabet $\mathcal{S}$ have entropy $H(\mathcal{S})$ and produce symbols once every T_s seconds. Let a discrete memoryless channel have capacity C and be used once every T_c seconds. Then, if

---

* The channel coding theorem is also known as the *noisy coding theorem*. The original proof of the theorem is given in Shannon (1948). A proof of the theorem is also presented in Hamming (1980, Chapters 9 and 10) in sufficient detail, so that a general appreciation of relevant results is developed. The second part of the theorem is referred to in the literature as *the converse to the coding theorem*. A proof of this theorem is presented in the following references: Viterbi and Omura (1979, pp. 28–34) and Gallager (1968, pp. 76–82).

$$\frac{H(\mathcal{S})}{T_s} \leqslant \frac{C}{T_c} \tag{2.63}$$

there exists a coding scheme for which the source output can be transmitted over the channel and be reconstructed with an arbitrarily small probability of error. The parameter C/T_c is called the critical rate. *When Eq. 2.63 is satisfied with the equality sign, the system is said to be* signaling at the critical rate.

(ii) Conversely, if

$$\frac{H(\mathcal{S})}{T_s} > \frac{C}{T_c}$$

it is not possible to transmit information over the channel and reconstruct it with an arbitrarily small probability of error.

The channel coding theorem is the single most important result of information theory. The theorem specifies the channel capacity C as a *fundamental limit* on the rate at which the transmission of reliable error-free messages can take place over a discrete memoryless channel.

It is important to note that the channel coding theorem does not show us how to construct a good code. Rather, the theorem can be characterized as an *existence proof* in the sense that it tells us that if the condition of Eq. 2.63 is satisfied, then good codes do exist. In Chapter 8, we describe several good codes suitable for use with discrete memoryless channels.

(1)   Application of the Channel Coding Theorem to Binary Symmetric Channels

Consider a discrete memoryless source that emits equally likely binary symbols (0s and 1s) once every T_s seconds. With the source entropy equal to one bit per source symbol (see Example 1), the information rate of the source is $(1/T_s)$ bits per second. The source sequence is applied to a binary channel encoder with *code rate r*. The encoder produces a symbol once every T_c seconds. Hence, the *encoded symbol transmission rate* is $(1/T_c)$ symbols per second. The encoder engages the use of a binary symmetric channel once every T_c seconds. Hence, the channel capacity per unit time is (C/T_c) bits per second, where C is determined by the prescribed channel transition probability p in accordance with Eqs. 2.61 and 2.62. Accordingly, the channel coding theorem [part (i)] implies that if

$$\frac{1}{T_s} \leqslant \frac{C}{T_c} \tag{2.64}$$

the probability of error can be made arbitrarily low by the use of a suitable encoding scheme. But the ratio T_c/T_s equals the code rate of the encoder:

$$r = \frac{T_c}{T_s} \tag{2.65}$$

Hence, we may restate the condition of Eq. 2.64 as

$$r \leqslant C \tag{2.66}$$

That is, for $r \leqslant C$, there exists a code (with code rate less than or equal to C) capable of achieving an arbitrarily low probability of error.

EXAMPLE 7

In this example, we present a graphical interpretation of the channel coding theorem. We also bring out a surprising aspect of the theorem by taking a look at a simple coding scheme.

Consider first a binary symmetric channel with transition probability $p = 10^{-2}$. For this value of p, we find from Eqs. 2.60 and 2.61 that the channel capacity $C = 0.9192$. Hence, from the channel coding theorem, we may state that for any $\varepsilon > 0$ and $r \leqslant 0.9192$, there exists a code of large enough length n and code rate r, and an appropriate decoding algorithm, such that when the code is used on the given channel, the average probability of decoding error is less than ε. This result is illustrated in Fig. 2.12, where we have plotted the average probability of error versus the code rate r. In this figure, we have arbitrarily set the limiting value $\varepsilon = 10^{-8}$.

To put the significance of this result in perspective, consider next a simple coding scheme that involves the use of a *repetition code,* in which each bit of the message is repeated several times. Let each bit (0 or 1) be repeated n times, where $n = 2m + 1$ is an odd integer. For example, for $n = 3$, we transmit 0 and 1 as 000 and 111, respectively. Intuitively, it would seem logical to use a *majority rule* for decoding, which operates as follows:

If in a block of n received bits (representing one bit of the message), the number of 0s exceeds the number of 1s, the decoder decides in favor of a 0. Otherwise it decides in favor of a 1.

Hence, an error occurs when $m + 1$ or more bits out of $n = 2m + 1$ bits are received incorrectly. Because of the assumed symmetric nature of the channel, the *average probability of error, P_e,* is independent of the a priori probabilities of 0 and 1. Accordingly, we find that P_e is given by (see Problem 2.7.1)

$$P_e = \sum_{i=m+1}^{n} \binom{n}{i} p^i (1 - p)^{n-i} \tag{2.67}$$

where p is the transition probability of the channel.

Table 2.3 gives the average probability of error P_e for a repetition code, which is calculated by using Eq. 2.67 for different values of the code rate r. The values given here assume the use of a binary symmetric channel with transition probability $p = 10^{-2}$. The improvement in reliability displayed in Table 2.3 is achieved at the cost of decreasing code rate. The results of this table are also shown plotted as the curve labeled "repetition code" in Fig. 2.12. This curve illustrates the *exchange of code rate for message reliability,* which is a characteristic of repetition codes.

Figure 2.12 highlights the unexpected result presented to us by the channel coding theorem. The result is that it is not necessary to have the code rate r approach zero (as in the case of repetition codes) so as to achieve more and more reliable operation of the communication link. The theorem merely requires that the code rate be less than the channel capacity C.

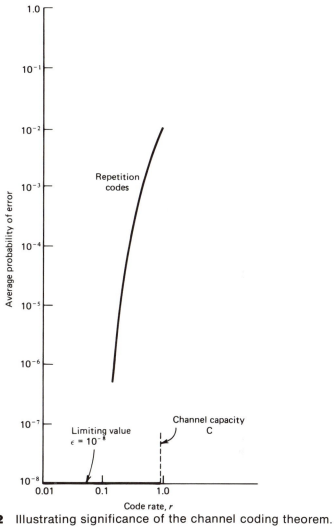

Figure 2.12 Illustrating significance of the channel coding theorem.

Table 2.3 Average Probability of Error for Repetition Code

Code Rate, $r = 1/n$	Average Probability of Error, P_e
1	10^{-2}
$\frac{1}{3}$	3×10^{-4}
$\frac{1}{5}$	10^{-6}
$\frac{1}{7}$	4×10^{-7}
$\frac{1}{9}$	10^{-8}
$\frac{1}{11}$	5×10^{-10}

2.8 DIFFERENTIAL ENTROPY AND MUTUAL INFORMATION FOR CONTINUOUS ENSEMBLES

The sources and channels considered in our discussion of information-theoretic concepts thus far have involved ensembles of random variables that are *discrete* in amplitude. In this section, we extend some of these concepts to *continuous* random variables and random vectors. The motivation for doing so is to pave the way for the description of another fundamental limit in information theory, which we take up in the next section.

Consider, a continuous random variable X with the *probability density function* $f_X(x)$. By analogy with the entropy of a discrete random variable, we introduce the following definition

$$h(X) = \int_{-\infty}^{\infty} f_X(x) \log_2\left[\frac{1}{f_X(x)}\right] dx \qquad (2.68)$$

We refer to $h(X)$ as the *differential entropy of* X to distinguish it from the ordinary or *absolute entropy*. We do so in recognition of the fact that although $h(X)$ is a useful mathematical quantity to know, it is *not* in any sense a measure of the randomness of X. Nevertheless, we justify the use of Eq. 2.68 as follows. We begin by viewing the continuous random variable X as the limiting form of a discrete random variable that assumes the value $x_k = k\Delta x$, where $k = 0, \pm 1, \pm 2, \ldots$, and Δx approaches zero. By definition, the continuous random variable X assumes a value in the interval $[x_k, x_k + \Delta x]$ with probability $f_X(x_k)\Delta x$. Hence, permitting Δx to approach zero, the ordinary entropy of the continuous random variable X may be written in the limit as follows

$$
\begin{aligned}
H(X) &= \lim_{\Delta x \to 0} \sum_{k=-\infty}^{\infty} f_X(x_k)\, \Delta x\, \log_2\left(\frac{1}{f_X(x_k)\Delta x}\right) \\
&= \lim_{\Delta x \to 0}\left[\sum_{k=-\infty}^{\infty} f_X(x_k)\log_2\left(\frac{1}{f_X(x_k)}\right)\Delta x - \log_2\Delta x \sum_{k=-\infty}^{\infty} f_X(x_k)\Delta x\right] \\
&= \int_{-\infty}^{\infty} f_X(x)\log_2\left(\frac{1}{f_X(x)}\right) dx - \lim_{\Delta x \to 0} \log_2\Delta x \int_{-\infty}^{\infty} f_X(x)dx \\
&= h(X) - \lim_{\Delta x \to 0} \log_2\Delta x \qquad (2.69)
\end{aligned}
$$

where, in the last line, we have made use of Eq. 2.68 and the fact that the total area under the curve of the probability density function $f_X(x)$ is unity. In the limit as Δx approaches zero, $\log_2\Delta x$ approaches infinity. This means that the entropy of a continuous random variable is infinitely large. Intuitively, we would expect this to be true, because a continuous random variable may assume a value anywhere in the interval $[-\infty,\infty]$ and the uncertainty associated with the variable is of the order of infinity. We avoid the problem associated with the term $\log_2\Delta x$ by adopting $h(X)$ as a differential entropy, with the term $-\log_2\Delta x$ serving as reference. Moreover, since the information transmitted over a channel is actually the difference between two entropy terms that have a common reference, the information will be the same as the difference between the corresponding differential entropy terms. We are therefore perfectly justified in using the term $h(X)$, defined in Eq. 2.68, as the differential entropy of the continuous random variable X.

When we have a continuous random vector $\mathbf{X}$ consisting of n random variables $X_1, X_2, \ldots, X_n$, we define the differential entropy of $\mathbf{X}$ as the *n-fold integral*

$$h(\mathbf{X}) = \int_{-\infty}^{\infty} f_{\mathbf{X}}(\mathbf{x}) \log_2 \left[\frac{1}{f_{\mathbf{X}}(\mathbf{x})} \right] d\mathbf{x} \qquad (2.70)$$

where $f_{\mathbf{X}}(\mathbf{x})$ is the *joint probability density function* of $\mathbf{X}$.

EXAMPLE 8   MAXIMUM DIFFERENTIAL ENTROPY FOR SPECIFIED VARIANCE

In this example, we solve an important *constrained optimization problem*. We determine the form that the probability density function of a random variable X must have for the differential entropy of X to assume its largest value for some prescribed variance. In mathematical terms, we may restate the problem as follows:

With the differential entropy of a random variable X defined by

$$h(X) = - \int_{-\infty}^{\infty} f_X(x) \log_2 f_X(x) dx,$$

find the probability density function $f_X(x)$ for which $h(X)$ is maximum, subject to the two constraints

$$\int_{-\infty}^{\infty} f_x(x) dx = 1 \qquad (2.71)$$

and

$$\int_{-\infty}^{\infty} (x - \mu)^2 f_X(x) \, dx = \sigma^2 = \text{constant} \qquad (2.72)$$

where μ is the mean of X and σ^2 is its variance.

The formula for $h(X)$ is that of Eq. 2.68, reproduced here with a minor modification. The first constraint, Eq. 2.71 simply states that the area under $f_X(x)$, a probability density function, must equal unity. The second constraint, Eq. 2.72, recognizes that the variance of X has a prescribed value. The second constraint is significant, because σ^2 is a measure of average power, and so maximization of the differential entropy $h(X)$ is performed subject to a constraint of constant power. The result of this constrained optimization will be exploited later on in Section 2.9.

We use the *method of Lagrange multipliers** to solve this constrained optimization problem. Specifically, the differential entropy $h(X)$ will attain its maximum value only when the integral

---

* The method of Lagrange multipliers is described in Kaplan (1952, pp. 128–130).

$$\int_{-\infty}^{\infty} [-f_X(x) \log_2 f_X(x) + \lambda_1 f_X(x) + \lambda_2 (x - \mu)^2 f_X(x)] dx$$

is *stationary*. The parameters λ_1 and λ_2 are known as *Lagrange multipliers*. That is to say, $h(X)$ is maximum only when the derivative of the integrand

$$-f_X(x) \log_2 f_X(x) + \lambda_1 f_X(x) + \lambda_2 (x - \mu)^2 f_X(x)$$

with respect to $f_X(x)$ is zero. This yields the result

$$-\log_2 e + \lambda_1 + \lambda_2 (x - \mu)^2 = \log_2 f_X(x)$$

$$= (\log_2 e) \ln f_X(x)$$

where e is the base of the natural logarithm. Solving for $f_X(x)$, we get

$$f_X(x) = \exp\left[-1 + \frac{\lambda_1}{\log_2 e} + \frac{\lambda_2}{\log_2 e}(x - \mu)^2\right] \tag{2.73}$$

Note that λ_2 has to be negative if the integrals of $f_X(x)$ and $(x - \mu)^2 f_X(x)$ with respect to x are to converge. Substituting Eq. 2.73 in Eqs. 2.71 and 2.72, and then solving for λ_1 and λ_2, we get

$$\lambda_1 = \frac{1}{2} \log_2 \left(\frac{e}{2\pi\sigma^2}\right)$$

and

$$\lambda_2 = -\frac{\log_2 e}{2\sigma^2}$$

The desired form for $f_X(x)$ is therefore described by

$$f_X(x) = \frac{1}{\sqrt{2\pi}\sigma} \exp\left(-\frac{(x - \mu)^2}{2\sigma^2}\right) \tag{2.74}$$

which is recognized as the probability density of a *Gaussian random variable X of mean μ and variance σ^2*. The maximum value of the differential entropy of such a random variable is obtained by substituting Eq. 2.74 in Eq. 2.68. The result of this substitution is given by

$$h(X) = \frac{1}{2} \log_2(2\pi e \sigma^2) \tag{2.75}$$

We may thus summarize the results of this example, as follows:

1. *For a given variance σ^2, the Gaussian random variable has the largest differential entropy attainable by any random variable.* That is, if X is a Gaussian random variable and Y is any other random variable with the same mean and variance, then for all Y

$$h(X) \geq h(Y) \tag{2.76}$$

where the equality holds if, and only if, $Y = X$.

2. *The entropy of a Gaussian random variable X is uniquely determined by the variance of X* (i.e., it is independent of the mean of X).

Indeed, it is because of Property 1 that the Gaussian channel model is so widely used in the study of digital communication systems.

(1) Mutual Information

Consider next a pair of continuous random variables X and Y. By analogy with Eq. 2.44, we define the *mutual information* between the random variables X and Y as follows

$$I(X;Y) = \int_{-\infty}^{\infty} \int_{-\infty}^{\infty} f_{X,Y}(x,y) \, \log_2 \left[\frac{f_X(x|y)}{f_X(x)} \right] dxdy \qquad (2.77)$$

where $f_{X,Y}(x,y)$ is the *joint probability density function of X* and Y, and $f_X(x|y)$ is the *conditional* probability density function of X, given that $Y = y$. Also, by analogy with Eqs. 2.42, 2.47, 2.41, and 2.51, we find that the mutual information $I(X;Y)$ has the following properties respectively,

1. $I(X;Y) = I(Y;X)$ $\qquad\qquad\qquad\qquad\qquad\qquad\qquad\qquad$ (2.78)

2. $I(X;Y) \geq 0$ $\qquad\qquad\qquad\qquad\qquad\qquad\qquad\qquad\qquad\quad$ (2.79)

3. $I(X;Y) = h(X) - h(X|Y)$ $\qquad\qquad\qquad\qquad\qquad\qquad\quad$ (2.80)

4. $I(X;Y) = h(Y) - h(Y|X)$ $\qquad\qquad\qquad\qquad\qquad\qquad\quad$ (2.81)

The parameter $h(X)$ is the differential entropy of X; likewise for $h(Y)$. The parameter $h(X|Y)$ is the *conditional differential entropy* of X, given Y; it is defined by the double integral

$$h(X|Y) = \int_{-\infty}^{\infty} \int_{-\infty}^{\infty} f_{X,Y}(x,y) \, \log_2 \left[\frac{1}{f_X(x|y)} \right] dxdy \qquad (2.82)$$

The parameter $h(Y|X)$ is the conditional differential entropy of Y, given X; it is defined in a manner similar to $h(X|Y)$.

2.9 CHANNEL CAPACITY THEOREM

In Section 2.2 we presented the source coding theorem, which defines the fundamental limit on the faithful representation of a discrete memoryless source. Then, in Section 2.7 we presented the channel coding theorem, which defines the fundamental limit on error-free transmission over a discrete memoryless channel. In this section, we present our third (and final) limit in information theory, which is defined by the *capacity theorem* for *band-limited, power-limited Gaussian channels*.

To be specific, consider a zero-mean stationary process $X(t)$ that is band-limited to B hertz. Let X_k, $k = 1,2, \ldots ,n$, denote the *continuous* random variables obtained by uniform *sampling* of the process $X(t)$ at the rate of $2B$ samples per second. (In Chapter 4, it is shown that if a random process $X(t)$, limited to the frequence interval $-B \leq f \leq B$, is sampled at a rate equal to or greater than $2B$ samples per second, then the original process can be reconstructed from its samples with zero mean-squared error). These samples are transmitted in T seconds over a noisy channel, also band-limited to B Hertz. Hence, the number of samples, n, is given by

$$n = 2BT \qquad (2.83)$$

We refer to X_k as a sample of the *transmitted signal*. The channel output is perturbed by *additive white Gaussian noise* of zero mean and power spectral density $N_0/2$. The noise is band-limited to B hertz. Let the continuous random variables Y_k, $k = 1, 2, \ldots, n$ denote samples of the *received signal*, as shown by

$$Y_k = X_k + N_k \qquad k = 1, 2, \ldots, n \qquad (2.84)$$

The *noise* sample N_k is Gaussian with zero mean and variance given by

$$\sigma^2 = N_0 B \qquad (2.85)$$

We assume that the samples Y_k, $k = 1, 2, \ldots, n$, are *statistically independent*.

A channel for which the noise and the received signal are as described is called a *discrete-time, memoryless Gaussian channel*. It is modeled as shown in Fig. 2.13.

To make meaningful statements about the channel, we have to assign a *cost* to each channel input. Typically, the transmitter is *power-limited*. It is therefore reasonable to define the cost as

$$E[X_k^2] = P \qquad k = 1, 2, \ldots, n \qquad (2.86)$$

where P is the *average transmitted power*. Accordingly, we define the *channel capacity* as

$$C = \max_{f_{X_k}(x)} \{ I(X_k; Y_k) : E[X_k^2] = P \} \qquad (2.87)$$

where $I(X_k; Y_k)$ is the average mutual information between a sample of the transmitted signal, X_k, and the corresponding sample of the received signal, Y_k. The maximization indicated in Eq. 2.87 is performed with respect to $f_{X_k}(x)$, the probability density function of X_k.

The average mutual information $I(X_k; Y_k)$ can be expressed in one of two equivalent forms, in accordance with Eqs. 2.80 and 2.81. For the purpose at hand, we use the form of Eq. 2.81, restated as

$$I(X_k; Y_k) = h(Y_k) - h(Y_k|X_k) \qquad (2.88)$$

Since X_k and N_k are independent random variables, and their sum equals Y_k, as in Eq. 2.84, we find that the conditional differential entropy of Y_k, given X_k, is equal to the conditional entropy of N_k (see Problem 2.8.4):

$$h(Y_k|X_k) = h(N_k) \qquad (2.89)$$

Hence, we may rewrite Eq. 2.88 as

$$I(X_k; Y_k) = h(Y_k) - h(N_k) \qquad (2.90)$$

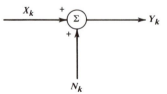

Figure 2.13 Model of discrete-time, memoryless Gaussian channel.

Since $h(N_k)$ is independent of the distribution of X_k, maximizing $I(X_k;Y_k)$ in accordance with Eq. 2.87 requires maximizing $h(Y_k)$, the differential entropy of sample Y_k of the received signal. For $h(Y_k)$ to be maximum, Y_k has to be a Gaussian random variable (see Example 8). That is, the samples of the received signal represent a noise-like process. Next, we observe that since N_k is Gaussian by assumption, the sample X_k of the transmitted signal must be Gaussian too. We may therefore state that the maximization specified in Eq. 2.87 is attained by choosing the samples of the transmitted signal from a noise-like process of average power P. Correspondingly, we may reformulate Eq. 2.87 as

$$C = I(X_k;Y_k) : X_k \text{ Gaussian}, \qquad E[X_k^2] = P \qquad (2.91)$$

where the mutual information $I(X_k;Y_k)$ is given in Eq. 2.90.

For the evaluation of the channel capacity C, we proceed in three stages:

1. The variance of sample Y_k of the received signal equals $P + \sigma^2$. Hence, the use of Eq. 2.75 yields the differential entropy of Y_k as

$$h(Y_k) = \frac{1}{2} \log_2[2\pi e(P + \sigma^2)] \qquad (2.92)$$

2. The variance of the noise sample N_k equals σ^2. Hence, the use of Eq. 2.75 yields the differential entropy of N_k as

$$h(N_k) = \frac{1}{2} \log_2(2\pi e\sigma^2) \qquad (2.93)$$

3. Substituting Eqs. 2.92 and 2.93 in Eq. 2.90 and recognizing the definition of channel capacity given in Eq. 2.91, we get the desired result:

$$C = \frac{1}{2} \log_2\left(1 + \frac{P}{\sigma^2}\right) \text{ bits/channel use} \qquad (2.94)$$

With the channel used n times for the transmission of n samples of the process $X(t)$ in T seconds, we find that the *channel capacity per unit time* is (n/T) times the result given in Eq. 2.94. The number n equals $2BT$, as in Eq. 2.83. Accordingly, we may express the channel capacity per unit time as

$$C = B \log_2\left(1 + \frac{P}{N_0 B}\right) \text{ bits/s} \qquad (2.95)$$

where we have used Eq. 2.85 for the noise variance σ^2.

Based on the formula of Eq. 2.95, we may now state Shannon's third (and most famous) theorem, the *channel capacity theorem*,* as follows:

---

* Shannon's channel capacity theorem is also referred to in the literature as the *Shannon–Hartley law* in recognition of early work by Hartley on information transmission (Hartley, 1928). In particular, Hartley showed that the amount of information that can be transmitted over a given channel is proportional to the product of the channel bandwidth and the time of operation.

The capacity of a channel of bandwidth B hertz, perturbed by additive white Gaussian noise of power spectral density $N_0/2$ and limited in bandwidth to B, is given by

$$C = B \log_2\left(1 + \frac{P}{N_0 B}\right) \text{ bits/s}$$

where P is the average transmitted power.

The channel capacity theorem is one of the most remarkable results of information theory for, in a single formula, it highlights most vividly the interplay between three key system parameters: channel bandwidth, average transmitted power (or, equivalently, average received signal power), and noise power spectral density at the channel output.

The theorem implies that, for given average transmitted power P and channel bandwidth B, we can transmit information at the rate C bits per second, as defined in Eq. 2.95, with arbitrarily small probability of error by employing sufficiently complex encoding systems. It is not possible to transmit at a rate higher than C bits per second by any encoding system without a definite probability of error. Hence, the channel capacity theorem defines the *fundamental limit* on the rate of error-free transmission for a power-limited, band-limited Gaussian channel. To approach this limit, however, the transmitted signal must have statistical properties approximating those of white Gaussian noise.

(2) Ideal System

We define an *ideal system* as one that transmits data at a bit rate R_b equal to the channel capacity C. We may then express the average transmitted power as

$$P = E_b C \tag{2.96}$$

where E_b is the transmitted energy per bit. Accordingly, the ideal system is defined by the equation

$$\frac{C}{B} = \log_2\left(1 + \frac{E_b}{N_0}\frac{C}{B}\right) \tag{2.97}$$

Equivalently, we may define the *energy-per-bit to noise power spectral density ratio*, E_b/N_0, in terms of the *bandwidth efficiency*, C/B, for the ideal system as

$$\frac{E_b}{N_0} = \frac{2^{C/B} - 1}{C/B} \tag{2.98}$$

This equation is displayed as the curve labeled "capacity boundary" in Fig. 2.14.

A plot of bandwidth efficiency R_b/B versus E_b/N_0 as in Fig. 2.14 is called the *bandwidth-efficiency diagram*. Based on this diagram, we can make the following *observations:*

1. For *infinite bandwidth*, the signal energy-to-noise density ratio E_b/N_0 approaches the limiting value

$$\left(\frac{E_b}{N_0}\right)_\infty = \lim_{B\to\infty} \left(\frac{E_b}{N_0}\right)$$

$$= \ln 2 = 0.693 \tag{2.99}$$

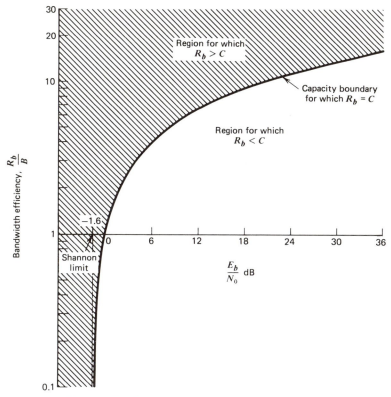

Figure 2.14 Bandwidth-efficiency diagram.

This value is called the *Shannon limit*. Expressed in decibels, it equals -1.6 dB. The corresponding limiting value of the channel capacity is obtained by letting the channel bandwidth B in Eq. 2.95 approach infinity; we thus find that

$$C_\infty = \lim_{B \to \infty} C$$

$$= \frac{P}{N_0} \log_2 e \qquad (2.100)$$

2. The *capacity boundary*, defined by the curve for the critical bit rate $R_b = C$, separates combinations of system parameters that have the potential for supporting error-free transmission ($R_b < C$) from those for which error-free transmission is not possible ($R_b > C$). The latter region is shown shaded in Fig. 2.14.

3. The diagram highlights potential *trade-offs* among E_b/N_0, R_b/B, and probability of symbol error P_e. In particular, we may view movement of the operating point along a horizontal line as trading P_e versus E_b/N_0 for a fixed R_b/B. On the other hand, we may view movement of the operating point along a vertical line as trading P_e versus R_b/B, for a fixed E_b/N_0.

The Shannon limit may also be defined in terms of the E_b/N_0 required by the ideal system for error-free transmission to be possible. Specifically, we may write for the ideal system:

$$P_e = \begin{cases} 0 & (E_b/N_0) \geq \ln 2 \\ 1 & (E_b/N_0) < \ln 2 \end{cases} \qquad (2.101)$$

This viewpoint of the ideal system is described in Fig. 2.15. The boundary between error-free transmission and unreliable transmission with possible errors, defined by the Shannon limit in Fig. 2.15, is analogous to the capacity boundary in Fig. 2.14.

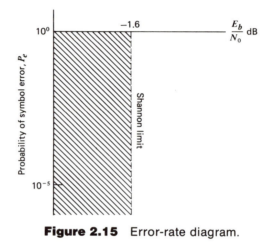

Figure 2.15 Error-rate diagram.

2.10 SUMMARY AND DISCUSSION

In this chapter we have established three fundamental limits on different aspects of a digital communication system. The limits are embodied in the source coding theorem, the channel coding theorem, and the channel capacity theorem. All three theorems are due to Shannon.

The *source coding theorem,* Shannon's first theorem, provides us with a yardstick by which we can measure the information emerging from a *discrete memoryless source.* The theorem tells us that we can make the average number of binary code elements (bits) per source symbol as small as, but no smaller than, the entropy of the source measured in bits. The *entropy* of a source is a function of the probabilities of the source symbols that constitute the alphabet of the source. Since entropy is a measure of uncertainty, the entropy is maximum when the associated probability distribution generates maximum uncertainty.

The *channel coding theorem,* Shannon's second theorem, is the most surprising as well as the single most important result of information theory. For a *binary symmetric channel,* the channel coding theorem tells us that for any *code rate r* less than or equal to the *channel capacity C*, codes do exist such that the average probability of error is as small as we want it. A binary symmetric channel is the simplest form of a discrete memoryless channel. It is symmetric because the probability of receiving a 1 if a 0 is sent is the same as the probability of receiving a 0 if a 1 is sent. This probability, the probability that an error will occur, is termed a *transition probability*. The transition probability, p, is determined not only by the additive noise at the channel output but also the kind of receiver used. The value of p uniquely defines the channel capacity C.

Shannon's third remarkable theorem, the *channel capacity theorem,* tells us that there is a maximum to the rate at which any communication system can operate reliably (i.e., free of errors) when the system is constrained in power. This maximum rate is called the *channel capacity,* measured in bits per second. When the system operates at a rate greater than the channel capacity, it is condemned to a high probability of error, regardless of the choice of signal set used for transmission or the receiver used for processing the received signal.

The subject of channel coding, inspired by the channel coding theorem, is treated in detail in Chapter 8. Finally, the channel capacity theorem is applied in Chapters 5 and 7 to compare the performance of different modulation techniques to that of the ideal system.

PROBLEMS

The problems are divided into sections that correspond to the major sections in the chapter. For example, the problems in Section P2.1 relate to the material in Section 2.1, and so on. This practice is followed in *all* subsequent chapters of the book.

P2.1 UNCERTAINTY, INFORMATION, AND ENTROPY

Problem 2.1.1 Let p denote the probability of some event. Plot the amount of information gained by the occurrence of this event for $0 \leqslant p \leqslant 1$.

Problem 2.1.2 A source emits one of four possible symbols during each signaling interval. The symbols occur with the probabilities:

$$p_0 = 0.4$$
$$p_1 = 0.3$$
$$p_2 = 0.2$$
$$p_3 = 0.1$$

Find the amount of information gained by observing the source emitting each of these symbols.

Problem 2.1.3 A source emits one of four symbols s_0, s_1, s_2, and s_3 with probabilities $\frac{1}{3}$, $\frac{1}{6}$, $\frac{1}{4}$, and $\frac{1}{4}$, respectively. The successive symbols emitted by the source are statistically independent. Calculate the entropy of the source.

Problem 2.1.4 Let X represent the outcome of a single roll of a fair die. What is the entropy of X?

Problem 2.1.5 The sample function of a Gaussian process of zero mean and unit variance is uniformly sampled, and then applied to a uniform quantizer having the input–output amplitude characteristic shown in Fig. P2.1. Calculate the entropy of the quantizer output.

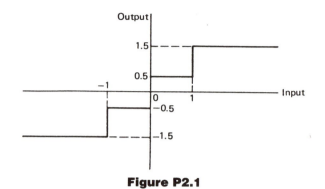

Figure P2.1

Problem 2.1.6 Consider a discrete memoryless source with source alphabet $\mathcal{S} = \{s_0, s_1, \ldots, s_{K-1}\}$ with source statistics $\{p_0, p_1, \ldots, p_{K-1}\}$. The nth extension of this source is another discrete memoryless source with source alphabet $\mathcal{S}^n = \{\sigma_0, \sigma_1, \ldots, \sigma_{M-1}\}$, where $M = K^n$. Let σ_i correspond to the sequence $\{s_{i_1}, s_{i_2}, \ldots, s_{i_n}\}$. Let $P(\sigma_i)$ denote the probability of σ_i.
(a) Show that

$$\sum_{i=0}^{M-1} P(\sigma_i) = 1$$

which is to be expected.
(b) Show that

$$\sum_{i=0}^{M-1} P(\sigma_i) \log_2\left(\frac{1}{p_{i_k}}\right) = H(\mathcal{S}) \qquad k = 1, 2, \ldots, n$$

where p_{i_k} is the probability of symbol s_{i_k}, and $H(\mathcal{S})$ is the entropy of the original source.
(c) Hence, show that

$$H(\mathcal{S}^n) = \sum_{i=0}^{M-1} P(\sigma_i) \log_2 \frac{1}{P(\sigma_i)}$$

$$= nH(\mathcal{S})$$

Problem 2.1.7 Consider a discrete memoryless source with source alphabet $\mathcal{S} = \{s_0, s_1, s_2\}$ and source statistics $\{0.7, 0.15, 0.15\}$.
(a) Calculate the entropy of the source.
(b) Calculate the entropy of the second-order extension of the source.

P2.2 SOURCE CODING THEOREM

Problem 2.2.1 Consider a discrete memoryless source whose alphabet consists of K equiprobable symbols.
(a) Explain why the use of a fixed-length code for the representation of such a source is about as efficient as any code can be.
(b) What conditions have to be satisfied by K and the code length for the coding efficiency to be 100 percent?

Problem 2.2.2 An inventor claims to have developed a device for storing arbitrary English text in computer memory; the device requires 1 bit of memory for every 10

Latin letters. The entropy of English has been estimated by researchers to be in the range 1–2 bits per letter. The estimate takes account of the fact that some English letters are more likely than others and that successive letters in English text are highly dependent. Given this estimate of the entropy of English, is the claim made by the inventor correct? Justify your answer.

Problem 2.2.3   Consider the four codes listed below:

Symbol	Code I	Code II	Code III	Code IV
s_0	0	0	0	00
s_1	10	01	01	01
s_2	110	001	011	10
s_3	1110	0010	110	110
s_4	1111	0011	111	111

Two of these four codes are prefix codes. Identify them, and construct their individual decision trees.

P2.3   HUFFMAN CODING

Problem 2.3.1   A discrete memoryless source has an alphabet of five symbols with their probabilities for its output, as given here:

Symbol	s_0	s_1	s_2	s_3	s_4
Probability	0.55	0.15	0.15	0.10	0.05

Compute two different Huffman codes for this source. Hence, for each of the two codes, find
(a)   The average code-word length.
(b)   The variance of the average code-word length over the ensemble of source symbols.

Problem 2.3.2   A discrete memoryless source has an alphabet of seven symbols with probabilities for its output, as described here:

Symbol	s_0	s_1	s_2	s_3	s_4	s_5	s_6
Probability	0.25	0.25	0.125	0.125	0.125	0.0625	0.0625

Compute the Huffman code for this source, moving a "combined" symbol as high as possible. Explain why the computed source code has an efficiency of 100 percent.

Problem 2.3.3   Consider a discrete memoryless source with alphabet $\{s_0, s_1, s_2\}$ and statistics $\{0.7, 0.15, 0.15\}$ for its output.
(a)   Apply the Huffman algorithm to this source. Hence, show that the average code-word length of the Huffman code equals 1.3 bits/symbol.
(b)   Let the source be extended to order two. Apply the Huffman algorithm to the resulting extended source, and show that the average code-word length of the new code equals 1.1975 bits/symbol.
(c)   Compare the average code-word length calculated in part (b) with the entropy of the original source.

P2.4   DISCRETE MEMORYLESS CHANNELS

Problem 2.4.1   Consider the transition probability diagram of a binary symmetric channel shown in Fig. 2.8. The input binary symbols 0 and 1 occur with equal probability. Find the probabilities of the binary symbols 0 and 1 appearing at the channel output.

Problem 2.4.2   Repeat the calculation in Problem 2.4.1, assuming that the input binary symbols 0 and 1 occur with probabilities 1/4 and 3/4, respectively.

P2.5   MUTUAL INFORMATION

Problem 2.5.1   Consider a binary symmetric channel characterized by the transition probability p. Plot the mutual information of the channel as a function of p_1, the a priori probability of symbol 1 at the channel input, for the transition probability $p = 0, 0.1, 0.2, 0.3, 0.5$.

P2.6   CHANNEL CAPACITY

Problem 2.6.1   Figure 2.10 depicts variation of the channel capacity of a binary symmetric channel with the transition probability p. Use the results of Problem 2.5.1 to explain this variation.

Problem 2.6.2   Consider the binary symmetric channel shown in Fig. 2.8. Let p_0 denote the probability of choosing binary symbol $x_0 = 0$, and let $p_1 = 1 - p_0$ denote the probability of choosing binary symbol $x_1 = 1$. Let p denote the transition probability of the channel.

(a)   Show that the average mutual information between the channel input and channel output is given by

$$I(\mathscr{X};\mathscr{Y}) = \mathscr{H}(z) - \mathscr{H}(p)$$

where

$$\mathscr{H}(z) = z \log_2\left(\frac{1}{z}\right) + (1 - z) \log_2\left(\frac{1}{1 - z}\right)$$

$$z = p_0 p + (1 - p_0)(1 - p)$$

and

$$\mathscr{H}(p) = p \log_2\left(\frac{1}{p}\right) + (1 - p) \log_2\left(\frac{1}{1 - p_0}\right)$$

(b)   Show that the value of p_0 that maximizes $I(\mathscr{X};\mathscr{Y})$ is equal to 1/2.
(c)   Show that the channel capacity equals

$$C = 1 - \mathscr{H}(p)$$

Problem 2.6.3   Two binary symmetric channels are connected in cascade, as shown in Fig. P2.2. Find the overall channel capacity of the cascaded connection, assuming that both channels have the same transition probability diagram shown in Fig. 2.8.

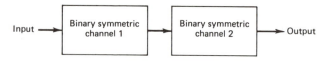

Figure P2.2

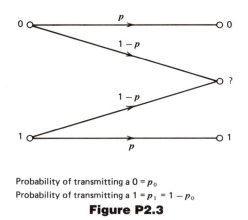

Probability of transmitting a 0 = p_0
Probability of transmitting a 1 = p_1 = 1 − p_0
Figure P2.3

Problem 2.6.4 The *binary erasure channel* is described in Fig. P2.3. The inputs are labeled "0" and "1," and the outputs are labeled "0", "1", and "?". Find the capacity of the channel.

P2.7 CHANNEL CODING THEOREM

Problem 2.7.1 Consider a digital communication system that uses a *repetition code* for the channel encoding/decoding. In particular, each transmission is repeated n times, where $n = 2m + 1$ is an odd integer. The decoder operates as follows. If in a block of n received bits, the number of 0s exceeds the number of 1s, the decoder decides in favor of a "0". Otherwise, it decides in favor of a "1". An error occurs when $m + 1$ or more transmissions out of $n = 2m + 1$ are incorrect. Assume a binary symmetric channel.
(a) For $n = 3$, show that the average probability of error is given by

$$P_e = 3p^2(1 - p) + p^3$$

where p is the transition probability of the channel.
(b) For $n = 5$, show that the average probability of error is given by

$$P_e = 10p^3(1 - p)^2 + 5p^4(1 - p) + p^5$$

(c) For the general case, show that the average probability of error is given by

$$P_e = \sum_{i=m+1}^{n} \binom{n}{i} p^i (1 - p)^{n-i}$$

P2.8 DIFFERENTIAL ENTROPY AND MUTUAL INFORMATION FOR CONTINUOUS RANDOM ENSEMBLES

Problem 2.8.1 Let $X_1, X_2, \ldots, X_n$ denote the elements of a Gaussian vector **X**. The X_i are independent with means μ_i and variances σ_i^2, $i = 1, 2, \ldots, n$. Show that the differential entropy of the vector **X** equals

$$h(\mathbf{X}) = \frac{n}{2} \log_2[2\pi e(\sigma_1^2 \, \sigma_2^2 \ldots \, \sigma_n^2)^{1/n}]$$

What does $h(\mathbf{X})$ reduce to if the variances are equal?

Problem 2.8.2   A continuous random variable X is constrained to a peak magnitude M; that is, $-M < X < M$.
(a) Show that the differential entropy of X is maximum when it is uniformly distributed, as shown by

$$f_X(x) = \begin{cases} 1/2M, & -M < x \leqslant M \\ 0, & \text{otherwise} \end{cases}$$

(b) Show that the maximum differential entropy of X is $\log_2 2M$.

Problem 2.8.3   Prove the properties given in Eqs. 2.78–2.81 for the mutual information $I(X;Y)$.

Problem 2.8.4   Consider the continuous random variable Y defined by

$$Y = X + N$$

where X and N are statistically independent. Show that the conditional differential entropy of Y, given X, equals

$$h(Y|X) = h(N)$$

where $h(N)$ is the differential entropy of N.

P2.9   CHANNEL CAPACITY THEOREM

Problem 2.9.1   A voice-grade channel of the telephone network has a bandwidth of 3.4 kHz.
(a) Calculate the channel capacity of the telephone channel for a signal-to-noise ratio of 30 dB.
(b) Calculate the minimum signal-to-noise ratio required to support information transmission through the telephone channel at the rate of 4800 bits per second.

Problem 2.9.2   Alphanumeric data are entered into a computer from a remote terminal through a voice-grade telephone channel. The channel has a bandwidth of 3.4 kHz, and output signal-to-noise ratio of 20 dB. The terminal has a total of 128 symbols. Assume that the symbols are equiprobable, and the successive transmissions are statistically independent.
(a) Calculate the channel capacity.
(b) Calculate the maximum symbol rate for which error-free transmission over the channel is possible.

Problem 2.9.3   A black-and-white television picture may be viewed as consisting of approximately 3×10^5 elements, each one of which may occupy one of 10 distinct brightness levels with equal probability. Assume (a) the rate of transmission is 30 picture frames per second, and (b) the signal-to-noise ratio is 30 dB.
   Using the channel capacity theorem, calculate the minimum bandwidth required to support the transmission of the resultant video signal.

NOTE   As a matter of interest, commercial television transmissions actually employ a bandwidth of 4 MHz.

CHAPTER THREE

DETECTION AND ESTIMATION

In the previous chapter, we used information-theoretic concepts to define fundamental limits on the efficiency of a source and the rate of reliable information transmission over a channel. In this chapter, we consider two other fundamental issues in digital communications, namely, detection and estimation in the presence of additive noise. Both of these topics are of particular interest in determining the optimum design of a receiver.

Detection theory deals with the design and evaluation of a decision-making processor that observes the received signal and guesses which particular symbol was transmitted according to some set of rules. *Estimation theory,* on the other hand, deals with the design and evaluation of a processor that uses information in the received signal to extract estimates of physical parameters or waveforms of interest. The results of detection and estimation are always subject to *errors.* The nature of the errors made in detection is of course different from those in estimation. Nevertheless, in both cases, the challenge is to control the errors so as to ensure an acceptable quality of performance.

We begin the study by describing a mathematical model of digital communication systems, which provides a framework for the development of detection theory. Later in the chapter, we consider some aspects of estimation theory.

3.1 MODEL OF DIGITAL COMMUNICATION SYSTEM

In Fig. 3.1, we depict the *conceptualized model* of a digital communication system. At the transmitter input, we have a *message source* that emits one *symbol* every T seconds, with the symbols belonging to an alphabet of M symbols which we denote by $m_1, m_2, \ldots, m_M$. For example, in the remote connection of two digital computers, we have one computer acting as an information source that calculates digital outputs based on observations and inputs fed into it. The resulting computer output is expressed as a sequence of 0s and 1s, which are transmitted to the second computer. In this example, the alphabet consists simply of the two binary symbols 0 and 1. A second example is that of a quaternary signaling scheme with an alphabet consisting of four possible symbols: 00, 01, 10, and 11. In the sequel, we assume that all M symbols of the alphabet are equally likely. Then we may write the *a priori probability* of the message source output as

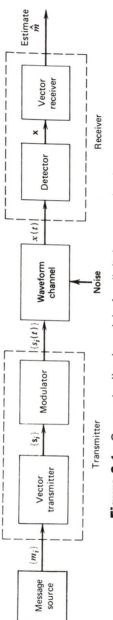

Figure 3.1 Conceptualized model of a digital communication system.

$$p_i = P(m_i \text{ emitted})$$

$$= \frac{1}{M} \quad \text{for all } i \tag{3.1}$$

The output of the message source is presented to a *vector transmitter*, producing a *vector* of real numbers. In particular, when the source output $m = m_i$, the vector transmitter output takes on the value

$$\mathbf{s}_i = \begin{bmatrix} s_{i1} \\ s_{i2} \\ \vdots \\ s_{iN} \end{bmatrix} \quad i = 1, 2, \ldots, M \tag{3.2}$$

where the *dimension* $N \leq M$. With this vector as input, the *modulator* then constructs a *distinct* signal $s_i(t)$ of duration T seconds. The signal $s_i(t)$ is necessarily of finite energy, as shown by

$$E_i = \int_0^T s_i^2(t) dt \quad i = 1, 2, \ldots, M \tag{3.3}$$

Note that the signal $s_i(t)$ is *real-valued*. One such signal is transmitted every T seconds. The particular signal chosen for transmission depends in some fashion on the incoming message and possibly on the signals transmitted in preceding time slots. Its characterization also depends on the nature of the *physical channel* available for communication. The channel is assumed to have two characteristics:

1. The channel is linear, with a bandwidth that is large enough to accommodate the transmission of the modulator output $s_i(t)$ without distortion.
2. The transmitted signal $s_i(t)$ is perturbed by an *additive, zero-mean, stationary, white, Gaussian noise* process, denoted by $W(t)$. The reasons for this assumption are that it makes calculations tractable, and also it is a reasonable description of the type of *receiver noise** present in many communication systems.

We refer to such a channel as an *additive white Gaussian noise (AWGN) channel*. Accordingly, we express the received random process, $X(t)$, as

$$X(t) = s_i(t) + W(t) \quad \begin{matrix} 0 \leq t \leq T \\ i = 1, 2, \ldots, M \end{matrix} \tag{3.4}$$

We may thus model the channel as in Fig. 3.2, where the noise process $W(t)$ is represented by the sample function $w(t)$, and the received random process $X(t)$ is correspondingly represented by the sample function $x(t)$. From here on, we refer to $x(t)$ as the *received signal*.

The receiver has the task of observing the received signal $x(t)$, for a duration of T seconds, and making a best *estimate* of the transmitted signal $s_i(t)$ or,

---

* "Front-end receiver noise" and "channel noise" mean one and the same thing. We use them interchangeably.

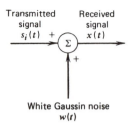

Figure 3.2 Model of an additive white Gaussian noise channel.

equivalently, the symbol m_i. This task is accomplished in two stages. The first stage is a *detector* that operates on the received random process $X(t)$ to produce a vector of random variables, **X**. By using an *observation vector* **x** (which is a sample value of **X**), prior knowledge of the s_i, and the a priori probabilities p_i, the second stage of the receiver, called the *vector receiver,* produces the estimate $\hat{m}$. However, owing to the presence of the additive noise at the receiver input, this decision-making process is statistical in nature, with the result that the receiver will make occasional errors. The requirement is to design the vector receiver so as to minimize the *average probability of symbol error* defined as

$$P_e = P(\hat{m} \neq m_i) \tag{3.5}$$

where m_i is the transmitted symbol and $\hat{m}$ is the estimate produced by the vector receiver. The resulting receiver is said to be *optimum in the minimum probability of error sense.*

It is customary to assume that the receiver is *time-synchronized* with the transmitter, which means that the receiver knows the instants of time when the modulation changes state. Sometimes, it is also assumed that the receiver is *phase locked* to the transmitter. In such a case, we speak of *coherent detection,* and we refer to the receiver as a *coherent receiver.* On the other hand, there may be no phase synchronism between the transmitter and receiver. In this second case, we speak of *noncoherent detection,* and we refer to the receiver as a *noncoherent receiver.* In this chapter, we shall always assume the existence of time synchronism; however, we shall distinguish between coherent and noncoherent detection.

The model just described provides a basis for the design of the optimum receiver, for which we will use the *geometric representation* of the known set of transmitted signals, $\{s_i(t)\}$. This method provides a great deal of insight, with considerable simplification of detail.* Application of the method in this book is confined to *real-valued signals.*

3.2 GRAM–SCHMIDT ORTHOGONALIZATION PROCEDURE

According to the model of Fig. 3.1, the task of transforming an incoming message $m_i, i = 1, 2, \ldots, M$, into a modulated wave $s_i(t)$ may be divided into separate discrete-time and continuous-time operations. The justification for

---

* For detailed treatments of the geometric representation of signals, see the following references: Wozencraft and Jacobs (1965, pp. 211–284), Sakrison (1968, pp. 219–271), Franks (1969, pp. 1–65), and Viterbi and Omura (1979, pp. 47–127).

this separation lies in the *Gram–Schmidt orthogonalization procedure*, which permits the representation of any set of M energy signals, $\{s_i(t)\}$, as linear combinations of N *orthonormal basis functions*, where $N \leq M$. That is to say, we may represent the given set of real-valued energy signals $s_1(t)$, $s_2(t)$, . . . , $s_M(t)$, each of duration T seconds, in the form

$$s_i(t) = \sum_{j=1}^{N} s_{ij}\phi_j(t) \qquad \begin{matrix} 0 \leq t \leq T \\ i = 1, 2, \ldots, M \end{matrix} \qquad (3.6)$$

where the coefficients of the expansion are defined by

$$s_{ij} = \int_0^T s_i(t)\phi_j(t)\,dt \qquad \begin{matrix} i = 1, 2, \ldots, M \\ j = 1, 2, \ldots, N \end{matrix} \qquad (3.7)$$

The real-valued basis functions $\phi_1(t)$, $\phi_2(t)$, . . . , $\phi_N(t)$ are *orthonormal*, by which we mean

$$\int_0^T \phi_i(t)\phi_j(t)\,dt = \begin{cases} 1 & \text{if } i = j \\ 0 & \text{if } i \neq j \end{cases} \qquad (3.8)$$

The first condition of Eq. 3.8 states that each basis function is *normalized* to have unit energy. The second condition states that the basis functions $\phi_1(t)$, $\phi_2(t)$, . . . , $\phi_N(t)$ are *orthogonal* with respect to each other over the interval $0 \leq t \leq T$.

Given the set of coefficients $\{s_{ij}\}$, $j = 1, 2, \ldots, N$, operating as input, we may use the scheme shown in Fig. 3.3a to generate the signal $s_i(t)$, $i = 1, 2,$ . . . , M, which follows directly from Eq. 3.6. It consists of a bank of N multipliers, with each multiplier supplied with its own basis function, followed by a summer. This scheme may be viewed as performing a similar role to that of the second stage or modulator in the transmitter of Fig. 3.1. Conversely, given the set of signals $\{s_i(t)\}$, $i = 1, 2, \ldots, M$, operating as input, we may use the scheme shown in Fig. 3.3b to calculate the set of coefficients $\{s_{ij}\}$, $j = 1, 2,$ . . . , N, which follows directly from Eq. 3.7. This second scheme consists of a bank of N *product-integrators* or *correlators* with a common input, and with each one supplied with its own basis function. As will be shown later, such a bank of correlators may be used as the first stage or detector in the receiver of Fig. 3.1.

To prove the Gram–Schmidt orthogonalization procedure, we proceed in two stages, as indicated next.

(1) Stage 1

First, we have to establish whether or not the given set of signals $s_1(t)$, $s_2(t)$, . . . , $s_M(t)$ is *linearly independent*. If not, then (by definition) there exists a set of coefficients $a_1, a_2, \ldots, a_M$, not all equal to zero, such that we may write

$$a_1 s_1(t) + a_2 s_2(t) + \cdots + a_M s_M(t) = 0 \qquad 0 \leq t \leq T \qquad (3.9)$$

Suppose, in particular, that $a_M \neq 0$. Then, we may express the corresponding signal $s_M(t)$ as

$$s_M(t) = -\left[\frac{a_1}{a_M} s_1(t) + \frac{a_2}{a_M} s_2(t) + \cdots + \frac{a_{M-1}}{a_M} s_{M-1}(t) \right] \qquad (3.10)$$

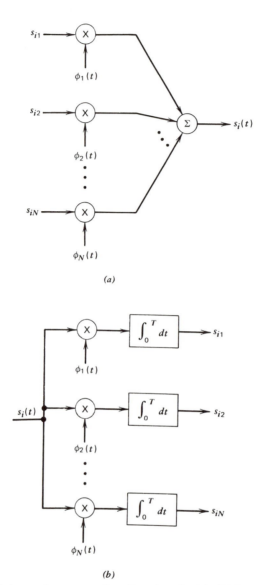

Figure 3.3 (a) Scheme for generating the signal $s_i(t)$. (b) Scheme for generating the set of coefficients $\{s_i\}$.

which implies that $s_M(t)$ may be expressed in terms of the remaining $(M - 1)$ signals.

Consider, next, the set of signals $s_1(t)$, $s_2(t)$, . . . , $s_{M-1}(t)$. Either this set of signals is linearly independent, or it is not. If not, then there exists a set of numbers b_1, b_2, . . . , b_{M-1}, not all equal to zero, such that

$$b_1 s_1(t) + b_2 s_2(t) + \cdots + b_{M-1} s_{M-1}(t) = 0 \qquad 0 \leqslant t \leqslant T \qquad (3.11)$$

Suppose that $b_{M-1} \neq 0$. Then, we may express $s_{M-1}(t)$ as a linear combination of the remaining $M - 2$ signals, as shown by

$$s_{M-1}(t) = - \left[\frac{b_1}{b_{M-1}} s_1(t) + \frac{b_2}{b_{M-1}} s_2(t) + \cdot \cdot \cdot + \frac{b_{M-2}}{b_{M-1}} s_{M-2}(t) \right] \quad (3.12)$$

Now, testing the set of signals $s_1(t), s_2(t), \ldots, s_{M-2}(t)$ for linear independence, and continuing in this fashion, it is clear that we will eventually end up with a linearly independent subset of the original set of signals. Let $s_1(t), s_2(t), \ldots, s_N(t)$ denote this subset of linearly independent signals, where $N \leq M$. The important point to note is that each member of the original set of signals $s_1(t), s_2(t), \ldots, s_M(t)$ may be expressed as a linear combination of this subset of N signals.

(2) Stage 2

Next, we wish to show that it is possible to construct a set of N orthonormal basis functions $\phi_1(t), \phi_2(t), \ldots, \phi_N(t)$ from the linearly independent signals $s_1(t), s_2(t), \ldots, s_N(t)$. As a starting point, define the first basis function as

$$\phi_1(t) = \frac{s_1(t)}{\sqrt{E_1}} \quad (3.13)$$

where E_1 is the energy of the signal $s_1(t)$. Then, clearly, we have

$$s_1(t) = \sqrt{E_1}\phi_1(t) \quad (3.14)$$
$$= s_{11}\phi_1(t)$$

where the coefficient $s_{11} = \sqrt{E_1}$ and $\phi_1(t)$ has unit energy, as required.

Next, using the signal $s_2(t)$, we define the coefficient s_{21} as

$$s_{21} = \int_0^T s_2(t)\phi_1(t)dt \quad (3.15)$$

We may thus define a new intermediate function

$$g_2(t) = s_2(t) - s_{21}\phi_1(t) \quad (3.16)$$

which is orthogonal to $\phi_1(t)$ over the interval $0 \leq t \leq T$. Now, we are ready to define the second basis function as

$$\phi_2(t) = \frac{g_2(t)}{\sqrt{\int_0^T g_2^2(t)dt}} \quad (3.17)$$

Substituting Eq. 3.16 in 3.17, and simplifying, we get the desired result

$$\phi_2(t) = \frac{s_2(t) - s_{21}\phi_1(t)}{\sqrt{E_2 - s_{21}^2}} \quad (3.18)$$

where E_2 is the energy of the signal $s_2(t)$. It is clear from Eq. 3.17 that,

$$\int_0^T \phi_2^2(t)dt = 1$$

and from Eq. 3.18 that

$$\int_0^T \phi_1(t)\phi_2(t)dt = 0$$

That is to say, $\phi_1(t)$ and $\phi_2(t)$ form an orthonormal set.

Continuing in this fashion, we may define

$$g_i(t) = s_i(t) - \sum_{j=1}^{i-1} s_{ij}\phi_j(t) \tag{3.19}$$

where the coefficients $s_{ij}, j = 1, 2, \ldots, i - 1$, are themselves defined by

$$s_{ij} = \int_0^T s_i(t)\phi_j(t)\,dt \tag{3.20}$$

Then it follows readily that the set of functions

$$\phi_i(t) = \frac{g_i(t)}{\sqrt{\int_0^T g_i^2(t)\,dt}} \qquad i = 1, 2, \ldots, N \tag{3.21}$$

forms an orthonormal set.

Since we have shown that each one of the derived subset of linearly independent signals $s_1(t), s_2(t), \ldots, s_N(t)$ may be expressed as a linear combination of the orthonormal basis functions $\phi_1(t), \phi_2(t), \ldots, \phi_N(t)$, it follows that each one of the original set of signals $s_1(t), s_2(t), \ldots, s_M(t)$ may be expressed as a linear combination of this set of basis functions, as described in Eq. 3.6. This completes the proof of the Gram–Schmidt orthogonalization procedure.

Note that the conventional Fourier series expansion of a periodic signal is an example of a particular expansion of this type. There are, however, two important distinctions that should be made:

1. The form of the basis functions $\phi_1(t), \phi_2(t), \ldots, \phi_N(t)$ has not been specified. That is to say, unlike the Fourier series expansion of a periodic signal, we have not restricted the Gram–Schmidt orthogonalization procedure to be in terms of sinusoidal functions.
2. The expansion of the signal $s_i(t)$ into a finite number of terms is not an approximation wherein only the first k terms are significant but rather an exact expression where N and only N terms are significant.

EXAMPLE 1

Consider the signals $s_1(t), s_2(t), s_3(t)$, and $s_4(t)$ shown in Fig. 3.4a. We wish to use the Gram–Schmidt orthogonalization procedure to find an orthonormal basis for this set of signals.

We observe immediately that $s_4(t) = s_1(t) + s_3(t)$, which means that this set of signals is not linearly independent. Accordingly, we base the Gram–Schmidt orthogonalization procedure on a subset consisting of the signals $s_1(t), s_2(t)$, and $s_3(t)$, which are linearly independent.

Step 1 We note that the energy of signal $s_1(t)$ is

$$\begin{aligned} E_1 &= \int_0^T s_1^2(t)\,dt \\ &= \int_0^{T/3} (1)^2\,dt \\ &= \frac{T}{3} \end{aligned}$$

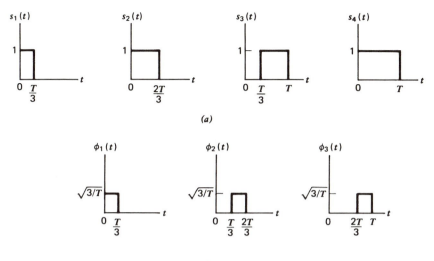

Figure 3.4 (a) Set of signals to be orthonormalized. (b) The resulting set of orthonormal functions.

The first basis function $\phi_1(t)$ is therefore (see Eq. 3.13)

$$\phi_1(t) = \frac{s_1(t)}{\sqrt{E_1}}$$
$$= \begin{cases} \sqrt{3/T} & 0 \leq t \leq T/3 \\ 0 & \text{elsewhere} \end{cases}$$

Step 2   From Eq. 3.15, we find that the coefficient s_{21} equals

$$s_{21} = \int_0^T s_2(t)\phi_1(t)\,dt$$
$$= \int_0^{T/3} (1) \left(\sqrt{\frac{3}{T}} \right) dt$$
$$= \sqrt{\frac{T}{3}}$$

The energy of signal $s_2(t)$ is

$$E_2 = \int_0^T s_2^2(t)$$
$$= \int_0^{2T/3} (1)^2 dt$$
$$= \frac{2T}{3}$$

The second basis function $\phi_2(t)$ is therefore (see Eq. 3.18)

$$\phi_2(t) = \frac{s_2(t) - s_{21}\phi_1(t)}{\sqrt{E_2 - s_{21}^2}}$$
$$= \begin{cases} \sqrt{3/T} & T/3 \leq t \leq 2T/3 \\ 0 & \text{elsewhere} \end{cases}$$

Step 3 Using Eq. 3.20, we find that the coefficient s_{31} equals

$$s_{31} = \int_0^T s_3(t)\phi_1(t)\,dt$$
$$= 0$$

and the coefficient s_{32} equals

$$s_{32} = \int_0^T s_3(t)\phi_2(t)\,dt$$
$$= \int_{T/3}^{2T/3} (1)\left(\sqrt{\frac{3}{T}}\right) dt$$
$$= \sqrt{\frac{T}{3}}$$

The pertinent value of the intermediate function $g_i(t)$, with $i = 3$, is therefore (see Eq. 3.19)

$$g_3(t) = s_3(t) - s_{31}\phi_1(t) - s_{32}\phi_2(t)$$
$$= \begin{cases} 1 & 2T/3 \leq t \leq T \\ 0 & \text{elsewhere} \end{cases}$$

Finally, using Eq. 3.21, we find that the third basis function $\phi_3(t)$ is

$$\phi_3(t) = \frac{g_3(t)}{\sqrt{\int_0^T g_3^2(t)\,dt}}$$
$$= \begin{cases} \sqrt{T/3} & 2T/3 \leq t \leq T \\ 0 & \text{elsewhere} \end{cases}$$

The resulting basis functions $\phi_1(t)$, $\phi_2(t)$, and $\phi_3(t)$ are shown in Fig. 3.4b. It is clear that these three basis functions are orthonormal, and that any of the original signals $s_1(t)$, $s_2(t)$, $s_3(t)$, and $s_4(t)$ may be expressed as a linear combination of them.

3.3 GEOMETRIC INTERPRETATION OF SIGNALS

Once we have adopted a convenient set of orthonormal basis functions $\{\phi_j(t)\}$, $j = 1, 2, \ldots, N$, then each signal in the set $\{s_i(t)\}$, $i = 1, 2, \ldots, M$ may be expanded as in Eq. 3.6, reproduced here for convenience:

$$s_i(t) = \sum_{j=1}^{N} s_{ij}\phi_j(t) \qquad \begin{matrix} 0 \leq t \leq T \\ i = 1, 2, \ldots, M \end{matrix} \tag{3.22}$$

The coefficients of the expansion, s_{ij}, are themselves defined by Eq. 3.7, also reproduced here for convenience:

$$s_{ij} = \int_0^T s_i(t)\phi_j(t)\,dt \qquad \begin{matrix} i = 1, 2, \ldots, M \\ j = 1, 2, \ldots, N \end{matrix} \tag{3.23}$$

Accordingly, each signal in the set $\{s_i(t)\}$ is completely determined by the *vector* of its coefficients, as shown by

$$\mathbf{s}_i = \begin{bmatrix} s_{i1} \\ s_{i2} \\ \cdot \\ \cdot \\ \cdot \\ s_{iN} \end{bmatrix} \qquad i = 1, 2, \ldots, M \tag{3.24}$$

The vector $\mathbf{s}_i$ is called the *signal vector*. Furthermore, if we conceptually extend our conventional notion of two- and three-dimensional Euclidean spaces to an *N-dimensional Euclidean space*, we may visualize the set of signal vectors $\{\mathbf{s}_i\}$, $i = 1, 2, \ldots, M$, as defining a corresponding set of M points in an N-dimensional Euclidean space, with N mutually perpendicular axes labeled ϕ_1, $\phi_2, \ldots, \phi_N$. This N-dimensional Euclidean space is called the *signal space*.

The idea of visualizing a set of energy signals geometrically, as described above, is of fundamental importance. For example, Fig. 3.5 illustrates the case of a two-dimensional signal space with three signals, that is, $N = 2$ and $M = 3$.

In an N-dimensional Euclidean space, we may define *lengths* of vectors and *angles* between vectors.* It is customary to denote the *length* or *norm* of a signal vector $\mathbf{s}_i$ by the symbol $\|\mathbf{s}_i\|$. The squared-length of any signal vector $\mathbf{s}_i$ is defined to be the *inner product* or *dot product* of $\mathbf{s}_i$ with itself. In the familiar case of $N = 2$ or 3, we have

$$\|\mathbf{s}_i\|^2 = (\mathbf{s}_i, \mathbf{s}_i)$$
$$= \sum_{j=1}^{N} s_{ij}^2 \tag{3.25}$$

where the s_{ij} are the elements of $\mathbf{s}_i$. For larger values of N, length is defined in the same way, and Eq. 3.25 remains valid.

The *cosine of the angle* between two vectors is defined as the inner product of the two vectors divided by the product of their individual norms. That is, the

* For further details on signal spaces, see Franks (1969).

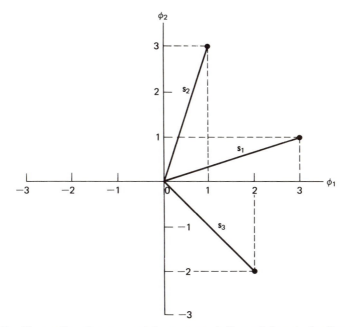

Figure 3.5 Illustrating the geometric representation of signals for the case when $N = 2$ and $M = 3$.

cosine of the angle between the vectors $\mathbf{s}_i$ and $\mathbf{s}_j$ equals the ratio $(\mathbf{s}_i, \mathbf{s}_j)/\|\mathbf{s}_i\| \|\mathbf{s}_j\|$. The two vectors $\mathbf{s}_i$ and $\mathbf{s}_j$ are thus *orthogonal* or *perpendicular* if their inner product $(\mathbf{s}_i, \mathbf{s}_j)$ is zero.

There is an interesting relationship between the energy content of a signal and its representation as a vector. By definition, the energy of a signal $s_i(t)$ of duration T seconds is equal to

$$E_i = \int_0^T s_i^2(t)\,dt \tag{3.26}$$

Therefore, substituting Eq. 3.22 in 3.26, we get

$$E_i = \int_0^T \left[\sum_{j=1}^N s_{ij}\phi_j(t) \right]\left[\sum_{k=1}^N s_{ik}\phi_k(t) \right] dt \tag{3.27}$$

Interchanging the order of summation and integration

$$E_i = \sum_{j=1}^N \sum_{k=1}^N s_{ij}s_{ik} \int_0^T \phi_j(t)\phi_k(t)\,dt \tag{3.28}$$

But since the $\phi_j(t)$ form an orthonormal set, then, in accordance with the two conditions of Eq. 3.8, we find that Eq. 3.28 reduces simply to

$$E_i = \sum_{j=1}^N s_{ij}^2 \tag{3.29}$$

Thus Eqs. 3.25 and 3.29 show that the energy of a signal $s_i(t)$ is equal to the squared-length of the signal vector $\mathbf{s}_i$ representing it.

In the case of a pair of signals $s_i(t)$ and $s_k(t)$, represented by the signal vectors $\mathbf{s}_i$ and $\mathbf{s}_k$, respectively, we may similarly show that

$$\|\mathbf{s}_i - \mathbf{s}_k\|^2 = \sum_{j=1}^N (s_{ij} - s_{kj})^2$$
$$= \int_0^T [s_i(t) - s_k(t)]^2 dt \tag{3.30}$$

where $\|\mathbf{s}_i - \mathbf{s}_k\|$ is the Euclidean distance between the points represented by the signal vectors $\mathbf{s}_i$ and $\mathbf{s}_k$.

3.4 RESPONSE OF BANK OF CORRELATORS TO NOISY INPUT

Suppose that the input to the bank of N product-integrators or correlators in Fig. 3.3b is not the transmitted signal $s_i(t)$ but rather the received random process $X(t)$ of the idealized AWGN channel of Fig. 3.2. That is to say,

$$X(t) = s_i(t) + W(t) \qquad 0 \leq t \leq T \tag{3.31}$$
$$i = 1, 2, \ldots, M$$

where $W(t)$ is a white Gaussian noise process of zero mean and power spectral density $N_0/2$. Correspondingly, we find that the output of each correlator is a random variable defined by

$$X_j = \int_0^T X(t)\phi_j(t)\,dt$$
$$= s_{ij} + W_j \qquad j = 1, 2, \ldots, N \qquad (3.32)$$

The first component, s_{ij}, is a deterministic quantity contributed by the transmitted signal $s_i(t)$; it is defined by

$$s_{ij} = \int_0^T s_i(t)\phi_j(t)\,dt \qquad (3.33)$$

The second component, W_j, is a random variable that arises because of the presence of noise at the input; it is defined by

$$W_j = \int_0^T W(t)\phi_j(t)\,dt \qquad (3.34)$$

Consider next a new random process, $X'(t)$, that is related to the received random process $X(t)$ as follows:

$$X'(t) = X(t) - \sum_{j=1}^N X_j\phi_j(t) \qquad (3.35)$$

Substituting Eqs. 3.31 and 3.32 into Eq. 3.35, we get

$$X'(t) = s_i(t) + W(t) - \sum_{j=1}^N (s_{ij} + W_j)\phi_j(t)$$

$$= W(t) - \sum_{j=1}^N W_j\phi_j(t)$$

$$= W'(t) \qquad (3.36)$$

which depends only on the noise $W(t)$ at the front end of the receiver and not at all on the transmitted signal $s_i(t)$. Thus we may express the received random process as

$$X(t) = \sum_{j=1}^N X_j\phi_j(t) + X'(t)$$

$$= \sum_{j=1}^N X_j\phi_j(t) + W'(t) \qquad (3.37)$$

Accordingly, we may view $W'(t)$ as a sort of remainder term that must be included on the right in order to preserve the equality in Eq. 3.37. It is of interest to contrast the expansion of the received random process $X(t)$, as in Eq. 3.37, with the corresponding expansion of the transmitted deterministic signal $s_i(t)$, as in Eq. 3.6.

We now wish to characterize the set of correlator outputs, $\{X_j\}$, $j = 1, 2, \ldots, N$. Since the received random process $X(t)$ is Gaussian, we deduce that each X_j is a Gaussian random variable. Hence, each X_j is characterized completely by its mean value and variance.

The noise process $W(t)$ has zero mean. Hence, the random variable W_j extracted from $W(t)$ in accordance with Eq. 3.34 has zero mean too. This implies that the mean value of the jth correlator output X_j depends only on s_{ij}, as shown by

$$
\begin{aligned}
m_{X_j} &= E[X_j] \\
&= E[s_{ij} + W_j] \\
&= s_{ij} + E[W_j] \\
&= s_{ij}
\end{aligned}
\tag{3.38}
$$

To find the variance of X_j, we note that

$$
\begin{aligned}
\sigma_{X_j}^2 &= \text{Var}[X_j] \\
&= E[(X_j - s_{ij})^2] \\
&= E[W_j^2]
\end{aligned}
\tag{3.39}
$$

Substituting Eq. 3.34 into Eq. 3.39, we get

$$
\sigma_{X_j}^2 = E\left[\int_0^T W(t)\phi_j(t)dt \int_0^T W(u)\phi_j(u)du \right]
$$
$$
= E\left[\int_0^T \int_0^T \phi_j(t)\phi_j(u)W(t)W(u)dt\,du \right]
\tag{3.40}
$$

Interchanging the order of integration and expectation:

$$
\sigma_{X_j}^2 = \int_0^T \int_0^T \phi_j(t)\phi_j(u)E[W(t)W(u)]dt\,du
$$
$$
= \int_0^T \int_0^T \phi_j(t)\phi_j(u)R_W(t,\,u)dt\,du
\tag{3.41}
$$

where $R_W(t,\,u)$ is the autocorrelation function of the noise process $W(t)$. Since this noise is stationary, $R_W(t,\,u)$ depends only on the time difference $t - u$. Furthermore, since the noise process $W(t)$ is white with power spectral density $N_0/2$, we may express $R_W(t,\,u)$ as follows:

$$
R_W(t,\,u) = \frac{N_0}{2}\,\delta(t - u)
\tag{3.42}
$$

where $\delta(t - u)$ is a *Dirac delta function* (impulse of unit area) at $t - u$. Therefore, substituting Eq. 3.42 in Eq. 3.41, we get

$$
\sigma_{X_j}^2 = \frac{N_0}{2} \int_0^T \int_0^T \phi_j(t)\phi_j(u)\delta(t - u)dt\,du
$$
$$
= \frac{N_0}{2} \int_0^T \phi_j^2(t)dt
\tag{3.43}
$$

Since the $\phi_j(t)$ have unit energy, we finally get the simple result

$$
\sigma_{X_j}^2 = \frac{N_0}{2} \qquad \text{for all } j
\tag{3.44}
$$

This shows that all the correlator outputs, $X_j, j = 1, 2, \ldots N$, have a variance equal to the power spectral density $N_0/2$ of the additive noise process $W(t)$.

Similarly, we find that since the $\phi_j(t)$ form an orthogonal set, then the X_j are mutually uncorrelated, as shown by

$$
\begin{aligned}
\text{Cov}[X_j X_k] &= E[(X_j - m_{X_j})(X_k - m_{X_k})] \\
&= E[(X_j - s_{ij})(X_k - s_{ik})] \\
&= E[W_j W_k] \\
&= E\left[\int_0^T W(t)\phi_j(t)dt \int_0^T W(u)\phi_k(u)du\right] \\
&= \int_0^T \int_0^T \phi_j(t)\phi_k(u) R_W(t, u)dt\ du \\
&= \frac{N_0}{2} \int_0^T \int_0^T \phi_j(t)\phi_k(u)\delta(t - u)dt\ du \\
&= \frac{N_0}{2} \int_0^T \phi_j(t)\phi_k(t)dt \\
&= 0 \qquad j \neq k
\end{aligned} \tag{3.45}
$$

Since the X_j are Gaussian random variables, Eq. 3.45 implies that they are also statistically independent.

Define the vector of N random variables at the correlator outputs as

$$
\mathbf{X} = \begin{bmatrix} X_1 \\ X_2 \\ \vdots \\ X_N \end{bmatrix} \tag{3.46}
$$

whose elements are independent Gaussian random variables with mean values equal to s_{ij} and variances equal to $N_0/2$. Since the elements of the vector $\mathbf{X}$ are statistically independent, we may express the conditional probability density function of the vector $\mathbf{X}$, given that the signal $s_i(t)$ or correspondingly the symbol m_i was transmitted, as the product of the conditional probability density functions of its individual elements. We may thus write

$$
f_{\mathbf{X}}(\mathbf{x}|m_i) = \prod_{j=1}^{N} f_{X_j}(x_j|m_i) \qquad i = 1, 2, \ldots, M \tag{3.47}
$$

where the vector $\mathbf{x}$ and scalar x_j are sample values of the random vector $\mathbf{X}$ and random variable X_j, respectively. The conditional probability density functions, $f_{\mathbf{X}}(\mathbf{x}|m_i)$, for each transmitted message m_i, $i = 1, 2, \ldots, M$ are called *likelihood functions*. These likelihood functions are, in fact, the channel characterization. Any channel whose likelihood functions satisfy Eq. 3.47 is called a *memoryless* channel.

Since each X_j is a Gaussian random variable with mean s_{ij} and variance $N_0/2$, we have

$$
f_{X_j}(x_j|m_i) = \frac{1}{\sqrt{\pi N_0}} \exp\left[-\frac{1}{N_0}(x_j - s_{ij})^2\right] \qquad \begin{aligned} j &= 1, 2, \ldots, N \\ i &= 1, 2, \ldots, M \end{aligned} \tag{3.48}
$$

Therefore, substituting Eq. 3.48 in Eq. 3.47, we find that the likelihood functions of an AWGN channel are defined by

$$f_{\mathbf{X}}(\mathbf{x}|m_i) = (\pi N_0)^{-N/2} \exp\left[-\frac{1}{N_0} \sum_{j=1}^{N} (x_j - s_{ij})^2\right] \qquad i = 1, 2, \ldots, M$$

(3.49)

Returning to Eq. 3.37, while it is now clear that the elements of the random vector $\mathbf{X}$ completely characterize the summation term $\sum X_j \phi_j(t)$, there remains the term $W'(t)$, which depends only on the original noise $W(t)$. Since the noise process $W(t)$ is Gaussian with zero mean, it follows that $W'(t)$ is also a zero-mean Gaussian process. Finally, we note that any random variable $W'(t_k)$, say, derived from the noise process $W'(t)$ by sampling it at time t_k, is in fact independent of the set of random variables $\{X_j\}$. That is to say,

$$E[W'(t_k)X_j] = 0 \qquad \begin{array}{l} j = 1, 2, \ldots, N \\ 0 \le t_k \le T \end{array}$$

(3.50)

Since any random variable based on the remainder noise term $W'(t)$ is independent of the set of random variables $\{X_j\}$ and the set of transmitted signals $\{s_i(t)\}$, we conclude that such a random variable is *irrelevant* to the decision as to which signal was transmitted. In other words, the set of random variables $\{X_j\}$ at the correlator outputs, based on the received random process $X(t)$, are the only data that are useful for the decision-making process and, hence, represent *sufficient statistics*.

3.5 DETECTION OF KNOWN SIGNALS IN NOISE

Assume that, in each time slot of duration T seconds, one of the M possible signals $s_1(t)$, $s_2(t)$, $\ldots$, $s_M(t)$ is transmitted with equal probability, namely, $1/M$. Then, for an AWGN channel, a possible realization or sample function, $x(t)$, of the received random process $X(t)$ may be described by

$$x(t) = s_i(t) + w(t) \qquad \begin{array}{l} 0 \le t \le T \\ i = 1, 2, \ldots, M \end{array}$$

(3.51)

where $w(t)$ is a sample function of the white Gaussian noise process $W(t)$, assumed to be of zero mean and power spectral density $N_0/2$. The receiver has to observe the signal $x(t)$ and make a "best estimate" of the transmitted signal $s_i(t)$ or equivalently the symbol m_i.

We note that when the transmitted signal $s_i(t)$, $i = 1, 2, \ldots, M$, is applied to a bank of correlators, with a common input and supplied with an appropriate set of N orthonormal basis functions, the resulting correlator outputs define the *signal vector* $\mathbf{s}_i$ (see Eq. 3.7). Since knowledge of the signal vector $\mathbf{s}_i$ is as good as knowing the transmitted signal $s_i(t)$ itself, and vice versa, we may represent $s_i(t)$ by a point in a Euclidean space of dimension $N \le M$. We refer to such a point as the *transmitted signal point* or *message point*. The collection of M message points in the N-dimensional Euclidean space is called a *signal constellation*.

However, the representation of the received signal $x(t)$ is complicated by the presence of the additive noise $w(t)$. We note that when the received signal $x(t)$ is applied to the same bank of N correlators as above, the correlator outputs define a new vector, $\mathbf{x}$, called the *observation* or *received vector*. The vector $\mathbf{x}$ differs from the signal vector $\mathbf{s}_i$ by the *noise vector* $\mathbf{w}$ whose orientation is completely random. That is to say,

$$\mathbf{x} = \mathbf{s}_i + \mathbf{w} \qquad i = 1, 2, \ldots, M \qquad (3.52)$$

The vectors $\mathbf{x}$ and $\mathbf{w}$ are sample values of the random vectors $\mathbf{X}$ and $\mathbf{W}$, respectively, that were described in Section 3.4. Note also that the noise vector $\mathbf{w}$ is completely characterized by the noise $w(t)$; the converse of this, however, is not true. The noise vector $\mathbf{w}$ represents that portion of the noise $w(t)$ which will interfere with the detection process. The remaining portion of this noise may be thought of as being tuned out by the bank of correlators.

Now, based on the observation vector $\mathbf{x}$, we may represent the received signal $x(t)$ by a point in the same Euclidean space used to represent the transmitted signal. We refer to this second point as the *received signal point*. The relationship between the observation vector, $\mathbf{x}$, the signal vector, $\mathbf{s}_i$, and the noise vector, $\mathbf{w}$, is illustrated in Fig. 3.6 for the case of $N = 3$.

We are now ready to state the detection problem. Given the observation vector $\mathbf{x}$, we have to perform a *mapping* from $\mathbf{x}$ to an estimate $\hat{m}$ of the transmitted symbol, m_i, in a way that would minimize the average probability of symbol error in the decision. The so-called maximum-likelihood detector provides the solution to this basic problem.

(1) Maximum-likelihood Detector

Suppose that, when the observation vector has the value $\mathbf{x}$, we make the decision $\hat{m} = m_i$. The average probability of symbol error in this decision,

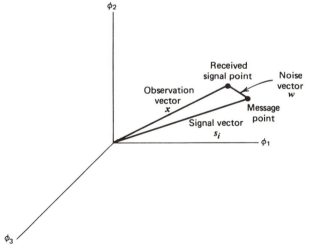

Figure 3.6   Illustrating the effect of noise perturbation on location of the received signal point.

which we denote by $P_e(m_i, \mathbf{x})$, is simply

$$P_e(m_i, \mathbf{x}) = P(m_i \text{ not sent}|\mathbf{x})$$
$$= 1 - P(m_i \text{ sent}|\mathbf{x}) \tag{3.53}$$

Now, since our criterion is to minimize the probability of error in mapping each given observation vector, $\mathbf{x}$, into a decision, we deduce from Eq. 3.53 that the *optimum decision rule* is as follows:

set $\hat{m} = m_i$ if

$$P(m_i \text{ sent}|\mathbf{x}) \geqslant P(m_k \text{ sent}|\mathbf{x}), \text{ for all } k \neq i \tag{3.54}$$

where $k = 1, 2, \ldots, M$. This decision rule is referred to as *maximum a posteriori probability*.

The condition of Eq. 3.54 may be expressed more explicitly in terms of the a priori probabilities of the transmitted signals and in terms of the likelihood functions. Applying Bayes' rule to Eq. 3.54, and for the moment ignoring possible ties in the decision-making process, we may restate this decision rule as

set $\hat{m} = m_i$ if

$$\frac{p_k f_{\mathbf{X}}(\mathbf{x}|m_k)}{f_{\mathbf{X}}(\mathbf{x})} \text{ is maximum for } k = i \tag{3.55}$$

where p_k is the a priori probability of occurrence of symbol m_k, $f_{\mathbf{X}}(\mathbf{x}|m_k)$ is the likelihood function that results when symbol m_k is transmitted, and $f_{\mathbf{X}}(\mathbf{x})$ is the unconditional joint probability density function of the random vector $\mathbf{X}$. However, in Eq. 3.55, we note that (1) the denominator term, $f_{\mathbf{X}}(\mathbf{x})$, is independent of the transmitted signal, and (2) the a priori probability $p_k = p_i$ when all the signals are transmitted with equal probability. Therefore, we may simplify the decision rule of Eq. 3.55 as

set $\hat{m} = m_i$ if

$$f_{\mathbf{X}}(\mathbf{x}|m_k) \text{ is maximum for } k = i \tag{3.56}$$

Ordinarily, we find it more convenient to work with the natural logarithm of the likelihood function rather than the likelihood function itself. For a memoryless channel, the logarithm of the likelihood function is commonly called the *metric*. Since the likelihood function $f_{\mathbf{X}}(\mathbf{x}|m_k)$ is always nonnegative, and since if $A > B > 0$ then $\ln A > \ln B$, we may restate the decision rule of Eq. 3.56 in terms of the metric as

set $\hat{m} = m_i$ if

$$\ln[f_{\mathbf{X}}(\mathbf{x}|m_k)] \text{ is maximum for } k = i \tag{3.57}$$

This decision rule is referred to as *maximum likelihood,* and the device for its implementation is correspondingly referred to as the *maximum-likelihood detector.* According to Eq. 3.57, a maximum-likelihood detector computes the metric for each transmitted message, compares them, and then decides in favor of the maximum.

It is useful to have a graphical interpretation of the maximum-likelihood decision rule. Let Z denote the N-dimensional space of all possibly observed vectors $\mathbf{x}$. We refer to this space as the *observation space*. Because we have assumed that the decision rule must say $\hat{m} = m_i$, where $i = 1, 2, \ldots, M$, the total observation space Z is correspondingly partitioned into M *decision regions*, denoted by $Z_1, Z_2, \ldots, Z_M$. Accordingly, we may restate the decision rule of Eq. 3.57 as follows:

vector $\mathbf{x}$ lies inside region Z_i if

$$\ln[f_{\mathbf{X}}(\mathbf{x}|m_k)] \text{ is maximum for } k = i \qquad (3.58)$$

Aside from the boundaries between the decision regions $Z_1, Z_2, \ldots, Z_M$, it is clear that this set of regions covers the entire space of possibly observed vectors $\mathbf{x}$. We adopt the convention that all ties are resolved at random. Specifically, if the observation vector $\mathbf{x}$ falls on the boundary between any two decision regions, Z_i and Z_k, say, the choice between the two possible decisions $\hat{m} = m_i$ and $\hat{m} = m_k$ is resolved a priori by the flip of a fair coin. Clearly, the outcome of such an event does not affect the ultimate value of the average probability of error since, on this boundary, the condition of Eq. 3.54 is satisfied with the equality sign.

To illustrate the above concept, consider an AWGN channel for which the likelihood function is defined by Eq. 3.49. We thus have

$$f_{\mathbf{X}}(\mathbf{x}|m_k) = (\pi N_0)^{-N/2} \exp\left[-\frac{1}{N_0} \sum_{j=1}^{N} (x_j - s_{kj})^2\right] \qquad k = 1, 2, \ldots, M$$

$$(3.59)$$

The corresponding value of the metric is therefore

$$\ln[f_{\mathbf{X}}(\mathbf{x}|m_k)] = -\frac{N}{2} \ln(\pi N_0) - \frac{1}{N_0} \sum_{j=1}^{N} (x_j - s_{kj})^2 \qquad k = 1, 2, \ldots, M$$

$$(3.60)$$

The constant term $-(N/2)\ln(\pi N_0)$ on the right-hand side of Eq. 3.60 is of no consequence in so far as application of the decision rule is concerned. Therefore, ignoring this term, and then substituting the remainder of Eq. 3.60 into Eq. 3.58, we may formulate the maximum-likelihood decision rule for an AWGN channel as follows:

vector $\mathbf{x}$ lies inside region Z_i if

$$-\frac{1}{N_0} \sum_{j=1}^{N} (x_j - s_{kj})^2 \text{ is maximum for } k = i \qquad (3.61)$$

Equivalently, we may state:

vector $\mathbf{x}$ lies inside region Z_i if

$$\sum_{j=1}^{N} (x_j - s_{kj})^2 \text{ is minimum for } k = i \qquad (3.62)$$

However, we note that

$$\sum_{j=1}^{N} (x_j - s_{kj})^2 = \|\mathbf{x} - \mathbf{s}_k\|^2 \qquad (3.63)$$

where $\|\mathbf{x} - \mathbf{s}_k\|$ is the distance between the received signal point and message point, represented by the vectors $\mathbf{x}$ and $\mathbf{s}_k$, respectively. Accordingly, we may rewrite the decision rule of Eq. 3.62 as follows:

vector $\mathbf{x}$ lies inside region Z_i if

$\|\mathbf{x} - \mathbf{s}_k\|$ is minimum for $k = i$

(3.64)

This shows that the maximum-likelihood decision rule is simply to choose the message point closest to the received signal point, which is intuitively satisfying.

In practice the need for squarers, as in the decision rule of Eq. 3.64, is avoided by recognizing that

$$\sum_{j=1}^{N} (x_j - s_{kj})^2 = \sum_{j=1}^{N} x_j^2 - 2 \sum_{j=1}^{N} x_j s_{kj} + \sum_{j=1}^{N} s_{kj}^2 \qquad (3.65)$$

The first summation term is independent of the index k and may therefore be ignored. The second summation term is the inner product of the observation vector $\mathbf{x}$ and signal vector $\mathbf{s}_k$. The third summation term is the energy of the transmitted signal $s_k(t)$. Accordingly, a decision rule equivalent to that of Eq. 3.64 is as follows:

vector $\mathbf{x}$ lies inside region Z_i if

$$\sum_{j=1}^{N} x_j s_{kj} - \frac{1}{2} E_k \text{ is maximum for } k = i \qquad (3.66)$$

where E_k is the energy of the transmitted signal $s_k(t)$:

$$E_k = \sum_{j=1}^{N} s_{kj}^2 \qquad (3.67)$$

From Eq. 3.66 we deduce that, for an AWGN channel, the decision regions are regions of the N-dimensional observation space Z, bounded by linear $(N-1)$-dimensional hyperplane boundaries. Figure 3.7 shows the example of decision regions for $M = 4$ signals and $N = 2$ dimensions, assuming that the signals are transmitted with equal energy E, and with equal probability.

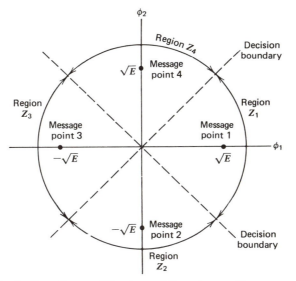

Figure 3.7   Illustrating the partitioning of the observation space into decision regions for the case when $N = 2$ and $M = 4$.

3.6   PROBABILITY OF ERROR

Suppose that the observation space Z is partitioned, in accordance with the maximum-likelihood decision rule, into a set of M regions $\{Z_i, i = 1, 2, \ldots, M\}$. Suppose also that symbol m_i (or, equivalently, signal vector $\mathbf{s}_i$) is transmitted, and an observation vector $\mathbf{x}$ is received. Then an error occurs whenever the received signal point represented by $\mathbf{x}$ does not fall inside region Z_i associated with the message point represented by $\mathbf{s}_i$. Averaging over all possibly transmitted symbols, we readily see that the *average probability of symbol error*, P_e, equals

$$P_e = \sum_{i=1}^{M} P(m_i \text{ sent}) P(\mathbf{x} \text{ does not lie inside } Z_i | m_i \text{ sent})$$

$$= \frac{1}{M} \sum_{i=1}^{M} P(\mathbf{x} \text{ does not lie inside } Z_i | m_i \text{ sent})$$

$$= 1 - \frac{1}{M} \sum_{i=1}^{M} P(\mathbf{x} \text{ lies inside } Z_i | m_i \text{ sent}) \qquad (3.68)$$

where we have used standard notation to denote the probability of an event and the conditional probability of an event. Since

$$P(\mathbf{x} \text{ lies inside } Z_i | m_i \text{ sent}) = \int_{Z_i} f_{\mathbf{X}}(\mathbf{x} | m_i) d\mathbf{x}$$

we may rewrite Eq. 3.68 in terms of the likelihood function (when m_i is sent) as follows:

$$P_e = 1 - \frac{1}{M} \sum_{i=1}^{M} \int_{Z_i} f_{\mathbf{X}}(\mathbf{x}|m_i)d\mathbf{x} \qquad (3.69)$$

For an N-dimensional observation vector $\mathbf{x}$, the integral in Eq. 3.69 is likewise N-dimensional.

The following example will illustrate the application of Eq. 3.69.

EXAMPLE 2   M-ARY SIGNALING

Consider a *quaternary signaling* scheme in which the received signal is defined by

$$x(t) = a_i + w(t) \qquad 0 \leqslant t \leqslant T$$
$$i = 1, 2, 3, 4$$

where the amplitude $a_i = \pm a/2, \pm 3a/2$, and $w(t)$ is the sample function of a white Gaussian noise process of zero mean and power spectral density $N_0/2$. The signal space, representing the signal component of $x(t)$, has two characteristic features:

1. It is one-dimensional in that the signal component is a scaled version of the time function

$$\phi_1(t) = \frac{1}{\sqrt{T}} \qquad 0 \leqslant t \leqslant T$$

2. There are four message points, whose locations on the $\phi_1(t)$-axis are as follows:

$$s_{i1} = \sqrt{T}\, a_i \qquad\qquad i = 1, 2, 3, 4$$
$$= \pm\sqrt{T}\, a/2, \pm 3\sqrt{T}\, a/2$$

Specifically, the signal-space diagram is as depicted in Fig. 3.8a. The diagram also includes the decision boundaries located at the points: $0, \pm a\sqrt{T}$, assuming that the four message points are equally likely. Figure 3.8b presents the transmitted *dibit* (pair of bits) corresponding to each of the four possible message points. Figure 3.8c presents the decision intervals on the $\phi_1(t)$-axis that correspond to the four possible received dibits.

Suppose we transmit the dibit 00, representing the message point $s_{11} = -3a\sqrt{T}/2$. Let X_1 denote the random variable derived from the received random process $X(t)$ with a sample function $x(t)$. From Eq. 3.48 we find that the conditional probability density function of the random variable X_1 is given by

$$f_{X_1}(x_1|m_1) = \frac{1}{\sqrt{\pi N_0}} \exp\left[-\frac{1}{N_0}(x_1 - s_{11})^2\right]$$

The conditional probability of choosing dibit 01 (representing the message point at s_{21}), given that dibit 00 was sent, is defined by

$$P(01|00) = \int_{-a\sqrt{T}}^{0} f_{X_1}(x_1|m_1)dx_1$$

$$= \frac{1}{\sqrt{\pi N_0}} \int_{-a\sqrt{T}}^{0} \exp\left[-\frac{1}{N_0}(x_1 - s_{11})^2\right]dx_1$$

$$= \frac{1}{\sqrt{\pi N_0}} \int_{-a\sqrt{T}}^{0} \exp\left[-\frac{1}{N_0}\left(x_1 + \frac{3a\sqrt{T}}{2}\right)^2\right]dx_1 \qquad (3.70)$$

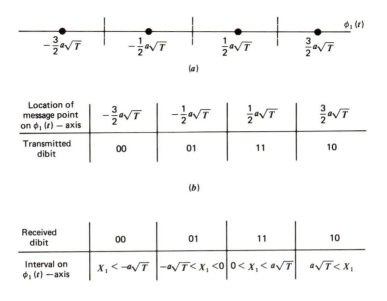

Figure 3.8 (*a*) Signal constellation (dashed lines represent decision boundaries). (*b*) Representation of transmitted dibits. (*c*) Decision intervals for received dibits.

From the definition of the *complementary error function*, we have (see Appendix E)

$$\text{erfc}(u) = \frac{2}{\sqrt{\pi}} \int_u^\infty \exp(-z^2)dz$$

The complementary error function erfc(u) equals twice the area under the tail of the curve of the probability density function of a Gaussian distributed variable of zero mean and variance 1/2, which lies above the value u. Hence, we may rewrite Eq. 3.70 as

$$P(01|00) = \frac{1}{2}\left[\text{erfc}\left(\frac{a}{2}\sqrt{\frac{T}{N_0}}\right) - \text{erfc}\left(\frac{3a}{2}\sqrt{\frac{T}{N_0}}\right)\right] \tag{3.71}$$

Similarly, we may show that the conditional probability of choosing dibit 11 (representing the message point at s_{31}), given that dibit 00 was sent, is defined by

$$P(11|00) = \frac{1}{2}\left[\text{erfc}\left(\frac{3a}{2}\sqrt{\frac{T}{N_0}}\right) - \text{erfc}\left(\frac{5a}{2}\sqrt{\frac{T}{N_0}}\right)\right] \tag{3.72}$$

Likewise, we may show that the conditional probability of choosing dibit 10 (representing the message point as s_{41}), given that dibit 00 was sent, is defined by

$$P(10|00) = \frac{1}{2}\text{erfc}\left(\frac{5a}{2}\sqrt{\frac{T}{N_0}}\right) \tag{3.73}$$

The *conditional probability of symbol error*, given that dibit 00 was sent, is defined by

$$P_e(00) = P(01|00) + P(11|00) + P(10|00)$$

$$= \frac{1}{2}\text{erfc}\left(\frac{a}{2}\sqrt{\frac{T}{N_0}}\right)$$

The conditional probabilities of errors, presented in Eqs. 3.71–3.73, are summarized in Table 3.1. This table also includes the conditonal probabilities of various other errors, given that the transmitted dibit was 01, 11, or 10. Note that for each transmitted dibit, the conditional probability of symbol error equals the sum of the three entries along the row corresponding to that dibit in Table 3.1.

Since all the four possible dibits 00,01,11, and 10 are assumed to occur with the same probability (equal to 1/4), we find that the *average probability of symbol error*, P_e, is equal to 1/4 times the sum of all 12 conditional probabilities of error given in Table 3.1. We thus get

$$P_e = \frac{3}{4} \operatorname{erfc}\left(\frac{a}{2} \sqrt{\frac{T}{N_0}}\right) \tag{3.74}$$

We may generalize this result as follows. Consider an *M-ary signaling* scheme, in which the received signal is defined by

$$x(t) = a_i + w(t) \qquad 0 \leqslant t \leqslant T$$

$$i = 1, 2, \ldots, M$$

where $a_i = \pm a/2, \pm 3a/2, \ldots, \pm(M - 1)a/2$. The average probability of symbol error for this scheme is given by

$$P_e = \left(\frac{M - 1}{M}\right) \operatorname{erfc}\left(\frac{a}{2} \sqrt{\frac{T}{N_0}}\right) \tag{3.75}$$

Putting $M = 4$ in Eq. 3.75, we get the result shown in Eq. 3.74. Note that for a fixed probability of symbol error P_e, increasing the number of levels M requires a corresponding increase in transmitted power.

(1)   Union Bound on the Probability of Error

For AWGN channels, the formulation of the average probability of symbol error, P_e, is conceptually straightforward. We simply write P_e in integral form by substituting Eq. 3.49 in Eq. 3.69. However, except for a few simple cases (e.g., the M-ary signaling scheme considered in Example 2 and some others considered in Chapters 5 and 7), numerical computation of the integral is impractical. To overcome this computational difficulty, we may resort to the use of *bounds,* which are usually adequate to predict the signal-to-noise ratio (within a decibel or so) required to maintain a prescribed error rate. The approximation to the integral defining P_e is made by simplifying the integral or simplifying the region of integration. In the sequel, we use the latter procedure to develop a simple and yet useful upper bound called the *union bound* as an approximation to the average probability of symbol error for any set of M equally likely signals (symbols) in AWGN.

Let A_{ik}, with $i, k = 1, 2, \ldots, M$, denote the event that the observation vector $\mathbf{x}$ is closer to the signal vector $\mathbf{s}_k$ than to $\mathbf{s}_i$, when the symbol m_i (vector $\mathbf{s}_i$) is sent. The conditional probability of symbol error when symbol m_i is sent, $P_e(m_i)$, is equal to the probability of the *union of events*, $A_{i1}, A_{i2}, \ldots, A_{i,i-1}$, $A_{i,i+1}, \ldots, A_{i,M}$. From probability theory we know that *the probability of a finite union of events is overbounded by the sum of the probabilities of the constituent events.* We may therefore write

Table 3.1 Conditional Probabilities of Error for Quaternary Signaling

Transmitted Dibit \ Received Dibit	00	01	11	10
00		$\frac{1}{2}\mathrm{erfc}\left(\frac{a}{2}\sqrt{\frac{T}{N_0}}\right)$ $-\frac{1}{2}\mathrm{erfc}\left(\frac{3a}{2}\sqrt{\frac{T}{N_0}}\right)$	$\frac{1}{2}\mathrm{erfc}\left(\frac{3a}{2}\sqrt{\frac{T}{N_0}}\right)$ $-\frac{1}{2}\mathrm{erfc}\left(\frac{5a}{2}\sqrt{\frac{T}{N_0}}\right)$	$\frac{1}{2}\mathrm{erfc}\left(\frac{5a}{2}\sqrt{\frac{T}{N_0}}\right)$
01	$\frac{1}{2}\mathrm{erfc}\left(\frac{a}{2}\sqrt{\frac{T}{N_0}}\right)$		$\frac{1}{2}\mathrm{erfc}\left(\frac{a}{2}\sqrt{\frac{T}{N_0}}\right)$ $-\frac{1}{2}\mathrm{erfc}\left(\frac{3a}{2}\sqrt{\frac{T}{N_0}}\right)$	$\frac{1}{2}\mathrm{erfc}\left(\frac{3a}{2}\sqrt{\frac{T}{N_0}}\right)$
11	$\frac{1}{2}\mathrm{erfc}\left(\frac{3a}{2}\sqrt{\frac{T}{N_0}}\right)$	$\frac{1}{2}\mathrm{erfc}\left(\frac{a}{2}\sqrt{\frac{T}{N_0}}\right)$ $-\frac{1}{2}\mathrm{erfc}\left(\frac{3a}{2}\sqrt{\frac{T}{N_0}}\right)$		$\frac{1}{2}\mathrm{erfc}\left(\frac{a}{2}\sqrt{\frac{T}{N_0}}\right)$
10	$\frac{1}{2}\mathrm{erfc}\left(\frac{5a}{2}\sqrt{\frac{T}{N_0}}\right)$	$\frac{1}{2}\mathrm{erfc}\left(\frac{3a}{2}\sqrt{\frac{T}{N_0}}\right)$ $-\frac{1}{2}\mathrm{erfc}\left(\frac{5a}{2}\sqrt{\frac{T}{N_0}}\right)$	$\frac{1}{2}\mathrm{erfc}\left(\frac{a}{2}\sqrt{\frac{T}{N_0}}\right)$ $-\frac{1}{2}\mathrm{erfc}\left(\frac{3a}{2}\sqrt{\frac{T}{N_0}}\right)$	

$$P_e(m_i) \leqslant \sum_{\substack{k=1 \\ k \neq i}}^{M} P(A_{ik}) \qquad i = 1, 2, \ldots, M \qquad (3.76)$$

This relationship is illustrated in Fig. 3.9 for the case of $M = 4$. In Fig. 3.9a, we show the four message points and associated decision regions, with the point s_1, representing the transmitted symbol. In Fig. 3.9b, we show the three constituent signal-space descriptions where, in each case, the transmitted message point s_1 and one other message point are retained. According to Fig. 3.9a, the conditional probability of symbol error, $P_e(m_i)$, is equal to the probability that the observation vector x lies in the shaded region of the two-dimensional signal

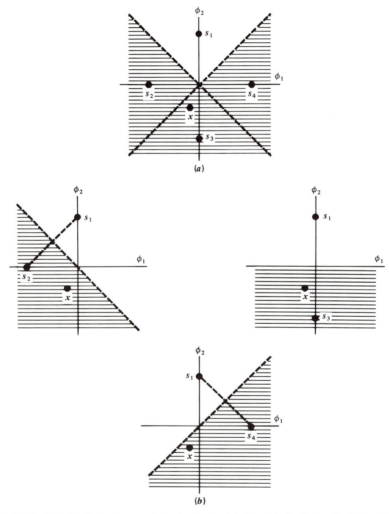

Figure 3.9 Illustrating the union bound. (a) Constellation of four message points. (b) Three constellations with a common message point and one other message point retained.

space. Clearly, this probability is less than the sum of the probabilities of the three individual events that $\mathbf{x}$ lies in the shaded regions of the constituent signal spaces depicted in Fig. 3.9b.

It is important to note that, in general, the probability $P(A_{ik})$ is different from the probability $P(\hat{m} = m_k|m_i)$. The latter is the probability that the observation vector $\mathbf{x}$ is closer to the signal vector $\mathbf{s}_k$ than every other, when $\mathbf{s}_i$ (or m_i) is sent. On the other hand, the probability $P(A_{ik})$ depends only on two signal vectors, $\mathbf{s}_i$ and $\mathbf{s}_k$. To emphasize this difference, we rewrite Eq. 3.76 by adopting $P_2(\mathbf{s}_i,\mathbf{s}_k)$ in place of $P(A_{ik})$. We thus have

$$P_e(m_i) \leqslant \sum_{\substack{k=1 \\ k \neq i}}^{M} P_2(\mathbf{s}_i,\mathbf{s}_k) \qquad i = 1, 2, \ldots, M \qquad (3.77)$$

Consider then a simplified digital communication system that involves the use of two equally likely messages represented by the vectors $\mathbf{s}_i$ and $\mathbf{s}_k$. Since white Gaussian noise is identically distributed along any set of orthogonal axes, we may temporarily choose the first axis in such a set as one that passes through the points $\mathbf{s}_i$ and $\mathbf{s}_k$; for examples, see Fig. 3.9b. The corresponding decision boundary is represented by the bisector that is perpendicular to the line joining the points $\mathbf{s}_i$ and $\mathbf{s}_k$. Accordingly, when the symbol m_i (vector $\mathbf{s}_i$) is sent, and if the observation vector $\mathbf{x}$ lies on the side of the bisector where $\mathbf{s}_k$ lies, an error is made. The probability of this event is given by

$$P_2(\mathbf{s}_i,\mathbf{s}_k) = P(\mathbf{x} \text{ is closer to } \mathbf{s}_k \text{ than } \mathbf{s}_i, \text{ when } \mathbf{s}_i \text{ is sent})$$

$$= \int_{d_{ik}/2}^{\infty} \frac{1}{\sqrt{\pi N_0}} \exp\left(-\frac{u^2}{N_0}\right) du \qquad (3.78)$$

where d_{ik} is the distance between $\mathbf{s}_i$ and $\mathbf{s}_k$; that is

$$d_{ik} = \|\mathbf{s}_i - \mathbf{s}_k\| \qquad (3.79)$$

From the definition of the complementary error function, we have

$$\text{erfc}(u) = \frac{2}{\sqrt{\pi}} \int_u^{\infty} \exp(-z^2) \, dz$$

Thus, in terms of this function, Eq. 3.78 takes on the compact form

$$P_2(\mathbf{s}_i,\mathbf{s}_k) = \frac{1}{2} \text{erfc}\left(\frac{d_{ik}}{2\sqrt{N_0}}\right) \qquad (3.80)$$

The complementary error function is a monotone decreasing function of its argument. We may therefore state that as the distance separating the message points $\mathbf{s}_i$ and $\mathbf{s}_k$ is increased, the probability of error is reduced, a result that is intuitively satisfying.

Substituting Eq. 3.80 into Eq. 3.77, we get

$$P_e(m_i) \leqslant \frac{1}{2} \sum_{\substack{k=1 \\ k \neq i}}^{M} \text{erfc}\left(\frac{d_{ik}}{2\sqrt{N_0}}\right) \qquad i = 1, 2, \ldots, M$$

Finally, with the M transmitted messages assumed equally likely, we find that the average probability of symbol error is overbounded as follows:

$$P_e = \frac{1}{M} \sum_{i=1}^{M} P_e(m_i)$$

$$\leq \frac{1}{2M} \sum_{i=1}^{M} \sum_{\substack{k=1 \\ k \neq i}}^{M} \text{erfc}\left(\frac{d_{ik}}{2\sqrt{N_0}}\right) \qquad (3.81)$$

where the distance d_{ik} is defined in Eq. 3.79. The second line of Eq. 3.81 defines the union bound on the average probability of symbol error for any set of M equally likely signals in additive white Gaussian noise.*

Equation 3.81 is the most general formulation of the *union bound*. A simpler result can be obtained by using the *minimum Euclidean distance* of a signal constellation. Let d_{min} denote this minimum distance; that is to say,

$$d_{min} = \min_{\substack{i,k \\ i \neq k}} d_{ik} \qquad (3.82)$$

Since the complementary error function $\text{erfc}(u)$ is a monotone decreasing function of the argument u, we may write

$$\sum_{\substack{k=1 \\ k \neq i}}^{M} \text{erfc}\left(\frac{d_{ik}}{2\sqrt{N_0}}\right) \leq (M-1)\,\text{erfc}\left(\frac{d_{min}}{2\sqrt{N_0}}\right)$$

Consequently, we may simplify the union bound on the probability of error given in Eq. 3.81 as

$$P_e \leq \frac{M-1}{2}\,\text{erfc}\left(\frac{d_{min}}{2\sqrt{N_0}}\right) \qquad (3.83)$$

This result represents a simplified form of the union bound that is easy to calculate.

3.7   CORRELATION RECEIVER

We note that for an AWGN channel and for the case when the transmitted signals $s_1(t)$, $s_2(t)$, . . . , $s_M(t)$ are equally likely, the optimum receiver consists of two subsystems, detailed in Fig. 3.10 and described next:

1.   The detector part of the receiver is shown in Fig. 3.10a. It consists of a bank of M *product-integrators* or *correlators* supplied with a corresponding set of coherent reference signals or orthonormal basis functions $\phi_1(t)$, $\phi_2(t)$, . . . , $\phi_N(t)$ that are generated locally. This bank of correlators operate on the received signal $x(t)$, $0 \leq t \leq T$, to produce the observation vector **x**.

---

* For the derivation of tighter bounds on the average probability of symbol error, see Viterbi and Omura (1979, pp. 58–59).

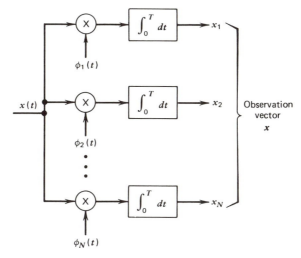

(a)

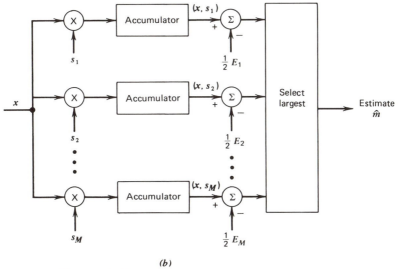

(b)

Figure 3.10 (a) Detector. (b) Vector receiver.

2. The second part of the receiver, namely, the vector receiver is shown in Fig. 3.10b. It is implemented in the form of a maximum-likelihood detector that operates on the observation vector $\mathbf{x}$ to produce an estimate $\hat{m}$ of the transmitted symbol m_i, $i = 1, 2, \ldots, M$, in a way that would minimize the average probability of symbol error. In accordance with Eq. 3.66, the N elements of the observation vector $\mathbf{x}$ are first multiplied by the corresponding N elements of each of the M signal vectors $\mathbf{s}_1, \mathbf{s}_2, \ldots, \mathbf{s}_M$, and the resulting products are successively summed in accumulators to form the corresponding set of inner products $\{(\mathbf{x}, \mathbf{s}_k)\}$, $k = 1, 2, \ldots, M$. Next, the inner products are corrected for the fact that the transmitted signal energies

may be unequal. Finally, the largest in the resulting set of numbers is selected, and a corresponding decision on the transmitted message is made.

The optimum receiver of Fig. 3.10 is commonly referred to as a *correlation receiver*.

3.8 MATCHED FILTER RECEIVER

Since each of the orthonormal basis functions $\phi_1(t)$, $\phi_2(t)$, . . . , $\phi_N(t)$ is assumed to be zero outside the interval $0 \leq t \leq T$, the use of multipliers shown in Fig. 3.10a may be avoided. This is desirable because analog multipliers are usually hard to build. Consider, for example, a linear filter with impulse response $h_j(t)$. With the received signal $x(t)$ used as the filter input, the resulting filter output, $y_j(t)$, is defined by the convolution integral:

$$y_j(t) = \int_{-\infty}^{\infty} x(\tau)h_j(t - \tau)d\tau \tag{3.76}$$

Suppose we now set the impulse response

$$h_j(t) = \phi_j(T - t) \tag{3.77}$$

Then the resulting filter output is

$$y_j(t) = \int_{-\infty}^{\infty} x(\tau)\phi_j(T - t + \tau)d\tau \tag{3.78}$$

Sampling this output at time $t = T$, we get

$$y_j(T) = \int_{-\infty}^{\infty} x(\tau)\phi_j(\tau)d\tau \tag{3.79}$$

and since $\phi_j(T)$ is zero outside the interval $0 \leq t \leq T$, we finally get

$$y_j(T) = \int_{0}^{T} x(\tau)\phi_j(\tau)d\tau \tag{3.80}$$

We note that $y_j(T) = x_j$, where x_j is the jth correlator output produced by the received signal $x(t)$ in Fig. 3.10a. Thus the detector part of the optimum receiver may also be implemented as in Fig. 3.11.

A filter whose impulse response is a time-reversed and delayed version of some signal $\phi_j(t)$, as in Eq. 3.77, is said to be *matched* to $\phi_j(t)$. Correspondingly, the optimum receiver based on the detector of Fig. 3.11 is referred to as the *matched filter receiver*.*

For a matched filter operating in real time to be physically realizable, it must be causal. That is to say, its impulse response must be zero for negative time, as shown by

$$h_j(t) = 0 \qquad t < 0$$

---

* For review papers on the matched filter and its properties see Turin (1960, 1976).

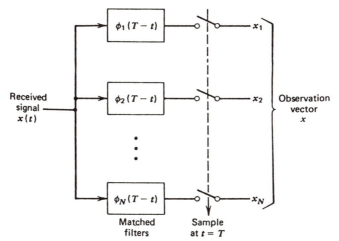

Figure 3.11   Detector part of matched filter receiver; the vector receiver part is as shown in Fig. 3.10*b*.

With $h_j(t)$ defined in terms of $\phi_j(t)$ as in Eq. 3.77, we see that the causality condition is satisfied provided that the signal $\phi_j(t)$ is zero outside the interval $0 \le t \le T$.

(1)   Maximization of Output Signal-to-noise Ratio

We may gain further insight into the operation of a matched filter by using output signal-to-noise ratio as the optimality criterion for deriving the matched filter from first principles.

Consider then a linear filter of impulse response $h(t)$, with an input that consists of a known signal, $\phi(t)$, and an additive noise component, $w(t)$, as shown in Fig. 3.12. We may thus write

$$x(t) = \phi(t) + w(t) \qquad 0 \le t \le T \tag{3.81}$$

where T is the observation instant. In particular, we may choose $\phi(t)$ to be one of the orthonormal basis functions. The $w(t)$ is the sample function of a white Gaussian noise process of zero mean and power spectral density $N_0/2$. Since the filter is linear, the resulting output, $y(t)$, may be expressed as

$$y(t) = \phi_o(t) + n(t) \tag{3.82}$$

where $\phi_o(t)$ and $n(t)$ are produced by the signal and noise components of the input $x(t)$, respectively. A simple way of describing the requirement that the

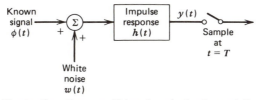

Figure 3.12   Illustrating the condition for derivation of the matched filter.

output signal component $\phi_o(t)$ be considerably greater than the output noise component $n(t)$ is to have the filter make the instantaneous power in the output signal $\phi_o(t)$, measured at time $t = T$, as large as possible compared with the average power of the output noise $n(t)$. This is equivalent to maximizing the *output signal-to-noise ratio* defined as

$$(\text{SNR})_O = \frac{|\phi_o(T)|^2}{E[n^2(t)]} \tag{3.83}$$

We now show that this maximization occurs when the filter is matched to the known signal $\phi(t)$ at the input.

Let $\Phi(f)$ denote the Fourier transform of the known signal $\phi(t)$, and $H(f)$ denote the transfer function of the filter. Then the Fourier transform of the output signal $\phi_o(t)$ is equal to $H(f)\Phi(f)$, and $\phi_o(t)$ is itself given by the inverse Fourier transform

$$\phi_o(t) = \int_{-\infty}^{\infty} H(f)\Phi(f)\exp(j2\pi ft)df \tag{3.84}$$

Hence, when the filter output is sampled at time $t = T$, we may write

$$|\phi_o(T)|^2 = \left| \int_{-\infty}^{\infty} H(f)\Phi(f)\exp(j2\pi fT)df \right|^2 \tag{3.85}$$

Consider next the effect of the noise $w(t)$ alone on the filter output. The power spectral density $S_N(f)$ of the output noise $n(t)$ is equal to the power spectral density of the input noise $w(t)$ times the squared magnitude of the transfer function $H(f)$. Since $w(t)$ is drawn from a process that is white with constant power spectral density $N_0/2$, it follows that

$$S_N(f) = \frac{N_0}{2} |H(f)|^2 \tag{3.86}$$

The average power of the output noise $n(t)$ is therefore

$$E[n^2(t)] = \int_{-\infty}^{\infty} S_N(f)df$$

$$= \frac{N_0}{2} \int_{-\infty}^{\infty} |H(f)|^2 df \tag{3.87}$$

Thus, substituting Eqs. 3.85 and 3.87 into Eq. 3.83, we may rewrite the expression for the output signal-to-noise ratio as

$$(\text{SNR})_O = \frac{\left| \int_{-\infty}^{\infty} H(f)\Phi(f)\exp(j2\pi fT)df \right|^2}{\dfrac{N_0}{2} \int_{-\infty}^{\infty} |H(f)|^2 df} \tag{3.88}$$

Our problem is to find, while holding the Fourier transform $\Phi(f)$ of the input signal fixed, the form of the transfer function $H(f)$ of the filter that makes $(\text{SNR})_O$ a maximum. To find the solution to this optimization problem, we apply a mathematical result known as Schwarz's* inequality to the numerator of Eq. 3.88. Accordingly, we may write

$$\left| \int_{-\infty}^{\infty} H(f)\Phi(f)\exp(j2\pi fT)df \right|^2 \leq \int_{-\infty}^{\infty} |H(f)|^2 df \int_{-\infty}^{\infty} |\Phi(f)|^2 df \qquad (3.89)$$

Using this relation in Eq. 3.88, we may simplify the output signal-to-noise ratio as

$$(\text{SNR})_O \leq \frac{2}{N_0} \int_{-\infty}^{\infty} |\Phi(f)|^2 df \qquad (3.90)$$

The right side of Eq. 3.90 is uniquely defined by two quantities:

1.   The signal energy given by (in accordance with Rayleigh's energy theorem)

$$\int_{-\infty}^{\infty} |\phi(t)|^2 \, dt = \int_{-\infty}^{\infty} |\Phi(f)|^2 \, dt$$

2.   The noise power spectral density $N_0/2$.

As such, the right side of Eq. 3.90 does not depend on the transfer function $H(f)$. Consequently, the output signal-to-noise ratio will be a maximum when $H(f)$ is chosen so that the equality holds; that is

$$(\text{SNR})_{O,\max} = \frac{2}{N_0} \int_{-\infty}^{\infty} |\Phi(f)|^2 df \qquad (3.91)$$

For this condition, $H(f)$ assumes its optimum value denoted as $H_{opt}(f)$. From Schwarz's inequality, we also find that, except for a scaling factor, the optimum value of this transfer function is defined by

$$H_{opt}(f) = \Phi^*(f)\exp(-j2\pi fT) \qquad (3.92)$$

where $\Phi^*(f)$ is the complex conjugate of the Fourier transform of the input signal $\phi(t)$. This relation states that, except for the necessary time delay factor $\exp(-j2\pi fT)$, the transfer function of the optimum filter is the same as the complex conjugate of the spectrum of the input signal.

Equation 3.92 specifies the matched filter in the frequency domain. To characterize it in the time domain, we take the inverse Fourier transform of $H_{opt}(f)$ in Eq. 3.92 to obtain the impulse response of the matched filter as

$$h_{opt}(t) = \int_{-\infty}^{\infty} \Phi^*(f)\exp[-j2\pi f(T - t)]df$$

Since for a real-valued signal $\phi(t)$ we have $\Phi^*(f) = \Phi(-f)$, we may also write

$$h_{opt}(t) = \int_{-\infty}^{\infty} \Phi(-f)\exp[-j2\pi f(T - t)]df$$

$$= \phi(T - t) \qquad (3.93)$$

---

* For a derivation of Schwarz's inequality, see Appendix G.

Equation 3.93 shows that the impulse response of the optimum filter is a time-reversed and delayed version of the input signal $\phi(t)$; that is, it is matched to the input signal. Note that the only assumptions we have made about the input noise $w(t)$ are that it is additive, stationary, and white with zero mean and power spectral density $N_0/2$.

The result given in Eq. 3.93 was derived by maximizing the output signal-to-noise ratio of the receiver. This result is identical, except for a minor change in notation, to that given in Eq. 3.77, which was derived earlier by minimizing the average probability of symbol error at the receiver output. The two derivations, however, were performed under somewhat different sets of conditions. Collecting these two sets of conditions together, we may now make the following important statement concerning these two criteria for optimum receiver design. Specifically, maximization of the output signal-to-noise ratio is equivalent to minimization of the average probability of symbol error under two assumptions:

1. The additive white noise at the receiver input is stationary with Gaussian statistics.
2. The a priori probabilities of the transmitted signals are known.

EXAMPLE 3 MATCHED FILTER FOR RF PULSE

Consider a rectangular RF pulse of duration T seconds and unit energy, as shown by (see Fig. 3.13a)

$$\phi(t) = \begin{cases} \sqrt{\dfrac{2}{T}} \cos(2\pi f_c t) & 0 \le t \le T \\ 0 & \text{elsewhere} \end{cases} \tag{3.94}$$

where f_c is a large integral multiple of $1/T$. The impulse response of a filter matched to $\phi(t)$ is therefore

$$\begin{aligned} h_{opt}(t) &= \phi(T - t) \\ &= \begin{cases} \sqrt{\dfrac{2}{T}} \cos(2\pi f_c t) & 0 \le t \le T \\ 0 & \text{elsewhere} \end{cases} \end{aligned} \tag{3.95}$$

which, in this example, works out to be the same as the signal $\phi(t)$ itself. The corresponding filter output, $\phi_o(t)$, is determined by convolving $h_{opt}(t)$ with $\phi(t)$; we thus obtain

$$\phi_o(t) = \begin{cases} (t/T) \cos(2\pi f_c t) & 0 \le t \le T \\ (2 - t/T) \cos(2\pi f_c t) & T \le t \le 2T \\ 0 & \text{elsewhere} \end{cases} \tag{3.96}$$

which is shown sketched in Fig. 3.13b. As expected, the filter output attains its maximum value at time $t = T$.

When a unit impulse of current is applied to the idealized parallel tuned circuit shown in Fig. 3.14a, the resulting voltage response is equal to the impulse response of the circuit, as shown by

$$h(t) = \frac{1}{C} \cos\left(\frac{t}{\sqrt{LC}}\right) \qquad 0 \le t \le \infty \tag{3.97}$$

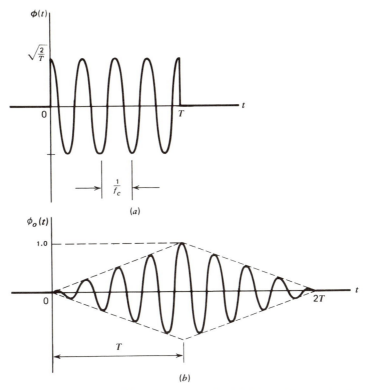

Figure 3.13 (a) RF pulse input. (b) Matched filter output.

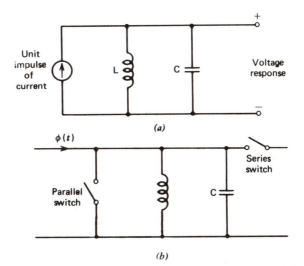

Figure 3.14 (a) Parallel tuned LC circuit. (b) Integrate-and-dump circuit.

where it is assumed that the initial energy in both L and C at time $t = 0$ is zero. Suppose we choose the circuit parameters L and C such that

$$\frac{1}{\sqrt{LC}} = 2\pi f_c$$

and

$$\frac{1}{C} = \sqrt{\frac{2}{T}}$$

We then find that the response $h(t)$ of the tuned circuit coincides with $h_{opt}(t)$ over the interval $0 \leqslant t \leqslant T$. However, this is not so for $t > T$. Thus, the matched filtering operation for the RF pulse $\phi(t)$ of Fig. 3.13a may be implemented by using the modification shown in Fig. 3.14b. The parallel switch closes briefly at time $t = 0$, dumping any residual energy in the filter. This ensures that signal energy received before $t = 0$ does not contribute to the output at time $t = T$. The series switch then closes briefly at time $t = T$, thereby sampling the filter output at the right time. This form of a matched filter is called an *integrate-and-dump filter*. Note that such a filter is not time-invariant; however, it does exhibit the desired impulse response as long as the timing of the two switches is properly synchronized with respect to the input RF pulse $\phi(t)$.

(2) Properties of Matched Filters

We note that a filter, which is matched to a known signal $\phi(t)$ of duration T seconds, is characterized by an impulse response that is a time-reversed and delayed version of the input $\phi(t)$, as shown by

$$h_{opt}(t) = \phi(T - t) \tag{3.98}$$

In the frequency domain, the matched filter is characterized by a transfer function that is, except for a delay factor, the complex conjugate of the Fourier transform of the input $\phi(t)$, as shown by

$$H_{opt}(f) = \Phi^*(f)\exp(-j2\pi fT) \tag{3.99}$$

Based on this fundamental pair of relations, we may derive some important properties of matched filters, which should help the reader develop an intuitive grasp of how a matched filter operates.

PROPERTY 1

The spectrum of the output signal of a matched filter with the matched signal as input is, except for a time delay factor, proportional to the energy spectral density of the input signal.

Let $\Phi_o(f)$ denote the Fourier transform of the filter output $\phi_o(t)$. Then

$$\begin{aligned} \Phi_o(f) &= H_{opt}(f)\Phi(f) \\ &= \Phi^*(f)\Phi(f)\exp(-j2\pi fT) \\ &= |\Phi(f)|^2 \exp(-j2\pi fT) \end{aligned} \tag{3.100}$$

which is the desired result.

PROPERTY 2

The output signal of a matched filter is proportional to a shifted version of the autocorrelation function of the input signal to which the filter is matched.

This property follows directly from Property 1, recognizing that the autocorrelation function and energy spectral density of a signal form a Fourier transform pair. Thus, taking the inverse Fourier transform of Eq. 3.100, we may express the matched-filter output as

$$\phi_o(t) = R_\phi(t - T) \tag{3.101}$$

where $R_\phi(\tau)$ is the autocorrelation function of the input $\phi(t)$ for lag τ. Equation 3.101 is the desired result. Note that at time $t = T$, we have

$$\phi_o(T) = R_\phi(0) = E \tag{3.102}$$

where E is the signal energy. That is, in the absence of noise, the maximum value of the matched-filter output, attained at time $t = T$, is proportional to the signal energy.

EXAMPLE 4

A possible exploitation of this property of a matched filter is illustrated in Fig. 3.15. Let us suppose that we have a signal lasting from $t = 0$ to $t = T$, which has the appearance and character of a sample function of a random process with a broad power spectrum, so that its autocorrelation function approximates a delta function. This signal may be generated by applying, at $t = 0$, a short pulse (short enough to approximate a delta function) to a linear filter with impulse response $\phi(t)$. The impulse-like input signal has components occupying a very wide frequency band, but their amplitudes and phases are such that they add constructively only at and near $t = 0$ and cancel each other out elsewhere. We may therefore view the signal-generating filter as an *encoder,* whereby the amplitudes and phases of the frequency components of the impulse-like input signal are coded in such a way that the filter output becomes noise-like in character, lasting from $t = 0$ to $t = T$, as in Fig. 3.16a. The signal $\phi(t)$ generated in this way is to be transmitted to a receiver via a distortionless but noisy channel. The requirement is to reconstruct at the receiver output a signal that closely approximates the original impulse-like signal at the transmitter input.

The optimum solution to such a requirement, in the presence of additive white Gaussian noise, is to employ a matched filter in the receiver, as in Fig. 3.15. We may view this matched filter as a *decoder,* whereby the useful signal component $\phi(t)$ of the receiver input is decoded in such a way that all frequency components at the filter output have zero phase at time $t = T$, and add constructively to produce a large pulse of

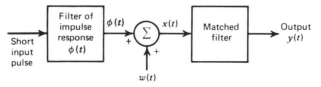

Figure 3.15 Viewing the matched filtering operation as an encoding-decoding process.

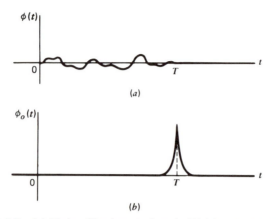

Figure 3.16 (a) Noise-like input signal. (b) Matched filter output.

nonzero width, as in Fig. 3.16b. Thus, in coding the impulse-like signal at the transmitter input, we have *spread* the signal energy out over a duration T, and in decoding the noise-like signal at the receiver input, we are able to concentrate this energy into a relatively narrow pulse. The extent to which the receiver output $\phi_o(t)$ approximates the original impulse-like signal is simply a reflection of the extent to which the autocorrelation function of the transmitted signal $\phi(t)$ approximates a delta function. The signal generating and reconstruction filters in Fig. 3.15 are said to constitute a *matched-filter pair*.*

PROPERTY 3

The output signal-to-noise ratio of a matched filter depends only on the ratio of the signal energy to the power spectral density of the white noise at the filter input.

To demonstrate this property, consider a filter matched to an input signal $\phi(t)$. From Property 2, the maximum value of the filter output, at time $t = T$, is proportional to the signal energy E. Substituting Eq. 3.92 in Eq. 3.87 gives the average output noise power as

$$E[n^2(t)] = \frac{N_0}{2} \int_{-\infty}^{\infty} |\Phi(f)|^2 df = \frac{N_0}{2} E \qquad (3.103)$$

where we have made use of Rayleigh's energy theorem. Therefore, the output signal-to-noise ratio has the maximum value

$$(\text{SNR})_{O,\text{max}} = \frac{E^2}{N_0 E/2} = \frac{2E}{N_0} \qquad (3.104)$$

This result is perhaps the most important parameter in the calculation of the performance of signal processing systems using matched filters. From Eq.

* The use of a matched-filter pair forms a basis of two important applications: (1) *spread-spectrum modulation,* discussed in Chapter 9, and (2) *pulse-compression radar,* discussed in Rihaczek (1969).

3.104 we see that dependence on the waveform of the input $\phi(t)$ has been completely removed by the matched filter. Accordingly, in evaluating the ability of a matched-filter receiver to combat additive white Gaussian noise, we find that all signals that have the same energy are equally effective. Note that the signal energy E is in joules and the noise spectral density $N_0/2$ is in watts per hertz, so that the ratio $2E/N_0$ is dimensionless; however, the two quantities have different physical meaning. We refer to E/N_0 as the *signal energy-to-noise density ratio*.

PROPERTY 4

The matched-filtering operation may be separated into two matching conditions; namely, spectral phase matching that produces the desired output peak at time T, and spectral amplitude matching that gives this peak value its optimum signal-to-noise density ratio.

In polar form, the spectrum of the signal $\phi(t)$ being matched may be expressed as

$$\Phi(f) = |\Phi(f)|\exp[j\theta(f)]$$

where $|\Phi(f)|$ is the amplitude spectrum and $\theta(f)$ is the phase spectrum of the signal. The filter is said to be *spectral phase matched* to the signal $\phi(t)$ if the transfer function of the filter is defined by*

$$H(f) = |H(f)|\exp[-j\theta(f) - j2\pi fT]$$

where $|H(f)|$ is real and nonnegative and T is a positive constant. The output of such a filter is

$$\phi_o'(t) = \int_{-\infty}^{\infty} H(f)\Phi(f)\exp(j2\pi ft)df$$

$$= \int_{-\infty}^{\infty} |H(f)||\Phi(f)|\exp[j2\pi f(t - T)]df$$

where the product $|H(f)||\Phi(f)|$ is real and nonnegative. The spectral phase matching ensures that all the spectral components of the output $\phi_o'(t)$ add constructively at time $t = T$, thereby causing the output to attain its maximum value, as shown by

$$\phi_o'(t) \le \phi_o'(T) = \int_{-\infty}^{\infty} |\Phi(f)||H(f)|df$$

For *spectral amplitude matching,* we choose the amplitude response $|H(f)|$ of the filter to shape the output for best signal-to-noise ratio at $t = T$ by using

$$|H(f)| = |\Phi(f)|$$

and the standard matched filter is the result.

* See Birdsall (1976).

3.9 DETECTION OF SIGNALS WITH UNKNOWN PHASE IN NOISE

Up to this point in our discussion, we have assumed that the information-bearing signal is completely known at the receiver. In practice, however, it is often found that in addition to the uncertainty due to the additive noise of a receiver, there is an additional uncertainty due to the randomness of certain signal parameters. The usual cause of this uncertainty is distortion in the transmission medium. Perhaps the most common random signal parameter is the phase, which is especially true for narrow-band signals. For example, transmission over a multiplicity of paths of different and variable lengths, or rapidly varying delays in the propagating medium from transmitter to receiver, may cause the phase of the received signal to change in a way that the receiver cannot follow. Synchronization with the phase of the transmitted carrier may then be too costly, and the designer may simply choose to disregard the phase information in the received signal at the expense of some degradation in the noise performance of the system.

Consider then a digital communication system in which the transmitted signal equals

$$s_i(t) = \sqrt{\frac{2E}{T}}\, \cos(2\pi f_i t) \qquad 0 \leqslant t \leqslant T \tag{3.105}$$

where E is the signal energy, T is the duration of the signaling interval, and the frequency f_i is an integral multiple of $1/2T$. When no provision is made to phase synchronize the receiver with the transmitter, the received signal will, for an AWGN channel, be of the form

$$x(t) = \sqrt{\frac{2E}{T}}\, \cos(2\pi f_i t + \theta) + w(t) \qquad 0 \leqslant t \leqslant T \tag{3.106}$$

where $w(t)$ is the sample function of a white Gaussian noise process of zero mean and power spectral density $N_0/2$. The phase θ is unknown, and it is usually considered to be the sample value of a random variable uniformly distributed between 0 and 2π radians. This implies a complete lack of knowledge of the phase. A digital communication system characterized in this way is said to be *noncoherent*.

We readily see that the detection schemes presented previously are inadequate for dealing with noncoherent systems, because if the received signal has the form described by Eq. 3.106, the output of the associated correlator in the receiver will be a function of the unknown phase θ. We shall now discuss, in a rather intuitive manner, the necessary modifications that may be introduced into the receiver in order to deal with this new situation.

Using a well-known trigonometric identity, we may rewrite Eq. 3.106 in the expanded form

$$x(t) = \sqrt{\frac{2E}{T}}\, \cos\theta \, \cos(2\pi f_i t) - \sqrt{\frac{2E}{T}}\, \sin\theta \, \sin(2\pi f_i t) + w(t) \qquad 0 \leqslant t \leqslant T$$

$$\tag{3.107}$$

Suppose that the received signal $x(t)$ is applied to a pair of correlators; we assume that one correlator is supplied with the reference signal $\sqrt{2/T} \cos(2\pi f_i t)$, and the other is supplied with the reference signal $\sqrt{2/T} \sin(2\pi f_i t)$. For both correlators, the observation interval is $0 \leq t \leq T$. Then in the absence of noise, we find that the first correlator output equals $\sqrt{E} \cos\theta$ and the second correlator output equals $-\sqrt{E} \sin\theta$. The dependence on the unknown phase θ may be removed by summing the squares of the two correlator outputs, and then taking the square root of the sum. Thus, when the noise $w(t)$ is zero, the result of these operations is simply $\sqrt{E}$, which is independent of the unknown phase θ. This suggests that for the detection of a sinusoidal signal of arbitrary phase, and which is corrupted by an additive white Gaussian noise, as in the model of Eq. 3.106, we may use the so-called *quadrature receiver* shown in Fig. 3.17a. Indeed, this receiver is optimum in the sense that it realizes this detection with the minimum probability of error.*

We next derive two equivalent forms of the quadrature receiver. The first form is easily obtained by replacing each correlator in Fig. 3.17a with a corresponding equivalent matched filter. We thus obtain the alternative form of quadrature receiver shown in Fig. 3.17b. In one branch of this receiver, we have a filter matched to the signal $\sqrt{2/T} \cos(2\pi f_i t)$, and in the other branch we have a filter matched to $\sqrt{2/T} \sin(2\pi f_i t)$, both defined for the time interval $0 \leq t \leq T$. The filter outputs are sampled at time $t = T$, squared, and added together.

To obtain the second equivalent form of the quadrature receiver, suppose we have a filter that is matched to $s(t) = \sqrt{2/T} \cos(2\pi f_i t + \theta)$, for $0 \leq t \leq T$. The envelope of the matched filter output is obviously unaffected by the value of phase θ. Therefore, for convenience, we may choose a matched filter with impulse response $\sqrt{2/T} \cos[2\pi f_i(T - t)]$, corresponding to $\theta = 0$. The output of such a filter in response to the received signal $x(t)$ is given by

$$y(t) = \sqrt{\frac{2}{T}} \int_0^T x(\tau)\cos[2\pi f_i(T - t + \tau)]d\tau$$

$$= \sqrt{\frac{2}{T}} \cos[2\pi f_i(T - t)] \int_0^T x(\tau)\cos(2\pi f_i\tau)d\tau$$

$$- \sqrt{\frac{2}{T}} \sin[2\pi f_i(T - t)] \int_0^T x(\tau)\sin(2\pi f_i\tau)d\tau \qquad (3.108)$$

The envelope of the matched filter output is proportional to the square root of the sum of the squares of the integrals in Eq. 3.108. The envelope, evaluated at time $t = T$, will therefore equal

$$l_i = \left\{ \left[\int_0^T x(\tau) \sqrt{\frac{2}{T}} \cos(2\pi f_i\tau)d\tau \right]^2 + \left[\int_0^T x(\tau) \sqrt{\frac{2}{T}} \sin(2\pi f_i\tau)d\tau \right]^2 \right\}^{1/2} \qquad (3.109)$$

But this is just the output of the quadrature receiver. Therefore, the output (at time T) of a filter matched to the signal $\sqrt{2/T} \cos(2\pi f_i t + \theta)$, of arbitrary phase

* For mathematical details of this derivation, see Whalen (1971, pp. 196–205).

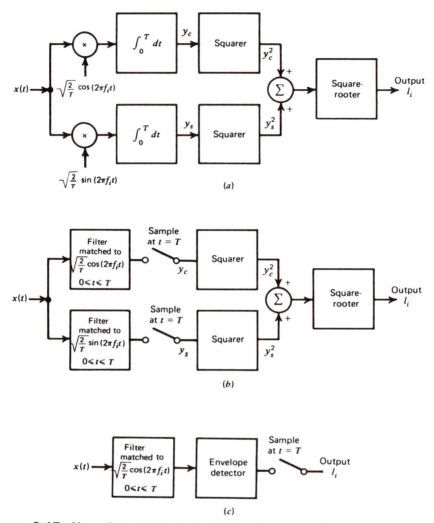

Figure 3.17 Noncoherent receivers. (a) Quadrature receiver using correlators. (b) Quadrature receiver using matched filters. (c) Noncoherent matched filter.

θ, followed by an envelope detector is the same as the corresponding output of the quadrature receiver of Fig. 3.17a. This form of receiver is shown in Fig. 3.17c. The combination of matched filter and envelope detector, shown in Fig. 3.17c, is called a *noncoherent matched filter*.

The need for an envelope detector following the matched filter in Fig. 3.17c may also be justified intuitively as follows. The output of a filter matched to a rectangular RF wave reaches a positive peak at the sampling instant $t = T$ (see Example 3). If, however, the phase of the filter is not matched to that of the signal, it is apparent that the peak will occur at a time different from the sampling instant. In actual fact, if the phases differ by π radians, we get a negative peak at the sampling instant. To avoid such poor sampling in the absence of prior information about the phase θ, it is reasonable to retain only the envelope of the matched filter output, since it is completely independent of the phase mismatch.

3.10 ESTIMATION: CONCEPTS AND CRITERIA

In the preceding sections of the chapter, our primary concern has been with the detection of a signal in additive noise. In the remaining part of the chapter, we consider some aspects of the related estimation problem.

To introduce the problem, suppose we are given a received signal $x(t)$ that is defined for a finite observation interval of duration T, as shown by

$$x(t) = s(t,\alpha) + w(t) \qquad 0 \leqslant t \leqslant T \tag{3.110}$$

where $s(t,\alpha)$ denotes a signal with an unknown parameter α, and $w(t)$ denotes a sample function of a white Gaussian noise process of zero mean and power spectral density $N_0/2$. For example, the parameter α may represent the amplitude or phase of a sinusoidal component of the received signal. *Parameter estimation* is the operation of assigning a value to the unknown parameter α, given the noise-corrupted received signal $x(t)$ of Eq. 3.110. The value assigned to the unknown parameter α is called an *estimate*. The algorithm yielding the estimate is called the *estimator*, the exact description of which is determined by the optimality criterion of interest.

Some commonly used criteria are as follows*:

1. *Minimum mean-square estimate.*   Let $\hat{\alpha}$ denote an estimate of the unknown parameter α. The *estimation error* is defined as $\alpha - \hat{\alpha}$. It is reasonable to associate with each estimate a *cost* that is a function of the estimation error. A cost function often used is a quadratic function of the estimation error, as depicted in Fig. 3.18*a*. Thus, treating α as the sample value of a random variable A, we may express the *average cost* incurred in using the estimate $\hat{\alpha}$ as

$$\mathscr{C}_1 = \int (\alpha - \hat{\alpha})^2 f_A(\alpha|\mathbf{x}) d\alpha$$

    where the integration is performed over all possible values of α. Also, $f_A(\alpha|\mathbf{x})$ is the *a posteriori* probability density function of the random variable A, given the observation vector $\mathbf{x}$; the vector $\mathbf{x}$ is a representation of the received signal $x(t)$ over the observation interval $0 \leqslant t \leqslant T$. The value of the estimate $\hat{\alpha}$ that minimizes the average cost $\mathscr{C}_1$ is called the *Bayes' minimum mean-square estimate.*

2. *Maximum a posteriori estimate.*   Another cost function of interest is the uniform cost function of Fig. 3.18*b*, in which Δ is an arbitrarily small but nonzero number. Based on this second cost function, the average cost incurred in using an estimate $\hat{\alpha}$ is given by (for small Δ)

$$\mathscr{C}_2 = 1 - f_A(\alpha|\mathbf{x})\Delta$$

---

* For detailed discussion of the estimation problem, see the following books: Van Trees (1968, pp. 52–86), Nahi (1969), Helstrom (1968, pp. 249–289), and Whalen (1971, pp. 321–362).

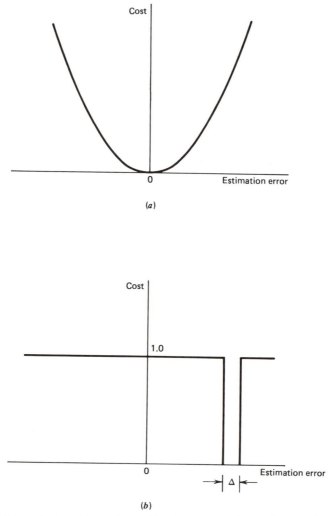

Figure 3.18   (a) Quadratic cost function. (b) Uniform cost function.

Accordingly, the best choice of the estimate $\hat{\alpha}$ is the value of α at which the a posteriori probability density function $f_A(\alpha|\mathbf{x})$ is maximum for then the average cost $\mathscr{C}_2$ is minimum. The estimate thus obtained is called the *maximum a posteriori (MAP) estimate*.

3. *Maximum likelihood estimate.*   After receiving the observation vector $\mathbf{x}$, it is reasonable to choose as an estimate of the unknown parameter α that value of α which gives the observation vector $\mathbf{x}$ a high probability of occurrence. Specifically, we first formulate the conditional probability density function $f_\mathbf{X}(\mathbf{x}|\alpha)$. With the observation vector $\mathbf{x}$ known, this density function represents a *likelihood function* that is a function of the unknown parameter α alone. The value of α for which the likelihood function $f_\mathbf{X}(\mathbf{x}|\alpha)$ is a maximum is called the *maximum likelihood (ML) estimate*. It is note-

worthy that the maximum likelihood estimate is the same as the maximum a posteriori estimate when the unknown parameter α is the sample value of a uniformly distributed random variable.

The ML estimate is popular because of its highly desirable properties. We emphasize its use in the next section.

3.11   MAXIMUM LIKELIHOOD ESTIMATION

We approach the development of the maximum likelihood estimate by choosing a set of orthonormal basis functions $\{\phi_i(t)\}$, $i = 1, 2, \ldots, N$. We approximate the received signal $x(t)$ in terms of a corresponding finite set of numbers $\{x_i\}$, $i = 1, 2, \ldots, N$, as shown by

$$x(t) \simeq \sum_{i=1}^{N} x_i \phi_i(t) \qquad 0 \le t \le T \tag{3.111}$$

where

$$x_i = \int_0^T x(t)\phi_i(t)dt \qquad i = 1, 2, \ldots, N \tag{3.112}$$

To represent $x(t)$ completely, we require an infinite set of numbers, due to the presence of the noise term $w(t)$. Substituting Eq. 3.110 into Eq. 3.112, we may thus write

$$x_i = s_i(\alpha) + w_i \tag{3.113}$$

where

$$s_i(\alpha) = \int_0^T s(t,\alpha)\phi_i(t)dt \qquad i = 1, 2, \ldots, N \tag{3.114}$$

and

$$w_i = \int_0^T w(t)\phi_i(t)dt \qquad i = 1, 2, \ldots, N \tag{3.115}$$

We note that the w_i are sample values of a set of independent Gaussian random variables of zero mean and variance $N_0/2$. Correspondingly, the x_i are sample values of a set of independent Gaussian random variables of mean values equal to $s_i(\alpha)$, $i = 1, 2, \ldots, N$, and variance $N_0/2$. Therefore, the likelihood function, given α, is defined by

$$f_\mathbf{X}(\mathbf{x}|\alpha) = (\pi N_0)^{-N/2} \prod_{i=1}^{N} \exp\left\{ -\frac{[x_i - s_i(\alpha)]^2}{N_0} \right\} \tag{3.116}$$

Now, if we let $N \to \infty$, this likelihood function is not well defined. Noting that we may divide a likelihood function by anything that does not depend on α and still have a likelihood function, we may avoid the convergence problem by dividing Eq. 3.116 by

$$f_{\mathbf{X}}(\mathbf{x}) = (\pi N_0)^{-N/2} \prod_{i=1}^{N} \exp\left(-\frac{x_i^2}{N_0}\right) \tag{3.117}$$

before letting $N \to \infty$. Clearly, we may do this, because this function does not depend on α. Define

$$\Lambda[x(t), \alpha] = \lim_{N \to \infty} \frac{f_{\mathbf{X}}(\mathbf{x}|\alpha)}{f_{\mathbf{X}}(\mathbf{x})} \tag{3.118}$$

Then, substituting Eqs. 3.116 and 3.117 into Eq. 3.118, cancelling common terms, and taking the logarithm, we get

$$\ln\Lambda[x(t), \alpha] = \frac{2}{N_0} \int_0^T x(t)s(t,\alpha)dt - \frac{1}{N_0} \int_0^T s^2(t,\alpha)dt \tag{3.119}$$

The *maximum likelihood estimate* $\hat{\alpha}$ is defined as that value of α that maximizes the likelihood function. Therefore, differentiating Eq. 3.119 with respect to α and setting the result equal to zero, we find that the maximum likelihood estimate $\hat{\alpha}$ is the solution of the equation

$$\int_0^T [x(t) - s(t,\hat{\alpha})] \frac{\partial s(t,\hat{\alpha})}{\partial \hat{\alpha}} dt = 0 \tag{3.120}$$

This equation is called the *likelihood equation*.

EXAMPLE 5   ESTIMATION OF PHASE

Consider a signal of known form

$$s(t,\theta) = a_c \sin(2\pi f_c t + \theta) \qquad 0 \le t \le T$$

where the amplitude a_c and frequency f_c are known. The problem is to estimate the unknown phase θ. From Eq. 3.120 we find that the maximum likelihood estimate $\hat{\theta}$ of the phase θ is the solution of

$$\int_0^T [x(t) - a_c \sin(2\pi f_c t + \hat{\theta})]\cos(2\pi f_c t + \hat{\theta})dt = 0 \tag{3.121}$$

We assume that $2f_c T = k$ where k is an integer. Then the integral of the second term in Eq. 3.121 is zero, so that the phase estimate $\hat{\theta}$ is the solution of

$$\int_0^T x(t)\cos(2\pi f_c t + \hat{\theta})dt = 0 \tag{3.122}$$

Expanding the cosine term in the integrand in Eq. 3.122 and rearranging terms, we get

$$\cos\hat{\theta} \int_0^T x(t)\cos(2\pi f_c t)dt = \sin\hat{\theta} \int_0^T x(t)\sin(2\pi f_c t)dt$$

Solving for $\hat{\theta}$ yields the desired estimate:

$$\hat{\theta} = \tan^{-1}\left[\frac{\int_0^T x(t)\cos(2\pi f_c t)dt}{\int_0^T x(t)\sin(2\pi f_c t)dt}\right] \tag{3.123}$$

The operations in Eq. 3.123 may be performed by using a pair of correlators or filters matched to $\cos(2\pi f_c t)$ and $\sin(2\pi f_c t)$, as indicated in Fig. 3.19.

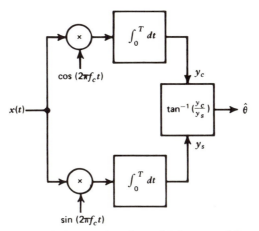

Figure 3.19 Estimator of phase of a sinusoidal wave of known amplitude and frequency.

From Eq. 3.122 we may also deduce the phase-locked loop realization shown in Fig. 3.20 consisting of a multiplier, an integrator, and a voltage-controlled oscillator (VCO), interconnected in the form of a feedback loop. The phase of the VCO output is denoted by $\hat{\hat{\theta}}$ to distinguish it from the maximum likelihood estimate $\hat{\theta}$. When the incoming signal $x(t)$ is noise-free, we have

$$x(t) = a_c \sin(2\pi f_c t + \theta)$$

The corresponding multiplier output is

$$e(t) = a_c \sin(2\pi f_c t + \theta)\cos(2\pi f_c t + \hat{\hat{\theta}})$$

$$= \frac{a_c}{2} \sin(\theta - \hat{\hat{\theta}}) + \frac{a_c}{2} \sin(4\pi f_c t + \theta + \hat{\hat{\theta}})$$

The integrator averages $e(t)$ with respect to time, thereby yielding an output $\bar{e}$ that is proportional to $\sin(\theta - \hat{\hat{\theta}})$. For a small phase difference, the integrator output $\bar{e}$ is therefore proportional to $\theta - \hat{\hat{\theta}}$. Also, the integrator acts to smooth the variations of $e(t)$ due to noise. When the integrator output $\bar{e}$ is applied to the VCO, the phase $\hat{\hat{\theta}}$ of the VCO output changes in such a way as to reduce $\bar{e}$ to zero. Then $\hat{\hat{\theta}}$ approaches the maximum likelihood estimate $\hat{\theta}$. This is the desired condition for the phase-locked loop and the condition suggested by Eq. 3.122. Thus a phase-locked loop may be viewed as a realization of the maximum likelihood estimation procedure.

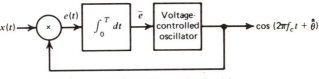

Figure 3.20 Phase-locked loop.

(1) Measures of an Estimator's Quality

The quality of an estimate $\hat{\alpha}$ is customarily assessed in terms of two specific measures, namely, the *expected value of the estimate,* and the *variance of the estimation error.*

Depending on the expected value, $E[\hat{\alpha}]$, three possible situations may arise:

1. The expected value of the estimate $\hat{\alpha}$ equals the true value of the parameter, α, being estimated; that is,

$$E[\hat{\alpha}] = \alpha \qquad \text{for all } \alpha$$

Under this condition, the estimate $\hat{\alpha}$ is said to be *unbiased*.

2. The expected value of the estimate $\hat{\alpha}$ differs from the true value of the unknown parameter, α, by a *fixed* amount; that is,

$$E[\hat{\alpha}] = \alpha + b$$

In this case, the estimate $\hat{\alpha}$ is said to have a *known bias*. We may then transform the biased estimate $\hat{\alpha}$ into an unbiased one simply by subtracting out the known bias b.

3. The expected value of the estimate $\hat{\alpha}$ differs from the true value of the parameter, α, by a *variable* amount; that is,

$$E[\hat{\alpha}] = \alpha + b(\alpha)$$

In this case, the estimate $\hat{\alpha}$ is said to have a *variable bias*. Since the bias $b(\alpha)$ now depends on the unknown parameter α, it cannot be subtracted out.

The *estimation error* equals the difference between the true value of the unknown parameter, α, and the estimate $\hat{\alpha}$. For the second measure of quality, we use the variance of the estimation error $\alpha - \hat{\alpha}$. This variance provides a measure of the *spread* of error.

A good estimate is one that has both a small bias and a small variance.

(2) Cramér-Rao Bound

In the case of the ML estimate, it is frequently difficult to compute the bias and variance. The problem is therefore tackled indirectly by first deriving a lower bound on the variance of *any* unbiased estimate, and then comparing the variance of the ML estimate with this lower bound.

Let $\hat{\alpha}$ denote *any* unbiased estimate of the real nonrandom parameter α. The *lower bound* on the variance of the estimation error is defined by

$$\text{Var}[\alpha - \hat{\alpha}] = E[(\alpha - \hat{\alpha})^2]$$

$$\geq \frac{-1}{E\left[\dfrac{\partial^2 \ln f_{\mathbf{X}}(\mathbf{x}|\alpha)}{\partial \alpha^2}\right]} \qquad (3.124)$$

where $\partial^2 f_{\mathbf{X}}(\mathbf{x}|\alpha)/\partial \alpha^2$ is assumed to exist and is absolutely integrable. The inequality of Eq. 3.124 is referred to as the *Cramér-Rao inequality*.* The quantity on the right side of the inequality is referred to as the *Cramér-Rao lower bound*.

---

* For a derivation of the Cramér-Rao inequality, see the following books: Van Trees (1968, pp. 66–72), Nahi (1969, pp. 246–251), Helstrom (1968, pp. 260–261), and Whalen (1971, pp. 325–331).

Any estimate that satisfies the Cramér-Rao bound with an equality is called an *efficient* estimate. Whenever an efficient estimate exists, the maximum likelihood estimation algorithm gives that estimate.

3.12   WIENER FILTER FOR WAVEFORM ESTIMATION

Another classic estimation problem in statistical communication theory is that of *linear estimation,* which involves the estimation of some desired response as a linear filtered version of an input signal. The reason for focusing on *linear estimators* is the simplicity in mechanization and construction, which is generally associated with linear filters. In this section, we present the *discrete-time version of the Wiener filter* for solving the linear estimation problem.

Let $X_n, X_{n-1}, \ldots, X_{n-M}$ denote random samples drawn from a *noisy* signal process $X(t)$, with a sampling interval of T seconds. The subscript n in the sample X_n signifies discrete time. Thus, X_n is the sample drawn from the process $X(t)$ at time $t = nT$, X_{n-1} is the sample drawn at time $t = (n - 1)T$, and so on for the other samples. That is, we have

$$X_n = X(nT)$$

For convenience of notation, the use of X_n is preferable to $X(nT)$.

The sequence of samples, $X_n, X_{n-1}, \ldots, X_{n-M}$ is applied to a *tapped-delay-line filter* (also called a *transversal filter*), as in Fig. 3.21. The filter consists of three sets of elements:

1. *Delay elements,* each of which produces a delay of T seconds.
2. *Multipliers,* each of which multiplies its respective *tap input* by a coefficient.
3. *Adders* for summing multiplier outputs.

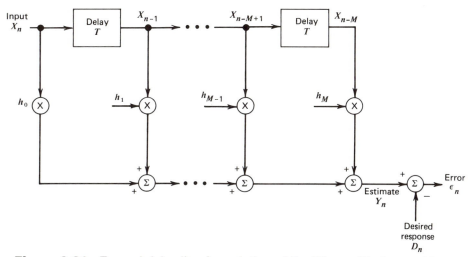

Figure 3.21   Tapped-delay-line formulation of the Wiener filtering problem.

The delay elements provide a means for the *storage of past samples of the input;* as such, they may be implemented by using a multistage *shift register.* We refer to the number of delay elements, M, as the *order* of the filter. We refer to the total delay, MT, as the *memory* of the filter.

Let $h_0, h_1, \ldots, h_M$ denote the filter coefficients used to weight the tap inputs $X_n, X_{n-1}, \ldots, X_{n-M}$, respectively. This sequence of coefficients defines the *impulse response* of the filter. Let Y_n denote the output of the filter, which is defined by the *convolution sum:*

$$Y_n = \sum_{k=0}^{M} h_k X_{n-k}\text{''} \tag{3.125}$$

Note that the filter output Y_n is a *linear* combination of the input samples. We wish to design the filter in such a way that the difference between a *desired response* D_n and the *filter output* Y_n is minimized in some statistical sense. In prediction, for example, D_n may equal the next sample X_{n+1} in the time sequence. Define*

$$\varepsilon_n = D_n - Y_n \tag{3.126}$$

The difference ε_n is called the *estimation error.* Clearly, ε_n is a random variable.

In the Wiener theory, the filter is optimized by minimizing the mean-square value of the estimation error ε_n. Let

$$\mathcal{E} = E[\varepsilon_n^2] \tag{3.127}$$

The *mean-squared error* $\mathcal{E}$ is a real and positive scalar quantity. The implication of minimizing $\mathcal{E}$ is that large values of the estimation error ε_n are considered much more serious than smaller values of ε_n. The use of the mean-squared error $\mathcal{E}$ as a *measure of waveform distortion* is a time-honored one.

To proceed with optimization of the filter, we substitute Eq. 3.126 in Eq. 3.127 and expand the squared expression. We thus get

$$\mathcal{E} = E[D_n^2] - 2E[D_n Y_n] + E[Y_n^2] \tag{3.128}$$

Next, substituting Eq. 3.125 in Eq. 3.128 and then interchanging the orders of summation and expectation in the last two terms, we get

$$\mathcal{E} = E[D_n^2] - 2 \sum_{k=0}^{M} h_k E[D_n X_{n-k}]$$

$$+ \sum_{k=0}^{M} \sum_{m=0}^{M} h_k h_m E[X_{n-k} X_{n-m}] \tag{3.129}$$

Assuming that the samples X_n and D_n are drawn from zero-mean jointly station-

* The letter "E" would have been an obvious symbol for denoting the estimation error, viewed as a random variable. Instead, we have used the Greek letter "ε" as the symbol for the estimation error so as to avoid confusion with the symbol for the expectation operator. A sample value of the estimation error will be denoted by the lowercase letter "e."

ary processes, we may interpret the three expectation terms on the right side of Eq. 3.129 as follows:

1. The expectation $E[D_n^2]$ is equal to the *variance of the desired response:*

$$\sigma_D^2 = E[D_n^2] \tag{3.130}$$

2. The expectation $E[D_n X_{n-k}]$ is equal to the *cross-correlation function between the desired response and the input signal for a lag of k samples:*

$$R_{DX}(k) = E[D_n X_{n-k}] \qquad k = 0, 1, \ldots, M \tag{3.131}$$

3. The expectation $E[X_{n-k} X_{n-m}]$ is equal to the *autocorrelation function of the input signal* for a lag of $(m - k)$ samples:

$$R_X(m - k) = E[X_{n-k} X_{n-m}] \qquad k, m = 0, 1, \ldots, M \tag{3.132}$$

Accordingly, we may rewrite the expression for the mean-squared error in the form

$$\mathcal{E} = \sigma_D^2 - 2 \sum_{k=0}^{M} h_k R_{DX}(k) + \sum_{k=0}^{M} \sum_{m=0}^{M} h_k h_m R_X(m - k) \tag{3.133}$$

With the cross-correlation function $R_{DX}(k)$ and the autocorrelation function $R_X(m - k)$ treated as known quantities, Eq. 3.133 states that *the mean squared error $\mathcal{E}$ is a second-order function of the filter coefficients $h_0, h_1, \ldots, h_M$.* Accordingly, we may visualize the dependence of the mean squared error $\mathcal{E}$ on these filter coefficients as a bowl-shaped surface with a unique minimum. We refer to this surface as the *error performance surface.* The requirement is to design the Wiener filter so that it operates at the minimum point of the error performance surface.

The mean squared error $\mathcal{E}$ attains its minimum value when its derivatives with respect to the filter coefficients are all zero. Differentiating Eq. 3.133 with respect to h_k, we get

$$\frac{\partial \mathcal{E}}{\partial h_k} = -2R_{DX}(k) + 2 \sum_{m=0}^{M} h_m R_X(m - k) \tag{3.134}$$

Setting this result equal to zero, we obtain the optimum values of the filter coefficients. Let these values be denoted by $h_{o0}, h_{o1}, \ldots, h_{oM}$. They are given by the solution of the set of simultaneous equations:

$$\sum_{m=0}^{M} h_{om} R_X(m - k) = R_{DX}(k) \qquad k = 0, 1, \ldots, M \tag{3.135}$$

This set of equations is called the *discrete-time version of the Wiener-Hopf equation,* the original form of which was developed for continuous-time signals.

Let $\mathcal{E}_{\min}$ denote the *minimum mean-squared error.* Substituting $h_{o0}, h_{o1}, \ldots, h_{oM}$ for the filter coefficients in Eq. 3.133, and then simplifying the expression for the minimum mean-squared error by the use of Eq. 3.135, we get the desired result:

$$\mathcal{E}_{min} = \sigma_D^2 - \sum_{k=0}^{M} h_{ok}R_X(k) \qquad (3.136)$$

A filter whose coefficients are defined by Eq. 3.135 is *optimum in the mean-square sense*. We refer to it as the *Wiener filter*. There is no other linear filter that we can design that produces a mean-squared error smaller in value than $\mathcal{E}_{min}$. In other words, for prescribed values of the correlations $R_X(k)$ and $R_{DX}(k)$, there is no other linear filter that produces a larger output signal-to-noise ratio than the Wiener filter, with the desired response and estimation error playing the roles of signal and noise, respectively.

(1) Principle of Orthogonality

The condition that defines the operation of the Wiener filter has a geometric interpretation that is intuitively satisfying. For this purpose, we rewrite Eq. 3.135 by using the definitions that we introduced earlier for the autocorrelation function $R_X(k)$ and the cross-correlation function $R_{DX}(m - k)$, as shown by

$$\sum_{m=0}^{M} h_{om}E[X_{n-k}X_{n-m}] = E[D_nX_{n-k}] \qquad k = 0, 1, \dots, M$$

Interchanging the order of expectation and summation, and then transposing terms, we get

$$E\left[\left(D_n - \sum_{m=0}^{M} h_{om}X_{n-m}\right)X_{n-k}\right] = 0 \qquad k = 0, 1, \dots, M \qquad (3.137)$$

However, the expression inside the inner brackets on the left side of this equation is recognized as the estimation error ε_{on} that results from the use of the Wiener filter. Hence, we may rewrite Eq. 3.137 as

$$E[\varepsilon_{on}X_{n-k}] = 0 \qquad k = 0, 1, \dots, M \qquad (3.138)$$

Equation 3.138 states that, for the Wiener filter, the *estimation error ε_{on} and any of the tap inputs $X_n, X_{n-1}, \dots, X_{n-M}$ are orthogonal*. This result is known as the *principle of orthogonality*.

As a corollary to the principle of orthogonality, we may state that the *estimation error ε_{on} and the filter output Y_{on} resulting from the use of a Wiener filter are also orthogonal*. That is to say,

$$E[\varepsilon_{on}Y_{on}] = 0 \qquad (3.139)$$

This result follows from the fact that

$$Y_{on} = \sum_{k=0}^{M} h_{ok}X_{n-k},$$

and consequently

$$E[\varepsilon_{on} Y_{on}] = E\left[\varepsilon_{on} \sum_{k=0}^{M} h_{ok} X_{n-k} \right]$$

$$= \sum_{k=0}^{M} h_{ok} E[\varepsilon_{on} X_{n-k}]$$

$$= 0,$$

where in the last line we have made use of the principle of orthogonality.

Equation 3.139 offers an interesting geometric interpretation of the variables that exist in a Wiener filter. If we view the three random variables representing the filter output, the desired response, and the estimation error as *vectors,* and recall that (by definition) the desired response equals the filter output plus the estimation error, we see that these three variables are related geometrically as illustrated in Fig. 3.22. In particular the vector $\mathbf{e}_o$ representing the estimation error is drawn perpendicular or "normal" to the vector $\mathbf{y}_o$ representing the filter output. For this reason, the equations that define the coefficients of the Wiener filter, labeled earlier as Eq. 3.135, are also called the *normal equations.* In the sequel, we will use this terminology to refer to this set of equations.

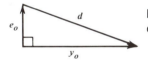

Figure 3.22 Geometric interpretation of the principle of orthogonality.

3.13 LINEAR PREDICTION

Prediction constitutes a special form of estimation. Specifically, the requirement is to use a finite set of present and past samples of a stationary process to predict a sample of the process in the *future*. We say that the prediction is *linear* if it is a linear combination of the given samples of the process. Our interest is confined to linear prediction. The filter designed to perform the prediction is called a *predictor*. The difference between the actual sample of the process at the (future) time of interest and the predictor output is called the *prediction error*. According to the Wiener filter theory, a predictor is designed to minimize the mean-square value of the prediction error.

Consider the random samples $X_{n-1}, X_{n-2}, \ldots, X_{n-M}$ drawn from a stationary process $X(t)$. Suppose the requirement is to make a prediction of the sample X_n. Let $\hat{X}_n$ denote the random variable resulting from this prediction. We thus write

$$\hat{X}_n = \sum_{k=1}^{M} h_{ok} X_{n-k} \tag{3.140}$$

where $h_{o1}, h_{o2}, \ldots, h_{oM}$ are the *optimum predictor coefficients*. Equation 3.140 is depicted in Fig. 3.23. We refer to M, the number of delay elements employed in the predictor, as its *order*.

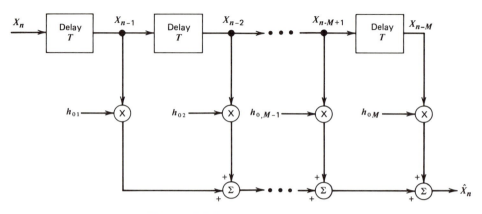

Figure 3.23 Linear predictor.

We may set up the normal equations for the predictor coefficients in Fig. 3.23 by minimizing the mean-square value of the prediction error. A more expedient approach, however, is to view the predictor as a special case of the Wiener filter. Specifically, we may proceed as follows:

1. The variance of the sample X_n, viewed as the desired response, equals

$$\sigma_X^2 = E[X_n^2]$$
$$= R_X(0) \tag{3.141}$$

where it is assumed that X_n has zero mean.

2. The cross-correlation function of X_n, acting as the desired response, and X_{n-k}, acting as the kth tap input of the predictor, is given by

$$E[X_n X_{n-k}] = R_X(k) \qquad k = 1, 2, \ldots, M \tag{3.142}$$

3. The autocorrelation function of the predictor's tap input X_{n-k} with another tap input X_{n-m} is given by

$$E[X_{n-k} X_{n-m}] = R_X(m - k) \qquad k, m = 1, 2, \ldots, M \tag{3.143}$$

Accordingly, bearing in mind the analogous roles of Eqs. 3.141, 3.142, and 3.143 for the predictor of Fig. 3.23 on the one hand, and Eqs. 3.130, 3.131, and 3.132 for the Wiener filter of Fig. 3.21 on the other hand, we may adapt the normal equations (Eq. 3.135) to fit the linear prediction problem (where we look into the future by one sample) as follows:

$$\sum_{m=1}^{M} h_{om} R_X(m - k) = R_X(k) \qquad k = 1, 2, \ldots, M \tag{3.144}$$

We see therefore that we need only know the autocorrelation function of the signal for different lags in order to solve the normal equations for the predictor coefficients.

(1) Prediction-error Process

The *prediction error,* denoted by ε_n, is defined by

$$\varepsilon_n = X_n - \hat{X}_n$$

$$= X_n - \sum_{k=1}^{M} h_{ok} X_{n-k} \tag{3.145}$$

Thus, given the present and past samples of a stationary process, namely, X_n, $X_{n-1}, \ldots, X_{n-M}$, and given the predictor coefficients $h_{o1}, h_{o2}, \ldots, h_{oM}$, we may compute the prediction error, ε_n, by using the structure shown in Fig. 3.24a. This structure is called a *prediction-error filter.*

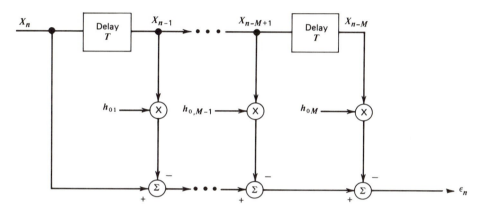

(a)

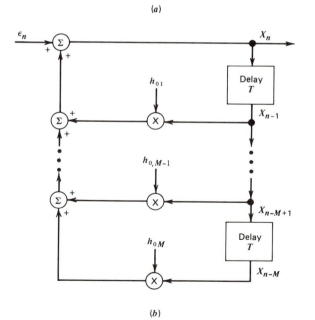

(b)

Figure 3.24 (a) Prediction-error filter. (b) Inverse filter.

The operation of prediction-error filtering is invertible. Specifically, we may rearrange Eq. 3.145 as

$$X_n = \sum_{k=1}^{M} h_{ok} X_{n-k} + \varepsilon_n \tag{3.146}$$

Hence, the "present" sample of the original process, X_n, may be computed as a linear combination of "past" samples of the process, $X_{n-1}, \ldots, X_{n-M}$, plus the "present" prediction error ε_n, where subscript n refers to the present. Figure 3.24b depicts the structure for performing the inverse operation so described. We refer to this second structure as the *inverse filter*. Note that the impulse response of the inverse filter has *infinite duration* because of feedback present in the filter, whereas the impulse response of the prediction-error filter has finite duration.

The structures of Figs. 3.24a and 3.24b confirm that there is a one-to-one correspondence between samples of a stationary process and those of the prediction-error process in that if we are given one, we can compute the other by means of a linear filtering operation.

At this point in our discussion, it is natural to question the reason for representing samples of a stationary process, $\{X_n\}$, by samples of the corresponding prediction-error process, $\{\varepsilon_n\}$. The answer to this question lies in the fact that the prediction-error variance is less than σ_X^2, the variance of X_n. If X_n has zero mean, so will ε_n. Then, the prediction-error variance equals

$$\sigma_E^2 = E[\varepsilon_n^2]$$
$$= R_X(0) - \sum_{k=1}^{M} h_{ok} R_X(k) \tag{3.147}$$

where we have adapted the formula of Eq. 3.136 to fit the predictor of Fig. 3.23. The summation term on the right side of Eq. 3.147, by virtue of the formula of Eq. 3.144 for the predictor coefficients, is less than $R_X(0)$. Since $\sigma_X^2 = R_X(0)$, we therefore have $\sigma_E^2 < \sigma_X^2$. This inequality is illustrated by a predictor of order one considered in Example 6, and a predictor of order two considered in Problem 3.13.2.

EXAMPLE 6 PREDICTOR OF ORDER ONE

Figures 3.25a and 6.25b show a predictor of order one and the corresponding prediction-error filter, respectively. There is a single predictor coefficient, h_{o1}, to be evaluated. From Eq. 3.144 we get

$$h_{o1} R_X(0) = R_X(1)$$

or

$$h_{o1} = \frac{R_X(1)}{R_X(0)} \tag{3.148}$$

Hence, the use of Eq. 3.147 yields the corresponding value of prediction-error variance:

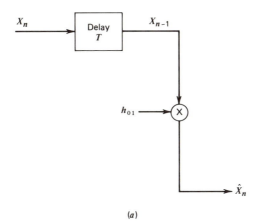

(a)

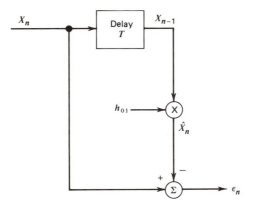

(b)

Figure 3.25 (a) Predictor of order one. (b) Prediction-error filter of order one.

$$\sigma_E^2 = R_X(0) - h_{o1}R_X(1)$$
$$= R_X(0) - \frac{R_X^2(1)}{R_X(0)}$$
$$= R_X(0)[1 - h_{o1}^2] \qquad (3.149)$$

Since $\sigma_X^2 = R_X(0)$, we therefore deduce from Eq. 3.149 that the prediction-error variance is less than the variance of the predictor input.

In Problem 3.13.2, we consider a predictor of order two. There it is shown that the corresponding prediction-error variance is almost always less than that for a predictor of order one.

We may formally present the significance of these results as two basic properties of linear prediction:

PROPERTY 1

The prediction-error variance decreases with increasing predictor order. In theory, this trend may go on indefinitely, or until a critical predictor order is reached, whereafter there is no further reduction.

PROPERTY 2

When the prediction-error variance reaches its minimum possible value, the prediction error process assumes the form of white noise. A prediction-error filter designed to whiten a stationary process is called a *whitening filter.* The resultant white noise sequence is known as the *innovations process* associated with the predictor input; the term "innovation" refers to "newness." Hence, only new information is retained in the innovations process.

Basically, prediction relies on the presence of correlation between adjacent samples of a stationary process. As we increase the predictor order, we successively reduce the correlation between adjacent samples of the process, until ultimately the prediction-error process consists of a sequence of uncorrelated samples. When this condition is reached, the prediction-error variance attains its minimum possible value, and the whitening of the original process is thereby accomplished.

Property 1 is exploited in the design of *waveform coders;* this application is considered in Chapter 5. Property 2 is exploited in the design of *source coders,* an example of which is considered in the next section.

3.14 LINEAR PREDICTIVE VOCODERS

Linear prediction provides the basis of an important source coding technique for the digitization of speech signals. The technique, known as *linear predictive vocoding,** relies on *parameterization* of speech signals according to a *physical model* for the speech production process. The term "vocoding" is a contraction of *voice coding;* the term "vocoder" refers to the device for performing it.

Figure 3.26 shows a classical model for the speech production process.† The model assumes that the sound-generating mechanism (i.e., source of *excitation*) is linearly separable from the intelligence-modulation mechanism (i.e., *vocal tract filter*). The precise form of the excitation depends on whether the speech sound is voiced or unvoiced.

Voiced sounds are produced by forcing air through the glottis with the tension of the vocal cords adjusted so that they vibrate in a relaxation oscillation, thereby producing quasi-periodic pulses of air that excite the vocal tract. *Fricative* or *unvoiced sounds,* on the other hand, are generated by forming a constriction at some point in the vocal tract (usually toward the mouth end), and forcing air through the constriction at a high enough velocity to produce turbu-

---

* For detailed discussions of linear predictive vocoders, see Rabiner and Schafer (1978, Chapter 8), Markel and Gray (1976), Flanagan, 1972.

† For a detailed discussion of digital models of the speech signal, see Rabiner and Schafer (1978, Chapter 7), Flanagan et al. (1979), and Flanagan (1972).

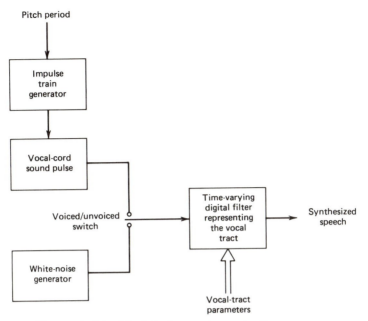

Figure 3.26   Model of speech-production process.

lence. Examples of voiced and unvoiced sounds are represented by utterances for the "A" and "S" segments in the word "salt." The speech waveform shown in Fig. 3.27a is the result of the utterance "every salt breeze comes from the sea" by a male subject. The waveform shown in Fig. 3.27b corresponds to the "A" segment in the word "salt," while the magnified waveform shown in Fig. 3.27c corresponds to the "S" segment. The generation of a voiced sound is modeled as the response of the vocal tract filter excited with a periodic sequence of impulses (very short pulses) spaced by a fundamental period equal to the *pitch* period. An unvoiced sound, on the other hand, is modeled as the response of the vocal filter excited with a white noise sequence. The vocal tract filter is time varying, so that its coefficients can provide an adequate representation for the input segment of voiced or unvoiced sound.

A linear predictive vocoder consists of a transmitter and a receiver having the block diagrams shown in Fig. 3.28. The transmitter, shown in Fig. 3.28a first performs *analysis* on the input speech signal, block by block. Typically, each block is 10–30 ms long, for which the speech-production process may be treated as essentially stationary. The parameters resulting from the analysis, namely, the prediction-error filter (analyzer) coefficients, a voiced/unvoiced parameter, and the pitch period, provide a complete description for the particular segment of the input speech signal. A digital representation of the parameters of this complete description constitutes the transmitted signal. The receiver, shown in Fig. 3.28b, first performs decoding, followed by *synthesis* of the speech signal; the latter operation utilizes the model of Fig. 3.26. The standard result of this analysis/synthesis is an artificial-sounding reproduction

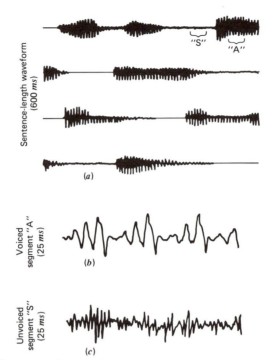

Figure 3.27 Time waveforms of speech for a long (sentence-length) segment and for short (25 ms) segments. This figure is reproduced from Flanagan et al. (1979) with permission.

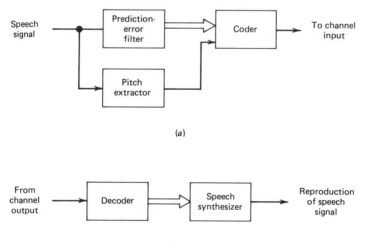

Figure 3.28 Block diagram of linear predictive vocoder. (a) Transmitter. (b) Receiver.

of the original speech signal. The somewhat poor reproduction quality of a linear predictive vocoder is tolerated for secure military communications where very low bit rates (4 kb/s or less) are required.

3.15 ADAPTIVE FILTERS

The design of a Wiener filter requires *a priori* information about the statistics of the data to be processed. In particular, to solve the normal equations (Eq. 3.135) for the filter coefficients, we require knowledge of two sets of correlation functions: (1) the autocorrelation function of the input signal for various lags, and (2) the cross-correlation function between a desired response and the input signal for a corresponding number of lags. However, when the filter operates in an environment in which these quantities are unknown or the environment is nonstationary, then we have a problem at our hands. In such situations, the use of an *adaptive filter* offers an attractive solution. By an adaptive filter we mean a *self-designing device* that has the following characteristics:

1. It contains a set of *adjustable* filter coefficients.
2. The coefficients are updated in accordance with an *algorithm*.
3. The algorithm operates with arbitrary *initial conditions;* each time new sample values are received for the input signal and the desired response, appropriate *corrections* are made to the previous values of the filter coefficients.
4. The adaptation is continued until the operating point of the filter on the error-performance surface moves close enough to the minimum point.

In this section, we develop an adaptive algorithm built around the tapped-delay-line filter. The development proceeds in two stages. First, we formulate the *steepest descent algorithm* that makes it possible to compute the Wiener filter in a *recursive* fashion. Second, we use estimates for unknown correlations to formulate an adaptive algorithm known as the *least-mean-square (LMS) algorithm.*

(1) Steepest Descent Algorithm

Consider a tapped-delay-line filter that is *time-varying.* Let $h_k(n)$ denote the value of the kth filter coefficient at time $t = nT$, where T is the sampling interval and $k = 0, 1, \ldots, M$. As a consequence of this time variation, the estimation error process becomes *nonstationary,* and so the mean-squared error (which involves ensemble averaging over the input signal and the desired response) assumes a time-varying form of its own. Let $\mathscr{E}(n)$ denote the mean-squared error that results from the use of this set of filter coefficients. Let $\nabla_k(n)$ denote the kth *gradient,* defined as the derivative of the mean-squared error with respect to the kth filter coefficient. That is,

$$\nabla_k(n) = \frac{\partial \mathscr{E}(n)}{\partial h_k(n)} \qquad k = 0, 1, \ldots, M \qquad (3.150)$$

Then, according to the steepest descent algorithm, we compute the updated value of the kth filter coefficient by making a correction in the present value, $h_k(n)$, in a direction opposite to that of the gradient $\nabla_k(n)$. It is intuitively reasonable that successive corrections to the filter coefficients in directions that are the negative of the respective gradients (i.e., in the direction of the steepest descent of the error performance surface) should eventually lead to the minimum point of the error-performance surface where the filter coefficients assume their individual optimum values (in accordance with the normal equations).

In mathematical terms, we express the steepest descent algorithm for the kth filter coefficient as

$$h_k(n + 1) = h_k(n) + \frac{1}{2}\,\mu[-\nabla_k(n)] \qquad k = 0, 1, \ldots, M \qquad (3.151)$$

where μ is a positive real-valued constant, and the factor $1/2$ is introduced to simplify the final form of the algorithm.

By definition, the mean-squared error at time $t = nT$ equals

$$\mathscr{E}(n) = E[\varepsilon_n^2] \qquad (3.152)$$

where ε_n is the corresponding estimation error. Hence, differentiating both sides of Eq. 3.152 with respect to the kth filter coefficient, and using the definition of Eq. 3.150, we get

$$\nabla_k(n) = \frac{\partial}{\partial h_k(n)}\, E[\varepsilon_n^2]$$

$$= E\left[\frac{\partial \varepsilon_n^2}{\partial h_k(n)}\right]$$

$$= 2E\left[\varepsilon_n \frac{\partial \varepsilon_n}{\partial h_k(n)}\right] \qquad (3.153)$$

The estimation error ε_n is defined by

$$\varepsilon_n = D_n - \sum_{k=0}^{M} h_k(n)X_{n-k}$$

This expression is the same as that defined in Eqs. 3.125 and 3.126, except for the use of $h_k(n)$ to denote the kth filter coefficient. Hence, differentiating ε_n with respect to $h_k(n)$, we get

$$\frac{\partial \varepsilon_n}{\partial h_k(n)} = -X_{n-k} \qquad k = 0, 1, \ldots, M \qquad (3.154)$$

where X_{n-k} is the tap input applied to the kth filter coefficient. Accordingly, we may rewrite Eq. 3.153 for the kth gradient as

$$\nabla_k(n) = -2E[\varepsilon_n X_{n-k}]$$
$$= -2R_{EX}(k) \qquad k = 0, 1, \ldots, M \qquad (3.155)$$

We may view $R_{EX}(k)$ as the correlation between the estimation error and the tap input applied to the kth filter coefficient or, equivalently, the cross-correlation function between the estimation error and the input signal for a lag of k samples.

Finally, substituting Eq. 3.155 in Eq. 3.151, we get the desired form of the steepest descent algorithm:

$$h_k(n + 1) = h_k(n) + \mu R_{EX}(k) \qquad k = 0, 1, \ldots, M \qquad (3.156)$$

There are no approximations made in the derivation of this algorithm, and so we find that as the number of iterations of the algorithm, n, becomes large, the kth filter coefficient $h_k(n)$ converges to the optimum value determined by the normal equations (Eq. 3.135). However, for this convergence to be achievable, the adaption constant μ has to be carefully chosen. Indeed, it can be shown that μ has to satisfy the following condition for convergence to occur:

$$0 < \mu < \frac{2}{\lambda_{\max}} \qquad (3.157)$$

where $\lambda_{\max}$ is the *largest eigenvalue of the correlation matrix* $\mathbf{R}_X$ of the tap inputs.*

(2)   Least-Mean-Square Algorithm

A shortcoming of the steepest descent algorithm is that it requires exact knowledge of the set of gradients, $\{\nabla_k(n), k = 0, \ldots, M\}$, at each iteration. In reality, exact measurements are not possible, and this set of gradients must be *estimated* from a limited number of input sample values, thereby introducing errors. We therefore need an algorithm that derives estimates for the set of

---

* The *correlation matrix* $\mathbf{R}_X$ of the tap inputs $X_n, X_{n-1}, \ldots, X_{n-M}$ in Fig. 3.20 is defined by the $(M + 1)$-by-$(M + 1)$ matrix:

$$\mathbf{R}_X = \begin{bmatrix} R_X(0) & R_X(1) & \ldots & R_X(M) \\ R_X(1) & R_X(0) & \ldots & R_X(M - 1) \\ \vdots & \vdots & & \vdots \\ R_X(M) & R_X(M - 1) & \ldots & R_X(0) \end{bmatrix}$$

where $R_x(l)$ is the autocorrelation of the input for a lag of l samples; that is

$$R_X(l) = E[X_n X_{n-l}]$$

The *eigenvalues* of the matrix $\mathbf{R}_X$ are roots of the *characteristic equation:*

$$\det(\mathbf{R}_X - \lambda \mathbf{I}) = 0$$

where $\det(\cdot)$ denotes determinant of the matrix in question, $\mathbf{I}$ is the $(M + 1)$-by-$(M + 1)$ identity matrix, and λ is a scalar. When the characteristic equation is expanded, it assumes the form of a polynomial equation of order M in λ. Hence, it has M roots that constitute the eigenvalues of $\mathbf{R}_X$.

For a proof of Eq. 3.157 that defines the condition necessary for convergence of the steepest descent algorithm, in terms of the eigenvalues of $\mathbf{R}_X$, see Haykin (1986, pp. 199–203) and Widrow and Stearns (1985, pp. 56–60).

gradients, $\{\nabla_k(n), k = 0, \ldots, M\}$, from the available data. One such algorithm is the *least-mean-square (LMS) algorithm*.* The attractive feature of the LMS algorithm is its simplicity: it does not require measurements of correlations, nor does it require the solution of the normal equations.

The LMS algorithm uses *instantaneous estimates* of the set of gradients $\{\nabla_k(n), k = 0, \ldots, M\}$ by using sample values of the estimation error ε_n and the corresponding set of tap inputs $\{X_{n-k}, k = 0, \ldots, M\}$. We note that both ε_n and X_{n-k} are random variables. Let e_n denote a *sample value* of the estimation error ε_n. Similarly, let x_{n-k} denote a *sample value* of the tap input X_{n-k}. From the first line of Eq. 3.155, we deduce the instantaneous estimate:

$$\hat{\nabla}_k(n) = -2e_n x_{n-k} \qquad k = 0, 1, \ldots, M \qquad (3.158)$$

Note that this estimate is *unbiased,* because its expected value is exactly the same as the actual value defined by Eq. 3.155.

We are now ready to formulate the LMS algorithm. By using the gradient estimate $\hat{\nabla}_k(n)$ in place of the actual gradient $\nabla_k(n)$ in the steepest descent algorithm of Eq. 3.151, we get the simple update equation:

$$\hat{h}_k(n + 1) = \hat{h}_k(n) + \mu e_n x_{n-k} \qquad k = 0, \ldots, M \qquad (3.159)$$

where we have used hats over the symbols for the filter coefficients to indicate that they are estimates.

The various terms that make up Eq. 3.159 deserve description: $\hat{h}_k(n)$ is the estimate of the kth filter coefficient before adaptation (i.e., *old* estimate), $\hat{h}_k(n + 1)$ is the estimate of the kth filter coefficient after adaptation (i.e., *updated* estimate), $e(n)$ is the value of the estimation error at the nth iteration, x_{n-k} is the value of the tap input applied to the kth filter coefficient at the nth iteration, and μ is the *adaptation constant*. Equation 3.159 states that the updated estimate of a coefficient in the tapped-delay-line filter is computed by incrementing the old estimate of the coefficient by an amount equal to the product of the estimation error and the tap input applied to the coefficient, with the correction scaled by the adaptation constant. Equation 3.159 constitutes the *adaptive process* of the LMS algorithm.

A sample value of the estimation error $\varepsilon(n)$ is defined by

$$e_n = d_n - \sum_{k=0}^{M} \hat{h}_k(n) x_{n-k} \qquad (3.160)$$

where d_n is a sample value of the desired response D_n, and the summation term (representing the filter output) is an estimate of d_n. Note that, at each iteration, all the filter coefficients enter the computation of the estimation error. Equation 3.160 constitutes the *filtering process* of the LMS algorithm.

For the *initial conditions,* it is customary to set all the filter coefficients to zero; thus

$$\hat{h}_k(0) = 0 \qquad k = 0, 1, \ldots, M \qquad (3.161)$$

---

* The LMS algorithm owes its name and first derivation to Widrow in 1966; see Widrow (1971). Its theory is discussed in detail in the following two books: Widrow and Stearns (1985, pp. 99–116) and Haykin (1986, pp. 216–259).

The combination of the initial conditions of Eq. 3.161, the filtering process of Eq. 3.160, and the adaptive process of Eq. 3.159 provides a complete description of the LMS algorithm. Figure 3.29 shows a structural interpretation of the LMS algorithm.

We may summarize the steps involved in the application of the LMS algorithm as follows:

1. Initialize the algorithm by setting all the filter coefficients to zero:

$$\hat{h}_k(0) = 0 \qquad k = 0, 1, \ldots, M$$

2. Using sample values of the desired response and the tap inputs available at iteration n, compute corresponding sample values of the estimation error:

$$e_n = d_n - \sum_{k=0}^{M} \hat{h}_k(n)x_{n-k}$$

3. For a prescribed adaptation constant μ, compute the updated estimate of the kth filter coefficient:

$$\hat{h}_k(n + 1) = \hat{h}_k(n) + \mu e_n x_{n-k} \qquad k = 0, 1, \ldots, M$$

4. Increment the iteration number n by one, go back to step 2, and repeat the computation.

Typically, the number of iterations required for the algorithm to converge close enough to the optimum Wiener condition is about 10 times the number of taps in the adaptive filter.

An LMS adaptive filter is in fact a *closed-loop feedback system* with time-varying coefficients (see Fig. 3.29). As such, the filter has a tendency to become *unstable* (*noncovergent*), unless the adaptation constant μ is properly chosen. A mathematical analysis of the stability (convergence) problem is much too complicated and certainly beyond the scope of this book. Nevertheless, we may make the following two statements for stationary inputs:

1. The LMS algorithm is *convergent in the mean* (i.e., the expected values of the filter coefficients approach their optimum Wiener values in the limit)

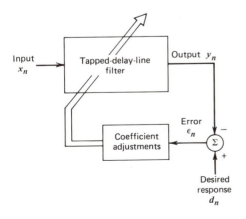

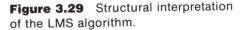

Figure 3.29 Structural interpretation of the LMS algorithm.

provided that μ satisfies the condition

$$0 < \mu < \frac{2}{\lambda_{\max}} \tag{3.162}$$

where $\lambda_{\max}$ is the largest eigenvalue of the correlation matrix $\mathbf{R}_x$ of the tap inputs.

2.   The LMS algorithm is *convergent in the mean square* (i.e., the average mean-squared error approaches a finite value in the limit) provided that μ satisfies the condition

$$0 < \mu < \frac{2}{\displaystyle\sum_{k=0}^{M} \lambda_k} \tag{3.163}$$

where $\sum_{k=0}^{M} \lambda_k$ is the sum of eigenvalues of the correlation matrix $\mathbf{R}_X$.

Since $\sum_{k=0}^{M} \lambda_k \geq \lambda_{\max}$, it is clear that if the LMS algorithm is convergent in the mean square, then it is convergent in the mean. Fortunately, computation of the upper limit on μ in Eq. 3.163 does not require explicit knowledge of the eigenvalues of the correlation matrix $\mathbf{R}_X$, since we have

$$\sum_{k=1}^{M} \lambda_k = (M + 1)R_X(0) \tag{3.164}$$
$$= \text{total average input power for all taps}$$

We may thus sum up this brief discussion of the stability problem of the LMS algorithm by stating that, for stationary inputs, convergence in both the mean and the mean-square is assured if

$$0 < \mu < \frac{2}{\text{total average input power}} \tag{3.165}$$

When the LMS algorithm operates in a stationary environment, the error performance surface has a constant shape as well as orientation. Then, starting with zero initial conditions, the algorithm proceeds in a step-by-step fashion closer and closer to the bottom of the error performance surface without prior knowledge of the statistical parameters of the environment. However, when the algorithm operates in a nonstationary environment, the bottom of the error-performance surface moves constantly, while the orientation and curvature of the surface may be changing too. In such a situation, the LMS algorithm has the added task of continually *tracking* the bottom of the error-performance surface. It turns out that the larger we make the adaptation constant μ, the faster the tracking capability of the LMS algorithm. However, a large adaptation constant μ may result in an unacceptably high *excess mean-squared error,* defined as that part of the mean-squared error in excess of the minimum attainable value $\varepsilon_{\min}$ (which results when the filter coefficients are set at their optimum values). We therefore find that in choosing a suitable value for the adaptation constant μ, a compromise must be made between fast tracking and low excess mean-squared error.

The LMS algorithm finds applications in various facets of communications. In Chapters 5 and 6, we consider two such applications.

3.16 SUMMARY AND DISCUSSION

In this chapter, we covered the two classical problems of *detection* and *estimation* in the context of digital communications.

To perform detection, the receiver observes the received signal for the duration of one symbol and guesses which particular symbol was sent according to some set of rules. In the first level of detection problems considered, the receiver is assumed to be in possession of exact replicas of the transmitted signals (representing the symbols of interest). This requires the receiver to be *coherent*, that is, synchronized with the transmitter not only in time but also in carrier *phase*. In coherent receivers, a source of errors in the decision-making process is *additive* thermal noise at the receiver input, which is modeled as *white Gaussian noise* of zero mean and constant power spectral density. We thus refer to the first level of detection problem as the *known signal in noise problem*. When all the signals are transmitted with equal probability, the *optimum* decision rule turns out to be *maximum likelihood; the criterion for optimality is minimization of the average probability of symbol error*. The optimum receiver is implemented in the form of a bank of *correlators* or *matched filters*, followed by a *maximum-likelihood detector*.

In the next level of detection problem considered, the receiver is *noncoherent* in that it has no knowledge of the carrier phase. The presence of phase uncertainty in the received signal represents another source of error in the decision-making process. Typically, the unknown phase is modeled as the sample value of a random variable uniformly distributed between 0 and 2π. We refer to the second level of detection problem as *the signal with unknown phase in noise problem*. The performance of a noncoherent receiver in noise is naturally inferior to that of the coherent version. However, the noncoherent receiver is simpler in implementation, since it does not require phase synchronization.

There is a third level of detection problem (not considered in the chapter), which is also encountered in practice. In the context of digital communications, the problem arises when the channel is inherently *random* in nature. A common technique is to transmit one of two signals centered at two different frequencies f_0 and f_1. The received signal has no deterministic component in that it consists of a sample function drawn from a random process centered at f_0 or f_1 (depending on the signal transmitted), plus additive white Gaussian noise. The decision rule as to which particular signal was transmitted is then based on the difference in the statistical characteristics of the two random processes centered at f_0 and f_1. We refer to the third level of detection problem as the *random signal in noise problem*. In our three-level heirarchy of detection problems, the random signal in noise problem is the most demanding in mathematical complexity.*

* For a mathematical treatment of the random signal in noise problem, see Van Trees (1971).

In the estimation part of the chapter, we described three basically different methods as follows.*

Maximum likelihood estimation is a nonlinear procedure involving maximization of a likelihood function. It is useful for estimating a real nonrandom parameter (e.g., phase of a sinusoidal signal) when there is no a priori knowledge of the parameter or any meaningful cost function available.

The *Wiener filter* is a linear estimator, optimized in the sense that it minimizes the mean-square value of the error between some desired response and the filter output. Its mathematical formulation requires knowledge of the autocorrelation function of the filter input and the cross-correlation function between the desired response and the filter output.

Finally, the *adaptive filter* is a nonlinear estimator that provides an estimate of some desired response without requiring knowledge of correlation functions; it is nonlinear in the sense that the filter coefficients are data dependent. The filter has the capability of not only responding to changes in the environmental conditions, but also tracking the changes. A popular adaptive filtering algorithm is the *least-mean-square LMS algorithm*. An important advantage of the LMS algorithm is that it is relatively simple to implement. However, this algorithm tends to *converge* to the optimum Wiener solution rather slowly. There is another class of algorithms known as the *least-squares algorithms*,† which minimize an index of performance defined as the sum of error squares. Because these least-squares algorithms make better use of all the past available information than stochastic gradient algorithms, their rate of convergence is faster. But this improvement in rate of convergence is achieved at the expense of increased complexity.

PROBLEMS

P3.2   GRAM–SCHMIDT ORTHOGONALIZATION PROCEDURE

Problem 3.2.1
(a) Using the Gram–Schmidt orthogonalization procedure, find a set of orthonormal basis functions to represent the three signals $s_1(t)$, $s_2(t)$, and $s_3(t)$ shown in Fig. P3.1
(b) Express each of these signals in terms of the set of basis functions found in part (a).

Problem 3.2.2   A periodic waveform $g_p(t)$ of fundamental period T_0 is represented by a truncated form of the Fourier series:

$$g'_p(t) = a_0 + 2 \sum_{n=1}^{N} \left[a_n \cos\left(\frac{2\pi nt}{T_0}\right) + b_n \sin\left(\frac{2\pi nt}{T_0}\right) \right]$$

where

$$a_0 = \frac{1}{T_0} \int_{-T_0/2}^{T_0/2} g_p(t)dt$$

---

* In parameter estimation theory, there is a set of problems similar in heirarchy to that in detection theory; for more details, see Van Trees (1968).
† For a detailed treatment of recursive least-squares algorithms, see Haykin (1986).

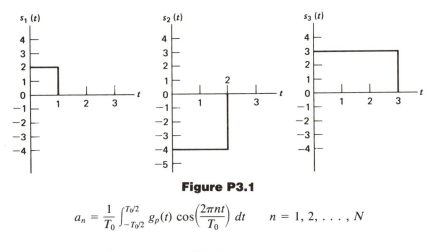

Figure P3.1

$$a_n = \frac{1}{T_0} \int_{-T_0/2}^{T_0/2} g_p(t) \cos\left(\frac{2\pi nt}{T_0}\right) dt \qquad n = 1, 2, \ldots, N$$

$$b_n = \frac{1}{T_0} \int_{-T_0/2}^{T_0/2} g_p(t) \sin\left(\frac{2\pi nt}{T_0}\right) dt \qquad n = 1, 2, \ldots, N$$

(a) Find a set of orthonormal basis functions to approximate $g_p(t)$.
(b) Set up schemes for generating the approximation to $g_p(t)$, and for computing the coefficients of the expansion (i.e., a_0, the a_n and the b_n).
(c) Determine the minimum value of the mean squared error

$$\mathcal{E} = \frac{1}{T_0} \int_0^{T_0} [g_p(t) - g_p'(t)]^2 dt$$

and show that it approaches zero as n approaches infinity.

P3.3 GEOMETRIC INTERPRETATION OF SIGNALS

Problem 3.3.1 Consider the set of signals

$$s_i(t) = \begin{cases} \sqrt{\frac{2E}{T}} \cos\left(2\pi f_c t + i\frac{\pi}{4}\right) & 0 \leq t \leq T \\ 0 & \text{elsewhere} \end{cases}$$

where $i = 1, 2, 3, 4$, and $f_c = n_c/T$ for some fixed integer n_c.

(a) What is the dimensionality, N, of the space spanned by this set of signals?
(b) Find a set of orthonormal basis functions to represent this set of signals.
(c) Using the expansion

$$s_i(t) = \sum_{j=1}^{N} s_{ij}\phi_j(t) \qquad i = 1, 2, 3, 4$$

find the coefficients s_{ij}.
(d) Plot the locations of $s_i(t)$, $i = 1, 2, 3, 4$, in the signal space, using the results of parts (b) and (c).

P3.4 RESPONSE OF BANK OF CORRELATORS TO NOISY INPUT

Problem 3.4.1 Consider Eq. 3.37, which shows that the received random process $X(t)$ may be expressed as

$$X(t) = \sum_{j=1}^{N} X_j \phi_j(t) + W'(t) \qquad 0 \leqslant t \leqslant T$$

where $W'(t)$ is a remainder noise term. The $\{\phi_j(t)\}, j = 1, 2, \ldots, N$ form an orthonormal set, and the X_j are defined by

$$X_j = \int_0^T X(t) \phi_j(t) dt$$

Let $W'(t_k)$ denote a random variable obtained by observing $W'(t)$ at time $t = t_k$. Show that

$$E[X_j W'(t_k)] = 0 \qquad \begin{aligned} & j = 1, 2, \ldots, N \\ & 0 \leqslant t \leqslant T \end{aligned}$$

P3.5   DETECTION OF KNOWN SIGNALS IN NOISE

Problem 3.5.1   Determine the optimum decision regions for the 16-signal constellations shown in Fig. P3.2a and P3.2b. Both constellations have the same peak power.

Problem 3.5.2   Let

$$s_1(t) = \sqrt{\frac{2}{T}} \cos\left(\frac{2\pi n t}{T}\right) \qquad 0 \leqslant t \leqslant T$$

$$s_2(t) = \sqrt{\frac{2}{T}} \cos\left(\frac{4\pi n t}{T}\right) \qquad 0 \leqslant t \leqslant T$$

$$s_3(t) = \sqrt{\frac{2}{T}} \cos\left(\frac{6\pi n t}{T}\right) \qquad 0 \leqslant t \leqslant T$$

where n is an arbitrary interger.

(a)   Sketch the signal space and decision boundaries for this set of signals.
(b)   Assuming that $s_1(t)$, $s_2(t)$, and $s_3(t)$ have equal a priori probabilities, show that the signal space can be reduced to two dimensions.

Problem 3.5.3   In the *Bayes' test*, applied to a binary hypothesis testing problem where we have to choose one of two possible hypotheses H_0 and H_1, we minimize the *risk R*

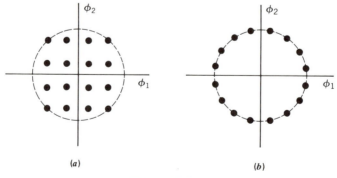

(a)   (b)

Figure P3.2

defined by

$$R = C_{00}p_0P(\text{say } H_0|H_0 \text{ is true})$$
$$+C_{10}p_0P(\text{say } H_1|H_0 \text{ is true})$$
$$+C_{11}p_1P(\text{say } H_1|H_1 \text{ is true})$$
$$+C_{01}p_1P(\text{say } H_0|H_1 \text{ is true})$$

The C_{00}, C_{10}, C_{11}, and C_{01} denote the costs assigned to the four possible outcomes of the experiment; the first subscript indicates the hypothesis chosen and the second the hypothesis that was true. Assume that $C_{10} > C_{00}$ and $C_{01} > C_{11}$. The p_0 and p_1 denote the a priori probabilities of hypotheses H_0 and H_1, respectively.

(a) Given the observation vector $\mathbf{x}$, show that the partitioning of the observation space so as to minimize the risk R leads to the *likelihood ratio test*:

$$\text{say } H_0 \text{ if } \Lambda(\mathbf{x}) < \eta$$
$$\text{say } H_1 \text{ if } \Lambda(\mathbf{x}) > \eta$$

where $\Lambda(\mathbf{x})$ is the *likelihood ratio*

$$\Lambda(\mathbf{x}) = \frac{f_{\mathbf{X}}(\mathbf{x}|H_1)}{f_{\mathbf{X}}(\mathbf{x}|H_0)}$$

and η is the *threshold* of the test defined by

$$\eta = \frac{p_0(C_{10} - C_{00})}{p_1(C_{01} - C_{11})}$$

(b) What are the cost values for which the Bayes' criterion reduces to the minimum probability of error criterion?

P3.6 PROBABILITY OF ERROR

Problem 3.6.1 Consider a signaling scheme whose signal space is symmetric in that its two message points are equispaced from the origin in a symmetric fashion. The symbols represented by the message points are equiprobable. Let $P(m_k|m_i)$ denote the conditional probability of choosing the symbol represented by message point m_k, given that the symbol represented by message point m_i was sent. The selection of m_k and m_i is random, with $m_k \neq m_i$. Hence, show that the average probability of error, P_e, equals the conditional probability of error, $P(m_k|m_i)$.

Problem 3.6.2 Consider the quanternary polar signaling scheme of Example 2. Find the conditional probability of correct reception, given that

(a) Dibit 00 was transmitted.
(b) Dibit 01 was transmitted.
(c) Dibit 11 was transmitted.
(d) Dibit 10 was transmitted.

Problem 3.6.3 The purpose of a *radar system* is basically to detect the presence of a target, and to extract useful information about the target. Suppose that in such a system, hypotheses H_0 is that there is no target present, so that the received signal $x(t) = w(t)$, where $w(t)$ is white Gaussian noise of zero mean and power spectral density $N_0/2$. For hypothesis H_1, a target is present, and $x(t) = w(t) + s(t)$, where $s(t)$ is an echo produced by the target. Assume that $s(t)$ is completely known. Evaluate the following probabilities:

(a) The *probability of false alarm* defined as the probability that the receiver decides a target is present when it is not.
(b) The *probability of detection* defined as the probability that the receiver decides a target is present when it is.

Problem 3.6.4 Consider an M-ary digital communication system, where $M = 2^N$. The M equally likely signal vectors lie at the corners of a *hypercube* centered at the origin. This is illustrated in Fig. P3.3 for $N = 3$ and $M = 8$. The signal energy per symbol is E.

(a) The transmitted signal $s_i(t)$ is defined in terms of a set of orthonormal functions, $\{\phi_i(t)\}$, by

$$s_i(t) = \sum_{j=1}^{N} s_{ij}\phi_i(t) \qquad i = 1, 2, \ldots, M$$

Hence, find the values of the coefficients s_{ij}.
(b) Show that the average probability of symbol error is given by

$$P_e = 1 - \left\{1 - \frac{1}{2}\operatorname{erfc}\left(\sqrt{\frac{E}{N_0\log_2 M}}\right)\right\}^N$$

where $N_0/2$ is the power spectral density of the additive white Gaussian noise (assumed to be of zero mean).

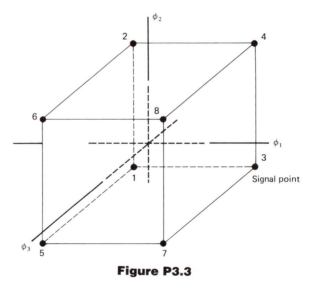

Figure P3.3

P3.8 MATCHED FILTER RECEIVER

Problem 3.8.1 Consider the signal $s(t)$ shown in Fig. P3.4
(a) Determine the impulse response of a filter matched to this signal and sketch it as a function of time.
(b) Plot the matched filter output as a function of time.
(c) What is the peak value of the output?

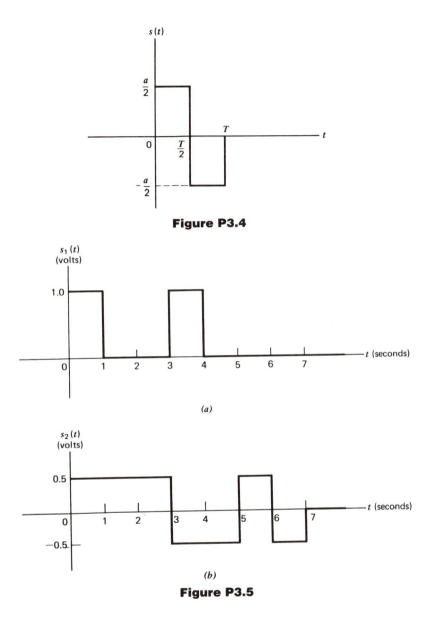

Figure P3.4

(a)

(b)

Figure P3.5

Problem 3.8.2

(a) Specify a matched filter for the signal $s_1(t)$ shown in Fig. P3.5a, and sketch the resulting filter output as a function of time.

(b) Repeat the problem for the signal $s_2(t)$ shown in Fig. 3.5b.

(c) Sketch the output of the filter matched to the signal $s_2(t)$ when the signal $s_1(t)$ is applied to the filter input.

Problem 3.8.3 It is proposed to implement a matched filter in the form of a tapped-delay-line filter with a set of tap weights $\{w_k\}$, $k = 0, 1, \ldots, K$. Given a signal $s(t)$ of duration T seconds to which the filter is matched, find the value of the w_k.

Problem 3.8.4 An *inverse filter*, with respect to a signal $s(t)$ of duration T and Fourier transform $S(f)$, is defined as having a transfer function equal to $\exp(-j2\pi fT)/S(f)$. Explain why an inverse filter is inferior to a matched filter when detecting the signal $s(t)$ in the presence of additive white noise.

Problem 3.8.5 Consider a pulse $s(t)$ defined by

$$s(t) = \begin{cases} 1 & 0 \leqslant t \leqslant T \\ 0 & \text{elsewhere} \end{cases}$$

It is proposed to approximate the matched filter for this pulse by a low-pass RC filter defined by the transfer function

$$H(f) = \frac{1}{1 + jf/f_0}$$

where $f_0 = 1/2\pi RC$ is the 3-dB bandwidth of the filter.

(a) Determine the optimum value of f_0 for which the RC filter provides the best approximation to the matched filter.
(b) Assuming an additive white noise of zero mean and power spectral density $N_0/2$, what is the peak output signal-to-noise ratio?
(c) By how many decibels must the transmitted energy be increased so at to realize the same performance as the perfectly matched filter?

Problem 3.8.6 Consider a signal $s(t)$ corrupted by additive *colored* noise $n(t)$ of power spectral density $S_N(f)$. To maximize the detection of the signal $s(t)$, the input is passed through a structure consisting of two components: (a) a *prewhitening filter* to transform the input noise $n(t)$ into a white noise, and (b) a filter matched to the signal component at the prewhitening filter output. Determine the overall transfer function of this structure.

Problem 3.8.7 In a binary digital communication system using on–off signaling, symbol 1 is represented by the pulse

$$s(t) = \begin{cases} a & 0 \leqslant t \leqslant T_b \\ 0 & \text{elsewhere} \end{cases}$$

and symbol 0 is represented by switching off the pulse. For predetection filtering, the receiver uses a matched filter, the maximum output of which is sampled and then applied to a decision device. Assume that the receiver noise is white, Gaussian, with zero mean and power spectral density $N_0/2$. Determine the average probability of error when symbols 1 and 0 occur with equal probability.

Problem 3.8.8 A binary signaling scheme uses the *Manchester code* to describe symbols 1 and 0, as illustrated in Fig. P3.6. The additive noise at the receiver input is white, Gaussian, with zero mean and power spectral density $N_0/2$. Assuming that symbols 1 and 0 occur with equal probability, find an expression for the average probability of error at the receiver output, using a matched filter.

Problem 3.8.9
(a) The two signals

$$s_1(t) = -s_2(t) = \begin{cases} \exp(-t) & t \geqslant 0 \\ 0 & \text{elsewhere} \end{cases}$$

are transmitted with equal probability over a channel with additive white Gaussian noise of zero mean and power spectral density $N_0/2$. The receiver bases its decision on the received signal over the interval $0 \leqslant t \leqslant 2$. Determine the minimum attainable probability of error at the receiver output.

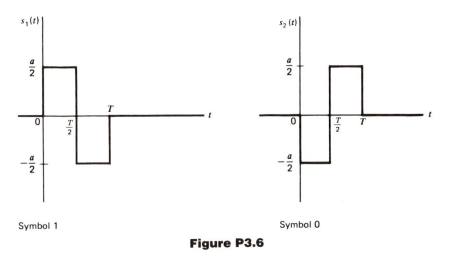

Symbol 1   Symbol 0

Figure P3.6

(b) Compare the result obtained in part (a) with the peformance of an optimum receiver that observes the received signal over the entire interval $-\infty < t < \infty$.

P3.9   DETECTION OF SIGNALS WITH UNKNOWN PHASE IN NOISE

Problem 3.9.1 The quadrature receiver shown in Fig. 3.17c consists of a noncoherent matched filter followed by a sampler. This receiver may also be viewed as an *energy detector*. Justify this statement.

Problem 3.9.2 Figure P3.7a shows a noncoherent receiver using a matched filter for the detection of a sinusoidal signal of known frequency but random phase, in the presence of additive white Gaussian noise. An alternative implementation of this receiver is its mechanization in the frequency domain as a *spectrum analyzer receiver*, as in Fig. P3.7b, where the correlator computes the finite time autocorrelation function $R_X(\tau)$ defined by

$$R_X(\tau) = \int_0^{T-\tau} x(t)x(t+\tau)dt \qquad 0 \leqslant \tau \leqslant T$$

Show that the square-law envelope detector output sampled at time $t = T$ in Fig. P3.7a is twice the spectral output of the Fourier transformer sampled at frequency $f = f_c$ in Fig. P3.7b.

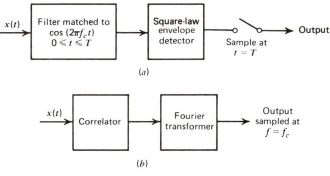

Figure P3.7

P3.10 ESTIMATION: CONCEPTS AND CRITERIA

Problem 3.10.1 Consider the problem of estimating the value of a uniformly distributed random variable. Show that the maximum a posteriori estimate and the maximum likelihood estimate are the same.

P3.11 MAXIMUM LIKELIHOOD ESTIMATION

Problem 3.11.1 Consider a signal of the form

$$s(t,a) = \begin{cases} as(t) & 0 \leq t \leq T \\ 0 & \text{elsewhere} \end{cases}$$

where $s(t)$ is completely known and the amplitude a is unknown. Find the maximum likelihood estimate of a in the presence of white Gaussian noise of zero mean and power spectral density $N_0/2$. What are the mean and variance of this estimate?

P3.12 WIENER FILTER FOR WAVEFORM ESTIMATION

Problem 3.12.1 The sample d_n of a stationary discrete time process is described by the first-order difference equation

$$d_n = a_1 d_{n-1} + w_{1,n}$$

where a_1 is a constant, and $w_{1,n}$ is the sample of a discrete-time white noise process $\{W_{1,n}\}$ of zero mean and variance σ_1^2. The sample d_n is transmitted through a noisy channel, yielding the received signal

$$x_n = d_n + w_{2,n}$$

where $w_{2,n}$ is the sample of another discrete-time white noise process of zero mean and variance σ_2^2, which is independent of $\{w_{1,n}\}$. The sequence $\{x_n\}$ is applied to a two-coefficient Wiener filter designed to produce an estimate of $\{d_n\}$.
(a) Find the coefficients h_{oo} and h_{o1} of the Wiener filter in terms of the constant a_1, and the variances σ_1^2 and σ_2^2.
(b) Find the minimum mean-squared error produced by the Wiener filter.

P3.13 LINEAR PREDICTION

Problem 3.13.1 Consider a linear predictor of order two, described by

$$\hat{x}_n = h_1 x_{n-1} + h_2 x_{n-2}$$

where the predictor coefficients h_1 and h_2 are selected to minimize the mean-square value of the prediction error:

$$e_n = x_n - \hat{x}_n$$

(a) Show that the optimum values of the predictor coefficients are given by

$$h_{o1} = \frac{\rho_1(1 - \rho_2)}{1 - \rho_1^2}$$

$$h_{o2} = \frac{\rho_2 - \rho_1^2}{1 - \rho_1^2}$$

where

$$\rho_1 = \frac{R_X(1)}{R_X(0)}$$

$$\rho_2 = \frac{R_X(2)}{R_X(0)}$$

(b) Show that the minimum mean-squared error equals

$$\mathscr{E}_{min} = \sigma_X^2 \left[(1 - \rho_1^2) - \frac{(\rho_1^2 - \rho_2)^2}{1 - \rho_1^2} \right]$$

where the variance $\sigma_X^2 = R_X(0)$.

(c) Show that $\mathscr{E}_{min}$ is less than the corresponding value for a predictor of order one, unless $\rho_2 = \rho_1^2$. What is the physical significance of this condition?

Problem 3.13.2 Consider a stationary discrete-time process $\{Y_n\}$. A sample of this process is described by

$$y_n = x_n + w_n$$

where x_n is the sample of another stationary discrete-time process $\{X_n\}$ of zero mean and variance σ_X^2, and w_n is the sample of a white Gaussian noise process $\{W_n\}$ of zero mean and variance σ_W^2. The sequence $\{y_n\}$ is applied to a first-order predictor designed to make a prediction of x_n based on y_{n-1}. The predictor is optimized in the minimum mean-square sense.

(a) Show that the predictor coefficient is given by

$$h_{o1} = \frac{\rho_1}{1 + (\sigma_W^2/\sigma_X^2)}$$

where

$$\rho_1 = \frac{R_X(1)}{R_X(0)}$$

(b) Find the minimum mean-squared error produced by the predictor.

P3.15   ADAPTIVE FILTERS

Problem 3.15.1

(a) Show that the steepest descent algorithm may be formulated as follows:

$$h_k(n + 1) = h_k(n) + \mu[R_{DX}(k) - \sum_{m=0}^{M} h_m(n) \, R_X(m - k)]$$

$$k = 0, 1, \ldots, M$$

where $R_{DX}(k)$ is the cross-correlation function of the desired response and the input signal for a lag of k samples, and $R_X(m - k)$ is the autocorrelation function of the input signal for a lag of $m - k$ samples.

(b) Using instantaneous estimates for the cross-correlation function $R_{DX}(k)$, show that this formulation of the steepest descent algorithm yields the same equation for the adaptive process in the LMS algorithm as that described in Section 3.15.

Problem 3.15.2   List the advantages and disadvantages of the LMS algorithm.

CHAPTER FOUR

SAMPLING PROCESS

As explained in the introduction, Chapter 1, a message signal may originate from a *digital* or *analog* source. If the message signal is derived from a digital source (e.g., digital computer), then from inception it is in the right form for processing by a digital communication system. If, however, the message signal happens to be analog in nature, as in a speech signal or video signal, then it has to be *converted* into digital form before it can be transmitted by digital means. The *sampling process* is the first process performed in *analog-to-digital conversion*. Two other processes, quantizing and encoding, are also involved in this conversion. The sampling process is considered in this chapter, and the other two are considered in the next chapter.

In the sampling process, a continuous-time signal is converted into a discrete-time signal by measuring the signal at periodic instants of time. For the sampling process to be of practical utility, it is necessary that we choose the sampling rate properly, so that the discrete-time signal resulting from the process uniquely defines the original continuous-time signal. This is the essence of the sampling theorem. In conceptual terms, the sampling theorem is easy to understand. However, related issues such as convergence and error analysis are mathematically difficult to handle. Moreover, practical application of the sampling theorem involves many subtleties that require careful attention.

In this chapter, we present an introductory treatment of the sampling process. We begin our study by deriving the sampling theorem for low-pass signals, and then extend its utility to band-pass signals. We also consider implications of the sampling theorem in the context of a stationary random process. In the remainder of the chapter, we consider the nature of signal distortion that may arise from improper application of the sampling theorem, practical aspects of sampling, and an important application of sampling in digital communications.

4.1 SAMPLING THEOREM

Consider an *analog signal* $g(t)$ that is *continuous in both time and amplitude*. We assume that $g(t)$ has *infinite duration* but *finite energy*. A segment of the signal $g(t)$ is depicted in Fig. 4.1a.

Let the *sample values* of the signal $g(t)$ at times $t = 0, \pm T_s, \pm 2T_s, \ldots$, be denoted by the series $\{g(nT_s), n = 0, \pm 1, \pm 2, \ldots\}$. We refer to T_s as the *sampling period* and $f_s = 1/T_s$ as the *sampling rate*.

We define the *discrete-time signal*, $g_\delta(t)$, that results from the sampling process as

$$g_\delta(t) = \sum_{n=-\infty}^{\infty} g(nT_s)\delta(t - nT_s) \tag{4.1}$$

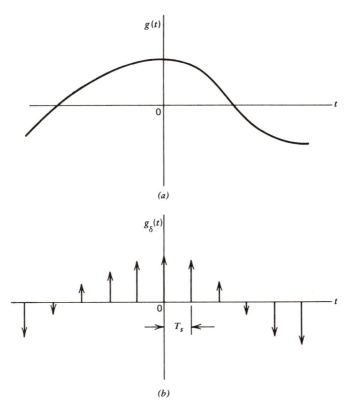

Figure 4.1 Illustration of the ideal sampling process. (a) Analog signal. (b) Discrete-time signal.

where $\delta(t - nT_s)$ is a *Dirac delta function* located at time $t = nT_s$. In Eq. 4.1, each delta function in the series is *weighted* by the corresponding sample value of the input signal $g(t)$. Figure 4.1b illustrates the construction of $g_\delta(t)$ from sample values of $g(t)$.

From the definition of a delta function, we have

$$g(nT_s)\, \delta(t - nT_s) = g(t)\, \delta(t - nT_s)$$

Hence, we may rewrite Eq. 4.1 in the equivalent form

$$g_\delta(t) = g(t) \sum_{n=-\infty}^{\infty} \delta(t - nT_s)$$

$$= g(t)\, \delta_{T_s}(t) \tag{4.2}$$

where $\delta_{T_s}(t)$ is the *Dirac comb* or *ideal sampling function*. According to Eq. 4.2, we may view the discrete-time signal $g_\delta(t)$ as the output of an *impulse modulator*, which operates with $g(t)$ as the modulating wave and $\delta_{T_s}(t)$ as the carrier wave. This circuit-theoretic interpretation of $g_\delta(t)$ is depicted in Fig. 4.2.

From the properties of the Fourier transform, we know that the multiplication of two time functions, as in Eq. 4.2, is equivalent to the convolution of

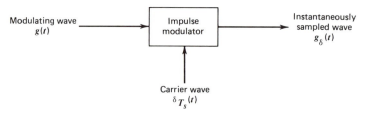

Figure 4.2 Circuit-theoretic interpretation of the ideal sampling process as impulse modulation.

their respective Fourier transforms. Let $G(f)$ and $G_\delta(f)$ denote the Fourier transforms of $g(t)$ and $g_\delta(t)$, respectively. For the Fourier transform of $\delta_{T_s}(t)$, we have

$$F[\delta_{T_s}(t)] = f_s \sum_{m=-\infty}^{\infty} \delta(f - mf_s) \tag{4.3}$$

where $F[\cdot]$ signifies the Fourier transform operation, and f_s is the sampling rate. Thus, transforming Eq. 4.2 into the frequency domain, we obtain

$$G_\delta(f) = G(f) \star \left[f_s \sum_{m=-\infty}^{\infty} \delta(f - mf_s) \right] \tag{4.4}$$

where the star $\star$ denotes convolution. Interchanging the orders of summation and convolution yields

$$G_\delta(f) = f_s \sum_{m=-\infty}^{\infty} G(f) \star \delta(f - mf_s) \tag{4.5}$$

From the properties of a delta function, we find that convolution of $G(f)$ and $\delta(f - mf_s)$ equals $G(f - mf_s)$. Hence, we may simplify Eq. 4.5 as follows:

$$G_\delta(f) = f_s \sum_{m=-\infty}^{\infty} G(f - mf_s) \tag{4.6}$$

From Eq. 4.6 we see that $G_\delta(f)$ represents a spectrum that is periodic in the frequency f with period f_s, but not necessarily continuous. In other words, the *process of uniformly sampling a signal in the time domain results in a periodic spectrum in the frequency domain with a period equal to the sampling rate.* Thus, $G_\delta(f)$ represents a *periodic extension* of the original spectrum $G(f)$.

Another useful expression for the Fourier transform $G_\delta(f)$ may be obtained by taking the Fourier transform of both sides of Eq. 4.1 and noting that the Fourier transform of the delta function $\delta(t - nT_s)$ is equal to $\exp(-j2\pi n f T_s)$. We may thus write

$$G_\delta(f) = \sum_{n=-\infty}^{\infty} g(nT_s) \exp(-j2\pi n f T_s) \tag{4.7}$$

This relation may be viewed as a complex Fourier series representation of the periodic frequency function $G_\delta(f)$, with the sequence of samples $\{g(nT_s)\}$ defining the coefficients of the expansion. Note that in the Fourier series defined by Eq. 4.7 the usual roles of time and frequency have been interchanged.

The relations, as just derived, apply to any continuous-time signal $g(t)$ of finite energy and infinite duration. Suppose, however, that the signal is *strictly band-limited,* with no frequency components higher than W hertz. That is, the Fourier transform $G(f)$ of the signal $g(t)$ has the property that $G(f)$ is zero for $|f| \geq W$, as illustrated in Fig. 4.3a; the shape of the spectrum shown in this figure is intended for the purpose of illustration only. The fact that the signal $g(t)$ has finite energy means that the area under the curve of the energy spectral density $|G(f)|^2$ is likewise finite. Suppose also that we choose the sampling

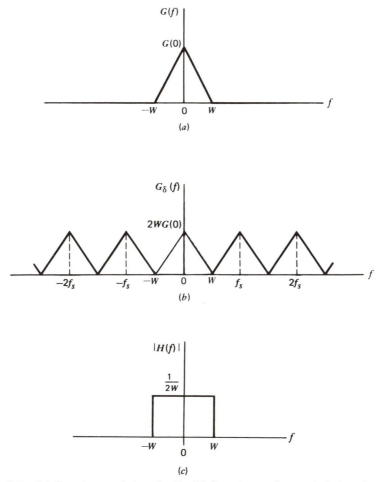

Figure 4.3 (a) Spectrum of signal $g(t)$. (b) Spectrum of sampled signal $g_\delta(t)$ for a sampling rate $f_s = 2W$. (c) Ideal amplitude response of reconstruction filter.

period $T_s = 1/2W$. Then the corresponding spectrum $G_\delta(f)$ of the sampled signal $g_\delta(t)$ is as shown in Fig. 4.3b. Putting $T_s = 1/2W$ in Eq. 4.7 yields

$$G_\delta(f) = \sum_{n=-\infty}^{\infty} g\left(\frac{n}{2W}\right) \exp\left(-\frac{j\pi n f}{W}\right) \qquad (4.8)$$

Putting $f_s = 2W$ in Eq. 4.6, we have

$$G(f) = \frac{1}{2W} G_\delta(f) \qquad -W < f < W \qquad (4.9)$$

It follows from Eq. 4.8 that we may also write

$$G(f) = \frac{1}{2W} \sum_{n=-\infty}^{\infty} g\left(\frac{n}{2W}\right) \exp\left(-\frac{j\pi n f}{W}\right) \qquad -W < f < W \qquad (4.10)$$

Therefore, if the sample values $g(n/2W)$ of the signal $g(t)$ are specified for all time, then the Fourier transform $G(f)$ of the signal is uniquely determined by using the Fourier series of Eq. 4.10. Because $g(t)$ is related to $G(f)$ by the inverse Fourier transform, it follows that the signal $g(t)$ is itself uniquely determined by the sample values $g(n/2W)$ for $-\infty \leqslant n \leqslant \infty$. In other words, the sequence $\{g(n/2W)\}$ contains all of the information of $g(t)$.

Consider next the problem of reconstructing the signal $g(t)$ from the sequence of sample values $\{g(n/2W)\}$. Substituting Eq. 4.10 in the formula for the inverse Fourier transform defining $g(t)$ in terms of $G(f)$, we get

$$g(t) = \int_{-\infty}^{\infty} G(f) \exp(j2\pi f t) df$$

$$= \int_{-W}^{W} \frac{1}{2W} \sum_{n=-\infty}^{\infty} g\left(\frac{n}{2W}\right) \exp\left(-\frac{j\pi n f}{W}\right) \exp(j2\pi f t) df$$

Interchanging the order of summation and integration:

$$g(t) = \sum_{n=-\infty}^{\infty} g\left(\frac{n}{2W}\right) \frac{1}{2W} \int_{-W}^{W} \exp\left[j2\pi f\left(t - \frac{n}{2W}\right)\right] df \qquad (4.11)$$

The integral term in Eq. 4.11 may be readily evaluated, yielding*

---

* The series

$$C(t) = \sum_{n=-\infty}^{\infty} g\left(\frac{n}{2W}\right) \frac{\sin(2\pi W t - n\pi)}{(2\pi W t - n\pi)}$$

or, equivalently

$$C(t) = \frac{\sin(2\pi W t)}{\pi} \sum_{n=-\infty}^{\infty} g\left(\frac{n}{2W}\right) \frac{(-1)^n}{(2W t - n)}$$

is called a *cardinal series* by mathematicians. The cardinal series $C(t)$ *converges absolutely if, and only if, the series*

$$g(t) = \sum_{n=-\infty}^{\infty} g\left(\frac{n}{2W}\right) \frac{\sin(2\pi Wt - n\pi)}{(2\pi Wt - n\pi)} \tag{4.12}$$

We may simplify the notation in Eq. 4.12 by using the *sinc function*, defined as

$$\operatorname{sinc} x = \frac{\sin(\pi x)}{\pi x} \tag{4.13}$$

where x is an independent variable. The sinc function exhibits an important property known as the *interpolatory property*, which is described as follows:

$$\operatorname{sinc} x = \begin{cases} 1 & x = 0 \\ 0 & x = \pm 1, \pm 2, \ldots \end{cases} \tag{4.14}$$

Using the definition of the sinc function, we may rewrite Eq. 4.12 as follows:

$$g(t) = \sum_{n=-\infty}^{\infty} g\left(\frac{n}{2W}\right) \operatorname{sinc}(2Wt - n) \tag{4.15}$$

Equation 4.15 provides an *interpolation formula* for reconstructing the original signal $g(t)$ from the sequence of sample values $\{g(n/2W)\}$, with the sinc function $\operatorname{sinc}(2Wt)$ playing the role of an *interpolation function*. Each sample is multiplied by a delayed version of the interpolation function, and all the resulting waveforms are added to obtain $g(t)$. It is noteworthy that Eq. 4.15 also represents the response of an ideal low-pass filter of bandwidth W and zero transmission delay, which is produced by an input signal consisting of the sequence of

---

$$\sum_{n=1}^{\infty} \frac{|g(n/2W)| + |g(-n/2W)|}{n}$$

converges. This result is known as the *absolute convergence principle*. It appears in the work of Whittaker (1929a, 1929b).

An improvement to the cardinal series convergence is reported in de Buda and de Buda (1987). Define

$$s(n) = \sum_{k=-\infty}^{n} (-1)^{k+n} g\left(\frac{k}{2W}\right)$$

so that

$$s(n) + s(n-1) = g\left(\frac{n}{2W}\right)$$

Then, the cardinal series

$$C(t) = \sum_{n=-\infty}^{\infty} s(n) \left[\operatorname{sinc}(2Wt + n) + \operatorname{sinc}(2Wt - n + 1)\right]$$

converges *absolutely* and *uniformly* to $g(t)$ for $-\infty < t < \infty$ if, and only if, $s(n)$ exists and converges to zero as n approaches infinity. Here $g(t)$ is not restricted to have finite energy.

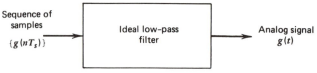

Figure 4.4 Reconstruction filter.

samples $\{g(n/2W)\}$ for $-\infty \leqslant n \leqslant \infty$. This is intuitively satisfying, since, by inspection of the spectrum of Fig. 4.3b, we see that the original signal $g(t)$ may be recovered exactly from the sequence of samples $\{g(n/2W)\}$ by passing it through an ideal low-pass filter of bandwidth W. This is illustrated in block diagrammatic form in Fig. 4.4. The ideal amplitude response of the *reconstruction filter* is shown in Fig. 4.3c.

(1) Signal Space Interpretation

We may develop another important interpretation of Eq. 4.15 by using the property that the function $\mathrm{sinc}(2Wt - n)$, where n is an integer, is one of a family of shifted sinc functions that are mutually orthogonal. To prove this property, we use the formula

$$\int_{-\infty}^{\infty} g_1(t)g_2^*(t)dt = \int_{-\infty}^{\infty} G_1(f)G_2^*(f)df \qquad (4.16)$$

where $g_1(t)$ and $G_1(f)$ form a Fourier transform pair; likewise for $g_2(t)$ and $G_2(f)$. This relation may be viewed as a generalization of Rayleigh's energy theorem. Put

$$g_1(t) = \mathrm{sinc}(2Wt - n) = \mathrm{sinc}\left[2W\left(t - \frac{n}{2W}\right)\right]$$

and

$$g_2(t) = \mathrm{sinc}(2Wt - m) = \mathrm{sinc}\left[2W\left(t - \frac{m}{2W}\right)\right]$$

For a sinc pulse, $\mathrm{sinc}(2Wt)$, we have the Fourier transform pair:

$$\mathrm{sinc}(2Wt) \rightleftharpoons \frac{1}{2W}\mathrm{rect}\left(\frac{f}{2W}\right) \qquad (4.17)$$

where, on the right side, we have made use of the definition of a *rectangular function*, namely

$$\mathrm{rect}(x) = \begin{cases} 1 & -\frac{1}{2} < x < \frac{1}{2} \\ 0 & |x| > \frac{1}{2} \end{cases} \qquad (4.18)$$

The functions $g_1(t)$ and $g_2(t)$ are time-shifted versions of the sinc pulse $\mathrm{sinc}(2Wt)$. Accordingly, using the time-shifting property of the Fourier transform, we may express the Fourier transforms of $g_1(t)$ and $g_2(t)$, as follows, respectively:

$$G_1(f) = \frac{1}{2W} \operatorname{rect}\left(\frac{f}{2W}\right) \exp\left(-\frac{j\pi n f}{W}\right)$$

and

$$G_2(f) = \frac{1}{2W} \operatorname{rect}\left(\frac{f}{2W}\right) \exp\left(-\frac{j\pi m f}{W}\right)$$

Hence, the use of these two Fourier transforms in Eq. 4.16 yields

$$\int_{-\infty}^{\infty} \operatorname{sinc}(2Wt - n)\operatorname{sinc}(2Wt - m)dt = \left(\frac{1}{2W}\right)^2 \int_{-W}^{W} \exp\left[-\frac{j\pi f}{W}(n - m)\right]df$$

$$= \frac{\sin[\pi(n - m)]}{2W\pi(n - m)}$$

$$= \frac{1}{2W} \operatorname{sinc}(n - m)$$

This result equals $1/2W$ when $n = m$, and zero when $n \neq m$ (see Eq. 4.14). We therefore have

$$\int_{-\infty}^{\infty} \operatorname{sinc}(2Wt - n)\operatorname{sinc}(2Wt - m)dt = \begin{cases} \dfrac{1}{2W} & n = m \\ 0 & n \neq m \end{cases} \qquad (4.19)$$

Accordingly, Eq. 4.15 represents the expansion of the signal $g(t)$ as an infinite sum of orthogonal functions with the coefficients of the expansion, $g(n/2W)$, defined by

$$g\left(\frac{n}{2W}\right) = 2W \int_{-\infty}^{\infty} g(t)\operatorname{sinc}(2Wt - n)dt \qquad (4.20)$$

We may thus view the coefficients $g(n/2W)$ of this expansion as coordinates in an infinite-dimensional signal space. In this space each signal corresponds to precisely one point and each point to one signal.

(2) Statement of the Sampling Theorem

We may now state the *sampling theorem** for band-limited signals of finite energy in two separate parts:

---

* The sampling theorem was introduced to communication theory in 1949 by Shannon. It is for this reason that the theorem is sometimes referred to in the literature as the "Shannon sampling theorem." However, the interest of communication engineers in the sampling theorem may be traced back to Nyquist (1928). Indeed, the sampling theorem was known to mathematicians much earlier. In particular, the corresponding mathematical theorem was first proved by E. T. Whittaker in 1915; the proof was based on interpolation between a given set of points (Whittaker, 1915). For a tutorial review of the sampling theorem, related issues, and historical notes, see Jerri (1977) and Higgins (1985).

1. *If a finite-energy signal g(t) contains no frequencies higher than W hertz, it is completely determined by specifying its ordinates at a sequence of points spaced 1/2W seconds apart.*
2. *If a finite energy signal g(t) contains no frequencies higher than W hertz, it may be completely recovered from its ordinates at a sequence of points spaced 1/2W seconds apart.*

Part 1 is a restatement of Eq. 4.10, and part 2 is a restatement of Eq. 4.15.

The minimum sampling rate of $2W$ samples per second, for a signal bandwidth of W hertz, is called the *Nyquist rate*. Correspondingly, the reciprocal, $1/2W$, is called the *Nyquist interval*. The sampling theorem serves as the basis for the interchangeability of analog signals and digital sequences, which is so valuable in digital communication systems.

4.2 QUADRATURE SAMPLING OF BAND-PASS SIGNALS

Up to this point in our discussion of the sampling process, we have focussed attention on low-pass signals. However, many of the signals encountered in practice are intrinsically band-pass in nature. In this section, we consider a scheme, called *quadrature sampling,* for the uniform sampling of band-pass signals.* The scheme represents a natural extension of the sampling of low-pass signals.

In the scheme described herein, rather than sample a band-pass signal directly, we precede the sampling operation by preparatory processing of the signal. The most obvious strategy is to represent the band-pass signal in terms of its *in-phase* and *quadrature components,* each of which may then be sampled separately.

Consider a band-pass signal $g(t)$ whose spectrum is limited to a bandwidth $2W$, centered around the frequency f_c, as illustrated in Fig. 4.5a. It is assumed that $f_c > W$. Let $g_I(t)$ denote the in-phase component of the band-pass signal $g(t)$ and $g_Q(t)$ denote its quadrature component. We may then express $g(t)$ in terms of $g_I(t)$ and $g_Q(t)$ as follows:†

$$g(t) = g_I(t)\cos(2\pi f_c t) - g_Q(t)\sin(2\pi f_c t) \qquad (4.21)$$

The in-phase component $g_I(t)$ and the quadrature component $g_Q(t)$ may be obtained (except for scaling factors) by multiplying the band-pass signal $g(t)$ by $\cos(2\pi f_c t)$ and $\sin(2\pi f_c t)$, respectively, and then suppressing the sum-frequency components by means of appropriate low-pass filters. Under the assumption that $f_c > W$, we find that $g_I(t)$ and $g_Q(t)$ are both low-pass signals limited to $-W < f < W$, as illustrated in Fig. 4.5b. Accordingly, each component may be sampled at the rate of $2W$ samples per second. This form of sampling is called *quadrature sampling.*

* The idea of quadrature sampling is discussed in Linden (1959).

† For the representation of band-pass signals in terms of their in-phase and quadrature components, see Appendix C.

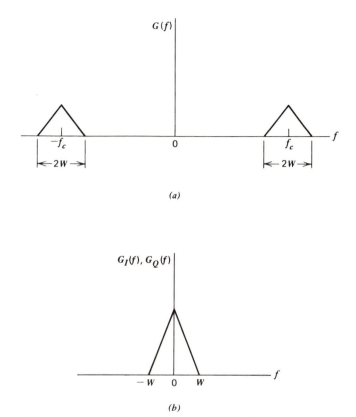

Figure 4.5 (a) Spectrum of bond-pass signal $g(t)$. (b) Spectrum of low-pass in-phase component $g_I(t)$ and quadrature component $g_Q(t)$.

To reconstruct the original band-pass signal from its quadrature-sampled version, we first reconstruct the in-phase component $g_I(t)$ and the quadrature component $g_Q(t)$ from their respective samples, multiply them respectively by $\cos(2\pi f_c t)$ and $\sin(2\pi f_c t)$, and then add the results.

The operations that constitute quadrature sampling are depicted in Fig. 4.6*a*, and those pertaining to the reconstruction process are depicted in Fig. 4.6*b*. Note that the samples of the quadrature component in the generator of Fig. 4.6*a* are the negative of those used as input in the reconstruction circuit of Fig. 4.6*b*.

4.3   RECONSTRUCTION OF A MESSAGE PROCESS FROM ITS SAMPLES

A study of the sampling process would be incomplete without considering the reconstruction of a *stationary message process* from its samples. By a "message process" we mean an ensemble of message waveforms, each one of which represents a realization of the process. In this section we show that when a wide-sense stationary message process (whose power spectrum is strictly band-limited) is reconstructed from the sequence of its samples taken at a rate

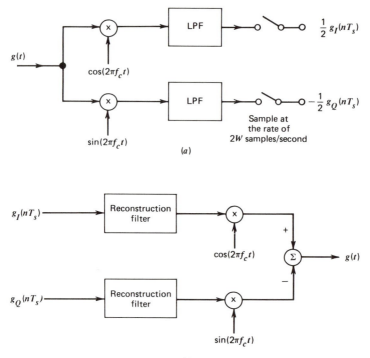

Figure 4.6 (a) Generation of in-phase and quadrative samples from band-pass signal $g(t)$. (b) Reconstruction of band-pass signal $g(t)$.

equal to twice the highest frequency component, the reconstructed process equals the original process in the *mean-square sense* for all time.

Consider then a wide-sense stationary message process $X(t)$ with autocorrelation function $R_X(\tau)$ and power spectral density $S_X(f)$. We assume that

$$S_X(f) = 0 \qquad \text{for } |f| \geq W \qquad (4.22)$$

Suppose we have available an infinite sequence of samples of the process taken at a uniform rate equal to $2W$, that is, twice the highest frequency component of the process. Using $X'(t)$ to denote the reconstructed process, based on this infinite sequence of samples, we may write (see Eq. 4.15)

$$X'(t) = \sum_{n=-\infty}^{\infty} X\left(\frac{n}{2W}\right) \text{sinc}(2Wt - n) \qquad (4.23)$$

where $X(n/2W)$ is the random variable obtained by sampling or observing the message process $X(t)$ at time $t = n/2W$. The *mean-square value of the error* between the original message process $X(t)$ and the reconstructed message process $X'(t)$ equals

$$\mathcal{E} = E[(X(t) - X'(t))^2]$$
$$= E[X^2(t)] - 2E[X(t)X'(t)] + E[(X'(t))^2] \qquad (4.24)$$

We recognize the first expectation term on the right side of Eq. 4.24 as the mean-square value of $X(t)$, which equals $R_X(0)$; thus

$$E[X^2(t)] = R_X(0) \tag{4.25}$$

For the second expectation term, we use Eq. 4.23 and so write

$$E[X(t)X'(t)] = E\left[X(t) \sum_{n=-\infty}^{\infty} X\left(\frac{n}{2W}\right) \text{sinc}(2Wt - n)\right]$$

Interchanging the order of summation and expectation:

$$E[X(t)X'(t)] = \sum_{n=-\infty}^{\infty} E\left[X(t)X\left(\frac{n}{2W}\right)\right] \text{sinc}(2Wt - n)$$

$$= \sum_{n=-\infty}^{\infty} R_X\left(t - \frac{n}{2W}\right) \text{sinc}(2Wt - n) \tag{4.26}$$

For a stationary process, the expectation $E[X(t)X'(t)]$ is independent of time t. Hence, putting $t = 0$ in the right side of Eq. 4.26 and recognizing that

$$R_X\left(-\frac{n}{2W}\right) = R_X\left(\frac{n}{2W}\right)$$

we may write

$$E[X(t)X'(t)] = \sum_{n=-\infty}^{\infty} R_X\left(\frac{n}{2W}\right) \text{sinc}(-n) \tag{4.27}$$

The term $R_X(n/2W)$ represents a sample of the autocorrelation function $R_X(\tau)$ taken at $\tau = n/2W$. Now, since the power spectral density $S_X(f)$ or equivalently the Fourier transform of $R_X(\tau)$ is zero for $|f| > W$, we may represent $R_X(\tau)$ in terms of its samples taken at $\tau = n/2W$ as follows (see Eq. 4.15):

$$R_X(\tau) = \sum_{n=-\infty}^{\infty} R_X\left(\frac{n}{2W}\right) \text{sinc}(2W\tau - n) \tag{4.28}$$

Accordingly, we deduce from Eqs. 4.27 and 4.28 that

$$E[X(t)X'(t)] = R_X(0) \tag{4.29}$$

For the third and final expectation term on the right side of Eq. 4.24, we again use Eq. 4.23 and so write

$$E[(X'(t))^2] = E\left[\sum_{n=-\infty}^{\infty} X\left(\frac{n}{2W}\right) \text{sinc}(2Wt - n) \sum_{k=-\infty}^{\infty} X\left(\frac{k}{2W}\right) \text{sinc}(2Wt - k)\right]$$

$$= E\left[\sum_{n=-\infty}^{\infty} \text{sinc}(2Wt - n) \sum_{k=-\infty}^{\infty} X\left(\frac{n}{2W}\right) X\left(\frac{k}{2W}\right) \text{sinc}(2Wt - k)\right]$$

Interchanging the order of expectation and inner summation:

$$E[(X'(t))^2] = \sum_{n=-\infty}^{\infty} \text{sinc}(2Wt - n) \sum_{k=-\infty}^{\infty} E\left[X\left(\frac{n}{2W}\right) X\left(\frac{k}{2W}\right)\right] \text{sinc}(2Wt - k)$$

$$= \sum_{n=-\infty}^{\infty} \text{sinc}(2Wt - n) \sum_{k=-\infty}^{\infty} R_X\left(\frac{n-k}{2W}\right) \text{sinc}(2Wt - k) \qquad (4.30)$$

However, in view of Eq. 4.28, we recognize that the inner summation on the right side of Eq. 4.30 equals $R_X(t - n/2W)$. Hence, we may simplify Eq. 4.30 as follows:

$$E[(X'(t))^2] = \sum_{n=-\infty}^{\infty} R_X\left(t - \frac{n}{2W}\right) \text{sinc}(2Wt - n)$$

$$= R_X(0) \qquad (4.31)$$

Finally, substituting Eqs. 4.25, 4.29, and 4.31 into Eq. 4.24, we get the result

$$\mathcal{E} = 0$$

as should be expected.

We may therefore state the sampling theorem for message processes as follows. If a stationary message process contains no frequencies higher than W hertz, it may be reconstructed from its samples at a sequence of points spaced $1/2W$ seconds apart with zero mean squared error (i.e., zero error power).

4.4   SIGNAL DISTORTION IN SAMPLING

In deriving the sampling theorem for a signal $g(t)$ in Section 4.1, we assumed that it is *strictly band-limited* with no frequencies higher than some value W. (We made a similar assumption in Section 4.3 when discussing the sampling theorem in the context of a stationary message process). However, *a signal cannot be finite in both time and frequency*. It follows therefore that the signal $g(t)$ must have infinite duration for its spectrum to be strictly band-limited. Indeed, in deriving the sampling theorem in Section 4.1, it was assumed that the signal has infinite duration in that time t may take on a value anywhere between $-\infty$ and ∞; see Eq. 4.12. In practice, we have to work with a *finite* segment of the signal, in which case the spectrum cannot be strictly band-limited. Consequently, when a signal of finite duration is sampled, an *error* in the reconstruction occurs as a result of the sampling process.

To develop an appreciation for this error, consider a signal $g(t)$ whose spectrum $G(f)$ decreases with increasing frequency f without limit, as shown in Fig. 4.7a. The spectrum $G_\delta(f)$ of the discrete-time signal $g_\delta(t)$, resulting from the use of idealized sampling, is the sum of $G(f)$ and an infinite number of frequency-shifted replicas of it, in accordance with Eq. 4.6. The replicas of $G(f)$ are shifted in frequency by multiples of the sampling rate f_s. Two replicas of $G(f)$ are shown in Fig. 4.7b, one at f_s and the other at $-f_s$. Clearly, the use of a low-pass reconstruction filter, with its pass-band extending from $-f_s/2$ to $f_s/2$, no longer yields an undistorted version of the original signal $g(t)$. Instead, we find that portions of the frequency-shifted replicas are *folded over* inside the desired spectrum. Specifically, high frequencies in $G(f)$ are reflected into low

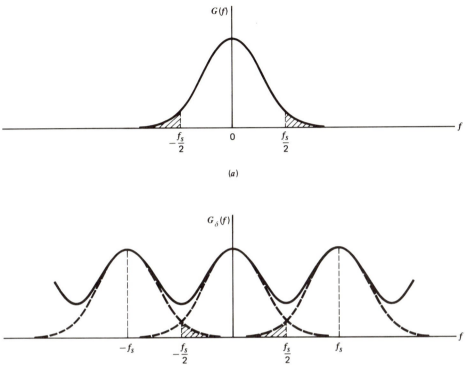

Figure 4.7 (a) Spectrum of a finite-energy signal $g(t)$ whose spectrum decreases with increasing frequency without limit. (b) Illustrating the composition of the spectrum $G_\delta(f)$ of the discrete-time signal $g_\delta(t)$; the dashed curves represent replicas of $G(f)$.

frequencies in $G_\delta(f)$. This is readily seen by comparing the shaded areas of the spectra shown in Fig. 4.7a and 4.7b. The phenomenon of a high-frequency in the spectrum of the original signal $g(t)$ seemingly taking on the identity of a lower frequency in the spectrum of the sampled signal $g(t)$ is called *aliasing* or *foldover*. When this effect occurs, information is inevitably lost in the sampling process.

The effect of aliasing on the output of the reconstruction filter depends on both the amplitude and phase components of the original spectrum $G(f)$, making an exact analysis of the effect difficult. Nevertheless, we may develop an appreciation for the problem by considering a *worst-case* analysis, as described next.

(1) Bound on Aliasing Error

Let $\{g(n/f_s)\}$ denote the sequence obtained by sampling an arbitrary signal $g(t)$ at the rate f_s samples per second. Let $g_i(t)$ denote the signal reconstructed from this sequence by *interpolation;* that is

$$g_i(t) = \sum_{n=-\infty}^{\infty} g\left(\frac{n}{f_s}\right) \text{sinc}(f_s t - n) \tag{4.32}$$

which is obtained from Eq. 4.15 by using f_s in place of $2W$. The absolute error

$$\varepsilon = |g(t) - g_i(t)| \tag{4.33}$$

is called the *aliasing error*.

It would be informative to have an *upper bound* on the aliasing error ε. We may develop such a bound by proceeding as follows. First, we express the signal $g(t)$ in terms of its spectrum $G(f)$ by using the inverse Fourier transform

$$g(t) = \int_{-\infty}^{\infty} G(f) \exp(j2\pi ft) df$$

or, equivalently

$$g(t) = \sum_{m=-\infty}^{\infty} \int_{(m-1/2)f_s}^{(m+1/2)f_s} G(f) \exp(j2\pi ft) df \tag{4.34}$$

Next, we use *Poisson's formula* written in the form*

$$\sum_{m=-\infty}^{\infty} G(f - mf_s) = \frac{1}{f_s} \sum_{n=-\infty}^{\infty} g\left(\frac{n}{f_s}\right) \exp\left(-\frac{j2\pi nf}{f_s}\right) \tag{4.35}$$

On the left side of this formula, we recognize spectrum repetition that results from sampling. On the right side, we have a series that has the appearance of a complex Fourier series, except for the fact that the roles of time and frequency have been interchanged. Indeed, multiplying this series by the exponential term $\exp(j2\pi ft)$ and integrating with respect to f from $-f_s/2$ to $f_s/2$, we get the reconstructed signal $g_i(t)$ as defined in Eq. 4.32. We may thus write

$$g_i(t) = \sum_{m=-\infty}^{\infty} \int_{-f_s/2}^{f_s/2} G(f - mf_s) \exp(j2\pi ft) df \tag{4.36}$$

Changing the variable of integration from f to $f - mf_s$, we get

$$g_i(t) = \sum_{m=-\infty}^{\infty} \exp(-j2\pi mf_s t) \int_{(m-1/2)f_s}^{(m+1/2)f_s} G(f) \exp(j2\pi ft) df \tag{4.37}$$

Substituting Eqs. 4.34 and 4.37 in Eq. 4.33, we may express the aliasing error as

$$\varepsilon = \left| \sum_{m=-\infty}^{\infty} \left[1 - \exp(-j2\pi mf_s t) \right] \int_{(m-1/2)f_s}^{(m+1/2)f_s} G(f) \exp(j2\pi ft) df \right| \tag{4.38}$$

We may now derive the desired bound on the aliasing error ε by combining

---

* The form of Poisson's formula given in Eq. 4.34 follows directly from Eqs. 4.6 and 4.8, with f_s used to denote the sampling rate in the second equation.

the following observations:

1. The term corresponding to $m = 0$ vanishes.
2. The absolute value of the sum of a set of terms is less than or equal to the sum of the absolute values of the individual terms.
3. The absolute value of the term $1 - \exp(-j2\pi m f_s t)$ is less than or equal to 2.
4. The absolute value of the integral in Eq. 4.38 is bounded as

$$\left| \int_{(m-1/2)f_s}^{(m+1/2)f_s} G(f)\exp(j2\pi ft)df \right| \leq \int_{(m-1/2)f_s}^{(m+1/2)f_s} |G(f)|df$$

Using these observations in Eq. 4.38 and reconstituting the integral, we find that the aliasing error is bounded as*

$$\varepsilon \leq 2 \int_{|f|>f_s/2} |G(f)|df \tag{4.39}$$

where $|G(f)|$ is the amplitude spectrum of the signal $g(t)$, and the integral is evaluated for frequencies defined by $|f| > f_s/2$. Thus, given an arbitrary signal $g(t)$ with amplitude spectrum $|G(f)|$, we may use Eq. 4.39 to perform a worst-case evaluation of the aliasing error G for varying sampling rate f_s.

The following simple example illustrates the validity of this bound on aliasing error.

EXAMPLE 1

Consider the time-shifted sinc pulse

$$g(t) = 2\,\text{sinc}(2t - 1)$$

which is shown plotted in Fig. 4.8a. Suppose $g(t)$ is sampled at times $t = 0, \pm1, \pm2,$ . . . , where it vanishes. Hence, with $f_s = 1$ sample per second, we have a sequence

$$g\left(\frac{n}{f_s}\right) = 0 \qquad \text{for } n = 0, \pm1, \pm2, \ldots.$$

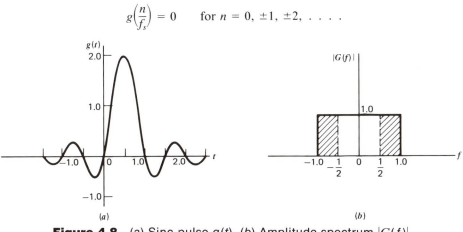

Figure 4.8   (a) Sinc pulse $g(t)$. (b) Amplitude spectrum $|G(f)|$.

* The bound given in Eq. 4.39 was first stated without proof in Weiss (1963). It is derived in Brown (1967). Our derivation follows Higgins (1985).

and a reconstructed signal

$$g_i(t) = 0 \qquad \text{for all } t$$

In Fig. 4.8a, we see that the sinc pulse $g(t)$ attains its maximum value of 2 at time t equal to 1/2. Hence, with $g_i(t) = 0$, we find from the defining Eq. 4.33 that the aliasing error cannot exceed $\max|g(t)| = 2$.

Consider next the use of Eq. 4.39. The amplitude spectrum of the sinc pulse $g(t)$ is defined by the rectangular function

$$|G(f)| = \text{rect}\left(\frac{f}{2}\right)$$

which is shown plotted in Fig. 4.8b. The total area under the curve of the amplitude spectrum $G(f)|$ that lies outside the frequency interval $-1/2 < f < 1/2$ is shown shaded in Fig. 4.8b. This area equals 1, and so we have

$$\int_{|f|>f_s/2} |G(f)|\,df = 1$$

where $f_s = 1$. Hence, from Eq. 4.39, we find that the upper bound on the aliasing error for this example is 2, confirming the previous result.

Signal-to-Distortion Ratio

We may also evaluate the effect of improper application of the sampling theorem by taking an *ensemble averaging* approach. Consider then a message process $X(t)$ that is wide-sense stationary but *not* strictly band-limited. Specifically, for a sampling rate f_s, we have

$$S_X(f) \neq 0 \qquad |f| > f_s/2 \tag{4.40}$$

where $S_X(f)$ is the power spectral density of the process. For the purpose of our evaluation, we may study the situation depicted in Fig. 4.9. The message process $X(t)$ is passed through an ideal low-pass filter with the transfer function

$$H_l(f) = \begin{cases} 1 & |f| < W \\ 0 & |f| > W \end{cases} \tag{4.41}$$

Let $X_l(t)$ denote the random process resulting at the low-pass filter output, which is strictly band-limited to $f_s/2$; that is

$$S_{X_l}(f) = \begin{cases} S_X(f) & |f| < f_s/2 \\ 0 & |f| \geq f_s/2 \end{cases} \tag{4.42}$$

where $S_{X_l}(f)$ is the power spectral density of $X_l(f)$. Hence, in accordance with the sampling theorem, the process $X_l(t)$ may be reconstructed from its samples exactly (in the sense of zero mean squared error). Let $X'(t)$ denote the process reconstructed from samples of $X_l(t)$. Naturally, $X'(t)$ is different from the original process $X(t)$, since $X_l(t)$ is different from $X(t)$.

To evaluate the average power of this difference, let the process $X(t)$ be passed through an ideal high-pass filter, also shown in the lower path of Fig. 4.9. This second filter has the transfer function

$$H_h(f) = \begin{cases} 0 & |f| < f_s/2 \\ 1 & |f| > f_s/2 \end{cases} \tag{4.43}$$

Let $X_h(t)$ denote the random process resulting at the high-pass filter output. Its power spectral density is defined by

$$S_{X_h}(f) = \begin{cases} 0 & |f| \le f_s/2 \\ S_X(f) & |f| > f_s/2 \end{cases} \tag{4.44}$$

Since

$$H_l(f) + H_h(f) = 1 \qquad \text{for all } f \tag{4.45}$$

it follows that

$$X_l(t) + X_h(t) = X(t) \qquad \text{for all } t$$

or, equivalently

$$X_h(t) = X(t) - X_l(t) \qquad \text{for all } t \tag{4.46}$$

We may therefore view $X_h(t)$ as the error between the process $X(t)$ and the process $X'(t)$ reconstructed from samples of $X_l(t)$. We now recognize that the mean-square value of a stationary random process equals the total area under the curve of its power spectral density. We may thus express the *mean squared error* as

$$\mathcal{E} = E[(X'(t) - X(t))^2]$$

$$= \int_{-\infty}^{\infty} S_{X_h}(f) \, df$$

$$= \int_{|f|>f_s/2} S_X(f) \, df \tag{4.47}$$

The total power (mean-square value) of the process $X(t)$ is given by

$$P = \int_{-\infty}^{\infty} S_X(f) \, df \tag{4.48}$$

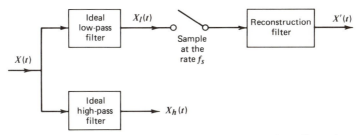

Figure 4.9   Conceptual model for evaluating signal-to-distortion ratio.

Hence, we may define the *signal-to-distortion* ratio (SDR) as

$$\text{SDR} = \frac{P}{\mathscr{E}}$$

$$= \frac{\int_{-\infty}^{\infty} S_X(f)\, df}{\int_{|f|>f_s/2} S_X(f)\, df} \tag{4.49}$$

Thus, given a message process $X(t)$ with power spectral density $S_X(f)$, we may use Eq. 4.49 to evaluate the SDR for varying sampling rates. Clearly, the SDR increases as the sampling rate f_s is increased.

EXAMPLE 2

Consider a message process $X(t)$ whose power spectral density is defined by the *Butterworth-shaped function*

$$S_X(f) = \frac{1}{1 + (f/f_0)^{2N}} \qquad N = 1, 2, 3, \ldots \tag{4.50}$$

where f_0 is the *3-dB cutoff frequency*, and the parameter N defines the *spectrum rolloff*. Let

$$\lambda = \frac{f}{f_0}$$

Then, substituting Eq. 4.50 in Eq. 4.49, we may express the SDR in terms of the normalized variable λ as

$$\text{SDR} = \frac{\int_0^{\infty} \dfrac{d\lambda}{1 + \lambda^{2N}}}{\int_{f_s/2f_0}^{\infty} \dfrac{d\lambda}{1 + \lambda^{2N}}} \tag{4.51}$$

The resulting SDR, expressed in decibels, is shown plotted versus the normalized sampling rate, f_s/f_0, in Fig. 4.10* for various values of rolloff parameter N.

From Fig. 4.10, we may make two observations:

1. To maintain an acceptable SDR (say, 30 dB), the sampling rate f_s has to be much greater than twice the 3-dB cutoff frequency f_0, especially when the rolloff parameter N is small.
2. For a fixed value of SDR, the required value of the sampling rate f_s decreases as the rolloff parameter N increases; in the limit, f_s approaches the theoretical minimum rate $2f_0$ as N approaches infinity.

These two points have practical implications of their own.

The implication of point 1 is that we must exercise care in the application of the sampling theorem. In particular, interpretation of the 3-dB cutoff frequency f_0 as the bandwidth extent of a signal and blind application of the sampling theorem can result in a large aliasing error.

The implication of point 2 is that we may reduce the aliasing error by shaping

* Figure 4.10 is adapted from Gagliardi (1978, p. 237).

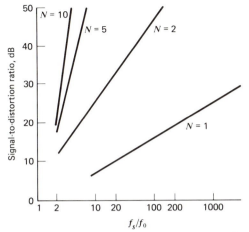

Figure 4.10 Signal-to-distortion ratio arising from the aliasing effect for But-terworth-shaped spectra.

the spectrum of a signal to roll off rapidly. This suggests the use of a *pre-alias* filter so as to realize the required spectrum rolloff.

(3) Sampling Procedure

We may now outline a practical procedure for the sampling of a signal whose spectrum is not strictly band-limited. The procedure involves the use of two corrective measures:

1. Prior to sampling, a low-pass pre-alias filter of high enough order is used to attenuate those high-frequency components of the signal that do *not* contribute significantly to the information content of the signal.
2. The filtered signal is sampled at a rate slightly higher than the Nyquist rate $2W$, where W is the cutoff frequency of the pre-alias filter.

It is of interest to note that the use of a sampling rate f_s higher than the Nyquist rate $2W$ has also the desirable effect of making it somewhat easier to design the low-pass reconstruction filter so as to recover the original analog signal from its sampled version. With such a sampling rate, we find that there are gaps, each of width $f_s - 2W$ between the frequency-shifted replicas of $G(f)$. Accordingly, we may design the reconstruction filter to satisfy the following characteristics:

1. The passband of the reconstruction filter extends from zero to W hertz.
2. The amplitude response of the reconstruction filter rolls off gradually from W to $f_s - W$ hertz. That is, the *guard band* (separating the passband from the stopband) has a width equal to $f_s - 2W$, which is nonzero for $f_s > 2W$.

Figure 4.11 illustrates these two characteristics of a reconstruction filter. Note that the nonzero width of the guard band ensures physical realizability of the reconstruction filter.

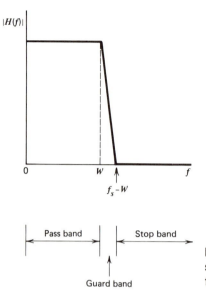

Figure 4.11 Illustrating the amplitude response of reconstruction filter for positive frequencies.

4.5 PRACTICAL ASPECTS OF SAMPLING AND SIGNAL RECOVERY

In practice, the sampling of an analog signal is accomplished by means of high-speed switching transistor circuits. Accordingly, we find that the resulting sampled waveform deviates from the ideal form of instantaneous sampling described in Section 4.1 because the operation of a physical switching circuit, however fast, still requires a nonzero interval of time. Furthermore, we often find that samples of an analog signal are lengthened intentionally for convenience in instrumentation or transmission. In this section we evaluate the effects of these practical deviations from the ideal condition. We also discuss a circuit for recovery of the original signal from its samples.

(1) Ordinary Samples of Finite Duration

Let an arbitrary analog signal $g(t)$ be applied to a switching circuit controlled by a sampling function $c(t)$ that consists of an infinite succession of rectangular pulses of amplitude A, duration T, and occurring with period T_s. The output of the switching circuit is denoted by $s(t)$. The waveforms $g(t)$, $c(t)$, and $s(t)$ are illustrated in parts (a), (b), and (c) of Fig. 4.12 respectively. We see that the switching operation merely extracts from the analog signal $g(t)$ successive portions of predetermined duration T, taken regularly at the rate $f_s = 1/T_s$. Accordingly, the sampled signal $s(t)$ consists of a sequence of positive and negative pulses, as in Fig. 4.12c. Mathematically, we have

$$s(t) = c(t)g(t) \qquad (4.52)$$

However, $c(t)$ may be expressed in the form of a complex Fourier series as

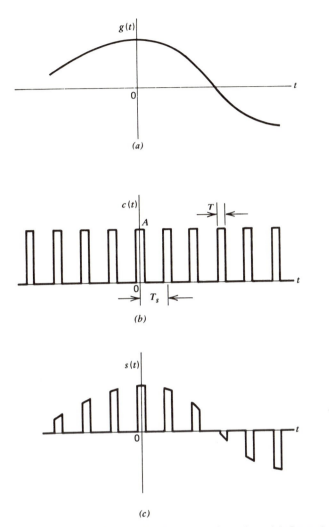

Figure 4.12   (a) Analog signal. (b) Sampling function. (c) Sampled signal.

follows:

$$c(t) = f_s TA \sum_{n=-\infty}^{\infty} \operatorname{sinc}(nf_s T) \exp(j2\pi nf_s t) \qquad (4.53)$$

where $T_s f_s = 1$ defines the sampling rate f_s. Therefore, substituting Eq. 4.53 in Eq. 4.52, we get

$$s(t) = f_s TA \sum_{n=-\infty}^{\infty} \operatorname{sinc}(nf_s T) \exp(j2\pi nf_s t) \, g(t) \qquad (4.54)$$

Taking the Fourier transform of both sides of Eq. 4.54 and using the frequency-shifting property of the Fourier transform, we get

$$S(f) = f_s T A \sum_{m=-\infty}^{\infty} \text{sinc}(m f_s T) \, G(f - m f_s) \tag{4.55}$$

where $S(f) = F[s(t)]$ and $G(f) = F[g(t)]$.

The relationship between the spectra $G(f)$ and $S(f)$ is illustrated in Fig. 4.13 assuming that $g(t)$ contains no frequencies lying outside the band $-W$ to W, and that the sampling rate f_s is greater than the Nyquist rate $2W$, so that there is no aliasing. We see that the effect of the finite duration of the sampling pulses is to multiply the nth lobe of the spectrum $S(f)$ by $(TA/T_s)\text{sinc}(n f_s T)$. The original signal $g(t)$ can be recovered from $s(t)$ with no distortion by passing $s(t)$ through an ideal low-pass filter whose bandwidth B, as before, satisfies the condition $W < B < f_s - W$. We conclude, therefore, that the use of sampling pulses of finite duration has no important effects on the sampling process.

Consider the case when $TA = 1$, so that each rectangular pulse of the sampling function $c(t)$ has unit area. Then, comparing Eq. 4.55 with Eq. 4.6, we see that as the pulse duration T approaches zero, $S(f)$ approaches $G_\delta(f)$. In other words, the discrete-time signal $g_\delta(t)$ produced by ideal sampling represents the limiting form of the sampled signal $s(t)$ as T approaches zero, with $TA = 1$.

(2) Flat-top Samples

Consider next the situation where the analog signal $g(t)$ is sampled instantaneously at the rate $f_s = 1/T_s$, and that the duration of each sample is lengthened to T, as illustrated in Fig. 4.14. (A practical reason for intentionally lengthening the duration of each sample is given later in the section.)

Using $s(t)$ to denote the sequence of flat-top pulses generated in this way, we may write

$$s(t) = \sum_{n=-\infty}^{\infty} g(nT_s) h(t - nT_s) \tag{4.56}$$

The $h(t)$ is a rectangular pulse of unit amplitude and duration T, defined as follows (see Fig. 4.15a):

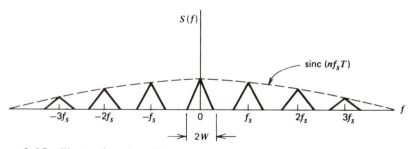

Figure 4.13 Illustrating the effect of using ordinary pulses of finite duration on the spectrum of a sampled signal.

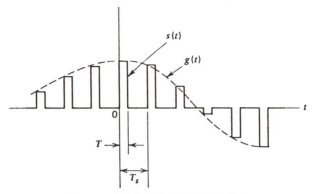

Figure 4.14   Flat-top samples.

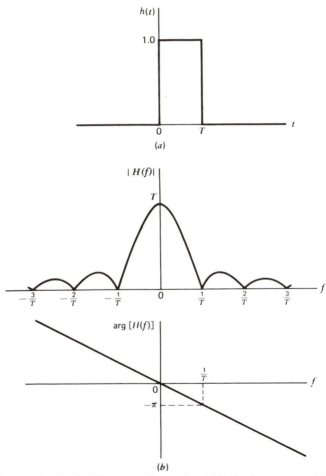

Figure 4.15   (a) Rectangular pulse $h(t)$. (b) Spectrum $H(f)$.

$$h(t) = \begin{cases} 1 & 0 < t < T \\ 0 & t < 0, \text{ and } t > T \end{cases}$$

$$= \text{rect}\left(\frac{t}{T} - \frac{1}{2}\right) \tag{4.57}$$

From Eq. 4.1 the discrete-time signal $g_\delta(t)$, obtained by instantaneously sampling $g(t)$, is given by

$$g_\delta(t) = \sum_{n=-\infty}^{\infty} g(nT_s)\delta(t - nT_s) \tag{4.58}$$

Convolving $g_\delta(t)$ with the pulse $h(t)$, we get

$$g_\delta(t) \star h(t) = \int_{-\infty}^{\infty} g_\delta(\tau)h(t - \tau)d\tau$$

$$= \int_{-\infty}^{\infty} \sum_{n=-\infty}^{\infty} g(nT_s)\delta(\tau - nT_s)h(t - \tau)d\tau$$

$$= \sum_{n=-\infty}^{\infty} g(nT_s) \int_{-\infty}^{\infty} \delta(\tau - nT_s)h(t - \tau)d\tau \tag{4.59}$$

From the sifting property of the delta function, we have

$$g_\delta(t) \star h(t) = \sum_{n=-\infty}^{\infty} g(nT_s)h(t - nT_s) \tag{4.60}$$

Therefore, from Eqs. 4.56 and 4.60 it follows that $s(t)$ is mathematically equivalent to the convolution of $g_\delta(t)$, the instantaneously sampled version of $g(t)$, and the pulse $h(t)$, as shown by

$$s(t) = g_\delta(t) \star h(t) \tag{4.61}$$

Taking the Fourier transform of both sides of Eq. 4.61 and recognizing that the convolution of two time functions is transformed into the multiplication of their respective Fourier transforms, we get

$$S(f) = G_\delta(f)H(f) \tag{4.62}$$

where $S(f) = F[s(t)]$, $G_\delta(f) = F[g_\delta(t)]$, and $H(f) = F[h(t)]$. Therefore, substitution of Eq. 4.6 into Eq. 4.62 yields

$$S(f) = f_s \sum_{m=-\infty}^{\infty} G\left(f - mf_s\right)H(f) \tag{4.63}$$

where $G(f) = F[g(t)]$.

Finally, suppose that $g(t)$ is strictly band-limited and that the sampling rate f_s is greater than the Nyquist rate. Then, passing $s(t)$ through a low-pass reconstruction filter, we find that the spectrum of the resulting filter output is equal to $G(f)H(f)$. This is equivalent to passing the original analog signal $g(t)$ through a low-pass filter of transfer function $H(f)$.

From Eq. 4.57 we find that

$$H(f) = T \text{ sinc}(fT)\exp(-j\pi fT) \tag{4.64}$$

which is shown plotted in Fig. 4.15b. Hence, we see that by using flat-top samples, we have introduced *amplitude distortion* as well as a *delay* of $T/2$. This effect is similar to the variation in transmission with frequency that is caused by the finite size of the scanning aperture in television and facsimile. Accordingly, the distortion caused by lengthening the samples, as in Fig. 4.14, is referred to as the *aperture effect*.

This distortion may be corrected by connecting an *equalizer* in cascade with the low-pass reconstruction filter. The equalizer has the effect of decreasing the in-band loss of the reconstruction filter as the frequency increases in such a manner as to compensate for the aperture effect. Ideally, the amplitude response of the equalizer is given by

$$\frac{1}{|H(f)|} = \frac{1}{T \text{ sinc}(fT)} = \frac{1}{T} \frac{\pi fT}{\sin(\pi fT)}$$

The amount of equalization needed in practice is usually small.

EXAMPLE 3

At the frequency $f = 0.5f_s$, which corresponds to the highest frequency component of the message signal for a sampling rate equal to the Nyquist rate, we find from Eq. 4.64 that the amplitude response of the equalizer normalized to that at zero frequency is equal to

$$\frac{1}{\text{sinc}(0.5f_sT)} = \frac{1}{\text{sinc}(0.5T/T_s)} = \frac{(\pi/2)(T/T_s)}{\sin[(\pi/2)(T/T_s)]}$$

where the ratio T/T_s is equal to the duty cycle of the sampling pulses. In Fig. 4.16 this

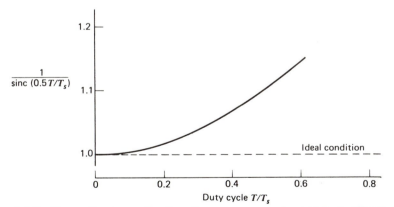

Figure 4.16   Normalized equalization (to compensate for aperture effect) plotted versus T/T_s.

result is plotted as a function of T/T_s. Ideally, it should be equal to one for all values of T/T_s. For a duty cycle of 10 percent, it is equal to 1.0041. It follows therefore that for duty cycles of less than 10 percent, the aperture effect becomes negligible, and the need for equalization may be omitted altogether.

(3) Sample-and-hold Circuit for Signal Recovery

In practice, the sampling process takes the form of natural sampling or flat-top sampling. In both cases, the spectrum of the sampled signal is scaled by the ratio T/T_s, where T is the sampling-pulse duration and T_s is the sampling period; see Eq. 4.55 for natural sampling and Eqs. 4.63 and 4.64 for flat-top sampling. Typically, this ratio is quite small, with the result that the signal power at the output of the low-pass reconstruction filter in the receiver is correspondingly small. We may obviously remedy this situation by the use of amplification, which can be quite large. A more attractive approach, however, is to use a simple *sample-and-hold* circuit.* The circuit, shown in Fig. 4.17a, consists of an amplifier of unity gain and low output impedance, a switch, and a capacitor; it is assumed that the load impedance is large. The switch is timed to close only for the small duration T of each sampling pulse, during which time the capacitor rapidly charges up to a voltage level equal to that of the input sample. When the

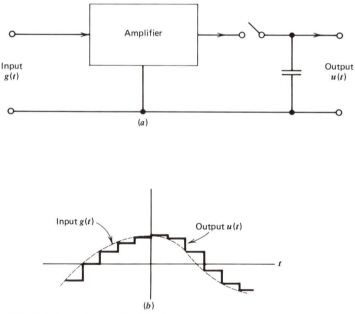

Figure 4.17 (a) Sample-and-hold circuit. (b) Idealized output waveform of the circuit.

---

* In the control literature, the sample-and-hold operation is referred to as the *zero-order hold* (Franklin and Powell, 1980, pp. 83–85).

switch is open, the capacitor retains its voltage level until the next closure of the switch. Thus the sample-and-hold circuit, in its ideal form, produces an output waveform that represents a *staircase interpolation* of the original analog signal, as illustrated in Fig. 4.17*b*.

From the previous discussion on flat-top sampling, we deduce that the output of a sample-and-hold circuit is defined by

$$u(t) = \sum_{n=-\infty}^{\infty} g(nT_s)h(t - nT_s) \tag{4.65}$$

where $h(t)$ is an impulse response representing the action of the sample-and-hold circuit; that is

$$h(t) = \begin{cases} 1 & 0 < t < T_s \\ 0 & t < 0, \text{ and } t > T_s \end{cases} \tag{4.66}$$

Correspondingly, the spectrum of the output of the sample-and-hold circuit is given by

$$U(f) = f_s \sum_{m=-\infty}^{\infty} H(f) \, G(f - mf_s) \tag{4.67}$$

where $G(f)$ is the Fourier transform of the original analog signal $g(t)$, and $H(f)$ is the transfer function of the sample-and-hold circuit, as shown by

$$H(f) = T_s \, \text{sinc}(fT_s)\exp(-j\pi fT_s) \tag{4.68}$$

In order to recover the original signal $g(t)$ without distortion, we need to pass the output of the sample-and-hold circuit through a low-pass filter designed to remove components of the spectrum $U(f)$ at multiples of the sampling rate f_s and an equalizer whose amplitude response equals $1/|H(f)|$. These operations are illustrated by the block diagram shown in Fig. 4.18.

4.6   PULSE-AMPLITUDE MODULATION

A discussion of the sampling process naturally leads to a type of pulse modulation known as pulse-amplitude modulation. In *pulse-amplitude modulation (PAM)*, *the amplitude of a carrier consisting of a periodic train of rectangular pulses is varied in proportion to sample values of a message signal.* In this type of modulation, the pulse duration is held *constant.** By making the amplitude of each rectangular pulse the same as the value of the message signal at the

---

* Pulse-amplitude modulation is a special form of *analog pulse modulation*. There are two other basic types of pulse modulation, namely, pulse-duration modulation and pulse-position modulation. In *pulse-duration modulation (PDM)*, the pulse amplitude is kept constant, but the pulse duration is varied in proportion to sample values of the message signal. In *pulse-position modulation (PPM)*, both the pulse amplitude and pulse duration are held constant, but the pulse position is varied in proportion to sample values of the message signal. For detailed treatment of pulse modulation, see Black (1953, pp. 237–298), Carlson (1986, Section 10.2), and Schwartz, Bennett, and Stein (1966, Chapter 6).

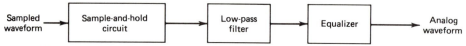

Figure 4.18 Components of a scheme for signal reconstruction.

leading edge of the pulse, we find that the version of PAM so defined is exactly the same as *flat-top sampling*. Indeed, we readily see this interpretation of pulse-amplitude modulation by looking at the waveform depicted in Fig. 4.14. Accordingly, the theory of flat-top sampling, described in Section 4.5, applies equally well to PAM.

What about the transmission bandwidth requirement of PAM? According to the definition given before in terms of rectangular pulses, we would require a very wide band of frequencies to transmit PAM. However, this need not be so if we were to formulate the definition of PAM in terms of a *standard* pulse, which the system is capable of transmitting. Let $v(t)$ denote such a pulse. We may then define a PAM wave, $s(t)$, as follows

$$s(t) = \sum_{n=-\infty}^{\infty} g(nT_s)v(t - nT_s) \tag{4.69}$$

where the $g(nT_s)$ are sample values of the message (modulating) signal $g(t)$, and T_s is the sampling period. Thus, the definition of PAM given in terms of rectangular pulses may be viewed as a special case of Eq. 4.69 by setting $v(t) = h(t)$, where $h(t)$ is itself defined by Eq. 4.57.

We may satisfy the transmission bandwidth requirement by finding a suitable pulse shape for $v(t)$. The issue of pulse shaping for the efficient transmission of PAM waves over band-limited channels is considered in Chapter 6.

4.7 TIME-DIVISION MULTIPLEXING

An important feature of pulse-amplitude modulation is a *conservation of time*. That is, for a given message signal, transmission of the associated PAM wave engages the communication channel for only a fraction of the sampling interval on a periodic basis. Hence, some of the time interval between adjacent pulses of the PAM wave is cleared for use by other independent message signals on a *time-shared basis*. By so doing, we obtain a *time-division multiplex system* (TDM), which enables the joint utilization of a common channel by a plurality of independent message signals without mutual interference.

The concept of TDM is illustrated by the block diagram shown in Fig. 4.19. Each input message signal is first restricted in bandwidth by a low-pass pre-alias filter to remove the frequencies that are nonessential to an adequate signal representation. The pre-alias filter outputs are then applied to a *commutator*, which is usually implemented using electronic switching circuitry. The function of the commutator is two-fold: (1) to take a narrow sample of each of the N input messages at a rate f_s that is slightly higher than $2W$, where W is the cutoff frequency of the pre-alias filter, and (2) to sequentially interleave these N samples inside a sampling interval $T_s = 1/f_s$. Indeed, this latter function is the

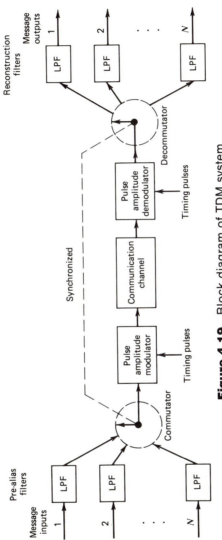

Figure 4.19 Block diagram of TDM system.

essence of the time-division multiplexing operation. Following the commutation process, the multiplexed signal is applied to a *pulse-amplitude modulator*, the purpose of which is to transform the multiplexed signal into a form suitable for transmission over the communication channel.

Suppose that the N message signals to be multiplexed have similar spectral properties. Then the sampling rate for each message signal is determined in accordance with the sampling theorem. Let T_s denote the sampling period so determined for each message signal. Let T_x denote the time spacing between adjacent samples in the time-multiplexed signal. It is rather obvious that we may set

$$T_x = \frac{T_s}{N} \tag{4.70}$$

Hence, the use of time-division multiplexing introduces a *bandwidth expansion factor N*, because the scheme must squeeze N samples derived from N independent message signals into a time slot equal to one sampling interval.

At the receiving end of the system, the received signal is applied to a *pulse-amplitude demodulator*, which performs the reverse operation of the pulse-amplitude modulator. The short pulses produced at the pulse demodulator output are distributed to the appropriate low-pass reconstruction filters by means of a *decommutator*, which operates in *synchronism* with the commutator in the transmitter. This synchronization is essential for a satisfactory operation of the TDM system, and provisions have to be made for it. The issue of synchronization is considered in the next chapter in the context of the T1 carrier, a 24-channel multiplexer used in digital telephony.

EXAMPLE 4

The waveforms shown in Fig. 4.20 illustrate the operation of a TDM system for $N = 2$. The dashed curves represent the waveforms of two message signals $g_1(t)$ and $g_2(t)$. The PAM waves corresponding to these two message signals are depicted as sequences of uniformly spaced rectangular pulses. The PAM wave corresponding to $g_1(t)$ is shown shaded, while the PAM wave corresponding to $g_2(t)$ is shown unmarked. The TDM wave is represented by the superposition of these two PAM waves.

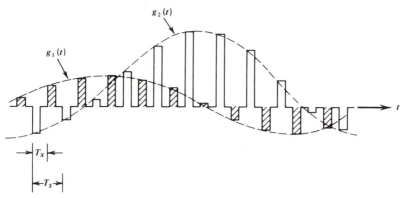

Figure 4.20 Waveforms illustrating TDM for two message signals.

The TDM system is highly sensitive to dispersion in the common channel, that is, to variations of amplitude with frequency or lack of proportionality of phase with frequency. Accordingly, accurate equalization of both the amplitude and pulse responses of the channel is necessary to ensure a satisfactory operation of the system; by "equalization" we mean the process of correcting channel-induced dispersion. To a first approximation, however, TDM is immune to amplitude nonlinearities in the channel as a source of *crosstalk*, because the different message signals are not simultaneously impressed on the channel. The term "crosstalk" refers to the signal from an adjacent channel spilling over into a desired time slot.

4.8 SUMMARY AND DISCUSSION

The *sampling theorem* is basic to the interchangeability of analog signals and digital sequences. In the context of low-pass signals whose energy is finite and whose spectrum is strictly band-limited, we may state the sampling theorem in two separate parts

1. The signal is completely described by specifying its samples taken at instants of time separated by $1/2W$ seconds.
2. The signal may be completely recovered from a knowledge of its samples taken at the rate of $2W$ samples per second.

The parameter W denotes the highest frequency component of the signal. The parameter $2W$ is called the *Nyquist rate*, and its reciprocal $1/2W$ is called the *Nyquist interval*.

Part 1 of the sampling theorem represents the basis for the first step in *analog-to-digital conversion*. It is applied in the *transmitter* of a digital communication system for the purpose of sending an analog signal from one point to another. Part 2, on the other hand, represents the basis for the last step in *digital-to-analog conversion*. It is applied in the *receiver* of the system for the purpose of recovering the original analog signal.

There is a third part to the sampling theorem. Any sequence of values $g(n/2W)$ can be interpolated to a band-limited function if, and only if,

$$\sum_{n=1}^{\infty} \frac{|g(n/2W)| + |g(-n/2W)|}{n} < \infty$$

This is known as the *absolute convergence principle*.

Two important phenomena arise in the practical application of the sampling theorem, which require careful attention:

1. *Aliasing* or *foldover*, which refers to some portions of the frequency-shifted replicas of the spectrum of the analog signal folding over inside the desired spectrum. Aliasing arises because an analog signal cannot be limited in both time and frequency simultaneously. The recipe for controlling its effect involves the use of a *pre-alias filter*, combined with the choice of a sampling rate in excess of the Nyquist rate.

2. *Aperture effect*, which results from the use of flat-top pulses to perform the sampling operation. The distortion due to the aperture effect is compensated for by cascading an *amplitude equalizer* with the reconstruction filter in the receiver.

Practical considerations also manifest themselves in other ways in matters relating to the sampling theorem. In particular, a cost-effective method for distortionless signal recovery consists of the cascade connection of three components: sample-and-hold-circuit, low-pass reconstruction filter, and amplitude equalizer. The inclusion of the *sample-and-hold-circuit* accounts for the low duty ratio of the periodic sequence of sampling pulses. The equalizer compensates for the amplitude distortion produced by the sample-and-hold-circuit.

In *pulse-amplitude modulation* (PAM), the amplitude of a periodic train of pulses (of fixed duration) is varied in proportion to a message signal. When a rectangular pulse is used in this modulation process, the resulting PAM wave is the same as that obtained from flat-top sampling.

Moreover, *time-division multiplexing* (representing a natural extension of pulse-amplitude modulation) provides an effective method for *sharing* a communication channel (representing a highly valuable resource) among multiple analog sources. Basically, this form of multiplexing involves the use of a *commutator* in the transmitter and a *decommutator* (the inverse of the commutator) in the receiver.

Two other issues of theoretical interest, albeit pertaining to different aspects of the sampling process, were also covered in the chapter:

1. *Quadrature sampling of a band-pass signal*, which involves sampling of the low-pass in-phase and quadrature components of the band-pass signal in accordance with the sampling theorem.
2. *Sampling of a message process*, which introduces a *mean-squared error* (between the message process and its reconstructed version) that is zero when the process is strictly band-limited and the sampling rate is twice the highest frequency component of the process.

In this chapter, we have taken an in-depth look at the sampling process and its various ramifications, from theoretical and practical viewpoints. We are thus prepared to take on a study of the remaining processes in analog-to-digital conversion, which we do in the next chapter.

PROBLEMS

$$\sin u \sin v = \frac{1}{2}[\cos(u-v) - \cos(u+v)]$$
$$\cos u \cos v = \frac{1}{2}[\cos(u-v) + \cos(u+v)]$$
$$\sin u \sin v = \frac{1}{2}[\sin(u+v) + \sin(u-v)]$$
$$\cos u \sin v = \frac{1}{2}[\sin(u+v) - \sin(u-v)]$$

P4.1   SAMPLING THEOREM

Problem 4.1.1

(a) Consider the cosine wave

$$g(t) = A \cos(2\pi f_0 t)$$

Plot the spectrum of the discrete-time signal $g_\delta(t)$ derived by sampling $g(t)$ at the

times $t_n = n/f_s$, where $n = 0, \pm 1, \pm 2, \ldots$, and
(i) $f_s = f_0$
(ii) $f_s = 2f_0$
(iii) $f_s = 3f_0$

(b) Repeat your calculations for the sine wave

$$g(t) = A \sin(2\pi f_0 t)$$

Comment on your results.

Problem 4.1.2 The signal

$$g(t) = 10 \cos(20\pi t)\cos(200\pi t)$$

is sampled at the rate of 250 samples per second.

(a) Determine the spectrum of the resulting sampled signal.
(b) Specify the cutoff frequency of the ideal reconstruction filter so as to recover $g(t)$ from its sampled version.
(c) What is the Nyquist rate for $g(t)$?

Problem 4.1.3 A signal $g(t)$ consists of two frequency components $f_1 = 3.9$ kHz and $f_2 = 4.1$ kHz in such a relationship that they just cancel each other out when the signal $g(t)$ is sampled at the instants $t = 0, T, 2T, \ldots$, where $T = 125 \ \mu s$. The signal $g(t)$ is defined by

$$g(t) = \cos\left(2\pi f_1 t + \frac{\pi}{2}\right) + A \cos(2\pi f_2 t + \phi)$$

Find the values of amplitude A and phase ϕ of the second frequency component.

Problem 4.1.4 Let E denote the energy of a strictly band-limited signal $g(t)$. Show that E may be expressed in terms of the sample values of $g(t)$, taken at the Nyquist rate, as follows:

$$E = \frac{1}{2W} \sum_{n=-\infty}^{\infty} \left| g\left(\frac{n}{2W}\right) \right|^2$$

where W is the highest frequency component of $g(t)$.

Problem 4.1.5 Consider a continuous-time signal $g(t)$ of finite energy, with a continuous spectrum $G(f)$. Assume that $G(f)$ is sampled uniformly at the discrete frequencies $f = kF_s$, thereby obtaining the sequence of frequency samples $G(kF_s)$, where k is an integer in the entire range $-\infty < k < \infty$, and F_s is the frequency sampling interval. Show that if $g(t)$ is duration-limited, so that it is zero outside the interval $-T < t < T$, then the signal is completely defined by specifying $G(f)$ at frequencies spaced $1/2T$ hertz apart.

Problem 4.1.6 Consider a sequence of samples $x(nT_s)$ obtained by sampling a continuous-time signal $x(t)$ at the rate $f_s = 1/T_s$. It is required to increase the sampling period to a new value $T_s' = MT_s$, where M is an integer. Such a process is called *decimation*. Let $y(nT_s')$ denote the new sequence with a sampling period T_s'.

(a) Find the relationship between the Fourier transforms of the new sequence $y(nT_s')$ and the original sequence $x(nT_s)$.
(b) What is the condition required to avoid the occurrence of aliasing in the decimation process?

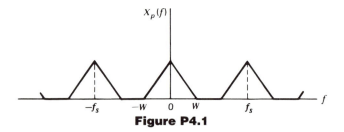

Figure P4.1

(c)   For a sequence $x(nT_s)$ with the spectrum shown in Fig. P4.1, find the spectrum of $y(nT_s')$, assuming that $1/4W < T_s < 1/2W$, and $M = 2$. How would you avoid the occurrence of aliasing in this example?

Problem 4.1.7   In the *interpolation* process, which is the dual of the decimation process, we increase the sampling rate by an integer ratio. Consider a sequence $x(nT_s)$, the sampling period of which is to be reduced to a new value $T_s' = T_s/L$, where L is an integer. This requirement is achieved by using the arrangement shown in Fig. P4.2a, where we first insert $(L - 1)$ zero-valued samples between each sample of $x(nT_s)$, and then pass the resulting sequence $v(nT_s')$ through a low-pass filter with a periodic frequency response, as in Fig. P4.2b.

(a)   Determine the relation between the Fourier transform of $v(nT_s')$ and that of $x(nT_s)$.
(b)   Specify the zero-frequency response $H(0)$ and bandwidth B of the low-pass filter so that the output sequence $y(nT_s')$ has a spectrum that is periodic with period f_s' equal to $1/T_s'$, and a dc value equal to that of the input sequence $x(nT_s)$.

P4.2   QUADRATURE SAMPLING OF BAND-PASS SIGNALS

Problem 4.2.1   The spectrum of a band-pass signal occupies a band of width 0.5 kHz, centered around ± 10 kHz. Find the Nyquist rate for quadrature sampling the in-phase and quadrature components of the signal.

Problem 4.2.2   A discrete-time signal, obtained by uniformly sampling a band-pass signal, has a periodic spectrum. This property offers the possibility of recovering not

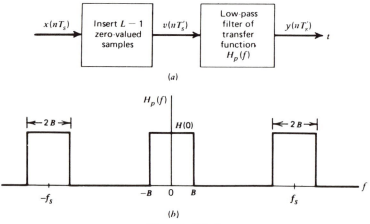

(a)

(b)

Figure P4.2

only the original signal but also frequency-translated versions of it. Demonstrate such an operation.

P4.3   RECONSTRUCTION OF A MESSAGE PROCESS FROM ITS SAMPLES

Problem 4.3.1   The power spectral density of a message process $X(t)$ is strictly band-limited to 3.4 kHz.
(a)   Find the minimum sampling rate for exact reconstruction of the process $X(t)$ from its samples to be attainable.
(b)   What is the corresponding value of the sampling period?
(c)   Specify the impulse response of the reconstruction filter.

P4.4   DISTORTION IN SAMPLING

Problem 4.4.1   The signals

$$g_1(t) = 10 \cos(100\pi t)$$

and

$$g_2(t) = 10 \cos(50\pi t)$$

are both sampled at times $t_n = n/f_s$, where $n = 0, \pm 1, \pm 2, \ldots$, and $f_s = 75$ samples per second. Show that the two sequences of samples thus obtained are identical. What is this phenomenon called?

Problem 4.4.2   Figure P4.3 shows the spectrum of a low-pass signal $g(t)$. The signal is sampled at the rate of 1.5 Hz, and then applied to a low-pass reconstruction filter with cutoff frequency at 1 Hz. Plot the spectrum of the resulting signal.

Problem 4.4.3   The spectrum of a signal $g(t)$ is defined by the Fourier transform

$$G(f) = \frac{1}{\sqrt{1 + f^2}}$$

(a)   Plot the spectrum of the discrete-time signal $g_\delta(t)$ derived from $g(t)$ by sampling it at the rate $f_s = 2.5$.
(b)   The signal $g(t)$ is passed through a pre-alias filter prior to sampling. The pre-alias filter consists of a Butterworth filter of order 4 whose amplitude response is defined by

$$H(f) = \frac{1}{\sqrt{1 + f^8}}$$

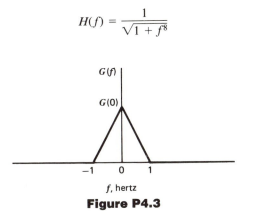

Figure P4.3

Plot the spectrum of the discrete-time signal obtained by sampling the pre-alias filter output at the rate $f_s = 2.5$.

Comment on your results.

Problem 4.4.4 Calculate the signal-to-distortion ratios for cases (a) and (b) specified in Problem 4.4.3.

P4.5 PRACTICAL ASPECTS OF SAMPLING AND SIGNAL RECOVERY

Problem 4.5.1 The spectrum of a signal $g(t)$ is shown in Fig. P4.4. This signal is sampled with a periodic train of rectangular pulses of duration 50/3 milliseconds. Plot the spectrum of the sampled signal for frequencies up to 50 hertz for the following two conditions:

(a) The sampling rate is equal to the Nyquist rate.
(b) The sampling rate is equal to 10 samples per second.

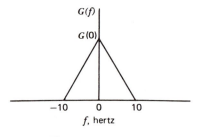

Figure P4.4

Problem 4.5.2 This problem is aimed at investigating the fact that practical electronic switching circuits will not produce a sampling function that consists of exactly rectangular pulses. Let $h(t)$ denote some arbitrary pulse shape, so that the sampling function $c(t)$ may be expressed as

$$c(t) = \sum_{n=-\infty}^{\infty} h(t - nT_s)$$

where T_s is the sampling period. The sampled version of an incoming analog signal $g(t)$ is defined by

$$s(t) = c(t)g(t)$$

(a) Show that the Fourier transform of $s(t)$ is given by

$$S(f) = f_s \sum_{n=-\infty}^{\infty} G(f - nf_s)H(nf_s)$$

where $G(f) = F[g(t)]$, $H(f) = F[h(t)]$, and $f_s = 1/T_s$.
(b) What is the effect of using the arbitrary pulse shape $h(t)$?

P4.6 PULSE-AMPLITUDE MODULATION

Problem 4.6.1 A sine wave, of amplitude 10 volts and frequency 4 hertz, is applied to a pulse-amplitude modulator. The modulator has the following parameters:

Sampling rate = 10 samples per second

Pulse duration = 25 milliseconds

Also, it samples the input at times $t = 0, \pm T_s, \pm 2T_s, \ldots$, where T_s is the sampling period.

Plot two full cycles of the modulator output.

P4.7 TIME-DIVISION MULTIPLEXING

Problem 4.7.1 Two signals $g_1(t)$ and $g_2(t)$ are to be transmitted over a common channel by means of time-division multiplexing. The highest frequency of $g_1(t)$ is 1 kHz, and that of $g_2(t)$ is 1.5 kHz. What is the minimum value of the permissible sampling rate? Justify your answer.

Problem 4.7.2 Six independent message sources of bandwidths W, W, $2W$, $2W$, $3W$, and $3W$ hertz are to be transmitted on a time-division multiplexed basis using a common communication channel.

(a) Set up a scheme for accomplishing this multiplexing requirement, with each message signal sampled at its Nyquist rate.
(b) Determine the minimum transmission bandwidth of the channel.

Problem 4.7.3 Twenty-four voice signals are sampled uniformly and then time-division multiplexed. The sampling operation uses flat-top samples with 1 microsecond duration. The multiplexing operation includes provision for synchronization by adding an extra pulse of sufficient amplitude and also 1 microsecond duration. The highest frequency component of each voice signal is 3.4 kHz.

(a) Assuming a sampling rate of 8 kHz, calculate the spacing between successive pulses of the multiplexed signal.
(b) Repeat your calculation assuming the use of Nyquist rate sampling.

CHAPTER FIVE

WAVEFORM CODING TECHNIQUES

Having sampled an analog message signal, the next step in its digital transmission is the generation of a *coded version* (*digital representation*) of the signal. *Pulse-code modulation* (PCM) provides one method for accomplishing such a requirement. In this method of signal coding, the message signal is sampled and the amplitude of each sample is *rounded off* (*approximated*) to the nearest one of a finite set of discrete levels, so that *both time and amplitude are represented in discrete form*. This allows the message to be transmitted by means of a digital (coded) waveform, thereby distinguishing pulse-code modulation from all analog modulation techniques.

In conceptual terms, pulse-code modulation is simple to understand. Moreover, it was the first method to be developed for the digital coding of waveforms. Indeed, it is the most applied of all digital coding systems in use today. Accordingly, pulse-code modulation is widely accepted as the *standard* against which other digital coders of analog signals are calibrated in performance.

The use of digital representation of analog signals (e.g., voice, video) offers the following advantages: (1) ruggedness to transmission noise and interference, (2) efficient regeneration of the coded signal along the transmission path, (3) the potential for communication privacy and security through use of *encryption,* and (4) the possibility of a uniform format for different kinds of baseband signals. These advantages, however, are attained at the cost of increased transmission bandwidth requirement and increased system complexity. With the increasing availability of wide-band communication channels, coupled with the emergence of the requisite device technology, the use of PCM has indeed become a practical reality.

PCM belongs to a class of signal coders known as *waveform coders,* in which an analog signal is *approximated* by mimicking the amplitude-versus-time waveform; hence, the name. *Waveform coders are* (*in principle*) *designed to be signal-independent.* As such, they are basically different from *source coders* (e.g., linear predictive vocoders considered in Chapter 3), which rely on a parameterization of the analog signal in accordance with an appropriate *model* for the generation of the signal.

In this chapter, we study various types of waveform coders. We begin the chapter by discussing the basic elements of pulse-code modulation.

5.1 PULSE-CODE MODULATION

Pulse-code modulation systems are complex in that the message signal is subjected to a large number of operations. The essential operations in the transmitter of a PCM system are *sampling, quantizing,* and *encoding,* as shown in the top part of Fig. 5.1. The sampling, quantizing, and encoding operations are usually performed in the same circuit, which is called an *analog-to-digital converter*. *Regeneration* of impaired signals occurs at intermediate points along the transmission path (channel) as indicated in the middle part of Fig. 5.1. At the receiver, the essential operations consist of one last stage of regeneration followed by *decoding,* then *demodulation* of the train of quantized samples, as in the bottom part of Fig. 5.1. The operations of decoding and reconstruction are usually performed in the same circuit, called a *digital-to-analog converter*. When time-division multiplexing is used, it becomes necessary to synchronize the receiver to the transmitter for the overall system to operate satisfactorily.

It is noteworthy that pulse-code modulation is not modulation in the conventional sense. The term "modulation" usually refers to the variation of some characteristic of a carrier wave in accordance with an information-bearing signal. The only part of pulse-code modulation that conforms to this definition is sampling. The subsequent use of quantization, which is basic to pulse-code modulation, introduces a signal distortion that has no counterpart in conventional modulation.

In the sequel, the basic signal-processing operations involved in PCM are considered, one by one.

(1) Sampling

The incoming message wave is sampled with a train of narrow rectangular pulses so as to closely approximate the instantaneous sampling process. In

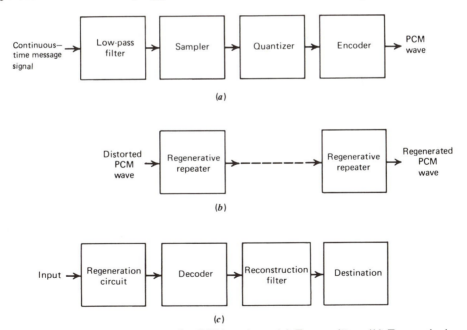

Figure 5.1 Basic elements of a PCM system. (*a*) Transmitter. (*b*) Transmission path. (*c*) Receiver.

order to ensure perfect reconstruction of the message at the receiver, the sampling rate must be greater than twice the highest frequency component W of the message wave (in accordance with the sampling theorem). In practice, a low-pass pre-alias filter is used at the front end of the sampler in order to exclude frequencies greater than W before sampling. Thus, the application of sampling permits the reduction of the continuously varying message wave to a limited number of discrete values per second.

(2)   Quantizing

An analog signal, such as voice, has a continuous range of amplitudes and therefore its samples cover a continuous amplitude range. In other words, within the finite amplitude range of the signal we find an infinite number of amplitude levels. However, it is not necessary in fact to transmit the exact amplitudes of the samples. Any human sense (the ear or the eye), as ultimate receiver, can detect only finite intensity differences. This means that the original analog signal may be approximated by a signal constructed of discrete amplitudes (selected on a minimum error basis from an available set). The existence of a finite number of discrete amplitude levels is a basic condition of PCM. Clearly, if we assign the discrete amplitude levels with sufficiently close spacing, we may make the approximated signal practically indistinguishable from the original analog signal.

The conversion of an analog (continuous) sample of the signal into a digital (discrete) form is called the *quantizing* process. Graphically, the quantizing process means that a straight line representing the relation between the input and the output of a linear analog system is replaced by a transfer characteristic that is staircase-like in appearance. Figure 5.2a depicts one such characteristic. The quantizing process has a *two-fold effect:* (1) the peak-to-peak range of input sample values is subdivided into a finite set of *decision levels* or *decision thresholds* that are aligned with the "risers" of the staircase, and (2) the output is assigned a discrete value selected from a finite set of *representation levels* or *reconstruction values* that are aligned with the "treads" of the staircase. In the case of a *uniform quantizer,* characterized as in Fig. 5.2a, the separation between the decision thresholds and the separation between the representation levels of the quantizer have a common value called the *step size.*

According to the staircase-like transfer characteristic of Fig. 5.2a, the decision thresholds of the quantizer are located at $\pm\Delta/2, \pm3\Delta/2, \pm5\Delta/2, \ldots$, and the representation levels are located at $0, \pm\Delta, \pm2\Delta, \ldots$, where Δ is the step size. A uniform quantizer characterized in this way is referred to as a *symmetric quantizer of the midtread type,* because the origin lies in the middle of a tread of the staircase.

Figure 5.3a shows another staircase-like transfer characteristic, in which the decision thresholds of the quantizer are located at $0, \pm\Delta, \pm2\Delta, \ldots$, and the representation levels are located at $\pm\Delta/2, \pm3\Delta/2, \pm5\Delta/2, \ldots$, where Δ is again the step size. A uniform quantizer having this second characteristic is referred to as a *symmetric quantizer of the midriser type,* because in this case the origin lies in the middle of a riser of the staircase.

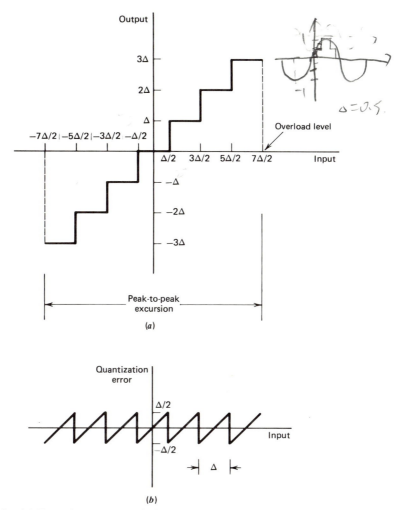

Figure 5.2   (a) Transfer characteristic of quantizer of midtread type. (b) Variation of the quantization error with input.

A quantizer of the midtread or midriser type, as defined, is *memoryless* in that the quantizer output is determined only by the value of a corresponding input sample, independently of earlier (or later) analog samples applied to the input.* The memoryless quantizer is the simplest and most often used quantizer.

In the transfer characteristics of Figs. 5.2a and 5.3a, we have included a parameter labeled the *overload level,* the absolute value of which is one half of the peak-to-peak range of input sample values. Moreover, the number of inter-

---

* A memoryless quantizer is inefficient if the input samples are statistically dependent; such dependencies would have to be removed either prior to quantizing or as part of the quantizing process. For a discussion of this issue, see Jayant and Noll (1984, Chapters 6 and 9).

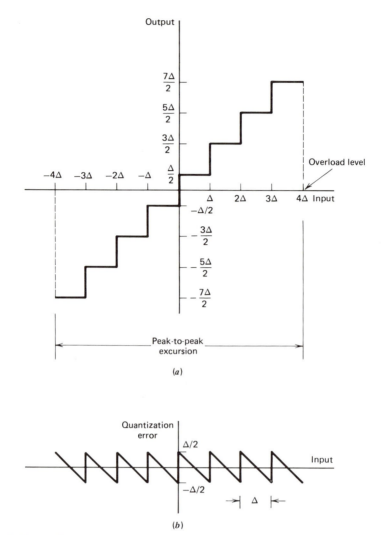

Figure 5.3 (a) Transfer characteristic of quantizer of midriser type. (b) Variation of the quantization error with input.

vals into which the peak-to-peak excursion is divided, or equivalently the number of representation levels, is equal to twice the absolute value of the overload level divided by the step size. Accordingly, for an analog input sample that lies anywhere inside an interval of either transfer characteristic, the quantizer produces a discrete output equal to the midvalue of the pair of decision thresholds in question. In so doing, however, a *quantization error* is introduced, the value of which equals the difference between the output and input values of the quantizer. Figures 5.2b and 5.3b show the variations of the quantization error with the input for the two uniform quantizer types. In both cases, we see that the maximum instantaneous value of this error is half of one step size, and the total range of variation is from minus half a step to plus half a step.

(3) Encoding

In combining the processes of sampling and quantizing, the specification of a continuous baseband signal becomes limited to a discrete set of values, but not in the form best suited for transmission over a line, radio path, or optical fiber. To exploit the advantages of sampling and quantizing, we require the use of an *encoding process* to translate the discrete set of sample values to a more appropriate form of signal. Any plan for representing each member of this discrete set of values as a particular arrangement of discrete events is called a *code*. One of the discrete events in a code is called a *code element* or *symbol*. For example, the presence or absence of a pulse is a symbol. A particular arrangement of symbols used in a code to represent a single value of the discrete set is called a *code-word* or *character*.

In a *binary code*, each symbol may be either of two distinct values or kinds, such as the presence or absence of a pulse. The two symbols of a binary code are customarily denoted as 0 and 1. In a *ternary code*, each symbol may be one of three distinct values or kinds, and so on for other codes. However, the maximum advantage over the effects of noise in a transmission medium is obtained by using a binary code, because a binary symbol withstands a relatively high level of noise and is easy to regenerate.

Suppose that, in a binary code, each code-word consists of n bits. Then, using such a code, we may represent a total of 2^n distinct numbers. For example, a sample quantized into one of $2^4 = 16$ levels may be represented by a 4-bit code-word. There are several ways of establishing a one-to-one correspondence between representation levels and code-words. A convenient one is to express the ordinal number of the representation level as a binary number. In the binary number system, each bit has a place-value that is a power of 2, as illustrated in Table 5.1 for the case of $n = 4$.

Table 5.1

Ordinal Number of Representation Level	Level Number Expressed as Sum of Powers of 2	Binary Number
0		0000
1	2^0	0001
2	2^1	0010
3	$2^1 + 2^0$	0011
4	2^2	0100
5	$2^2 \quad + 2^0$	0101
6	$2^2 + 2^1$	0110
7	$2^2 + 2^1 + 2^0$	0111
8	2^3	1000
9	$2^3 \quad + 2^0$	1001
10	$2^3 \quad + 2^1$	1010
11	$2^3 \quad + 2^1 + 2^0$	1011
12	$2^3 + 2^2$	1100
13	$2^3 + 2^2 \quad + 2^0$	1101
14	$2^3 + 2^2 + 2^1$	1110
15	$2^3 + 2^2 + 2^1 + 2^0$	1111

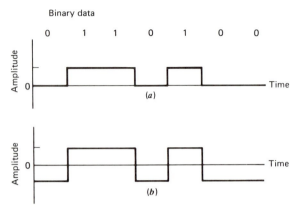

Figure 5.4 Two binary waveforms. (a) Nonreturn-to-zero unipolar. (b) Nonreturn-to-zero polar.

There are several formats (waveforms) for the representation of binary sequences produced by analog-to-digital conversion (or by other sources). Figure 5.4 depicts two such formats.* In Fig. 5.4a, binary symbol 1 is represented by a pulse of constant amplitude for the duration of one bit, and symbol 0 is represented by switching off the pulse for the same duration. This format is called a *nonreturn-to-zero unipolar signal,* or *on–off signal*. In Fig. 5.4b, symbols 1 and 0 are represented by pulses of positive and negative amplitude, respectively, with each pulse occupying one complete bit duration. This second format is called a *nonreturn-to-zero polar signal*.

(4) Regeneration
The most important feature of PCM systems lies in the ability to control the effects of distortion and noise produced by transmitting a PCM wave through a channel. This capability is accomplished by reconstructing the PCM wave by means of a chain of *regenerative repeaters* located at sufficiently close spacing along the transmission route. As illustrated in Fig. 5.5, three basic functions are performed by a regenerative repeater, namely, *equalization, timing,* and *decision making*. The equalizer shapes the received pulses so as to compensate for the effects of amplitude and phase distortions produced by imperfections in the transmission characteristics of the channel. The timing circuit provides a periodic pulse train, derived from the received pulses, for sampling the equalized pulses at the instants of time where the signal-to-noise ratio is a maximum. The decision device is enabled when, at the sampling time determined by the timing circuit, the amplitude of the equalized pulse plus noise exceeds a predetermined voltage level. Thus, for example, in a PCM system with on–off signaling, the repeater makes a decision in each bit interval as to whether or not a pulse is

---

* For a more complete list of formats (waveforms) for the representation of binary data, and a discussion of their characteristics, see Chapter 6. In that chapter, the issue of pulse shaping (and its effect on baseband transmission) of PCM waves is also discussed.

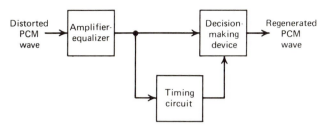

Figure 5.5 Block diagram of a regenerative repeater.

present. If the decision is "yes," a clean new pulse is transmitted to the next repeater. If, on the other hand, the decision is "no," a clean base line is transmitted. In this way, the accumulation of distortion and noise in a repeater span is completely removed, provided that the disturbance is not too large to cause an error in the decision-making process. Ideally, except for delay, the regenerated signal is exactly the same as the signal originally transmitted. In practice, however, the regenerated signal departs from the original signal for two main reasons:

1. The presence of channel noise and interference causes the repeater to make wrong decisions occasionally, thereby introducing *bit errors* into the regenerated signal; we shall have more to say on this issue in Section 5.2.
2. If the spacing between received pulses deviates from its assigned value, a *jitter* is introduced into the regenerated pulse position, thereby causing distortion.

(5) Decoding

The first operation in the receiver is to regenerate (i.e., reshape and clean up) the received pulses. These clean pulses are then regrouped into code-words and decoded (i.e., mapped back) into a quantized PAM signal. The *decoding* process involves generating a pulse the amplitude of which is the linear sum of all the pulses in the code-word, with each pulse weighted by its place-value (2^0, 2^1, 2^2, 2^3, . . .) in the code.

(6) Reconstruction

The final operation in the receiver is to recover the analog signal. This is done by passing the decoder output through a low-pass reconstruction filter whose cutoff frequency is equal to the message bandwidth W. Assuming that the transmission path is error-free, the recovered signal includes no noise with the exception of the initial distortion introduced by the quantization process.

(7) Multiplexing

In applications using PCM, it is natural to multiplex different message sources by time-division, whereby each source keeps its individuality throughout the journey from the transmitter to the receiver. This individuality accounts for the comparative ease with which message sources may be dropped or reinserted in a time-division multiplex system. As the number of independent message sources is increased, the time interval that may be allotted to each source has to be reduced, since all of them must be accommodated into a time interval equal to the reciprocal of the sampling rate. This, in turn, means that the allowable duration of a code-word representing a single sample is reduced. However, pulses tend to become more difficult to generate and to transmit as their duration is reduced. Furthermore, if the pulses become too short, impairments in the transmission medium begin to interfere with the proper operation of the system. Accordingly, in practice, it is necessary to restrict the number of independent message sources that can be included within a time-division group.

(8) Synchronization

For a PCM system with time-division multiplexing to operate satisfactorily, it is necessary that the timing operations at the receiver, except for the time lost in transmission and regenerative repeating, follow closely the corresponding operations at the transmitter. In a general way, this amounts to requiring a local clock at the receiver to keep the same time as a distant standard clock at the transmitter, except that the local clock is somewhat slower by an amount corresponding to the time required to transport the message signals from the transmitter to the receiver. One possible procedure to synchronize the transmitter and receiver clocks is to set aside a code element or pulse at the end of a *frame* (consisting of a code-word derived from each of the independent message sources in succession) and to transmit this pulse every other frame only. In such a case, the receiver includes a circuit that would search for the pattern of 1s and 0s alternating at half the frame rate, and thereby establish synchronization between the transmitter and receiver.

When the transmission is interrupted, it is highly unlikely that the transmitter and receiver clocks will continue to indicate the same time for long. Accordingly, in carrying out a synchronization process, we must set up an orderly procedure for detecting the synchronizing pulse. The procedure consists of observing the code elements one by one until the synchronizing pulse is detected. That is, after observing a particular code element long enough to establish the absence of the synchronizing pulse, the receiver clock is set back by one code element and the next code element is observed. This searching process is repeated until the synchronizing pulse is detected. Clearly, the time required for synchronization depends on the epoch at which proper transmission is reestablished.

5.2 CHANNEL NOISE AND ERROR PROBABILITY

The performance of a PCM system is influenced by two major sources of noise that are independent:

1. Channel noise, which may be introduced anywhere along the transmission path.
2. Quantizing noise, which is introduced in the transmitter and is carried along to the receiver output.

Although these two sources of noise appear simultaneously when the system is operating, we consider them separately, so that we may develop some insight into their individual effects on the receiver performance. Channel noise is considered in this section, and quantizing noise is considered in the next section.

The effect of channel noise is to introduce *transmission errors* in reconstruction of the original PCM wave at the receiver output. Specifically, a symbol 0 is occasionally mistaken for a symbol 1, or vice versa. Clearly, the more frequently such transmission errors occur, the more dissimilar the receiver output becomes compared with the original PCM wave. The fidelity of information transmission by PCM in the presence of channel noise is conveniently measured in terms of the *error rate* or *probability of error,* that is, the probability that the symbol at the receiver output differs from that transmitted.

Consider a binary-encoded PCM wave $s(t)$ that uses the nonreturn-to-zero unipolar (i.e., on–off) format. When symbol 1 is sent, $s(t)$ equals $s_1(t)$ defined by

$$s_1(t) = \sqrt{\frac{E_{max}}{T_b}} \qquad 0 \leq t \leq T_b \tag{5.1}$$

where T_b is the *bit duration*, and E_{max} is the *maximum* or *peak signal energy.* When symbol 0 is sent, the transmitter is switched off, and so $s(t)$ equals $s_2(t)$ defined by

$$s_2(t) = 0 \qquad 0 \leq t \leq T_b \tag{5.2}$$

The channel noise $w(t)$ is modeled as *additive white Gaussian noise* (AWGN) with zero mean and power spectral density $N_0/2$. Correspondingly, the received signal equals

$$x(t) = s(t) + w(t) \qquad 0 \leq t \leq T_b \tag{5.3}$$

where the transmitted PCM wave $s(t)$ equals $s_1(t)$ or $s_2(t)$, depending on whether symbol 1 or 0 was sent.

From Chapter 3, we know that for an AWGN channel, the optimum receiver uses a matched filter or, equivalently, correlator. The receiver structure shown in Fig. 5.6 for a binary encoded PCM system uses a *matched filter*. The matched filter output is sampled at time $t = T_b$, where T_b is the *bit duration*. The resulting sample value is compared with a threshold by means of a decision device. If the threshold is exceeded, the receiver decides in favor of symbol 1; if the threshold is not exceeded, it decides in favor of symbol 0. In the case of ties, the decision is made by flipping a fair coin.

To calculate the probability of error incurred by this receiver, we use the signal-space approach described in Chapter 3. From Eqs. 5.1 and 5.2, we see

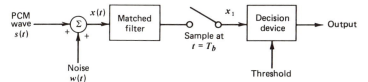

Figure 5.6 Receiver for baseband transmission of binary-encoded PCM wave.

that for the problem at hand there is only one basis function of unit energy, namely,

$$\phi_1(t) = \sqrt{\frac{1}{T_b}} \qquad 0 \leq t \leq T_b \tag{5.4}$$

Then, we may express the transmitted waveform $s_1(t)$ in terms of $\phi_1(t)$ as follows:

$$s_1(t) = \sqrt{E_{\max}}\, \phi_1(t) \qquad 0 \leq t \leq T_b \tag{5.5}$$

An on–off PCM system is therefore characterized by having a signal space that is one-dimensional and with two message points, as shown in Fig. 5.7a. The coordinates of the two message points equal

$$s_{11} = \int_0^{T_b} s_1(t)\, \phi_1(t)dt$$
$$= \sqrt{E_{\max}} \tag{5.6}$$

and

$$s_{21} = \int_0^{T_b} s_2(t)\, \phi_1(t)dt$$
$$= 0 \tag{5.7}$$

The message point corresponding to $s_1(t)$ or symbol 1 is located at s_{11} equal to $\sqrt{E_{\max}}$, and the message point corresponding to $s_2(t)$ or symbol 0 is located at the origin.

We assume that binary symbols 0 and 1 occur with equal probability. Correspondingly, the threshold used by the decision device in Fig. 5.6 is set at $\sqrt{E_{\max}}/2$, the halfway point between the two message points of Fig. 5.7a. To realize the decision rule as to whether symbol 1 or 0 was sent, we must partition the one-dimensional signal space of Fig. 5.7a into two decision regions:

1. The set of points closest to the message point at $\sqrt{E_{\max}}$.
2. The set of points closest to the second message point at the origin.

The corresponding decision regions are shown marked in Fig. 5.7a as Z_1 and Z_2, respectively.

The decision rule is now simply to guess symbol 1 or signal $s_1(t)$ was sent if the received signal point falls in region Z_1, and guess symbol 0 or signal $s_2(t)$ was sent if the received signal point falls in region Z_2. Naturally, two kinds of erroneous decisions are likely to be made by the receiver. The first kind of error is: Symbol 0 is sent, but the channel noise $w(t)$ is such that the received signal point falls inside region Z_1, and so the receiver decides in favor of symbol 1. The second kind of error is: Symbol 1 is sent but the channel noise is such that the received signal point falls inside region Z_2 and so the receiver decides in favor of symbol 0. We refer to these conditional errors as *errors of the first* and *second kind,* respectively.

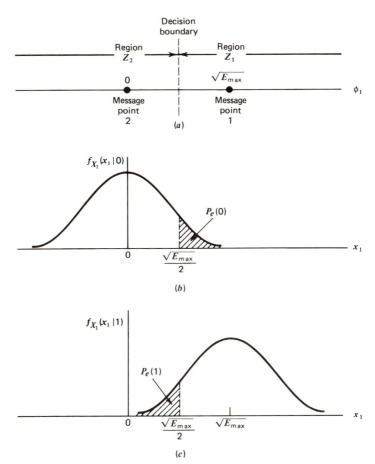

Figure 5.7 (a) Signal space diagram for on–off PCM system. (b) Likelihood function, given that symbol 0 was sent. (c) Likelihood function, given that symbol 1 was sent.

The received signal (observation) point is calculated by sampling the matched filter output at time $t = T_b$. We thus write

$$x_1 = \int_0^{T_b} x(t) \, \phi_1(t)dt \tag{5.8}$$

The received signal point x_1 may lie anywhere along the ϕ_1-axis in Fig. 5.7a. We say so because x_1 is the sample value of a Gaussian-distributed random variable X_1. When symbol 0 is sent, the mean of X_1 is zero. When, however, symbol 1 is sent, the mean of X_1 equals $\sqrt{E_{\max}}$. Regardless of which of the two symbols is sent, the variance of X_1 equals $N_0/2$, where $N_0/2$ is the power spectral density of the channel noise.

To calculate the probability of making an error of the first kind, assume that symbol 0 is sent. Under this condition, an error is made if the receiver decides in favor of symbol 1. Now from Fig. 5.7a we note that the decision region associated with symbol 1 is defined by

$$Z_1: \frac{\sqrt{E_{max}}}{2} < x_1 < \infty$$

Since the random variable X_1, with sample value x_1, has a Gaussian distribution with zero mean and variance $N_0/2$, the likelihood function (under the assumption that symbol 0 was sent) is defined by

$$f_{X_1}(x_1|0) = \frac{1}{\sqrt{\pi N_0}} \exp\left(-\frac{x_1^2}{N_0}\right) \tag{5.9}$$

This function is shown plotted in Fig. 5.7b. Let $P_e(0)$ denote the *conditional probability* of deciding in favor of symbol 1, given that symbol 0 was sent; that is, $P_e(0)$ is the probability of an error of the first kind. The probability $P_e(0)$ is the total area under the shaded part of the curve in Fig. 5.7b, which lies above $\sqrt{E_{max}}/2$. Hence, we may write

$$P_e(0) = \int_{\sqrt{E_{max}}/2}^{\infty} f_{X_1}(x_1|0)dx_1$$

$$= \frac{1}{\sqrt{\pi N_0}} \int_{\sqrt{E_{max}}/2}^{\infty} \exp\left(-\frac{x_1^2}{N_0}\right) dx_1 \tag{5.10}$$

Let

$$z = \frac{x_1}{\sqrt{N_0}}$$

Then we may rewrite Eq. 5.10 as

$$P_e(0) = \frac{1}{\sqrt{\pi}} \int_{\frac{1}{2}\sqrt{E_{max}/N_0}}^{\infty} \exp(-z^2)dz$$

By definition, the *complementary error function* equals (see Appendix E)

$$\text{erfc}(u) = \frac{2}{\sqrt{\pi}} \int_{u}^{\infty} \exp(-z^2)dz \tag{5.11}$$

Accordingly, we may express the conditional probability of error $P_e(0)$ as

$$P_e(0) = \frac{1}{2} \text{erfc}\left(\frac{1}{2}\sqrt{\frac{E_{max}}{N_0}}\right) \tag{5.12}$$

To calculate the probability of making an error of the second kind, assume that symbol 1 is sent. Under this condition, an error is made when the receiver decides in favor of symbol 0. From Fig. 5.7a we note that the decision region associated with symbol 0 is defined by

$$Z_2: -\infty < x_1 < \frac{\sqrt{E_{max}}}{2}$$

In this case, the Gaussian distributed random variable X_1 with sample value x_1

has a mean of $\sqrt{E_{\max}}$ and variance of $N_0/2$. The corresponding likelihood function $f_{X_1}(x_1|1)$ is therefore as shown plotted in Fig. 5.7c. Let $P_e(1)$ denote the second conditional probability of error, given that symbol 1 was sent; that is, $P_e(1)$ is the probability of making an error of the second kind. The probability $P_e(1)$ equals the area under the shaded part of the curve in Fig. 5.7c, which lies below $\sqrt{E_{\max}}/2$. Accordingly, we also find that

$$P_e(1) = \frac{1}{2} \operatorname{erfc}\left(\frac{1}{2} \sqrt{\frac{E_{\max}}{N_0}}\right) \tag{5.13}$$

The fact that $P_e(0) = P_e(1)$ is confirmation of the *symmetric* nature of the channel.

To determine the average probability of error in the receiver, we now note that the two possible kinds of error considered before are mutually exclusive events in that if the receiver chooses symbol 1, then symbol 0 is excluded from appearing, and vice versa. Furthermore, $P_e(0)$ and $P_e(1)$ are conditional probabilities. Thus, assuming that the a priori probability of sending a 0 is p_0, and the a priori probability of sending a 1 is p_1, we find that the *average probability of error* in the receiver is given by

$$P_e = p_0 P_e(0) + p_1 P_e(1) \tag{5.14}$$

Since $P_e(1) = P_e(0)$ and $p_0 + p_1 = 1$, we obtain

$$P_e = P_e(1) = P_e(0)$$

or

$$P_e = \frac{1}{2} \operatorname{erfc}\left(\frac{1}{2} \sqrt{\frac{E_{\max}}{N_0}}\right) \tag{5.15}$$

The ratio $E_{\max}/N_0$ represents the *peak signal energy-to-noise spectral density ratio*. This ratio may also be equated to the *peak signal-to-noise power ratio* as follows. The peak signal energy $E_{\max}$ may be written as

$$E_{\max} = P_{\max} T_b$$

where $P_{\max}$ is the *maximum* or *peak signal power,* and T_b is the bit duration. Hence, we may express the ratio $E_{\max}/N_0$ as

$$\frac{E_{\max}}{N_0} = \frac{P_{\max}}{N_0/T_b} \tag{5.16}$$

The ratio N_0/T_b may be viewed as the average noise power contained in a transmission bandwidth equal to the bit rate $1/T_b$. We may thus view $E_{\max}/N_0$ as the peak signal-to-noise power ratio, as previously stated.

The important point that has emerged from this analysis is that *the average probability of error in a PCM receiver depends solely on the ratio of the peak signal energy to the noise power spectral density measured at the receiver input.* As shown in Fig. 5.8, the error probability P_e decreases very rapidly as this ratio is increased, so that eventually a very small increase in transmitted

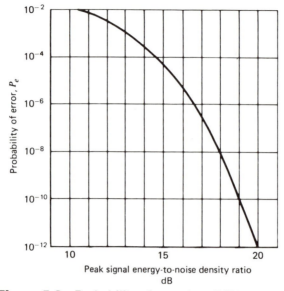

Figure 5.8 Probability of error in a PCM receiver.

signal energy (power) will make the reception of binary pulses almost error free. The nature of this improvement is further emphasized in Table 5.2 where, in the last column, we have assumed a bit rate of 10^5 bits per second.

Clearly, there is an *error threshold** (at about 17 dB, say) below which the receiver performance may involve significant numbers of errors, and above which the effect of channel noise is practically negligible. In other words, provided that the peak signal energy-to-noise density ratio E_{max}/N_0 exceeds the error threshold, channel noise has virtually no effect on the receiver performance, which is precisely the goal of PCM. When, however, this ratio drops below the error threshold, there is a sharp increase in the rate at which errors occur in the receiver. Because decision errors result in the construction of incorrect code-words, we find that when the errors are frequent, the reconstructed message at the receiver output bears little resemblance to the original message.

In most transmission systems, the effects of noise and distortion from the individual links cumulate. For a given quality of overall transmission, the longer the system, the more severe are the requirements on each link. In a PCM system, however, because the signal can be regenerated as often as necessary, the effects of amplitude, phase, and nonlinear distortions in one link (if not too severe) have practically no effect on the regenerated input signal to the next

---

* The basic idea of a threshold in PCM was first described in Oliver, Pierce, and Shannon (1948). In that paper, the probability of error calculation was based on the processing of samples derived directly from the received signal. On the other hand, the results shown plotted in Fig. 5.8 are based on a matched filter receiver. For the situation described herein, the use of a matched filter produces a processing gain of 3 decibels. This accounts for the discrepancy between the results plotted in Fig. 5.8 and those presented in the paper by Oliver, Pierce, and Shannon.

Table 5.2

Peak Signal Energy-to-Noise Density Ratio, E_{max}/N_0	Probability of Error P_e	For a bit rate of 10^5 bits per second, this is about one error every
10.3 dB	10^{-2}	10^{-3} second
14.4	10^{-4}	10^{-1} second
16.6	10^{-6}	10 seconds
18	10^{-8}	20 minutes
19	10^{-10}	1 day
20	10^{-12}	3 months

link. We have also seen that the effect of channel noise can be made practically negligible by using a peak signal energy-to-noise spectral density ratio above threshold. For all practical purposes, then, the transmission requirements for a PCM link are almost independent of the total length of the system.

Another important characteristic of a PCM system is its *ruggedness to interference*. We have seen that channel noise in a PCM system, using on–off signaling, produces no effect unless the peak amplitude is greater than half the pulse height. Similarly, interference caused by stray impulses or crosstalk will produce no effect unless the peak amplitude of this interference plus noise is greater than half the pulse height. Thus the combined presence of channel noise and interference causes the error threshold necessary for satisfactory operation to increase. If an adequate margin over the error threshold is provided in the first place, the system can withstand the presence of relatively large amounts of interference. In other words, a PCM system is *rugged*. However, for the ruggedness to be realizable in practice, the probability of error would have to be low enough (10^{-5} or less).

EXAMPLE 1   AVERAGE TRANSMITTED POWER FOR THE PROVISION OF NOISE MARGIN

Consider an *M-ary PCM system* that uses a code-word consisting of n code elements, each having one of M possible discrete amplitude values; hence, the name "M-ary." In order to provide an adequate noise margin and thereby maintain a negligibly small error rate, there must be a certain separation between these M discrete amplitude levels. Call this separation $k\sigma_N$, where k is a constant and $\sigma_N^2 = N_0 B$ is the noise variance measured in a channel bandwidth B. From Table 5.2, we see that, for on–off signaling, k is about 7 for an average error rate of 1 in 10^6 bits. The number of amplitude levels M is usually an integer power of 2. The average transmitted power will be least if the amplitude range is symmetrical about zero. Then the discrete amplitude levels, normalized with respect to the separation $k\sigma_N$, will have the values $\pm 1/2, \pm 3/2, \ldots, \pm(M-1)/2$. We assume that these M different amplitude levels are equally likely. Accordingly, we find that the average transmitted power is given by

$$P = \frac{2}{M}\left[\left(\frac{1}{2}\right)^2 + \left(\frac{3}{2}\right)^2 + \ldots + \left(\frac{M-1}{2}\right)^2\right](k\sigma_N)^2$$

$$= k^2\sigma_N^2\left(\frac{M^2-1}{12}\right) \tag{5.17}$$

Based on the result of Eq. 5.17, we can make two observations:

1. For a prescribed noise variance σ_N^2, the average transmitted power P (required to operate above the error threshold) increases rapidly with the number of discrete amplitude levels M.
2. For the special case of $M = 2$, which corresponds to nonreturn-to-zero polar signaling, we have $P = k^2\sigma_N^2/4$. That is, for the same noise margin, the use of nonreturn-to-zero polar signaling requires one half of the *average* transmitted power needed for the nonreturn-to-zero unipolar signaling. Note that, for the latter case, the average power is one half of the peak power.

EXAMPLE 2 M-ARY PCM VIEWED IN THE LIGHT OF THE CHANNEL CAPACITY THEOREM

In this example, we look at an M-ary PCM system in the light of Shannon's channel capacity theorem under the assumption that the system operates above the error threshold. That is, the average probability of error due to transmission noise is assumed to be negligible.

Consider then the M-ary PCM system of Example 1, which is used to transmit a message signal with its highest frequency component equal to W hertz. The signal is sampled at the Nyquist rate of $2W$ samples per second. We assume that the system uses a quantizer of the midriser type, with L equally likely representation levels. Hence, the probability of occurrence of any one of the L representation levels is $1/L$. Correspondingly, the amount of information carried by a single sample of the signal is $\log_2 L$ bits. With a maximum sampling rate of $2W$ samples per second, the maximum rate of information transmission of the PCM system, measured in bits per second, is given by

$$R_b = 2W \log_2 L \text{ bits/s} \tag{5.18}$$

As in Example 1, we assume that the PCM system uses a code-word consisting of n code elements, each having one of M possible discrete amplitude values. By using such a code, we have M^n different possible code-words. For a unique encoding process, we require

$$L = M^n \tag{5.19}$$

Clearly, the rate of information transmission in the system is unaffected by the use of an encoding process. We may therefore eliminate L between Eqs. 5.18 and 5.19 to obtain

$$R_b = 2Wn \log_2 M \text{ bits/s} \tag{5.20}$$

Equation 5.17, derived in Example 1, defines the average transmitted power required to maintain an M-ary PCM system operate above the error threshold. Hence, solving this equation for the number of discrete amplitude levels, M, we get

$$M = \left(1 + \frac{12P}{k^2N_0B}\right)^{1/2} \tag{5.21}$$

where $\sigma_N^2 = N_0B$ is the variance of the channel noise measured in a bandwidth B. Therefore, substituting Eq. 5.21 in Eq. 5.20, we obtain

$$R_b = Wn \log_2\left(1 + \frac{12P}{k^2N_0B}\right) \tag{5.22}$$

In Chapter 6, we show that the channel bandwidth B required to transmit a rectangular

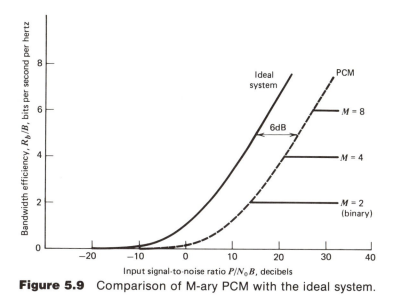

Figure 5.9 Comparison of M-ary PCM with the ideal system.

pulse of duration $1/2nW$ is given by

$$B = \kappa nW \tag{5.23}$$

where κ is a constant with a value between 1 and 2. Using the minimum possible value $\kappa = 1$, we find that the channel bandwidth $B = nW$. We may thus rewrite Eq. 5.22 as

$$R_b = B \log_2\left(1 + \frac{12P}{k^2 N_0 B}\right) \tag{5.24}$$

The *ideal system* is described by Shannon's channel capacity theorem. This theorem is given in Eq. 2.94, reproduced here for convenience:

$$C = B \log_2\left(1 + \frac{P}{N_0 B}\right) \tag{5.25}$$

where C is the channel capacity. Hence, comparing Eq. 5.24 with Eq. 5.25, we see that they are identical if the average transmitted power in the PCM system is increased by the factor $k^2/12$, compared with the ideal system. Perhaps the most interesting thing to note about Eq. 5.24 is that the form is right. Power and bandwidth are exchanged on a logarithmic basis, and the information capacity C is proportional to the channel bandwidth B.

The ideal system operates under two constraints, namely, limited bandwidth and limited power. An M-ary PCM system, on the other hand, operates under the additional constraint of a finite number of amplitude levels, M. In Fig. 5.9, we have plotted the bandwidth efficiency R_b/B in bits per second per hertz versus the input signal-to-noise ratio $P/N_0 B$ in decibels for the ideal system by using Eq. 5.25 with $R_b = C$, and for an M-ary PCM system with varying M by using Eq. 5.24 for $k = 7$. The constraint of a finite M has the effect of driving the M-ary PCM system into *saturation* when the bandwidth efficiency satisfies the condition

$$\frac{R_b}{B} \geq 2 \log_2 M$$

This condition is obtained by setting the constant $\kappa = 1$ in Eq. 5.23 and eliminating nW between Eqs. 5.20 and 5.23. Clearly, the higher the value of M, the higher will be the saturation level of the M-ary PCM system be. In any event, whenever the PCM system enters saturation, the probability of error assumes the limiting value of unity.

5.3 QUANTIZATION NOISE AND SIGNAL-TO-NOISE RATIO

As explained in Section 5.1, *quantization noise* is produced in the transmitter end of a PCM system by rounding off sample values of an analog baseband signal to the nearest permissible representation levels of the quantizer. As such, quantization noise differs from channel noise in that it is *signal dependent*. In this section, we evaluate statistical characteristics of quantization noise by making certain assumptions that permit a mathematical analysis of the problem.

Consider a memoryless quantizer that is both uniform and symmetric, with a total of L representation levels. Let x denote the quantizer input, and y denote the quantizer output. These two variables are related by the transfer characteristic of the quantizer, as shown by

$$y = Q(x) \tag{5.26}$$

which is a staircase function that befits the type of midtread or midriser quantizer of interest. Suppose that the input x lies inside the interval

$$\mathcal{I}_k = \{x_k < x \leq x_{k+1}\} \qquad k = 1, 2, \ldots, L \tag{5.27}$$

where x_k and x_{k+1} are decision thresholds of the interval $\mathcal{I}_k$ as depicted in Fig. 5.10. Correspondingly, the quantizer output y takes on a discrete value y_k, $k = 1, 2, \ldots, L$. That is,

$$y = y_k, \qquad \text{if } x \text{ lies in the interval } \mathcal{I}_k \tag{5.28}$$

Let q denote the quantization error, with values in the range $-\Delta/2 \leq q \leq \Delta/2$. We may then write

$$y_k = x + q, \qquad \text{if } x \text{ lies in the interval } \mathcal{I}_k \tag{5.29}$$

We assume that the quantizer input x is the sample value of a random variable X of zero mean and variance σ_X^2. When the quantization is fine enough (say, the number of representation levels L is greater than 64), the distortion produced by quantization noise affects the performance of a PCM system as

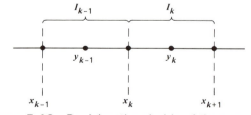
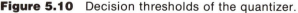

Figure 5.10 Decision thresholds of the quantizer.

though it were an additive independent source of noise with zero mean and mean-square value determined by the quantizer step size Δ. The reason for this is that the power spectral density of the quantization noise in the receiver output is practically independent of that of the baseband signal over a wide range of input signal amplitudes. Furthermore, for a baseband signal of a root mean-square value that is large compared to a quantum step, it is found that the power spectral density of the quantization noise has a large bandwidth compared with the signal bandwidth. Thus, with the quantization noise uniformly distributed throughout the signal band, its interfering effect on a signal is similar to that of thermal noise.*

Let the random variable Q denote the quantization error, and let q denote its sample value. (The symbol used for this random variable should not be confused with that for the transfer characteristic of the quantizer.) Lacking information to the contrary, we assume that the random variable Q is uniformly distributed over the possible range $-\Delta/2$ to $\Delta/2$, as shown by

$$f_Q(q) = \begin{cases} \dfrac{1}{\Delta} & -\dfrac{\Delta}{2} \le q \le \dfrac{\Delta}{2} \\ 0 & \text{otherwise} \end{cases} \tag{5.30}$$

where $f_Q(q)$ is the probability density function of the quantization error. For this to be justifiable, we must ensure that the incoming signal does not overload the quantizer. Then the mean of the quantization error is zero, and its variance σ_Q^2 is the same as the mean-square value, as shown by

$$\sigma_Q^2 = E[Q^2]$$

$$= \int_{-\infty}^{\infty} q^2 f_Q(q) dq \tag{5.31}$$

Substituting Eq. 5.30 in Eq. 5.31, we get

$$\sigma_Q^2 = \frac{1}{\Delta} \int_{-\Delta/2}^{\Delta/2} q^2 dq$$

$$= \frac{\Delta^2}{12} \tag{5.32}$$

Thus, the variance of the quantization noise, produced by a uniform quantizer, grows as the square of the step size. This is perhaps the most often used result in quantization.

Let the variance of the baseband signal $x(t)$ at the quantizer input be denoted by σ_X^2. When the baseband signal is reconstructed at the receiver output, we obtain the original signal plus quantization noise. We may therefore define an *output signal-to-quantization noise ratio* (SNR) as

$$(\text{SNR})_0 = \frac{\sigma_X^2}{\sigma_Q^2} = \frac{\sigma_X^2}{\Delta^2/12} \tag{5.33}$$

* See Bennett (1948).

Clearly, the smaller we make the step size Δ, the larger will the SNR be.

Equation 5.33 defines the performance of a quantizing noise-limited PCM system that uses a uniform quantizer. In the sequel, we illustrate the usefulness of this result by considering an example.

EXAMPLE 3 SIGNAL-TO-QUANTIZATION NOISE RATIO

Suppose that the quantizer input x represents the sample value of a random variable X with zero mean and variance σ_X^2. The quantizer is assumed to be uniform, symmetric, and of the midtread type. Let x_{max} denote the absolute value of the overload level of the quantizer, and Δ denote its step size. We may then express the number of representation levels in the quantizer as

$$L = 1 + \frac{2x_{max}}{\Delta} \tag{5.34}$$

For a binary code with a code-word of n bits, up to 2^n representation levels can be represented. Since the number of representation levels is odd for a midtread quantizer, we write

$$L = 2^n - 1 \tag{5.35}$$

Eliminating L between Eqs. 5.34 and 5.35, and solving for Δ, we obtain

$$\Delta = \frac{x_{max}}{2^{n-1} - 1} \tag{5.36}$$

The ratio x_{max}/σ_x is called the *loading factor*. To avoid significant overload distortion, we let the amplitude range of the quantizer input x extend from $-4\sigma_X$ to $4\sigma_X$, which corresponds to a loading factor of 4. We then find that the *probability of overload* (i.e., the probability that a sample value of the signal falls outside the total amplitude range of the quantizer, $8\sigma_X$) is less than 10^{-4}. Thus with $x_{max} = 4\sigma_X$, we may rewrite Eq. 5.31 as

$$\Delta = \frac{4\sigma_X}{2^{n-1} - 1} \tag{5.37}$$

The use of Eq. 5.37 in Eq. 5.33 yields the corresponding value of the signal-to-quantization noise ratio as

$$(\text{SNR})_0 = \frac{3}{4}(2^{n-1} - 1)^2 \tag{5.38}$$

For large n (typically, $n > 6$), we may approximate this result as

$$(\text{SNR})_0 \simeq \frac{3}{16}(2^{2n}) \tag{5.39}$$

Hence, expressing the SNR in decibels, we may write*

$$10 \log_{10}(\text{SNR})_0 \simeq 6n - 7.2 \tag{5.40}$$

This formula states that each bit in the code-word of a PCM system contributes 6 dB to

---

* The linear increase of SNR with the code-word length n is reported in Oliver, Pierce, and Shannon (1948). This reference also includes a derivation of the channel capacity of a PCM system, which was presented in Example 2.

the signal-to-noise ratio. It gives a good description of the noise performance of a PCM system, provided that the following conditions are satisfied:

1. The system operates with an average signal power above the error threshold so that the effect of channel noise is made negligible, and performance is thereby limited essentially by quantization noise alone.
2. The quantization error is uniformly distributed.
3. The quantization is fine enough (say, $n > 6$) to prevent signal-correlated patterns in the quantization error waveform.
4. The quantizer is aligned with the input for a loading factor of 4.

Note that a change in the loading factor merely modifies the constant term 7.2 in Eq. 5.40, but does not alter the rate of increase of SNR (expressed in decibels) with n.

In Section 5.2, we indicated that the channel bandwidth B for a binary PCM system is typically nW. We may therefore rewrite Eq. 5.40 as

$$10 \log_{10}(\text{SNR})_0 \simeq 6 \left(\frac{B}{W}\right) - 7.2 \tag{5.41}$$

This result shows that in a PCM system limited by quantizing noise, doubling the channel bandwidth permits twice the number of bits in a code-word and therefore increases the SNR by $6n$ dB.

(1)   Idle Channel Noise

A discussion of noise in PCM systems would be incomplete without a description of *idle channel noise*. As the name implies, idle channel noise is the coding noise measured at the receiver output with *zero* transmitter input. The zero-input condition arises, for example, during silences in speech. The average power of this form of noise depends on the type of quantizer used. In a quantizer of the midriser type, as in Fig. 5.3a, zero input amplitude is encoded into one of the two innermost representation levels $\pm\Delta/2$. Assuming that these two representation levels are equiprobable, the idle channel noise for midriser quantizer has zero mean and an average power of $\Delta^2/4$. On the other hand, in a quantizer of the midtread type, as in Fig. 5.2a, the output is zero for zero input, and the idle channel noise is correspondingly zero. In practice, however, the idle channel noise is never exactly zero due to the inevitable presence of background noise or interference. Moreover, the characterization of a quantizer exhibits deviations from its idealized form. Accordingly, we find that the average power of idle channel noise in a midtread quantizer is also in the order of, although less than, $\Delta^2/4$.

5.4   ROBUST QUANTIZATION

In the previous section, it was shown that for a uniform quantizer with a step size Δ the variance of the quantization noise is $\sigma_Q^2 = \Delta^2/12$, provided that the input signal does not overload the quantizer. Hence, under this condition, the variance of quantization noise is independent of the variance of the input signal. The implication of this result is that the SNR decreases with a decrease in input power level relative to the overload point of the quantizer. However, in certain applications, notably in the use of PCM for the transmission of speech signals, the same quantizer has to accommodate input signals with widely varying power levels. For example, the range of voltages covered by speech signals, from the peaks of loud talk to the weak passages of weak talk, is on the order of 1000 to 1. It would therefore be highly desirable from a practical viewpoint for the signal-to-quantization noise ratio to remain essentially constant for a wide range of input power levels. A quantizer that satisfies this requirement is said to be *robust*.

The provision for such a robust performance necessitates the use of a *nonuniform quantizer,* characterized by a step size that increases as the separation from the origin of the transfer characteristic is increased. Accordingly, in the case of speech signals, the weak passages (which generally occur with high probability and therefore require extra protection) are assigned more representation levels and thereby favored at the expense of the loud passages (which occur relatively infrequently). The result is that a nearly uniform percentage precision is achieved through the greater part of the amplitude range of the input signal with a smaller number of representation levels than would be possible by means of a uniform quantizer. Also, the nonuniform quantizer exploits a characteristic of human hearing, namely that large amplitudes mask quantization noise to some extent.

The desired form of nonuniform quantization can be achieved by using a *compressor followed by a uniform quantizer.* By cascading this combination, with an *expander* complementary to the compressor, the original signal samples are restored to their correct values except for quantization errors. Ideally, the compression and expansion laws are exactly the *inverse* of each other, as illustrated in Fig. 5.11. This figure depicts the transfer characteristics of the compressor, uniform quantizer, and expander. In particular, it shows the relationship between the decision thresholds at the compressor input and the representation (reconstruction) levels at the expander output. Thus all sample values of the compressor input, which lie inside an interval $\mathcal{I}_k$ (say), are assigned the discrete value defined by the kth representation level at the expander output.

The combination of a *comp*ressor and an *exp*ander is called a *compander.* Naturally, in an actual PCM system, the combination of compressor and uniform quantizer is located in the transmitter, while the expander is located in the receiver.

(1) Variance of the Quantization Error

Consider the *model* shown in Fig. 5.12. Indeed, any nonuniform quantizer can be represented by such a model. The transfer characteristic of the compressor is represented by a memoryless nonlinearity $c(x)$, where x is the sample value of a random variable X denoting the compressor input. The characteristic $c(x)$ is a monotonically increasing function that has *odd symmetry:*

$$c(-x) = -c(x) \tag{5.42}$$

With the sample value x bounded in the range $-x_{\max}$ to $x_{\max}$, the function $c(x)$

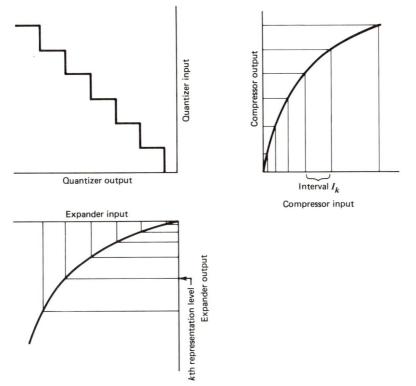

Figure 5.11   Transfer characteristics of compressor, uniform quantizer, and expander (shown for positive amplitudes only).

similarly ranges from $-x_{\max}$ to $x_{\max}$, as shown by

$$c(x) = \begin{cases} x_{\max} & x = x_{\max} \\ 0 & x = 0 \\ -x_{\max} & x = -x_{\max} \end{cases} \tag{5.43}$$

The monotonic property of $c(x)$ ensures that it is completely invertible, which means that the inverse function $c^{-1}(x)$ is uniquely defined for a prescribed $c(x)$. We may therefore write

$$c(x)\, c^{-1}(x) = 1 \tag{5.44}$$

where $c(x)$ defines the compressor, and $c^{-1}(x)$ defines the expander. Thus the sample value x of the compressor input is reproduced exactly at the expander output (in the absence of quantization).

The compression characteristic $c(x)$ relates nonuniform intervals (between decision thresholds) at the compressor input to uniform intervals at the com-

Figure 5.12   Model of nonuniform quantizer.

pressor output. The uniform intervals are of width $2x_{max}/L$ each, where L is the number of representation levels of the quantizer. To proceed with the analysis, it is assumed that L is large. The compressor characteristic $c(x)$ in the kth interval, $\mathcal{I}_k$, may then be approximated by a straight-line segment with a slope equal to $2x_{max}/L\Delta_k$, where Δ_k is the width of interval $\mathcal{I}_k$. Correspondingly, we may write for large L and x inside the interval $\mathcal{I}_k$:

$$\frac{dc(x)}{dx} \simeq \frac{2x_{max}}{L\Delta_k} \qquad k = 0, 1, \ldots, L-1 \qquad (5.45)$$

where $dc(x)/dx$ is the derivative of $c(x)$ with respect to x.

Two other assumptions are made, concerning the probability density function $f_X(x)$ of the random variable X at the compressor input:

1.   The probability density function $f_X(x)$ is symmetric.
2.   In each interval $\mathcal{I}_k$, $k = 1, 2, \ldots, L$, the probability density function $f_X(x)$ is approximately constant.

As a direct consequence of the second assumption, we may write

$$f_X(x) \simeq f_X(y_k) \qquad x_k < X \leq x_{k+1} \qquad (5.46)$$

where the representation level y_k lies in the middle of interval $\mathcal{I}_k$; that is

$$y_k = \frac{1}{2}(x_k + x_{k+1}) \qquad k = 0, 1, \ldots, L-1 \qquad (5.47)$$

The width of interval $\mathcal{I}_k$ equals

$$\Delta_k = x_{k+1} - x_k \qquad k = 0, 1, \ldots, L-1 \qquad (5.48)$$

Accordingly, the probability that the random variable X lies in the interval $\mathcal{I}_k$ equals

$$\begin{aligned} p_k &= P(x_k < X \leq x_{k+1}) \\ &= f_X(y_k)\,\Delta_k \qquad k = 0, 1, \ldots, L-1 \end{aligned} \qquad (5.49)$$

with the constraint

$$\sum_{k=0}^{L-1} p_k = 1 \qquad (5.50)$$

Let the random variable Q denote the quantization error, as shown by

$$Q = y_k - X \qquad x_k < X \leq x_{k+1} \qquad (5.51)$$

We may therefore express the variance of Q as

$$\begin{aligned} \sigma_Q^2 &= E[Q^2] \\ &= E[(X - y_k)^2] \\ &= \int_{-x_{max}}^{x_{max}} (x - y_k)^2 f_X(x)\,dx \end{aligned} \qquad (5.52)$$

Dividing up the region of integration into L intervals, and using Eq. 5.49, we may rewrite Eq. 5.52 as

$$\sigma_Q^2 = \sum_{k=0}^{L-1} \frac{p_k}{\Delta_k} \int_{x_k}^{x_{k+1}} (x - y_k)^2 dx \qquad (5.53)$$

Substituting Eq. 5.47 in Eq. 5.53, and carrying out the integration with respect to x, we finally get the result

$$\sigma_Q^2 = \frac{1}{12} \sum_{k=0}^{L-1} p_k \, \Delta_k^2 \qquad (5.54)$$

In this formula, we may view $\Delta_k^2/12$ as the variance of quantization error conditional on interval $\mathcal{I}_k$. Note that for a uniform quantizer, $\Delta_k = \Delta$ for all k and so Eq. 5.54 reduces to the simple form given in Eq. 5.32.

Earlier, we developed an approximation for the derivative (slope) of the compressor characteristic $c(x)$ in interval $\mathcal{I}_k$. We rewrite this approximate relation for x in the interval $\mathcal{I}_k$ as (see Eq. 5.45)

$$\Delta_k \simeq \frac{2x_{\max}}{L} \left[\frac{dc(x)}{dx} \right]^{-1} \qquad k = 0, 1, \ldots, L - 1 \qquad (5.55)$$

Hence, substituting Eq. 5.55 in Eq. 5.54, we obtain the following approximate formula for the variance of the error produced by nonuniform quantization:

$$\sigma_Q^2 \simeq \frac{x_{\max}^2}{3L^2} \sum_{k=0}^{L-1} p_k \left[\frac{dc(x)}{dx} \right]^{-2} \qquad (5.56)$$

In view of Eqs. 5.46 to 5.49, we may equivalently write*

$$\sigma_Q^2 \simeq \frac{x_{\max}^2}{3L^2} \int_{-x_{\max}}^{x_{\max}} f_X(x) \left[\frac{dc(x)}{dx} \right]^{-2} dx \qquad (5.57)$$

The approximate formula of Eq. 5.57 for the error variance σ_Q^2 in nonuniform quantization is based on two assumptions: (1) the number of representation levels L of the quantizer is large, and (2) the overload distortion is negligible. Given a compressor characteristic $c(x)$, we may use the formula to evaluate σ_Q^2 for prescribed overload levels $\pm x_{\max}$.

With the aid of this formula, we may now establish the condition for robust quantization. The output signal-to-quantization ratio is defined by

$$(\text{SNR})_0 = \frac{\sigma_X^2}{\sigma_Q^2}$$

---

* The approximate formula of Eq. 5.57 is due to Bennett (1948). A derivation of this formula and discussion of related issues can also be found in Jayant and Noll (1984, pp. 129–146) and Gersho (1977).

By definition, the variance of the quantizer input x equals

$$\sigma_X^2 = \int_{-x_{max}}^{x_{max}} x^2 f_X(x) \, dx$$

Hence, using this formula and that of Eq. 5.57, we may write

$$(SNR)_0 = \frac{3L^2}{x_{max}^2} \frac{\int_{-x_{max}}^{x_{max}} x^2 f_X(x) \, dx}{\int_{-x_{max}}^{x_{max}} f_X(x)[dc(x)/dx]^{-2} \, dx} \tag{5.58}$$

For a robust performance, the output signal-to-noise ratio should ideally be independent of the probability density function of the input random variable X. From Eq. 5.58 we see that this requirement is met if the compressor characteristic $c(x)$ satisfies the first-order differential equation

$$\frac{dc(x)}{dx} = \frac{K}{x} \qquad -x_{max} \leq x \leq x_{max} \tag{5.59}$$

where K is a constant. Integrating Eq. 5.59 with respect to x, and using the boundary condition that $c(x_{max}) = x_{max}$ to evaluate the constant of integration, we get

$$c(x) = x_{max} + K \ln\left(\frac{x}{x_{max}}\right) \qquad x > 0 \tag{5.60}$$

where ln denotes the natural logarithm.

The compressor characteristic of Eq. 5.60 is unrealizable since $c(0)$ is not finite. In practice, this difficulty is eliminated by modifying the compressor characteristic in such a way that it is well-behaved for small x; elsewhere, it retains the logarithmic behavior. Two widely used solutions to the problem are as follows:

1. *μ-law companding*   In the *μ-law companding,* the compressor characteristic $c(x)$ is *continuous,* approximating a linear dependence on x for low input levels and a logarithmic one for high input levels. Specifically, it is described by*

$$\frac{c(|x|)}{x_{max}} = \frac{\ln(1 + \mu|x|/x_{max})}{\ln(1 + \mu)} \qquad 0 \leq \frac{|x|}{x_{max}} \leq 1 \tag{5.61}$$

The special case of uniform quantization corresponds to the parameter μ equal to 0. The normalized form of μ-law compressor characteristic, $c(x)/x_{max}$, is shown plotted in Fig. 5.13a for $0 \leq x/x_{max} \leq 1$, and three different values of μ. A practical value for μ is 255. The μ-law is used for PCM telephone systems in the United States, Canada, and Japan.

2. *A-law companding*   In the *A-law companding,* the compressor character-

---

* The μ-law companding was first described in the literature by Holzwarth (1949). For a detailed treatment of the μ-law, see Smith (1957).

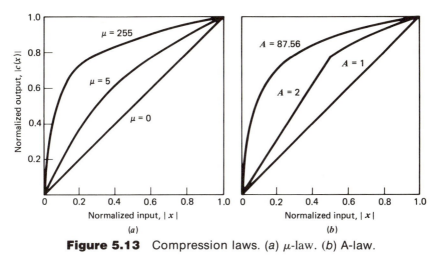

Figure 5.13   Compression laws. (a) μ-law. (b) A-law.

istic $c(x)$ is *piecewise,* made up of a linear segment for low-level inputs and a logarithmic segment for high-level inputs. It is described by*

$$\frac{c(|x|)}{x_{max}} = \begin{cases} \dfrac{A|x|/x_{max}}{1 + \ln A} & 0 \leqslant \dfrac{|x|}{x_{max}} \leqslant \dfrac{1}{A} \\[3mm] \dfrac{1 + \ln(A|x|/x_{max})}{1 + \ln A} & \dfrac{1}{A} \leqslant \dfrac{|x|}{x_{max}} \leqslant 1 \end{cases} \qquad (5.62)$$

The special case of uniform quantization corresponds to the parameter $A = 1$. The normalized form of the A-law compressor characteristic, $c(x)/x_{max}$, is shown plotted in Fig. 5.13b for $0 \leqslant x/x_{max} \leqslant 1$ and three different values of A. A practical value for A is 87.56. The A-law companding is used for PCM telephone systems in Europe.

EXAMPLE 4

To illustrate the advantage gained by the use of robust quantization, Fig. 5.14 shows curves† of output signal-to-noise ratio versus input signal power for both uniform and nonuniform quantizers, the latter based on the μ-law companding. The operating conditions are as follows:

1. The number of representation levels $L = 256$.
2. The parameter $\mu = 255$.
3. The random variable X, denoting the compressor input, is *Laplacian distributed,* as shown by the probability density function

$$f_X(x) = \frac{1}{\sqrt{2}\sigma_X} \exp\left(-\frac{\sqrt{2}|x|}{\sigma_X}\right) \qquad (5.63)$$

where σ_X^2 is the variance of X. There is experimental evidence to justify the use of a

---

* The A-law companding was originated by Cattermole; for details, see Cattermole (1969).
† The curves shown in Fig. 5.14 are adapted, with permission, from Jayant and Noll (1984).

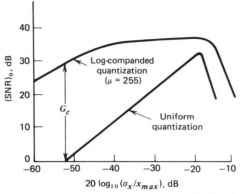

Figure 5.14 Output signal-to-noise ratio characteristics for uniform and log-companded forms of quantization.

Laplacian model as a first approximation to the *long-term averaged* probability density function of speech signals.

4. The input is bounded to the range $-x_{max}$ to x_{max}.

The two curves shown in Fig. 5.14 take overload distortion into account, which rapidly becomes dominant when the input reaches a critical level.

From Fig. 5.14, we see that the performance of the uniform quantizer is highly dependent on the input. On the other hand, the nonuniform quantizer based on the μ-law companding achieves a dynamic range of about 30 dB (from -15 dB to -45 dB on the abscissa) for which the SNR remains within 3 dB of the maximum value of 38 dB.

Another useful parameter for assessing the improvement that results from the use of a compressor/expander combination is the *companding gain*, G_c, defined as

$$G_c = \left. \frac{dc(x)}{dx} \right|_{x \to 0} \tag{5.64}$$

where $c(x)$ is the compressor characteristic. For the μ-law described in Eq. 5.61, we have

$$G_c = \frac{\mu}{\ln(1 + \mu)} \tag{5.65}$$

With $\mu = 255$, the use of Eq. 5.65 yields $20\log_{10}G_c = 33.3$ dB. In Fig. 5.14, the companding gain (expressed in decibels) is shown for low-level inputs. The implication of this companding gain is that the smallest step size resulting from the use of the nonuniform quantizer described herein is smaller than the step size of a corresponding uniform quantizer (with identical values for the overload level x_{max} and the number of representation levels L) by a factor equal to $G_c = 32$.

5.5 DIFFERENTIAL PULSE-CODE MODULATION

In the use of PCM for the digitization of a voice or video signal, the signal is sampled at a rate slightly higher than the Nyquist rate. The resulting sampled signal is then found to exhibit a high correlation between adjacent samples. The meaning of this high correlation is that, in an average sense, the signal does not change rapidly from one sample to the next with the result that the difference between adjacent samples has a variance that is smaller than the variance of the signal itself. When these highly correlated samples are encoded, as in a standard PCM system, the resulting encoded signal contains redundant information. In particular, symbols that are not absolutely essential to the transmission of information are generated as a result of the encoding process. By removing this redundancy before encoding, we obtain a more efficient coded signal.

Now, if we know a sufficient part of a redundant signal, we may use linear prediction to infer the rest, or at least make the most probable estimate. (Linear prediction was studied in Chapter 3.) In particular, if we know the past behavior of a signal up to a certain point in time, it is possible to make some inference about its future values. Suppose then a baseband signal $x(t)$ is sampled at the rate $f_s = 1/T_s$ to produce a sequence of correlated samples T_s seconds apart. Let the sequence be denoted by $\{x(nT_s)\}$, where n takes on integer values. The fact that it is possible to predict future values of the signal $x(t)$ provides motivation for the differential quantization scheme shown in Fig. 5.15a. In this scheme the input to the quantizer is a signal

$$e(nT_s) = x(nT_s) - \hat{x}(nT_s) \tag{5.66}$$

which is the difference between the unquantized input sample $x(nT_s)$ and a prediction of it, denoted by $\hat{x}(nT_s)$. This predicted value is produced by using a

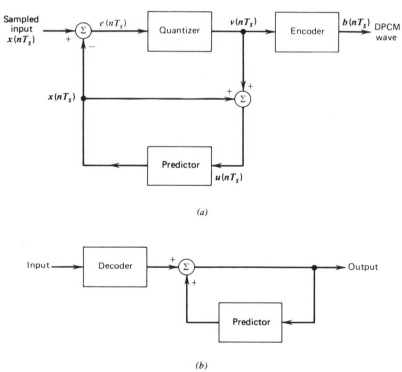

(a)

(b)

Figure 5.15 DPCM system. (a) Transmitter. (b) Receiver.

predictor whose input, as we will see, consists of a quantized version of the input signal $x(nT_s)$. The difference signal $e(nT_s)$ is called a *prediction error*, since it is the amount by which the predictor fails to predict the input exactly.

By encoding the quantizer output, as in Fig. 5.15a, we obtain an important variation of PCM, which is known as *differential pulse-code modulation* (DPCM).* It is the encoded signal that is used for transmission.

Let the nonlinear function $Q(\cdot)$ define the input–output characteristic of the quantizer. The quantizer output may then be represented as

$$v(nT_s) = Q[e(nT_s)]$$
$$= e(nT_s) + q(nT_s) \tag{5.67}$$

where $q(nT_s)$ is the quantization error. According to Fig. 5.15a, the quantizer output $v(nT_s)$ is added to the predicted value $x(nT_s)$ to produce the predictor input

$$u(nT_s) = \hat{x}(nT_s) + v(nT_s) \tag{5.68}$$

Substituting the second line of Eq. 5.67 in Eq. 5.68, we get

$$u(nT_s) = \hat{x}(nT_s) + e(nT_s) + q(nT_s) \tag{5.69}$$

However, from Eq. 5.66 we observe that $\hat{x}(nT_s) + e(nT_s)$ is equal to the input signal $x(nT_s)$. Therefore, we may rewrite Eq. 5.69 as follows:

$$u(nT_s) = x(nT_s) + q(nT_s) \tag{5.70}$$

which represents a quantized version of the input signal $x(nT_s)$. That is, irrespective of the properties of the predictor, the quantized signal $u(nT_s)$ at the predictor input differs from the original input signal $x(nT_s)$ by the quantization error. Accordingly, if the prediction is good, the variance of the prediction error $e(nT_s)$ will be smaller than the variance $x(nT_s)$, so that a quantizer with a given number of representation levels can be adjusted to produce a quantizing error with a smaller variance than would be possible if the input signal $x(nT_s)$ were quantized directly as in a standard PCM system.

The receiver for reconstructing the quantized version of the input is shown in Fig. 5.15b. It consists of a decoder to reconstruct the quantized error signal. The quantized version of the original input is reconstructed from the decoder output using the same predictor as used in the transmitter of Fig. 5.15a. In the absence of channel noise, we find that the encoded signal at the receiver input is identical to the encoded signal at the transmitter output. Accordingly, the corresponding receiver output is equal to $u(nT_s)$, which differs from the original input $x(nT_s)$ only by the quantizing error $q(nT_s)$ incurred as a result of quantizing the prediction error $e(nT_s)$.

From the foregoing analysis we observe that, in a noise-free environment, the predictors in the transmitter and receiver operate on the same sequence of

---

* Differential pulse-code modulation was invented by Cutler; the invention is described in the patent (Cutler, 1952).

samples $u(nT_s)$. It is with this purpose in mind that a feedback path is added to the quantizer in the transmitter, as shown in Fig. 5.15a.

The output signal-to-quantization noise ratio of a signal coder (based on the use of PCM or DPCM) is defined in the first line of Eq. 5.33, which is reproduced here for convenience:

$$(\text{SNR})_0 = \frac{\sigma_X^2}{\sigma_Q^2}$$

where σ_X^2 is the variance of the original input $x(nT_s)$, assumed to be of zero mean, and σ_Q^2 is the variance of the quantization error $q(nT_s)$. We may rewrite this equation as

$$(\text{SNR})_0 = \left(\frac{\sigma_X^2}{\sigma_E^2}\right)\left(\frac{\sigma_E^2}{\sigma_Q^2}\right)$$
$$= G_P(\text{SNR})_P \tag{5.71}$$

where $(\text{SNR})_P$ is the prediction error-to-quantization noise ratio, defined by

$$(\text{SNR})_P = \frac{\sigma_E^2}{\sigma_Q^2} \tag{5.72}$$

and G_P is the *prediction gain* produced by the differential quantization scheme, defined by

$$G_P = \frac{\sigma_X^2}{\sigma_E^2} \tag{5.73}$$

The quantity G_P, when greater than unity, represents the gain in signal-to-noise ratio that is due to the differential quantization scheme of Fig. 5.15. Now, for a given baseband signal, the variance σ_X^2 is fixed, so that G_P is maximized by minimizing the variance σ_E^2 of the prediction error $e(nT_s)$. Accordingly, our objective should be to design the predictor so as to minimize σ_E^2.

5.6 DELTA MODULATION

The exploitation of signal correlations in DPCM suggests the further possibility of oversampling a baseband signal (i.e., at a rate much higher than the Nyquist rate) purposely to increase the correlation between adjacent samples of the signal, so as to permit the use of a simple quantizing strategy for constructing the encoded signal. *Delta modulation* (DM), which is the one-bit (or two-level) version of DPCM, is precisely such a scheme.*

In its basic form, DM provides a staircase approximation to the oversampled version of an input baseband signal, as illustrated in Fig. 5.16a. The difference between the input and the approximation is quantized into only two levels,

---

* For the original papers on delta modulation, see Schouten, DeJager, and Greefkes (1952) and DeJager (1952). The simplicity of delta modulation has inspired much detailed analysis, numerous refinements and variations; see Steele (1975), and Jayant and Noll (1984, pp. 372–427).

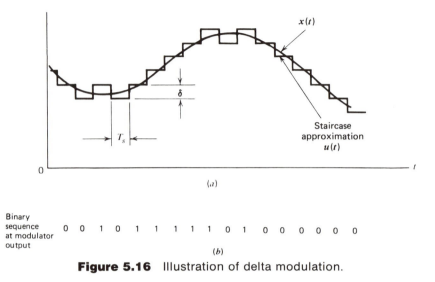

(a)

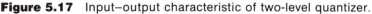

Binary
sequence   0 0 1 0 1 1 1 1 1 0 1 0 0 0 0 0 0 0
at modulator
output

(b)

Figure 5.16 Illustration of delta modulation.

namely, $\pm\delta$, corresponding to positive and negative differences, respectively. Thus, if the approximation falls below the signal at any sampling epoch, it is increased by δ. If, on the other hand, the approximation lies above the signal, it is diminished by δ. Provided that the signal does not change too rapidly from sample to sample, we find that the staircase approximation remains within $\pm\delta$ of the input signal.

Note that δ denotes the absolute value of the two representation levels of the one-bit quantizer used in the DM. These two levels are indicated in the transfer characteristic of Fig. 5.17. The step size Δ of the quantizer is therefore related to δ by

$$\Delta = 2\delta \tag{5.74}$$

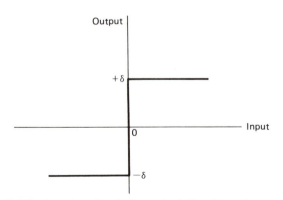

Figure 5.17 Input–output characteristic of two-level quantizer.

Denote the input signal as $x(t)$ and the staircase approximation to it as $u(t)$. Then, the basic principle of delta modulation may be formalized in the following set of discrete-time relations:

$$e(nT_s) = x(nT_s) - \hat{x}(nT_s)$$
$$= x(nT_s) - u(nT_s - T_s) \qquad (5.75)$$

$$b(nT_s) = \delta \, \text{sgn}[e(nT_s)] \qquad (5.76)$$

and

$$u(nT_s) = u(nT_s - T_s) + b(nT_s) \qquad (5.77)$$

where T_s is the sampling period; $e(nT_s)$ is a prediction error representing the difference between the present sample value $x(nT_s)$ of the input signal and the latest approximation to it, namely $\hat{x}(nT_s) = u(nT_s - T_s)$. The binary quantity $b(nT_s)$ is the algebraic sign ot the error $e(nT_s)$, except for the scaling factor δ. Indeed, $b(nT_s)$ is the one-bit word transmitted by the DM system.

Details of the modulator follow directly from Eqs. 5.75–5.77. It consists of a *summer*, a *two-level quantizer*, and an *accumulator* interconnected as shown in Fig. 5.18*a*. We assume that the accumulator is initially set to zero. Then, we may solve Eqs. 5.75–5.77 for the accumulator output, obtaining the result

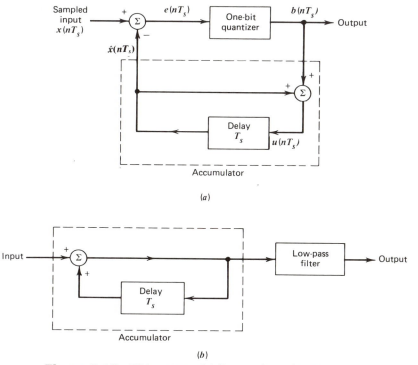

(a)

(b)

Figure 5.18 DM system. (a) Transmitter. (b) Receiver.

$$u(nT_s) = \delta \sum_{i=1}^{n} \text{sgn}[e(iT_s)]$$

$$= \sum_{i=1}^{n} b(iT_s) \tag{5.78}$$

Thus, at each sampling instant, the accumulator increments the approximation to the input signal by $\pm\delta$, depending on the binary output of the modulator. Indeed, the accumulator does the best it can to track the input by an increment $+\delta$ or $-\delta$ at a time. In the receiver, shown in Fig. 5.18b, the staircase approximation $u(t)$ is reconstructed by passing the incoming sequence of positive and negative pulses through an accumulator in a manner similar to that used in the transmitter. The out-of-band quantization noise in the high-frequency staircase waveform $u(t)$ is rejected by passing it through a low-pass filter with a bandwidth equal to the original signal bandwidth.

In comparing the DPCM and DM networks of Figs. 5.15 and 5.18 we see that, except for an output low-pass filter, delta modulation is a special case of differential pulse-code modulation.

Delta modulation offers two unique features: (1) a one-bit code-word for the output, which eliminates the need for word framing, and (2) simplicity of design for both the transmitter and the receiver. These two features make the use of delta modulation attractive for some types of digital communications and for digital voice storage.

(1) Quantization Noise

Delta modulation systems are subject to two types of quantization error: (1) slope-overload distortion, and (2) granular noise. We first discuss the cause of slope-overload distortion, and then we discuss granular noise.

Let $q(nT_s)$ denote the *quantizing error*. We may then write (see Eq. 5.70)

$$u(nT_s) = x(nT_s) + q(nT_s) \tag{5.79}$$

Accordingly, using Eq. 5.79 to eliminate $u(nT_s - T_s)$ from Eq. 5.75, we may express the prediction error $e(nT_s)$ as

$$e(nT_s) = x(nT_s) - x(nT_s - T_s) - q(nT_s - T_s) \tag{5.80}$$

Thus, except for the quantization error $q(nT_s - T_s)$, the quantizer input is a first backward difference of the input signal, which may be viewed as a digital approximation to the derivative of the input signal or, equivalently, as the inverse of a digital integration process. If we consider the maximum slope of the original input waveform $x(t)$, it is clear that in order for the sequence of samples $\{u(nT_s)\}$ to increase as fast as the input sequence of samples $\{x(nT_s)\}$ in a region of maximum slope of $x(t)$, we require that the condition

$$\frac{\delta}{T_s} \geq \max \left| \frac{dx(t)}{dt} \right| \tag{5.81}$$

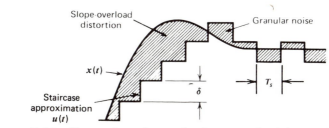

Figure 5.19 Illustration of quantization error in delta modulation.

be satisfied. Otherwise, we find that the step size $\Delta = 2\delta$ is too small for the staircase approximation $u(t)$ to follow a steep segment of the input waveform $x(t)$, with the result that $u(t)$ falls behind $x(t)$, as illustrated in Fig. 5.19. This condition is called *slope-overload*, and the resulting quantization error is called *slope-overload distortion* (*noise*). Note that since the maximum slope of the staircase approximation $u(t)$ is fixed by the step size Δ, increases and decreases in $u(t)$ tend to occur along straight lines. For this reason, a delta modulator using a fixed step size is often referred to as a *linear delta modulator* (LDM).

In contrast to slope-overload distortion, *granular noise* occurs when the step size Δ is too large relative to the local slope characteristics of the input waveform $x(t)$, thereby causing the staircase approximation $u(t)$ to hunt around a relatively flat segment of the input waveform; this phenomenon is also illustrated in Fig. 5.19. The granular noise is analogous to quantization noise in a PCM system.

We thus see that there is a need to have a large step size to accommodate a wide dynamic range, whereas a small step size is required for the accurate representation of relatively low-level signals. It is therefore clear that the choice of the optimum step size that minimizes the mean-square value of the quantizing error in a linear delta modulator will be the result of a compromise between slope overload distortion and granular noise.

EXAMPLE 5 MAXIMUM OUTPUT SIGNAL-TO-NOISE RATIO FOR SINUSOIDAL MODULATION

In this example we consider the effect of quantization noise under the simplifying assumption of no slope-overload. We further assume the use of sinusoidal modulation, as shown by

$$x(t) = a_0 \cos(2\pi f_0 t)$$

The maximum slope of the signal $x(t)$ is given by

$$\max \left| \frac{dx(t)}{dt} \right| = 2\pi f_0 a_0$$

The use of Eq. 5.81 constrains the choice of step size $\Delta = 2\delta$, so as to avoid slope-overload. In particular, it imposes the following condition on the value of δ:

$$\frac{\delta}{T_s} \geq 2\pi f_0 a_0 \qquad (5.82)$$

Equivalently, we may impose the following condition on the amplitude of the sinusoidal modulation:

$$a_0 \leq \frac{2\pi f_0 T_s}{\delta} \qquad (5.83)$$

Hence, the maximum permissible value of the output signal power equals

$$P_{max} = \frac{a_0^2}{2}$$

$$= \frac{\delta^2}{8\pi^2 f_0^2 T_s^2} \qquad (5.84)$$

When there is no slope-overload, the maximum quantization error is $\pm\delta$. We assume that the quantizing error is uniformly distributed (which is a reasonable approximation for small δ). We may then use Eq. 5.32 to evaluate the variance of the quantization error, σ_Q^2. Specifically, using Eq. 5.74 for the step size Δ of a delta modulator, and substituting it in Eq. 5.32, we get the desired result in terms of δ:

$$\sigma_Q^2 = \frac{\delta^2}{3} \qquad (5.85)$$

Typically, however, the receiver contains (at its output end) a low-pass filter whose bandwidth is set equal to the message bandwidth (i.e., highest possible frequency component of the message signal), denoted as W. We note that $f_0 \leq W$. Moreover, W is small compared to the sampling rate $1/T_s$. Hence, assuming that the average power of the quantization error is uniformly distributed over a frequency interval extending from $-1/T_s$ to $1/T_s$, we get the result:

$$\text{Average output noise power} = W T_s \left(\frac{\delta^2}{3} \right) \qquad (5.86)$$

Correspondingly, the maximum value of the output signal-to-noise ratio equals

$$(\text{SNR})_{0,max} = \frac{P_{max}}{W T_s (\delta^2/3)}$$

$$= \frac{3}{8\pi^2 W f_0^2 T_s^3} \qquad (5.87)$$

Equation 5.87 shows that, under the assumption of no slope-overload distortion, the maximum output signal-to-noise ratio of a delta modulator is proportional to the sampling rate cubed. This, therefore, indicates a 9 dB improvement with doubling of the sampling rate.

By comparison, in the case of standard PCM, if we double the bit rate by doubling the number of bits per sample, we achieve a 6 dB increase in SNR for each *added* bit. For example, by doubling the bit rate from 40 to 80 kilobits per second, the SNR is increased by 9 dB using DM. On the other hand, if PCM is employed and the bit rate is similarly doubled by increasing the number of bits per sample from 5 to 10 (keeping the sampling rate fixed at 8 kHz), the SNR is improved by 30 dB. The increase of SNR with bit rate is therefore much more dramatic for PCM than for DM.

(2) Adaptive Delta Modulation

The performance of a delta modulator can be improved significantly by making the step size of the modulator assume a time-varying form. In particular, during a steep segment of the input signal the step size is increased. Conversely, when the input signal is varying slowly, the step size is reduced. In this way, the step size is *adapted* to the level of the input signal. The resulting method is called *adaptive delta modulation* (ADM).

There are several types of ADM, depending on the type of scheme used for adjusting the step size. In one type, a *discrete* set of values is provided for the step size. In another type, a *continuous* range for step-size variation is provided. There are also other types. In the sequel, we describe an example of the first type of ADM.

Figure 5.20 shows the block diagrams of the transmitter and receiver of an ADM system. In practical implementations of the system, the step size $\Delta(nT_s)$ or $2\delta(nT_s)$ is constrained to lie between minimum and maximum values. In particular, we write

$$\delta_{min} \leq \delta(nT_s) \leq \delta_{max} \tag{5.88}$$

The upper limit, δ_{max}, controls the amount of slope-overload distortion. The lower limit, δ_{min}, controls the amount of idle channel noise. Inside these limits,

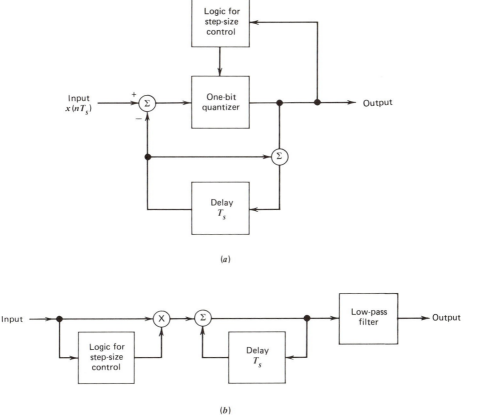

(a)

(b)

Figure 5.20 Adaptive delta modulator. (a) Transmitter. (b) Receiver.

the adaptation rule for $\delta(nT_s)$ is expressed in the general form

$$\delta(nT_s) = g(nT_s)\,\delta(nT_s - T_s) \tag{5.89}$$

where the time-varying multiplier $g(nT_s)$ depends on the present binary output $b(nT_s)$ of the delta modulator and the M previous values $b(nT_s - T_s), \ldots, b(nT_s - MT_s)$. The algorithm is initiated with a starting step size $\delta_{start} = \delta_{min}$.

A simple version of the formula in Eq. 5.89 involves the use of $b(nT_s)$ and $b(nT_s - T_s)$ only, as shown by

$$g(nT_s) = \begin{cases} K & \text{if } b(nT_s) = b(nT_s - T_s) \\ K^{-1} & \text{if } b(nT_s) \neq b(nT_s - T_s) \end{cases} \tag{5.90}$$

This adaptation algorithm is called a *constant factor ADM with one-bit memory*, where the term "one-bit memory" refers to the explicit utilization of the single previous bit $b(nT_s - T_s)$. The algorithm of Eq. 5.90, with $K = 1.5$, has been found to be well-matched to typical speech and image inputs alike, for a wide range of bit rates.* In particular, at bit rates of 20 to 60 kb/s, the use of the algorithm for speech coding realizes gains in SNR equal to 5 to 10 dB over the optimum LDM for which the constant K equals 1.

5.7 CODING SPEECH AT LOW BIT RATES

The use of PCM at the standard rate of 64 kb/s demands a high channel bandwidth for its transmission. In certain applications, however, channel bandwidth is at a premium, in which case there is a definite need for *speech coding at low bit rates, while maintaining acceptable fidelity or quality of reproduction*. A major motivation for bit rate reduction is for secure transmission over radio channels that are inherently of low capacity. The fundamental limits on bit rate suggested by speech perception and information theory show that high-quality speech coding is possible at rates considerably less than 64 kb/s (the rate may actually be as low as 2 kb/s). The price that has to be paid for attaining this advantage is increased processing *complexity* (and therefore increased cost of implementation). Also, in many coding schemes, increased complexity translates into increased processing *delay* time. (Delay is of no concern in applications that involve voice storage, as in "voice mail").

For coding speech at low bit rates, a waveform coder of prescribed configuration is optimized by exploiting both *statistical characterization of speech waveforms and properties of hearing*. In particular, the design philosophy has two aims in mind:

1. To remove redundancies from the speech signal as far as possible.
2. To assign the available bits to code the nonredundant parts of the speech signal in a perceptually efficient manner.

As we strive to reduce the bit rate from 64 kb/s (used in standard PCM) to 32,

---

* Jayant and Noll (1984, pp. 400–405). This reference also describes several other adaptation algorithms for ADM and discusses their performances for speech and image inputs.

16, 8, and 4 kb/s, the algorithms for redundancy removal and bit assignment become increasingly more sophisticated. As a rule of thumb, in the 64 to 8 kb/s range, the computational complexity (measured in terms of multiply-add operations) required to code speech increases by an order of magnitude when the bit rate is halved, for approximately equal speech quality.

In the sequel, we describe two schemes for coding speech, one at 32 kb/s and the other at 16 kb/s.*

(1)   Adaptive Differential Pulse-code Modulation

Reduction in the number of bits per sample from 8 (as used in standard PCM) to 4 involves the combined use of *adaptive quantization* and *adaptive prediction*. In this context, the term "adaptive" means being responsive to changing level and spectrum of the input speech signal. The variation of performance with speakers and speech material, together with variations in signal level inherent in the speech communication process, make the combined use of adaptive quantization and adaptive prediction necessary to achieve best performance over a wide range of speakers and speaking situations. A digital coding scheme that uses both adaptive quantization and adaptive prediction is called *adaptive differential pulse-code modulation* (ADPCM).

The term "adaptive quantization" refers to a quantizer that operates with a *time-varying* step size $\Delta(nT_s)$, where T_s is the sampling period. At any given time identified by the index n, the adaptive quantizer is assumed to have a uniform transfer characteristic. The step size $\Delta(nT_s)$ is varied so as to match the variance σ_X^2 of the input signal $x(nT_s)$. In particular, we write

$$\Delta(nT_s) = \phi \, \hat{\sigma}_X(nT_s) \tag{5.91}$$

where ϕ is a constant, and $\hat{\sigma}_X(nT_s)$ is an *estimate* of the standard deviation $\sigma_X(nT_s)$ (i.e., square root of the variance σ_X^2). For a nonstationary input, $\sigma_X(nT_s)$ is time-variable. Hence, the problem of adaptive quantization, according to Eq. 5.91, is one of estimating $\sigma_X(nT_s)$ continuously.

To proceed with the application of Eq. 5.91, we may compute the estimate $\hat{\sigma}_X(nT_s)$ in one of two ways:

1.   Unquantized samples of the input signal are used to derive forward estimates of $\sigma_X(nT_s)$, as in Fig. 5.21a.
2.   Samples of the quantizer output are used to derive backward estimates of $\sigma_X(nT_s)$, as in Fig. 5.21b.

The respective quantization schemes are referred to as *adaptive quantization with forward estimation* (AQF) and *adaptive quantization with backward estimation* (AQB).

The AQF scheme of Fig. 5.21a first goes through a learning period by buf-

---

* For a complete discussion of coding speech at low bit rates, see Jayant, (1986), Jayant and Noll (1984, pp. 188–210, 290–311), and Flanagan et al., (1979). Much of the material presented in Section 5.7 is based on these three references.

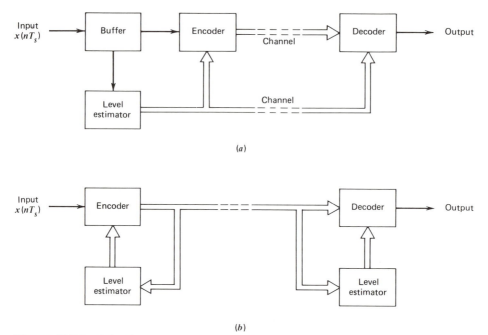

Figure 5.21 Adaptive quantization. (a) Adaptive quantization with forward estimation (AQF). (b) Adaptive quantization with backward estimation (AQB).

fering unquantized samples of the input speech signal. The samples are released after the estimate $\hat{\sigma}_X(nT_s)$ has been obtained. This estimate is obviously independent of quantizing noise. We therefore find that the step size $\Delta(nT_s)$ obtained from AQF is more reliable than that from AQB. However, the use of AQF requires the explicit transmission of level information (typically, about 5 to 6 bits per step size sample) to a remote decoder, thereby burdening the system with additional *side information* that has to be transmitted to the receiver. Also, a processing *delay* (on the order of 16 ms for speech) in the encoding operation results from the use of AQF, which is unacceptable in some applications. The problems of level transmission, buffering, and delay intrinsic to AQF are all avoided in the AQB scheme of Fig. 5.21*b* by using the recent history of the quantizer output to extract information for the computation of the step size $\Delta(nT_s)$. Accordingly, AQB is usually preferred over AQF in practice.

It is noteworthy that an adaptive quantizer with backward estimation, as in Fig. 5.21*b*, represents a nonlinear feedback system. As such, it is not obvious that the system will be stable. However, it has been shown that the system is indeed stable in the sense that if the quantizer input $x(nT_s)$ is *bounded*, then so are the backward estimate $\hat{\sigma}_X(nT_s)$ and the corresponding step size $\Delta(nT_s)$.

The use of adaptive prediction in ADPCM is justified because speech signals are inherently *nonstationary*, a phenomenon that manifests itself in the fact that the autocorrection function and power spectral density of speech signals are time-varying functions of their respective variables. This implies that the design of predictors for such inputs should likewise be time-varying, that is, adaptive. As with adaptive quantization, there are two schemes for performing adaptive prediction:

1. *Adaptive prediction with forward estimation* (APF), in which unquantized samples of the input signal are used to derive estimates of the predictor coefficients.
2. *Adaptive prediction with backward estimation* (APB), in which samples of the quantizer output and the prediction error are used to derive estimates of the predictor coefficients.

The respective schemes are shown in Fig. 5.22.

In the APF scheme of Fig. 5.22*a*, N unquantized samples of the input speech are first buffered and then released after computation of M predictor coefficients that are optimized for the buffered segment of input samples. The choice

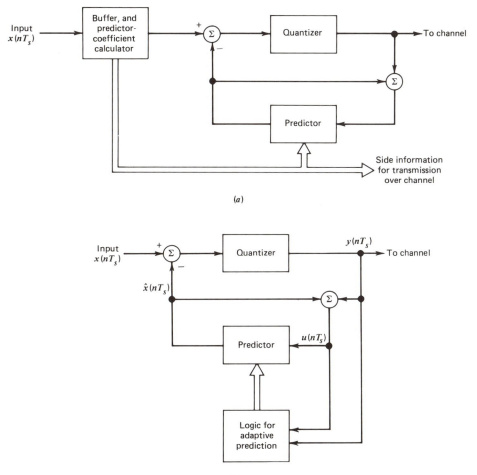

(*a*)

(*b*)

Figure 5.22 Adaptive prediction. (*a*) Adaptive prediction with forward estimation (APF). (*b*) Adaptive prediction with backward estimation (APB).

of M involves a compromise between an adequate prediction gain and an acceptable amount of side information. Likewise, the choice of learning period or buffer length N involves a compromise between the rate at which statistics of the input speech signal change and the rate at which information on predictor coefficients must be updated and transmitted to the receiver. For speech, a good choice of N corresponds to a 16 ms buffer for a sampling rate of 8 kHz, and a choice of $M = 10$ ensures adequate use of the short-term predictability of speech.

However, APF suffers from the same intrinsic disadvantages (side information, buffering, and delay) as AQF. These disadvantages are eliminated by using the APB scheme of Fig. 5.22b. Since, in the latter scheme, the optimum predictor coefficients are estimated on the basis of quantized and transmitted data, they can be updated as frequently as desired; for example, from sample to sample. Accordingly, APB is the preferred method of prediction. Continuing then with discussion of the adaptive prediction scheme shown in Fig. 5.22b, the box labeled "logic for adaptive prediction" in Fig. 5.22b is intended to represent the mechanism for updating the predictor coefficients. Let $y(nT_s)$ denote the quantizer output, where T_s is the sampling period and n is the time index. Then, from Fig. 5.22b we deduce that the corresponding sample value of the predictor input is given by

$$u(nT_s) = \hat{x}(nT_s) + y(nT_s) \tag{5.92}$$

where $\hat{x}(nT_s)$ is the prediction of the speech input sample $x(nT_s)$. We may rewrite Eq. 5.92 as

$$y(nT_s) = u(nT_s) - \hat{x}(nT_s) \tag{5.93}$$

Accordingly, with $u(nT_s)$ representing a sample value of the predictor input, and $\hat{x}(nT_s)$ representing a sample value of the predictor output, we may view $y(nT_s)$ as the corresponding value of the prediction error insofar as the adaptation process is concerned. The structure of the predictor, assumed to be of order M, is shown in Fig. 5.23. For adaptation of the predictor coefficients, we may use the *least-mean-square (LMS) algorithm** described in Chapter 3. Thus, reformulating the LMS algorithm for the problem at hand, we may write

$$\hat{h}_k(nT_s + T_s) = \hat{h}_k(nT_s) + \mu y(nT_s)u(nT_s - kT_s) \qquad k = 1, 2, \ldots, M \tag{5.94}$$

where μ is the adaptation constant. For the initial conditions, we set all of the predictor coefficients equal to zero at $n = 0$. The correction term in the update, Eq. 5.94, consists of the product $y(nT_s)u(nT_s - kT_s)$, scaled by the adaptation constant μ. By using a small value for μ, the correction term will (on the average) decrease with the number of iterations n. Indeed, for stationary

---

* A version of the LMS algorithm for the predictor and an adaptive algorithm for the quantizer have been combined in a synchronous fashion at both the encoder and decoder of a DPCM scheme. The performance of these algorithms is so impressive at 32 kb/s that ADPCM is now accepted internationally as a standard coding technique along with 64 kb/s using standard PCM. For details see Benvenuto, et al. (1986).

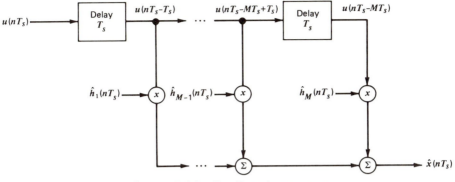

Figure 5.23 Predictor's structure.

speech inputs and small quantization effects, the correction term may assume a small enough value in the *steady-state* condition for the average mean-squared error to closely approach the minimum possible value. For *nonstationary* inputs, on the other hand, the choice of the adaptation constant μ requires extra care in that it has to be small enough for the fluctuations in the predictor coefficients about their optimum values to be acceptable, and yet large enough for the adaptation algorithm to track variations in the input statistics.

(2) Adaptive Sub-band Coding

PCM and ADPCM are both *time-domain coders* in that the speech signal is processed in the time-domain as a single full-band signal. We next describe a *frequency-domain coder,* in which the speech signal is divided into a number of sub-bands and each one is encoded separately. The coder is capable of digitizing speech at a rate of 16 kb/s with a quality comparable to that of 64 kb/s PCM. To accomplish this performance, it exploits the quasi-periodic nature of *voiced speech* and a characteristic of the hearing mechanism known as *noise masking*. (A description of voiced sounds and their distinction from unvoiced sounds is presented in Chapter 3).

Periodicity of voiced speech manifests itself in the fact that people speak with a characteristic *pitch frequency*. This periodicity permits pitch prediction, and therefore a further reduction in the level of the prediction error that requires quantization, compared to differential pulse-code modulation without pitch prediction. The number of bits per sample that needs to be transmitted is thereby greatly reduced, without a serious degradation in speech quality.

The number of bits per sample can be reduced further by making use of the *noise-masking phenomenon* in perception. That is, the human ear does not perceive noise in a given frequency band if the noise is about 15 dB below the signal level in that band. This means that a relatively large coding error (the equivalent of noise) can be tolerated near *formants,* and that the coding rate can be correspondingly reduced. In the context of speech production, the *formant frequencies* (or simply formants) are the resonance frequencies of the vocal tract tube. The formants depend on the shape and dimensions of the vocal tract.

In *adaptive sub-band coding* (ASBC), noise shaping is accomplished by *adaptive bit assignment*. In particular, the number of bits used to encode each sub-band is varied dynamically and shared with other sub-bands, such that the encoding accuracy is always placed where it is needed in the frequency-domain characterization of the signal. Indeed, sub-bands with little or no energy may not be encoded at all.

A block diagram of the adaptive sub-band coding scheme is shown in Fig. 5.24. Specifically, the speech band is divided into a number of contiguous bands by a bank of band-pass filters (typically four to eight). The output of each band-pass filter is frequency-translated to assume a low-pass form by a modulation process equivalent to single-sideband modulation. It is then sampled (or resampled) at a rate slightly higher than its Nyquist rate (twice the width of the pertinent sub-band), and then digitally encoded by using an ADPCM with fixed

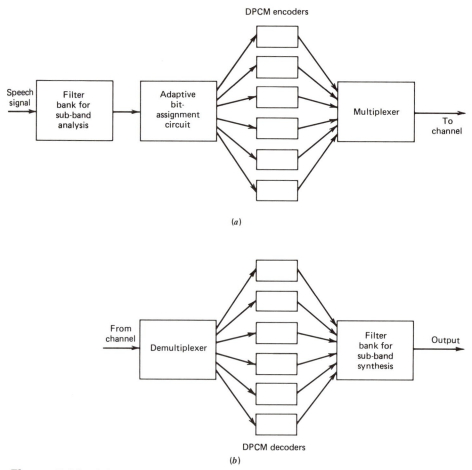

Figure 5.24 Adaptive sub-band coding scheme. (a) Transmitter. (b) Receiver.

(typically, first-order) prediction. A specific coding strategy is employed for each sub-band in accordance with perceptual criteria tailored to that band. Bit assignment information is transmitted to the receiver, enabling it to decode the sub-band signals individually and modulate them back to their original locations in the frequency band. Finally, they are summed to produce an output signal that provides a close replica of the original speech signal.

Let f_s denote the sampling rate for the (full band) input signal, and N the average number of bits used to encode a sample of the signal. The corresponding bit rate is therefore Nf_s bits per second. Clearly, we may write

$$Nf_s = (MN) \left(\frac{f_s}{M} \right)$$

where M is the number of sub-bands, assumed to be of equal widths. The sampling rate for each sub-band is recognized to be f_s/M. The implication is a total number of MN bits per sample for the M sub-bands. To illustrate the significance of this result, consider a scheme with $M = 4$ equal-width sub-bands, a standard sampling rate $f_s = 8$ kHz for normal speech, and $N = 2$ bits per sample. Then, the sampling rate for each sub-band is 2 kHz, and the total number of bits per sample for the 4 sub-bands is 8. For a speech segment with a predominance of low-frequency components, for example, we may use the bit assignment 5, 2, 1, 0 bits for the four sub-bands in increasing frequency. On the other hand, for a speech segment with a predominance of high-frequency components, the appropriate bit assignment might be 1, 1, 3, 3.

Thus, the adaptive sub-band coding scheme of Fig. 5.24 varies the assignment of available bits to the various sub-bands dynamically in accordance with the spectral content of the input speech signal, thereby helping control the shape of the overall quantizing noise spectrum as a function of frequency. Specifically, more representation levels (on the average) are used for the lower frequency bands where pitch and formant information have to be preserved. If, however, high-frequency energy is dominant in the input speech signal, the scheme automatically assigns a larger number of representation levels to the higher frequency components of the input. It is also noteworthy that the quantizing noise within any sub-band is kept within that band. That is, a low-level speech input in a sub-band of the scheme cannot be hidden by quantizing noise produced in another sub-band.

The complexity of a 16 kb/s adaptive sub-band coder is typically 100 times that of a 64 kb/s PCM coder for about the same reproduction quality. However, as a result of the large number of arithmetic operations involved in designing the adaptive sub-band coder, there is a processing delay of 25 ms; on the other hand, no such delay is encountered in the PCM coder.

(3) Subjective Quality

In coding speech, it is normal practice to supplement objective measures of performance such as signal-to-noise ratio (SNR) with subjective measures of quality, based on a *mean opinion score* (MOS). Indeed, in assessing the reproduction quality of digital speech coders (particularly at low bit rates), MOS ratings are often found to be more revealing than SNRs.

An MOS is obtained by conducting formal tests with human subjects.* An MOS of 5 represents *perfect* quality; however, such a score is hardly ever attained. A score of 4 or more means *high* quality. (In standard waveform coders, high quality is referred to by telephone engineers as "toll quality," when certain transmission specifications are also met.) An MOS exceeding 4 indicates that the reproduced speech is as intelligible to test subjects as the original and also free of distortion. A score between 3 and 4 represents *communication* quality, implying that intelligibility is still very high and that distortion is present but not obvious.

In subjective measurements, it is found that 64 kb/s PCM and 32 kb/s DPCM coders rate "high" for quality, and that the best 16 kb/s adaptive sub-band coders approach the higher bit-rate PCM coders in quality, attaining MOS ratings very close to 4. If, however, the comparison was to be made on the basis of SNR measurements, the adaptive sub-band coders (no matter how complex) perform poorly compared to the higher bit-rate PCM coders; this supports the earlier statement that SNR ratings are not always as revealing as MOS ratings.

In one respect, 16 kb/s adaptive sub-band coders fall short of 64 kb/s PCM and 32 kb/s ADPCM coders in that their quality drops sharply with tandem codings (*successive encoding–decoding stages*). On the other hand, the higher bit-rate coders maintain a high quality after as many as eight coding–decoding stages. This issue is of particular concern in a combined analog–digital transmission path. In an all-digital link, the multistage advantage of the higher bit-rate coders is not significant since the coded signals are decoded into analog form only once.

5.8 APPLICATIONS

In this section of the chapter, we describe two related applications: (1) hierarchy of *digital multiplexers,* whereby digitized voice and video signals as well as digital data are combined into one final data stream, and (2) *light wave transmission link* that is well-suited for use in a long-haul telecommunication network.

(1) Digital Multiplexers

In Chapter 4 we introduced the idea of time-division multiplexing whereby a group of analog signals (e.g., voice signals) are sampled sequentially in time at a *common* sampling rate and then multiplexed for transmission over a common line. In this section we consider the multiplexing of digital signals at different bit rates.† This enables us to combine several digital signals, such as computer

---

* The use of mean opinion score as a subjective measure of quality is discussed in the following references: Daumer (1982) and Jayant and Noll (1985, pp. 10–12). For a comparison of digital speech coders, based on MOS ratings, see Jayant (1986).

† For additional information on digital multiplexers, see Bell Telephone Laboratories (1970, Chapter 26).

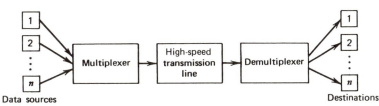

Figure 5.25 Conceptual diagram of multiplexing–demultiplexing.

outputs, digitized voice signals, digitized facsimile and television signals, into a single data stream (at a considerably higher bit rate than any of the inputs). Figure 5.25 shows a conceptual diagram of the digital multiplexing–demultiplexing operation.

The multiplexing of digital signals may be accomplished by using *a bit-by-bit interleaving procedure* with a selector switch that sequentially takes a bit from each incoming line and then applies it to the high-speed common line. At the receiving end of the system the output of this common line is separated out into its low-speed individual components and then delivered to their respective destinations.

Two major groups of digital multiplexers are used in practice:

1. One group of multiplexers is designed to combine relatively low-speed digital signals, up to a maximum rate of 4800 bits per second, into a higher speed multiplexed signal with a rate of up to 9600 bits per second. These multiplexers are used primarily to transmit data over voice-grade channels of a telephone network. Their implementation requires the use of *modems* in order to convert the digital format into an analog format suitable for transmission over telephone channels. The theory of a modem (*mo*dulator–*dem*odulator) is covered in Chapter 7.

2. The second group of multiplexers, designed to operate at much higher bit rates, forms part of the data transmission service generally provided by communication carriers. For example, Fig. 5.26 shows a block diagram of the digital hierarchy based on the T1 carrier, which has been developed by the Bell System. The T1 carrier system, described below, is designed to operate at 1.544 megabits per second, the T2 at 6.312 megabits per second, the T3 at 44.736 megabits per second, and the T4 at 274.176 megabits per second. The system is thus made up of various combinations of lower order T-carrier subsystems designed to accommodate the transmission of voice signals, Picturephone® service, and television signals by using PCM, as well as (direct) digital signals from data terminal equipment.

There are some basic problems involved in the design of a digital multiplexer, irrespective of its grouping:

1. Digital signals cannot be directly interleaved into a format that allows for their eventual separation unless their bit rates are locked to a common clock. Accordingly, provision has to be made for *synchronization* of the incoming digital signals, so that they can be properly interleaved.

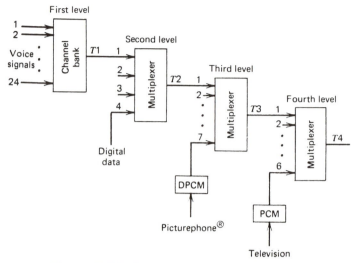

Figure 5.26 Digital hierarchy, Bell system.

2. The multiplexed signal must include some form of *framing,* so that its individual components can be identified at the receiver.
3. The multiplexer has to handle small variations in the bit rates of the incoming digital signals. For example, a 1000-kilometer coaxial cable carrying 3×10^8 pulses per second will have about one million pulses in transit, with each pulse occupying about one meter of the cable. A 0.01 percent variation in the propagation delay, produced by a 1°F decrease in temperature, will result in 100 fewer pulses in the cable. Clearly, these pulses must be absorbed by the multiplexer.

   In order to cater for the requirements of synchronization and rate adjustment to accommodate small variations in the input data rates, we may use a technique known as *bit stuffing.* The idea here is to have the outgoing bit rate of the multiplexer slightly higher than the sum of the maximum expected bit rates of the input channels by stuffing in additional non-information carrying pulses. All incoming digital signals are stuffed with a number of bits sufficient to raise each of their bit rates to equal that of a locally generated clock. To accomplish bit stuffing, each incoming digital signal or bit stream is fed into an *elastic store* at the multiplexer. The elastic store is a device that stores a bit stream in such a manner that the stream may be read out at a rate different from the rate at which it is read in. At the demultiplexer, the stuffed bits must obviously be removed from the multiplexed signal. This requires a method that can be used to identify the stuffed bits. To illustrate one such method, and also show one method of providing frame synchronization, we describe the signal format of the Bell System *M12 multiplexer,* which is designed to combine four T1 bit streams into one T2 bit stream. We begin the description by considering the T1 system first and then the M12 multiplexer.

T1 System

The *T1 carrier system* is designed to accommodate 24 voice channels primarily for short-distance, heavy usage in metropolitan areas. The T1 system was pioneered by the Bell System in the United States in the early 1960s; with its introduction the shift to digital communication facilities started.* The T1 system has been adopted for use throughout the United States, Canada, and Japan. It forms the basis for a complete hierarchy of higher order multiplexed systems that are used for either long-distance transmission or transmission in heavily populated urban centers.

A voice signal (male or female) is essentially limited to a band from 300 to 3400 Hz in that frequencies outside this band do not contribute much to articulation efficiency. Indeed, telephone circuits that respond to this range of frequencies give quite satisfactory service. Accordingly, it is customary to pass the voice signal through a low-pass filter with a cutoff frequency of about 3.4 kHz prior to sampling. Hence, with $W = 3.4$ kHz, the nominal value of the Nyquist rate is 6.8 kHz. The filtered voice signal is usually sampled at a slightly higher rate, namely, 8 kHz, which is the *standard* sampling rate in telephone systems.

For companding, the T1 system uses a *piecewise-linear* characteristic (consisting of 15 linear segments) to approximate μ-law companding of Eq. 5.61 with the constant $\mu = 255$. This approximation is constructed in such a way that the segment-end points lie on the compression curve computed from Eq. 5.61, and their projections onto the vertical axis are spaced uniformly. Table 5.3 gives the projections of the segment-end points onto the horizontal axis, and the step sizes of the individual segments. The table is normalized to 8159, so that all values are represented as integer numbers. Segment 0 of the approximation is a colinear segment, passing through the origin; it contains a total of 32 uniform quantizing levels. Linear segments 1a, 2a, . . . , 7a lie above the horizontal axis, whereas linear segments 1b, 2b, . . . , 7b lie below the horizontal axis; each of these 14 segments contains 16 uniform representation

* For a description of the original version of the T1 PCM system, see Fultz and Penick (1965). The description given here is based on an updated version of this system; see Henning and Pan (1972).

Table 5.3   The 15-Segment Companding Characteristic ($\mu = 255$)

Linear segment number	Step size	Projections of segment-end point onto the horizontal axis
0	2	±31
1a, 1b	4	±95
2a, 2b	8	±223
3a, 3b	16	±479
4a, 4b	32	±991
5a, 5b	64	±2015
6a, 6b	128	±4063
7a, 7b	256	±8159

levels. For colinear segment 0 the representation levels at the compressor input are ± 1, ± 3, . . . , ± 31, and the corresponding compressor output levels are 0, ± 1, . . . , ± 15. For linear segments 1a and 1b, the representation levels at the compressor input are ± 35, ± 39, . . . , ± 95, and the corresponding compressor output levels are ± 16, ± 17, . . . , ± 31, and so on for the other linear segments.

There are a total of $31 + 14 \times 16 = 255$ output levels associated with the 15-segment companding characteristic described above. To accommodate this number of output levels, each of the 24 voice channels uses a binary code with an 8-bit word. The first bit indicates whether the input voice sample is positive or negative; this bit is a 1 if positive and a 0 if negative. The next three bits of the code word identify the particular segment inside which the amplitude of the input voice sample lies, and the last four bits identify the actual quantizing step inside that segment.

With a sampling rate of 8 kHz, each frame of the multiplexed signal occupies a period of 125 μs. In particular, it consists of twenty-four 8-bit words, plus a single bit that is added at the end of the frame for the purpose of synchronization. Hence, each frame consists of a total of $24 \times 8 + 1 = 193$ bits. Correspondingly, the duration of each bit equals 0.647 μs, and the corresponding bit rate is 1.544 megabits per second.

In addition to the voice signal, a telephone system must also pass special supervisory signals to the far end. This *signaling information* is needed to transmit dial pulses, as well as telephone off-hook/on-hook signals. In the T1 system this requirement is accomplished as follows. Every sixth frame, the least significant (that is, the eighth) bit of each voice channel is deleted and a *signaling bit* is inserted in its place, thereby yielding on average $7\frac{5}{6}$-bit operation for each voice input. The sequence of signaling bits is thus transmitted at a rate equal to the sampling rate divided by six, that is, 1.333 kilobits per second.

For two reasons, namely, the assignment of the eighth digit in every sixth frame to signaling, and the need for two signaling paths for some switching systems, it is necessary to identify a super frame of 12 frames in which the sixth and twelfth frames contain two signaling paths. To accomplish this identification and still allow for rapid synchronization of the receiver framing circuitry, the frames are divided into *odd* and *even* frames. In the odd-numbered frames, the 193rd digit is made to alternate between 0 and 1. Accordingly, the framing circuit searches for the pattern 101010101010 . . . to establish frame synchronization. In the even-numbered frames, the 193rd digit is made to follow the pattern 000111000111 . . . . This makes it possible for the receiver to identify the sixth and twelfth frames as those that follow a 01 transition or 10 transition of this digit, respectively. Figure 5.27 depicts the signaling format of the T1 system.

M12 Multiplexer

Figure 5.28 illustrates the signal format of the M12 multiplexer. Each frame is subdivided into four subframes. The first subframe (first line in Fig. 5.28) is transmitted, then the second, the third, and the fourth, in that order.

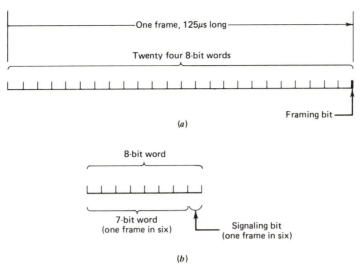

Figure 5.27 T1 bit stream format. (*a*) Coarse structure of a frame. (*b*) Frame structure of a word.

Bit-by-bit interleaving of the incoming four T1 bit streams is used to accumulate a total of 48 bits, 12 from each input. A *control bit* is then inserted by the multiplexer. Each frame contains a total of 24 control bits, separated by sequences of 48 data bits. Three types of control bits are used in the M12 multiplexer to provide synchronization and frame indication, and to identify which of the four input signals has been stuffed. These control bits are labeled as F, M, and C in Fig. 5.28. Their functions are as follows:

1. The F-control bits, two per subframe, constitute the *main* framing pulses. The subscripts on the F-control bits denote the actual bit (0 or 1) transmitted. Thus the main framing sequence is $F_0F_1F_0F_1F_0F_1F_0F_1$ or 01010101.

2. The M-control bits, one per subframe, form *secondary* framing pulses to identify the four subframes. Here again the subscripts on the M-control bits denote the actual bit (0 or 1) transmitted. Thus the secondary framing sequence is $M_0M_1M_1M_1$ or 0111.

3. The C-control bits, three per subframe are *stuffing indicators*. In particular, C_I refers to input channel I, C_{II} refers to input channel II, and so forth. For example, the three C-control bits in the first subframe following M_0 in the first subframe are stuffing indicators for the first T1 signal. The insertion of

M_0	[48]	C_I	[48]	F_0	[48]	C_I	[48]	C_I	[48]	F_1	[48]
M_1	[48]	C_{II}	[48]	F_0	[48]	C_{II}	[48]	C_{II}	[48]	F_1	[48]
M_1	[48]	C_{III}	[48]	F_0	[48]	C_{III}	[48]	C_{III}	[48]	F_1	[48]
M_1	[48]	C_{IV}	[48]	F_0	[48]	C_{IV}	[48]	C_{IV}	[48]	F_1	[48]

Figure 5.28 Signal format of Bell system M12 multiplexer.

a stuffed bit in this T1 signal is indicated by setting all three C-control bits to 1. To indicate no stuffing, all three are set to 0. If the three C-control bits indicate stuffing, the stuffed bit is located in the position of the first information bit associated with the first T1 signal that follows the F_1-control bit in the same subframe. In a similar way, the second, third, and fourth T1 signals may be stuffed, as required. By using *majority logic decoding* in the receiver, a single error in any of the three C-control bits can be detected. This form of decoding means simply that the majority of the C-control bits determine whether an all-one or all-zero sequence was transmitted. Thus three 1s or combinations of two 1s and a 0 indicate that a suffed bit is present in the information sequence, following the control bit F_1 in the pertinent subframe. On the other hand, three 0s or combinations of two 0s and a 1 indicate that no stuffing is used.

The demultiplexer at the receiving M12 unit first searches for the main framing sequence $F_0F_1F_0F_1F_0F_1F_0F_1$. This establishes identity for the four input T1 signals and also for the M- and C- control bits. From the $M_0M_1M_1M_1$ sequence, the correct framing of the C-control bits is verified. Finally, the four T1 signals are properly demultiplexed and destuffed.

The signal format just described has two safeguards:

1. It is possible, although unlikely, that with just the $F_0F_1F_0F_1F_0F_1F_0F_1$ sequence, one of the incoming T1 signals may contain a similar sequence. This could then cause the receiver to lock onto the wrong sequence. The presence of the $M_0M_1M_1M_1$ sequence provides verification of the genuine $F_0F_1F_0F_1F_0F_1F_0F_1$ sequence, thereby ensuring that the four T1 signals are properly demultiplexed.
2. The single-error correction capability built into the C-control bits ensures that the four T1 signals are properly destuffed.

The capacity of the M12 multiplexer to accommodate small variations in the input data rates can be calculated from the format of Fig. 5.28. In each M frame, defined as the interval containing one cycle of $M_0M_1M_1M_1$ bits, one bit can be stuffed into each of four input T1 signals. Each such signal has

$$12 \times 6 \times 4 = 288 \text{ positions in each } M \text{ frame}$$

Also the T1 signal has a bit rate equal to 1.544 megabits per second. Hence, each input can be incremented by

$$1.544 \times 10^3 \times \frac{1}{288} = 5.4 \text{ kilobits/s}$$

This result is much larger than the expected change in the bit rate of the incoming T1 signal. It follows therefore that the use of only one stuffed bit per input channel in each frame is sufficient to accommodate expected variations in the input signal rate.

The local clock that determines the outgoing bit rate also determines the nominal *stuffing rate S,* defined as the average number of bits stuffed per channel in any frame. The M12 multiplexer is designed for $S = 1/3$. Accordingly, the nominal bit rate of the T2 line is

$$1.544 \times 4 \times \frac{49}{48} \times \frac{288}{288\text{-}S} = 6.312 \text{ megabits/s}$$

This also ensures that the nominal T2 clock frequency is a multiple of 8 kHz (the nominal sampling rate of a voice signal), which is a desirable feature.

(2) Lightwave Transmission

Optical fiber waveguides have unique characteristics that make them highly attractive as a transmission medium. In particular, their low transmission losses and high bandwidths are important for long-haul, high-speed communications. Other advantages include small size, light weight, and immunity to electromagnetic interference.

Consider the basic optical fiber link shown in Fig. 5.29 consisting of a transmitter, an optical fiber waveguide, and a receiver. The binary data fed into the transmitter input may represent the output of a multiplexer in the digital heirarchy of Fig. 5.26. In any event, the transmitter emits pulses of optical *power,* with each pulse being "on" or "off" in accordance with the input data. Note that power is a baseband quantity that varies at the data (i.e., modulation) rate and *not* the optical frequency. For the *light source,* we may use a laser injection diode or a semiconductor light emitting diode (LED). In a system design, the choice of the light source determines the optical signal power available for transmission. The *driver* for the light source, typically, consists of a high-current–low-voltage device. The light source is thus turned on and off by switching the drive current on and off in a corresponding manner.

The on–off light pulses produced by the transmitter are launched into the optical fiber waveguide. The *collector efficiency* of the fiber depends on its core diameter and acceptance angle. We thus have to account for a *source-to-fiber coupling loss* that varies over a wide range, depending on the particular combination of light source and optical fiber selected. During the course of propagation along the fiber, a light pulse suffers an additional loss or attenuation that increases exponentially with distance; we refer to this as *fiber loss.* Another important phenomenon that occurs during propagation is *dispersion,* which causes light originally concentrated into a short pulse to spread out into a broader pulse as it propagates along the optical fiber waveguide.

At the receiver, the original input data are *regenerated* by performing three basic operations in the following order:

1. *Detection,* whereby light pulses impinging on the receiver input are converted back into pulses of electrical current.
2. *Pulse shaping and timing* that involves amplification, filtering, and equalization of the electrical pulses, as well as the extraction of timing information.
3. *Decision making,* according to which a particular received pulse is declared "on" or "off."

Typically, the detector consists of a photodiode that responds to light. The

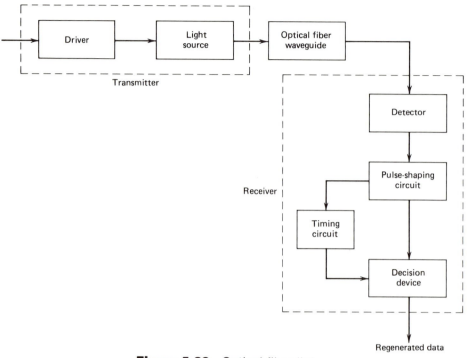

Figure 5.29 Optical fiber link.

choice of the detector and its associated circuitry determines the *receiver sensitivity*. It is also important to recognize that since the optical power is modulated at the transmitter, and since the optical fiber waveguide operates linearly on the propagation power, the detector behaves as a *linear device that converts power to current*.

From this discussion, we see that a lightwave transmission link differs from its coaxial cable counterpart in that power, rather than current (both baseband quantities), propagates through the optical fiber waveguide; otherwise, their individual block diagrams look very much alike. It differs from its microwave counterpart in that a lightwave receiver employs direct detection of the incoming signal rather than first performing a heterodyne downconversion.

In the design of a lightwave transmission link, two separate factors have to be considered: *transmission bandwidth* and *signal losses*. The transmission bandwidth of an optical fiber is determined by the dispersion phenomenon. This, in turn, limits the feasible data rate or the rate at which light pulses can propagate through the optical fiber waveguide. The signal losses are contributed by source-to-fiber coupling loss, fiber loss, fiber-to-fiber loss (due to joining the fiber by a permanent splice or the use of a demountable connector), and fiber-to-detector coupling loss. The fiber loss is insensitive to the input data rate, but it varies with wavelength. In any case, knowing these losses, and knowing the optical power available from the light source, we can determine the power available at the detector.

5.9 SUMMARY AND DISCUSSION

In this chapter, we have considered various coding techniques for analog signals, which may be summarized in point form as follows:

1. The standard *pulse-code modulation* (PCM) is conceptually straightforward. It can be highly demanding in bandwidth requirement. For speech coding, the use of PCM at 64 kb/s (corresponding to a sampling rate of 8 kHz and a code-word length of 8 bits) provides a high quality of reproduction. The only special algorithm used to attain it is that of a nonuniform memoryless quantization based on the μ-law companding with $\mu = 255$. Indeed, the quality of reproduction attained by this means is so high at the receiving end of a telephone line that few people can discern whether the voice signal at the other end has been transmitted digitally.

2. The use of *linear delta modulation* (LDM) offers simplicity of design. However, it is inferior to PCM in performance. Its performance can be improved by the use of an adaptive algorithm for step-size adjustment.

3. The use of *adaptive differential pulse-code modulation* (ADPCM) makes it possible to reduce the bit rate of coded speech by a factor of two, compared to the 64 kb/s PCM coder. This significant reduction in bit rate is attained by increasing system complexity through the combined use of adaptive quantization and adaptive prediction, both of which are based on backward estimation algorithms that utilize waveform memory.

4. In *adaptive sub-band coding,* the bit rate for speech coding is reduced yet again (down to 16 kb/s) by a further increase in system complexity. The reduction in bit rate is made possible by the use of an adaptive bit assignment algorithm, whereby the available number of bits is dynamically shared between a predetermined number of sub-bands in accordance with the frequency content of the input speech signal.

The ADPCM and adaptive sub-band coding techniques (both of which are examples of waveform coding) and the Huffman and linear predictive vocoders (both of which are examples of source coding that were considered in Chapters 2 and 3, respectively) share a common feature: they all achieve bit-rate reductions in their own individual ways by removing redundancy from the input signal.

In another important redundancy-removal technique, called *vector quantization,* a *block of samples* is quantized together instead of individual samples. A vector quantizer (like an adaptive quantizer) depends on utilization of waveform memory. It is of particular interest for low bit-rate speech applications.*

Another point that needs to be stressed is although much of the discussion on signal coding techniques has focused on speech signals, nevertheless, the use of PCM, DM, ADPCM, and linear predictive coding is equally applicable to the digitization of video and image signals.

---

* For tutorial papers on vector quantization, see the following two references: Makhoul, Roucos, and Gish (1985) and Gersho and Cuperman (1983).

PROBLEMS

P5.1   PULSE-CODE MODULATION

Problem 5.1.1

(a)   A sinusoidal signal, with an amplitude of 3.25 volts, is applied to a uniform quantizer of the midtread type whose output takes on the values 0, ±1, ±2, ±3 volts, as in Fig. P5.1a. Sketch the waveform of the resulting quantizer output for one complete cycle of the input.

(b)   Repeat this evaluation for the case when the quantizer is of the midriser type whose output takes on the values ±0.5, ±1.5, ±2.5, ±3.5 volts, as in Fig. P5.1b.

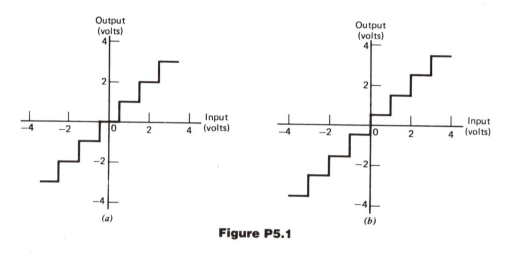

(a)   (b)

Figure P5.1

Problem 5.1.2   The signal

$$m(t) = 6 \sin(2\pi t) \text{ volts}$$

is transmitted using a 4-bit binary PCM system. The quantizer is of the midriser type, with a step size of 1 volt. Sketch the resulting PCM wave for one complete cycle of the input. Assume a sampling rate of four samples per second, with samples taken at $t = \pm 1/8, \pm 3/8, \pm 5/8, \ldots,$ seconds.

Problem 5.1.3   Figure P5.2 shows a PCM wave in which the amplitude levels of +1 volt and −1 volt are used to represent binary symbols 1 and 0, respectively. The code-word used consists of three bits. Find the sampled version of an analog signal from which this PCM wave is derived.

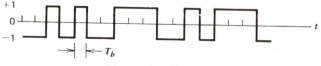

Figure P5.2

P5.2   CHANNEL NOISE AND ERROR PROBABILITY

Problem 5.2.1   A binary PCM system, using on–off signaling, operates just above the error threshold with an average probability of error equal to 10^{-6}. Suppose that the filter bandwidth at the receiver input is doubled. Find the new value of the average probability of error. You may use Table E.1 of Appendix E to evaluate the complementary error function.

Problem 5.2.2   Consider a binary PCM system that uses polar signaling with symbols 1 and 0 represented by $+a/2$ and $-a/2$ volts, respectively. The receiver is shown in Fig. 5.6 where the additive noise $w(t)$ is the sample function of a Gaussian process with zero mean and power spectral density $N_0/2$. Calculate the average probability of symbol error for this method of signaling.

Problem 5.2.3   A continuous-time signal is sampled and then transmitted as a PCM wave. The additive noise at the input of the decision device in the receiver has a variance of 0.01 volts^2.

(a)   Assuming the use of nonreturn-to-zero polar signaling, determine the pulse amplitude that must be transmitted for the average error rate not to exceed 1 bit in 10^8 bits.

(b)   If the added presence of interference causes the error rate to increase to 1 bit in 10^6 bits, what is the variance of the interference?

Problem 5.2.4   In a binary PCM system, symbols 0 and 1 have a priori probabilities p_0 and p_1, respectively. The conditional probability density function of a random variable X (with sample value x) obtained by observing (at some fixed time) the received signal, given that symbol 0 was transmitted, is denoted by $f_X(x|0)$. Similarly, $f_X(x|1)$ denotes the conditional probability density function of X, given that symbol 1 was transmitted. Let η denote the threshold used in the receiver, so that if the sample value x exceeds η, the receiver decides in favor of 1; otherwise, it decides in favor of 0.

(a)   Show that the optimum threshold η_{opt}, for which the average probability of error is a minimum, is given by the solution of the equation:

$$\frac{f_X(\eta_{\text{opt}}|1)}{f_X(\eta_{\text{opt}}|0)} = \frac{p_0}{p_1}$$

(b)   For the case of polar signals in additive Gaussian noise of zero mean and variance σ^2, show that the optimum threshold is given by

$$\eta_{\text{opt}} = \frac{\sigma^2}{a} \ln\left(\frac{p_0}{p_1}\right)$$

where $\pm a/2$ define the voltage levels corresponding to symbols 0 and 1.

(c)   For the case of on–off signaling, show that the corresponding result is

$$\eta_{\text{opt}} = \frac{a}{2} + \frac{\sigma^2}{a} \ln\left(\frac{p_0}{p_1}\right)$$

where a is the height of the pulse representing symbol 1.

Problem 5.2.5   Consider a chain of $(n - 1)$ regenerative repeaters, with a total of n sequential decisions made on a binary PCM wave, including the final decision made at the receiver. Assume that any binary symbol transmitted through the system has an independent probability p_1 of being inverted by any repeater. Let p_n represent the probability that a binary symbol is in error after transmission through the complete system.

(a) Show that

$$p_n = \tfrac{1}{2}[1 - (1 - 2p_1)^n]$$

(b) If p_1 is very small and n is not too large, what is the approximate value of p_n?

P5.3 QUANTIZATION NOISE AND SIGNAL-TO-NOISE RATIO

Problem 5.3.1 Consider a uniform quantizer characterized by the input–output relation illustrated in Fig. 5.2a. Assume that a Gaussian-distributed random variable with zero mean and unit variance is applied to this quantizer input. Find the probability density function of the discrete random variable at the quantizer output.

Problem 5.3.2 A PCM system uses a uniform quantizer followed by a 7-bit binary encoder. The bit rate of the system is equal to 50×10^6 bits per second.

(a) What is the maximum message bandwidth for which the system operates satisfactorily?
(b) Determine the output signal-to-quantizing noise ratio when a full-load sinusoidal modulating wave of frequency 1 MHz is applied to the output.

P5.4 ROBUST QUANTIZATION

Problem 5.4.1
(a) A sine wave is applied to the input of a compressor using the μ-law with $\mu = 255$. Plot the waveform of one complete cycle of the compressor output.
(b) Repeat your calculation for a compressor using the A-law with $A = 87.56$.

Problem 5.4.2 Show that for $\mu = A$ the μ-law and A-law have the same companding gain G_c.

Problem 5.4.3
(a) Show that for small signals the signal-to-noise ratios for a nonuniform quantizer using the A-law companding and a uniform quantizer are related by

$$10 \log_{10}[(\text{SNR})_{\text{A-law}}] = 10 \log_{10}[(\text{SNR})_{\text{uniform}}] + 20 \log_{10}G_c$$

where G_c is the companding gain.
(b) Given that $A = 87.56$, find the equivalent increase of SNR produced by the use of A-law companding. How many bits does this equivalent increase in SNR represent?

Problem 5.4.4 Show that the use of A-law companding provides a ratio of maximum step size to minimum step size equal to the parameter A.

Problem 5.4.5 Consider a Laplacian-distributed random variable X with probability density function

$$f_X(x) = \frac{1}{\sqrt{2}\sigma_X} \exp\left(- \frac{\sqrt{2}|x|}{\sigma_X}\right)$$

The variance σ_X^2 equals 0.02. The random variable X is applied to a compressor with transfer characteristic $c(x)$ and overload level $x_{\text{max}} = 1$.

(a) Find the probability density function of the compressor output, with $c(x)$ defined by the μ-law, $\mu = 100$.

(b) Repeat your calculation for a compressor that uses the A-law, $A = 87.56$.

P5.5   DIFFERENTIAL PULSE-CODE MODULATION

Problem 5.5.1   In the DPCM system of the open-loop type, depicted in Fig. P5.3, show that in the absence of transmission noise the transmitting and receiving prediction filters operate on slightly different input signals.

Problem 5.5.2

(a) Consider a DPCM system whose transmitter uses a first-order predictor optimized in the minimum mean-square sense. Calculate the prediction gain of the system for the following values of correlation coefficient for the message signal:

$$\text{(i)}\ \rho_1 = \frac{R_x(1)}{R_x(0)} = 0.825$$

$$\text{(ii)}\ \rho_1 = \frac{R_x(1)}{R_x(0)} = 0.950$$

The correlation coefficient ρ_1 equals the autocorrelation function of the message signal for lag T, normalized with respect to the mean-square value of the signal.

(b) Suppose that the predictor is made suboptimal by setting the coefficient $h_1 = 1$. Calculate the resulting values of the prediction gain for the two values of ρ_1 specified in part (a). What is the minimum value of ρ_1 for which the suboptimum system produces a prediction gain greater than one?

Problem 5.5.3   The predictor in a DPCM system operates on quantized samples of the message signal. Assume that the quantization is fine enough to justify the following:

(a) The quantizing error is orthogonal to the incoming message signal.

(b) The quantizing error is a stationary white noise process.

(c) The output signal-to-noise ratio is large compared to unity.

Show that, under these conditions, the performance is the same as if the predictor was operating on the original message signal.

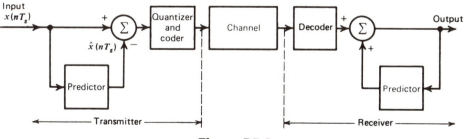

Figure P5.3

Problem 5.6.1 The ramp signal $x(t) = at$ is applied to a delta modulator that operates with a sampling period T_s and step size $\Delta = 2\delta$.

(a) Show that slope-overload distortion occurs if $\delta < aT_s$.
(b) Sketch the modulator output for the following three values of step size:

   (i) $\delta = 0.75aT_s$
   (ii) $\delta = aT_s$
   (iii) $\delta = 1.25aT_s$

Problem 5.6.2 Consider a speech signal with maximum frequency of 3.4 kHz and maximum amplitude of 1 volt. This speech signal is applied to a delta modulator whose bit rate is set at 20 kilobits per second. Discuss the choice of an appropriate step size for the modulator.

Problem 5.6.3 Consider a delta modulator whose receiver does *not* include a low-pass filter as in Example 5. Hence, show that the maximum output signal-to-noise ratio of such a modulator (assuming no slope-overload distortion) increases by 6 dB when the sampling rate is doubled. What is the improvement that results from the use of a low-pass filter at the receiver output?

Problem 5.6.4 The idle channel noise in a delta modulator (that includes a low-pass filter with a cutoff frequency just high enough to pass the message signal) is negligibly small. Justify the validity of this statement.

Problem 5.6.5 Consider an adaptive delta modulator, for which the input signal $x(nT_s)$ is given by

$$x(nT_s) = \begin{cases} -0.5 & n < 5 \\ 19.5 & 5 \le n \le 15 \\ -0.5 & 15 < n \end{cases}$$

The adaptation algorithm for the modulator is defined by Eq. 5.90 with the constant K equal to 2. The maximum and minimum permissible values of the step size are as follows:

$$\delta_{max} = 8$$

$$\delta_{min} = 1$$

The initial conditions are given by

$$\delta_{start} = 1$$

$$u(0) = 1$$

$$b(0) = -1$$

Plot the staircase approximation $u(nT_s)$ and the binary output $b(nT_s)$ for $0 \le n \le 25$.

P5.7 CODING SPEECH AT LOW BIT RATES

Problem 5.7.1 Equation 5.91 defines the formula for the step size $\Delta(n)$ of an AQB in terms of the standard deviation estimate $\hat{\sigma}_X(n)$. To obtain this estimate, it is proposed to use the following formula for the corresponding variance estimate $\hat{\sigma}_X^2(n)$ in terms of samples of the quantizer output:

$$\hat{\sigma}_X^2(n) = (1 - \lambda) \sum_{k=1}^{\infty} \lambda^{k-1} y^2(n - k)$$

where $y(n - k)$, $k = 1, 2, \ldots$, are previous quantizer outputs, and λ is an *exponential weighting factor* that lies in the range $0 < \lambda < 1$. Its inclusion accounts for the nonstationarity of the quantizer input, since the magnitude of the weighting decreases monotonically with the age of the y^2 term in question. The factor $(1 - \lambda)$ normalizes the sum of weights (i.e., $1, \lambda, \lambda^2, \ldots$) to unity.

Show that the formula for computing $\hat{\sigma}_X^2(n)$ can be written in the recursive form

$$\hat{\sigma}_X^2(n) = \lambda \, \hat{\sigma}_X^2(n - 1) + (1 - \lambda) \, y^2(n - 1)$$

Problem 5.7.2   A sub-band coding scheme involves a total of 7 sub-bands. The sampling rate for the full-band input signal is 8 kHz, and the average number of bits used to encode a sample of the input is 2. Calculate (a) the sampling rate for each sub-band, and (b) the total number of bits per sample for the group of 7 sub-bands.

P5.8   APPLICATIONS

Problem 5.8.1   In the first version of the T1-system, the code-word for each of the 24 voice channels has 7 bits. At the end of each code-word, a single bit is added for the purpose of signaling. Finally, a single bit is added at the end of each frame for synchronization. How much noisier is this T1 system compared to the version described in Section 5.8?

Problem 5.8.2   A block diagram of the overall T carrier TDM/PCM telephone system is shown in Fig. 5.26. It consists of various lower order T-carrier subsystems interconnected so as to accommodate the requirements of voice channels, Picturephone® service, and commercial television network programming. Given that the data rate for one Picturephone® service is 6.312 megabits per second, and that for one television service is 44.736 megabits per second, determine the capacity of each system level measured in terms of the number of (a) voice, (b) picturephone, or (c) television channels that it can accommodate.

Problem 5.8.3   Consider a lightwave transmission link with the following parameters:

Drive current	250 mA
LED light source	40 mW/A
Light source-to-fiber coupling loss	10 dB
Fiber loss	3.5 dB/km
Fiber-to-fiber loss	0.5 dB/splice
Fiber-to-detector coupling loss	2 dB
Detector	0.5 A/W

The receiver is separated from the transmitter by 6 km. The optical fiber is available in lengths of 2 km. Calculate (a) the available optical power at the detector input, and (b) the electrical current produced at the detector output.

CHAPTER SIX

BASEBAND SHAPING FOR DATA TRANSMISSION

In the previous chapter, we described various techniques for converting an analog signal into digital form. There is another way in which digital data can arise in practice; the data may represent the output of a source of information that is inherently discrete in nature (e.g., a digital computer). When digital data (of whatever origin) are transmitted through a band-limited channel, *dispersion* in the channel causes an overlap in time between successive symbols. This form of distortion, known as *intersymbol interference,* can pose a serious problem to the quality of reception if it is left uncontrolled. In this chapter, we present techniques that enable perfect reception in the absence of noise. The techniques are based on *shaping the baseband response* of the system. As such, they are applicable to baseband data transmission systems using coaxial cables or optical fiber waveguides.

We begin the discussion by expanding on the characteristics of *waveforms (formats)* for the baseband representation of binary data, an issue that was briefly considered in Section 5.1.

6.1 DISCRETE PAM SIGNALS

The use of an appropriate waveform for *baseband* representation of digital data is basic to its transmission from a source to a destination. Figure 6.1 shows four different formats for the representation of the binary data sequence 0110100011.

In the *unipolar format* (also known as *on–off signaling*), symbol 1 is represented by transmitting a pulse, whereas symbol 0 is represented by switching off the pulse. When the pulse occupies the full duration of a symbol, the unipolar format is said to be of the *nonreturn-to-zero* (NRZ) type. When it occupies a fraction (usually one-half) of the symbol duration, it is said to be of the *return-to-zero* (RZ) type. The NRZ version of the unipolar format (based on a rectangular pulse) is shown in Fig. 6.1a. The unipolar format offers simplicity of implementation. However, it contains a dc component that is often found to be objectionable.

In the *polar format,* a positive pulse is transmitted for symbol 1, and a negative pulse for symbol 0. It can be of the NRZ or RZ type. The NRZ version of the polar format (using rectangular pulses) is depicted in Fig. 6.1b. Unlike

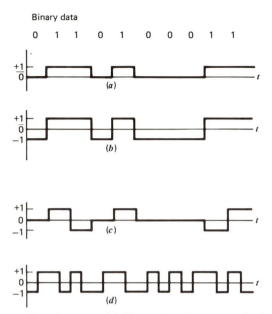

Figure 6.1   Binary data formats. (*a*) Nonreturn-to-zero unipolar format. (*b*) Non-return-to-zero polar format. (*c*) Nonreturn-to-zero bipolar format. (*d*) Manchester format.

the unipolar waveform, a polar waveform has no dc component, provided that the 0s and 1s in the input data occur in equal proportion.

In the *bipolar format* (also known as *pseudoternary signaling*), positive and negative pulses are used alternately for the transmission of 1s (with the alternation taking place at every occurrence of a 1), and no pulses for the transmission of 0s. It can be of the NRZ or RZ type. The NRZ version of the bipolar format (based on rectangular pulses) is depicted in Fig. 6.1*c*. Note that in this representation, there are *three* levels: +1, 0, and −1 (hence, the alternative name "pseudoternary"). An attractive feature of the bipolar format is the absence of a dc component, even though the input binary data may contain long strings of 0s and 1s. This property does not hold for the unipolar and polar formats. Also, the pulse alternation property of the bipolar format provides a capability for in-service performance monitoring in the sense that any isolated error, whether it causes the deletion or creation of a pulse, will violate this property. Moreover, the bipolar format eliminates ambiguity that may arise because of *polarity inversion* during the course of transmission (this problem is a characteristic of switched telephone networks). It is for these reasons that the bipolar format is adopted for use in the T1 carrier system for digital telephony (see Section 5.8). The absence of dc transmission permits repeaters, on the T1 carrier, to be transformer-coupled.

In the *Manchester format* (also known as *biphase baseband signaling*), symbol 1 is represented by transmitting a positive pulse for one-half of the symbol duration, followed by a negative pulse for the remaining half of the symbol

Table 6.1   Natural and Gray codes

Level	Natural code	Gray code
−3	00	00
−1	01	01
+1	10	11
+3	11	10

duration; for symbol 0, these two pulses are transmitted in reverse order. The Manchester format is depicted in Fig. 6.1*d*. Clearly, it has no dc component.

The NRZ versions of the unipolar, polar, and bipolar formats make efficient use of bandwidth. However, they do not offer a synchronization capability. On the other hand, the Manchester format has a built-in synchronization capability, because there is a predictable transition during each bit interval. But this capability is attained at the expense of a bandwidth requirement twice that of the NRZ unipolar, polar, and bipolar formats.

The utilization of bandwidth can be made more efficient by adopting an *M-ary format* for the representation of the input binary data. An example of this representation, namely, the *polar quaternary* format of the NRZ type, is shown in Fig. 6.2*a*. A quaternary code has four distinct symbols, referred to as *dibits* (pairs of bits). Each dibit is assigned a level in accordance with the *natural code* described in Table 6.1. To obtain the waveform of Fig. 6.2*a*, the input binary sequence 0110100011 is viewed as a new sequence of dibits, {01,10,10,00,11}, and each dibit in the sequence is represented by a level of its own. In Table 6.1, we also show a second coding scheme, called the *Gray code,* in which the adjacent bits are arranged in such a way that they differ by only *one bit.** In some applications, this behavior is a desirable attribute. Fig. 6.2*b* shows the NRZ polar quaternary format of the Gray-encoded version of the binary sequence 0110100011.

Earlier, we remarked that the bipolar format is immune from the polarity inversion-ambiguity problem. The only other one among those depicted in Fig. 6.1 that is also so immune is the unipolar format. In both cases, the immunity is a result of using no pulses for the representation of binary 0s. We may build a similar capability into the other formats by the inclusion of *differential encoding.* The use of this technique is illustrated in Fig. 6.3. Differential encoding starts with an arbitrary *initial* bit. In Fig. 6.3, a positive level is shown for the initial bit. The information in the input binary sequence itself is encoded in terms of *signal transitions.* For example, a signal transition is used to designate symbol 0, while no transition is used to designate symbol 1. Using such a procedure, the *differentially encoded polar waveform* for the representation of

---

* In general, the *Gray code* can be derived from the natural code as follows. Let b_k denote the kth bit in the natural code, where $k = 1, 2, \ldots, N$, with $k = N$ denoting the most significant bit. Given the natural code, we may determine the kth bit of the Gray code, by using the formula

$$g_k = \begin{cases} b_k & k = N \\ b_k + b_{k+1} & k = 1, 2, \ldots, N - 1 \end{cases}$$

where + denotes modulo-2 addition. For details of modulo-2 arithmetic, see Appendix H.

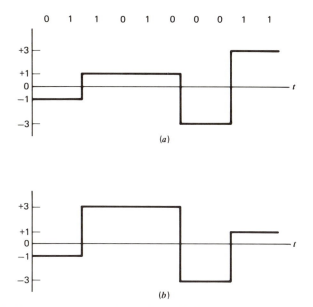

Figure 6.2   Polar quaternary format. (*a*) Natural-encoded. (*b*) Gray-encoded.

the input binary sequence 0110100011 is as shown in Fig. 6.3. It is apparent that the polarity of a differentially encoded waveform may be inverted without affecting its interpretation. The original binary information may be recovered by sampling the received wave and comparing the polarity of adjacent samples to establish whether or not a signal transition has occurred.

6.2   POWER SPECTRA OF DISCRETE PAM SIGNALS

The signaling formats shown in Figs. 6.1 and 6.2*a* may be viewed as different implementations of a *discrete amplitude-modulated pulse train*. In particular, we may describe them all as different realizations (sample functions) of a random process $X(t)$ defined by

$$X(t) = \sum_{k=-\infty}^{\infty} A_k v(t - kT) \tag{6.1}$$

where the coefficient A_k is a *discrete* random variable, $v(t)$ is a *basic pulse shape*, and T is the *symbol duration*. The basic pulse $v(t)$ is centered at the

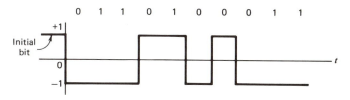

Figure 6.3   Differential encoding format.

Table 6.2 Data formats[a]

NRZ format	Coefficient A_k	Basic pulse $v(t)$
Unipolar	$A_k = \begin{cases} a, \text{ symbol } 1 \\ 0, \text{ symbol } 0 \end{cases}$	$v(t)$ consists of rectangular pulse of unit amplitude and duration T_b.
Polar	$A_k = \begin{cases} a, \text{ symbol } 1 \\ -a, \text{ symbol } 0 \end{cases}$	
Bipolar	$A_k = \begin{cases} a, -a, \text{ alternating 1s} \\ 0, \text{ symbol } 0 \end{cases}$	
Manchester	$A_k = \begin{cases} a, \text{ symbol } 1 \\ -a, \text{ symbol } 0 \end{cases}$	$v(t)$ consists of doublet pulse of heights ± 1, and total duration T_b.
Polar quaternary (natural coding)	$A_k = \begin{cases} 3a, \text{ dibit } 11 \\ a, \text{ dibit } 10 \\ -a, \text{ dibit } 01 \\ -3a, \text{ dibit } 00 \end{cases}$	$v(t)$ consists of rectangular pulse of unit amplitude and duration $2T_b$.

[a] In Figs. 6.1 and 6.2, the amplitude a is given as one.

origin, $t = 0$, and normalized such that

$$v(0) = 1 \tag{6.2}$$

In Table 6.2, we present a summary of sample values of the coefficient A_k and the basic pulse $v(t)$ for the various formats depicted in Figs. 6.1 and 6.2a.

The *data signaling rate* (or simply *data rate*) is defined as the rate, measured in *bits per second* (b/s), at which data are transmitted. It is also common practice to refer to the data signaling rate as the *bit rate*. This rate is denoted by

$$R_b = \frac{1}{T_b} \tag{6.3}$$

where T_b is the *bit duration*. In contrast, the *modulation rate* is defined as *the rate at which signal level is changed,* depending on the nature of the format used to represent the digital data. The modulation rate is measured in *bauds,** or *symbols per second*. For an M-ary format (with M an integer power of two) used to represent binary data, we find that the *symbol duration* of the M-ary format is related to the bit duration T_b by

$$T = T_b \log_2 M \tag{6.4}$$

Correspondingly, one baud equals $\log_2 M$ bits per second.

We are now ready to evaluate the *power spectra* of the various discrete PAM signals of interest. By so doing, we will be able to make a more complete assessment of their individual spectral characteristics.

To proceed with the analysis, we model the mechanism responsible for the generation of the sequence $\{A_k\}$, defining the coefficients in Eq. 6.1, as a *discrete stationary random source*. The source is characterized as having the

* The "baud" is named after Emile Baudot in honor of his pioneering work in telegraphy.

ensemble-averaged autocorrelation function:

$$R_A(n) = E[A_k A_{k-n}] \tag{6.5}$$

where E is the *expectation operator*. Accordingly, we find that the power spectral density of the discrete PAM signal $X(t)$ defined in Eq. 6.1 is given by (using the Wiener-Khintchine relations of Appendix D.6)

$$S_X(f) = \frac{1}{T} |V(f)|^2 \sum_{n=-\infty}^{\infty} R_A(n) \exp(-j2\pi nfT) \tag{6.6}$$

where $V(f)$ is the Fourier transform of the basic pulse $v(t)$. The values of the functions $V(f)$ and $R_A(n)$ depend on the type of discrete PAM signal being considered. In the sequel, we evaluate Eq. 6.6 for the unipolar, polar, and bipolar formats of the NRZ type, and also the Manchester format. The evaluations for other formats are posed as problems to the reader at the end of the chapter (see Problems 6.2.2 and 6.2.3).

(1)   NRZ Unipolar Format

Suppose the 0s and 1s of a random binary sequence occur with equal probability. Then, for a unipolar format of the NRZ type, we have

$$P(A_k = 0) = P(A_k = a) = \frac{1}{2}$$

Hence, for $n = 0$, we may write

$$E[A_k^2] = (0)^2 P(A_k = 0) + (a)^2 P(A_k = a)$$
$$= \frac{a^2}{2}$$

Consider next the product $A_k A_{k-n}$ for $n \neq 0$. This product has four possible values, namely, 0, 0, 0, and a^2. Assuming that the successive symbols in the binary sequence are statistically independent, these four values occur with a probability of 1/4 each. Hence, for $n \neq 0$, we may write

$$E[A_k A_{k-n}] = 3(0)(\tfrac{1}{4}) + (a^2)(\tfrac{1}{4})$$
$$= \frac{a^2}{4} \quad n \neq 0$$

We may thus express the autocorrelation function $R_A(n)$ as follows:

$$R_A(n) = \begin{cases} \dfrac{a^2}{2} & n = 0 \\ \dfrac{a^2}{4} & n \neq 0 \end{cases} \tag{6.7}$$

For the basic pulse $v(t)$, we have a rectangular pulse of unit amplitude and duration T_b. Hence, the Fourier transform of $v(t)$ equals

$$V(f) = T_b \operatorname{sinc}(fT_b) \tag{6.8}$$

where the *sinc function* is defined by

$$\text{sinc}(\lambda) = \frac{\sin(\pi\lambda)}{\pi\lambda}$$

Hence, the use of Eqs. 6.7 and 6.8 in Eq. 6.6, with $T = T_b$, yields the following result for the power spectral density of the NRZ unipolar format:

$$S_X(f) = \frac{a^2 T_b}{4} \text{sinc}^2(fT_b) + \frac{a^2 T_b}{4} \text{sinc}^2(fT_b) \sum_{n=-\infty}^{\infty} \exp(-j2\pi nfT_b) \quad (6.9)$$

We next use *Poisson's formula* written in the form

$$\sum_{n=-\infty}^{\infty} \exp(-j2\pi nfT_b) = \frac{1}{T_b} \sum_{m=-\infty}^{\infty} \delta\left(f - \frac{m}{T_b}\right) \quad (6.10)$$

where $\delta(f)$ denotes a *Dirac delta function* at $f = 0$. Accordingly, substituting Eq. 6.10 in Eq. 6.9 and recognizing that the sinc function $\text{sinc}(fT_b)$ has nulls at $f = \pm1/T_b, \pm2/T_b, \ldots$, we may simplify the expression for the power spectral density $S_X(f)$ as

$$S_X(f) = \frac{a^2 T_b}{4} \text{sinc}^2(fT_b) + \frac{a^2}{4} \delta(f) \quad (6.11)$$

The presence of the Dirac delta function $\delta(f)$ accounts for one half of the power contained in the unipolar waveform.

Curve a of Fig. 6.4 shows a *normalized* plot of Eq. 6.11. Specifically the power spectral density $S_X(f)$ is normalized with respect to $a^2 T_b$, and the frequency f is normalized with respect to the bit rate $1/T_b$. We see that most of the power of the NRZ unipolar format lies between dc and the bit rate of the input data.

(2) NRZ Polar Format

Consider next a polar format of the NRZ type, for which the input binary data consists of independent and equally likely symbols. Following a procedure similar to that described for the unipolar format, we find that for this second case

$$R_A(n) = \begin{cases} a^2 & n = 0 \\ 0 & n \neq 0 \end{cases} \quad (6.12)$$

The basic pulse $v(t)$ for the polar format is the same as that for the unipolar format. Hence, the use of Eqs. 6.8 and 6.12 in Eq. 6.6, with the symbol period $T = T_b$, yields the power spectral density of the NRZ polar format as

$$S_X(f) = a^2 T_b \text{sinc}^2(fT_b) \quad (6.13)$$

The normalized form of this equation is shown plotted as curve b in Fig. 6.4. Here again we see that most of the power of the NRZ polar format lies inside the main lobe of the sinc-shaped curve, which extends up to the bit rate $1/T_b$.

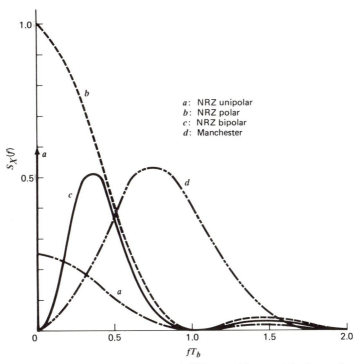

Figure 6.4   Power spectra of different binary data formats.

(3)   NRZ Bipolar Format

The *constraint* that successive 1s in the bipolar format be assigned pulses of alternating polarity may be viewed as a form of *precoding*. Indeed, in Section 6.4 we will show that the unipolar format can be transformed into the corresponding bipolar format by means of a special precoder. In any event, this constraint introduces intersymbol correlation that has to be taken into account.

We note that the bipolar format has three levels: a, 0, and $-a$. Then assuming that the 1s and 0s in the input binary data occur with equal probability, we find that the respective probabilities of occurrence of these three levels are as follows:

$$P(A_k = a) = \tfrac{1}{4}$$

$$P(A_k = 0) = \tfrac{1}{2}$$

$$P(A_k = -a) = \tfrac{1}{4}$$

Hence, for $n = 0$, we may write

$$E[A_k^2] = (a)^2 P(A_k = a) + (0)^2 P(A_k = 0) + (-a)^2 P(A_k = -a)$$

$$= \frac{a^2}{2}$$

For $n = 1$, the dibit represented by the sequence (A_{k-1}, A_k) can assume only four possible forms: (0,0), (0,1), (1,0), and (1,1). The respective values of the product A_kA_{k-1} are 0, 0, 0, and $-a^2$; the last value results from the fact that successive 1s alternate in polarity. Each of the dibits occurs with probability 1/4, on the assumption that successive symbols in the binary sequence occur with equal probability. Hence, we may write

$$E[A_kA_{k-1}] = 3(0)(\tfrac{1}{4}) + (-a^2)(\tfrac{1}{4})$$

$$= -\frac{a^2}{4}$$

For $n > 1$, we find that $E[A_kA_{k-n}] = 0$ (see Problem 6-2-1). Accordingly, for the NRZ bipolar format, we have

$$R_A(n) = \begin{cases} a^2/2 & n = 0 \\ -a^2/4 & n = \pm 1 \\ 0 & \text{otherwise} \end{cases} \tag{6.14}$$

where, in the second line on the right side, we have made note of the fact that $R_A(-n) = R_A(n)$.

The basic pulse $v(t)$ for the NRZ bipolar format has its Fourier transform defined by Eq. 6.8. Hence, substituting Eqs. 6.8 and 6.14 in Eq. 6.6, with $T = T_b$, we find that the power spectral density of the NRZ bipolar format is given by

$$S_X(f) = T_b \, \text{sinc}^2(fT_b) \left[\frac{a^2}{2} - \frac{a^2}{4} \left(\exp(j2\pi fT_b) + \exp(-j2\pi fT_b) \right) \right]$$

$$= \frac{a^2 T_b}{2} \, \text{sinc}^2(fT_b)[1 - \cos(2\pi fT_b)]$$

$$= a^2 T_b \, \text{sinc}^2(fT_b)\sin^2(\pi fT_b) \tag{6.15}$$

The normalized form of this equation is shown plotted in curve c in Fig. 6.4. We see that although most of the power lies inside a bandwidth equal to the bit rate $1/T_b$ (as with unipolar and polar waveforms of the NRZ type), the spectral content of the NRZ bipolar format is relatively small around zero frequency.

(4) Manchester Format

Consider next the Manchester format for which the input binary data consists of independent, equally likely symbols. Then, the autocorrelation function $R_A(n)$ for the Manchester format is the same as that given in Eq. 6.12 for the NRZ polar format. The basic pulse $v(t)$ for the Manchester format consists of a doublet pulse of unit amplitude and total duration T_b. Hence, the Fourier transform of the pulse equals

$$V(f) = jT_b \, \text{sinc}\left(\frac{fT_b}{2}\right) \sin\left(\frac{\pi fT_b}{2}\right) \tag{6.16}$$

Thus, substituting Eqs. 6.12 and 6.16 in Eq. 6.6, we find that the power spectral density of the Manchester format is given by

$$S_X(f) = a^2 T_b \, \text{sinc}^2\left(\frac{1}{2}fT_b\right) \sin^2\left(\frac{\pi fT_b}{2}\right) \tag{6.17}$$

The normalized form of this equation is shown plotted as curve *d* in Fig. 6.4. As anticipated, most of the power in the Manchester format lies inside a bandwidth equal to $2/T_b$ (i.e., twice that of unipolar, polar, and bipolar formats of the NRZ type).

6.3   INTERSYMBOL INTERFERENCE

Consider Fig. 6.5, which depicts the basic elements of a *baseband binary PAM system*. The input signal consists of a binary data sequence $\{b_k\}$ with a bit duration of T_b seconds. This sequence is applied to a pulse generator, producing the discrete PAM signal

$$x(t) = \sum_{k=-\infty}^{\infty} a_k v(t - kT_b) \qquad (6.18)$$

where $v(t)$ denotes the basic pulse, normalized such that $v(0) = 1$, as in Eq. 6.2. The coefficient a_k depends on the input data and the type of format used. The waveform $x(t)$ represents one realization of the random process $X(t)$ considered in Section 6.2. Likewise, a_k is a sample value of the random variable A_k.

The PAM signal $x(t)$ passes through a transmitting filter of transfer function $H_T(f)$. The resulting filter output defines the transmitted signal, which is modified as a result of transmission through the channel of transfer function $H_C(f)$. The channel may represent a coaxial cable or optical fiber, where the major source of system degradation is *dispersion* in the channel. In any event, for the present we assume that the channel is *noiseless* but dispersive. The channel output is passed through a receiving filter of transfer function $H_R(f)$. This filter output is sampled synchronously with the transmitter, with the sampling instants being determined by a clock or timing signal that is usually extracted from the receiving filter output. Finally, the sequence of samples thus obtained is used to reconstruct the original data sequence by means of a decision device. Each sample is compared to a threshold. We assume that symbols 1 and 0 are equally likely, and the threshold is set half way between their representation levels. If the threshold is exceeded, a decision is made in favor of symbol 1. If, on the other hand, the threshold is not exceeded, a decision is made in favor of symbol 0. If the sample value equals the threshold exactly, the flip of a fair coin will determine which symbol was transmitted.

The receiving filter output may be written as*

$$y(t) = \mu \sum_{k=-\infty}^{\infty} a_k p(t - kT_b) \qquad (6.19)$$

where μ is a scaling factor and the pulse $p(t)$ is *normalized* such that

$$p(0) = 1 \qquad (6.20)$$

---

* To be precise, an arbitrary time delay t_0 should be included in the argument of the pulse $p(t - kT_b)$ in Eq. 6.19 to represent the effect of transmission delay through the system. For convenience, we have put this delay equal to zero in Eq. 6.19.

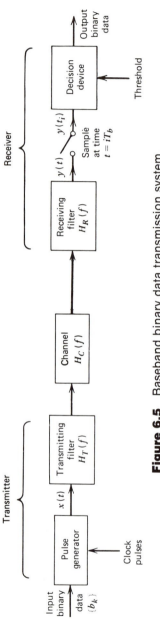

Figure 6.5 Baseband binary data transmission system.

244

The output $y(t)$ is produced in response to the binary data waveform applied to the input of the transmitting filter. Specifically, the pulse $\mu p(t)$ is the response of the cascade connection of the transmitting filter, the channel, and the receiving filter, which is produced by the pulse $v(t)$ applied to the input of this cascade connection. Therefore, we may relate $p(t)$ to $v(t)$ in the frequency domain by writing

$$\mu P(f) = V(f)H_T(f)H_C(f)H_R(f) \qquad (6.21)$$

where $P(f)$ and $V(f)$ are the Fourier transforms of $p(t)$ and $v(t)$, respectively. Note that the normalization of $p(t)$ as in Eq. 6.20 means that the total area under the curve of $P(f)$ equals unity.

The receiving filter output $y(t)$ is sampled at time $t_i = iT_b$ (with i taking on integer values), yielding

$$y(t_i) = \mu \sum_{k=-\infty}^{\infty} a_k p(iT_b - kT_b)$$

$$= \mu a_i + \mu \sum_{\substack{k=-\infty \\ k \neq i}}^{\infty} a_k p(iT_b - kT_b) \qquad (6.22)$$

In Eq. 6.22, the first term μa_i is produced by the ith transmitted bit. The second term represents the residual effect of all other transmitted bits on the decoding of the ith bit; this residual effect is called *intersymbol interference* (ISI).

In physical terms, ISI arises because of imperfections in the overall frequency response of the system. When a short pulse of duration T_b seconds is transmitted through a band-limited system, the frequency components constituting the input pulse are differentially attenuated and, more significantly, differentially delayed by the system. Consequently, the pulse appearing at the output of the system is *dispersed* over an interval longer than T_b seconds. Thus, when a sequence of short pulses (representing binary 1s and 0s) are transmitted through the system, one pulse every T_b seconds, the dispersed responses originating from different symbol intervals will interfere with each other, thereby resulting in intersymbol interference.

In the absence of ISI, we observe from Eq. 6.22 that

$$y(t_i) = \mu a_i$$

which shows that, under these conditions, the ith transmitted bit can be decoded correctly. The presence of ISI in the system, however, introduces errors in the decision device at the receiver output. Therefore, in the design of the transmitting and receiving filters, the objective is to minimize the effects of ISI, and thereby deliver the digital data to its destination with the smallest error rate possible.

6.4 NYQUIST'S CRITERION FOR DISTORTIONLESS BASEBAND BINARY TRANSMISSION

Typically, the transfer function of the channel and the transmitted pulse shape are specified, and the problem is to determine the transfer functions of the transmitting and receiving filters so as to reconstruct the transmitted data sequence $\{b_k\}$. The receiver does this by *extracting* and then *decoding* the corresponding sequence of weights, $\{a_k\}$, from the output $y(t)$. Except for a scaling factor, y(t) is determined by the a_k and the received pulse $p(t)$; see Eq. 6.19. The *extraction* involves sampling the output $y(t)$ at some time $t = iT_b$. The *decoding* requires that the weighted pulse contribution $a_k p(iT_b - kT_b)$ for $k = i$ be *free* from ISI due to the overlapping tails of all other weighted pulse contributions represented by $k \neq i$. This, in turn, requires that we *control* the received pulse $p(t)$, as shown by

$$p(iT_b - kT_b) = \begin{cases} 1 & i = k \\ 0 & i \neq k \end{cases} \tag{6.23}$$

where, by normalization, $p(0) = 1$. If $p(t)$ satisfies the condition of Eq. 6.23, the receiver output, given by Eq. 6.22, simplifies to

$$y(t_i) = \mu a_i$$

which implies zero intersymbol interference. Hence, the condition of Eq. 6.23 assures *perfect reception in the absence of noise*.

From a design point of view, it is informative to transform the condition of Eq. 6.23 into the frequency domain. Consider then the sequence of samples $\{p(nT_b)\}$, where $n = 0, \pm 1, \pm 2, \ldots$. From the discussion presented in Section 4.1 on the sampling process for a low-pass function, we recall that sampling in the time domain produces periodicity in the frequency domain. In particular, we may write (see Eq. 4.5)

$$P_\delta(f) = R_b \sum_{n=-\infty}^{\infty} P(f - nR_b) \tag{6.24}$$

where $R_b = 1/T_b$ is the *bit rate*; $P_\delta(f)$ is the Fourier transform of an infinite periodic sequence of delta functions of period T_b, and whose strengths are weighted by the respective sample values of $p(t)$. That is, $P_\delta(f)$ is given by

$$P_\delta(f) = \int_{-\infty}^{\infty} \sum_{m=-\infty}^{\infty} [p(mT_b) \, \delta(t - mT_b)] \exp(-j2\pi ft)dt \tag{6.25}$$

Let the integer $m = i - k$. Then, $i = k$ corresponds to $m = 0$, and likewise $i \neq k$ corresponds to $m \neq 0$. Accordingly, imposing the condition of Eq. 6.23 on the sample values of $p(t)$ in the integral of Eq. 6.25, we get

$$P_\delta(f) = \int_{-\infty}^{\infty} p(0) \, \delta(t) \exp(-j2\pi ft)dt$$
$$= p(0) \tag{6.26}$$

where we have made use of the sifting property of the delta function. Since $p(0) = 1$, by normalization, we thus see from Eqs. 6.24 and 6.26 that the condition for zero intersymbol interference is satisfied if

$$\sum_{n=-\infty}^{\infty} P(f - nR_b) = T_b \tag{6.27}$$

Equation 6.23 formulated in terms of the time function $p(t)$, or equivalently, Eq. 6.27 formulated in terms of the corresponding frequency function $P(f)$, constitutes the *Nyquist criterion for distortionless baseband transmission** in the absence of noise. It provides a method for constructing band-limited functions to overcome the effects of intersymbol interference. The method depends on sampling the received signal at midpoints of the signaling intervals.

(1) Ideal Solution

A frequency function $P(f)$, occupying the narrowest band, that satisfies Eq. 6.27 is obtained by permitting only one nonzero component in the series (on the left side of the equation) for each f in the range extending from $-B_o$ to B_o, where B_o denotes half the bit rate:

$$B_o = \frac{R_b}{2} \tag{6.28}$$

That is, we specify $P(f)$ as

$$P(f) = \frac{1}{2B_o} \text{rect}\left(\frac{f}{2B_o}\right) \tag{6.29}$$

In this solution, no frequencies of absolute value exceeding half the bit rate are needed. Hence, one signal waveform that produces zero intersymbol interference is defined by the *sinc function:*

$$p(t) = \frac{\sin(2\pi B_o t)}{2\pi B_o t}$$
$$= \text{sinc}(2B_o t) \tag{6.30}$$

Figures 6.6*a* and 6.6*b* show plots of $P(f)$ and $p(t)$, respectively. In Fig. 6.6*a*, the frequency function $P(f)$ is shown plotted for positive frequencies only. In Fig. 6.6*b*, we have also included the signaling intervals and the corresponding centered sampling instants. The function $p(t)$ can be regarded as the impulse response of an ideal low-pass filter with passband amplitude response $1/(2B_o)$ and bandwidth B_o. The function $p(t)$ has its peak value at the origin and goes through zero at integer multiples of the bit duration T_b. It is apparent that if the received waveform $y(t)$ is sampled at the instants of time $t = 0, \pm T_b, \pm 2T_b,$ . . . , then the pulses defined by $\mu p(t - iT_b)$ with arbitrary amplitude μ and index $i = 0, \pm 1, \pm 2, . . .$, will not interfere with each other.

Although this choice of pulse shape for $p(t)$ achieves economy in bandwidth in that it solves the problem of zero intersymbol interference with the minimum

---

* The criterion described by Eq. 6.23 or Eq. 6.27 was first formulated by Nyquist in the study of telegraph transmission theory (Nyquist, 1928). In the literature, it is referred to as *Nyquist's first criterion*. In his classic 1928 paper, Nyquist described another method, referred to in the literature as *Nyquist's second criterion*. The second method makes use of the instants of transition between unlike symbols in the received signal rather than centered samples. A discussion of both criteria is presented in Bennett (1970, pp. 78–92).

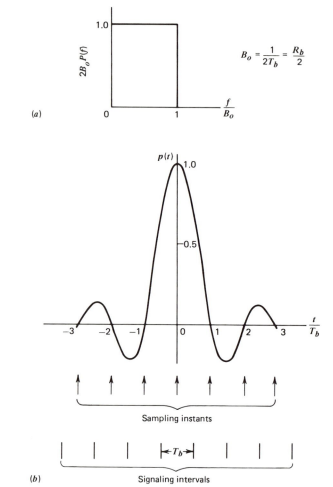

Figure 6.6 (a) Ideal amplitude response. (b) Ideal basic pulse shape.

bandwidth possible, there are two practical difficulties that make it an undesirable objective for system design:

1. It requires that the amplitude characteristic of $P(f)$ be flat from $-B_o$ to B_o, and zero elsewhere. This is physically unrealizable because of the abrupt transitions at $\pm B_o$.

2. The function $p(t)$ decreases as $1/|t|$ for large $|t|$, resulting in a slow rate of decay. This is caused by the discontinuity of $P(f)$ at $\pm B_o$. Accordingly, there is practically no margin of error in sampling times in the receiver.

To evaluate the effect of this *timing error*, consider the sample of $y(t)$ at $t = \Delta t$, where Δt is the timing error. To simplify the analysis we have put the correct sampling time t_i equal to zero. We thus obtain, in the absence of noise

$$y(\Delta t) = \mu \sum_k a_k p(\Delta t - kT_b)$$

$$= \mu \sum_k a_k \frac{\sin[2\pi B_o(\Delta t - kT_b)]}{2\pi B_o(\Delta t - kT_b)} \qquad (6.31)$$

Since $2B_o T_b = 1$, we may rewrite Eq. 6.31 as

$$y(\Delta t) = \mu \sum_k a_k \, \mathrm{sinc}(2B_o \Delta t - k)$$

$$= \mu a_0 \, \mathrm{sinc}(2B_o \Delta t) + \frac{\mu \sin(2\pi B_o \Delta t)}{\pi} \sum_{\substack{k \\ k \neq 0}} \frac{(-1)^k a_k}{2B_o \Delta t - k} \qquad (6.32)$$

The first term on the right side of Eq. 6.32 defines the desired symbol, whereas the remaining series represents the intersymbol interference caused by the timing error Δt in sampling the output $y(t)$. In certain cases, it is possible for this series to diverge, thereby causing erroneous decisions in the receiver.

(2) Practical Solution

We may overcome the practical difficulties posed by the ideal solution by extending the bandwidth from $B_o = R_b/2$ to an adjustable value between B_o and $2B_o$. In so doing, we permit three components in the series on the left side of Eq. 6.27 for $|f| \leq B_o$, as shown by

$$P(f) + P(f - 2B_o) + P(f + 2B_o) = \frac{1}{2B_o} \qquad -B_o \leq f \leq B_o \qquad (6.33)$$

We may devise several band-limited functions that satisfy Eq. 6.33. A particular form of $P(f)$ that embodies many desirable features is constructed by a *raised cosine spectrum*. This frequency characteristic consists of a flat portion and a *rolloff* portion that has a sinusoidal form, as follows

$$P(f) = \begin{cases} \dfrac{1}{2B_o} & |f| < f_1 \\[2mm] \dfrac{1}{4B_o}\left\{1 + \cos\left[\dfrac{\pi(|f| - f_1)}{2B_o - 2f_1}\right]\right\}, & f_1 \leq |f| < 2B_o - f_1 \\[2mm] 0 & |f| \geq 2B_o - f_1 \end{cases} \qquad (6.34)$$

The frequency f_1 and bandwidth B_o are related by

$$\alpha = 1 - \frac{f_1}{B_o} \qquad (6.35)$$

The parameter α is called the *rolloff factor*. For $\alpha = 0$, that is, $f_1 = B_o$, we get the minimum bandwidth solution described earlier.

The frequency response $P(f)$, normalized by multiplying it by $2B_o$, is shown plotted in Fig. 6.7a for three values of α namely, 0, 0.5, and 1. We see that for $\alpha = 0.5$ or 1, the rolloff characteristic of $P(f)$ cuts off gradually as compared

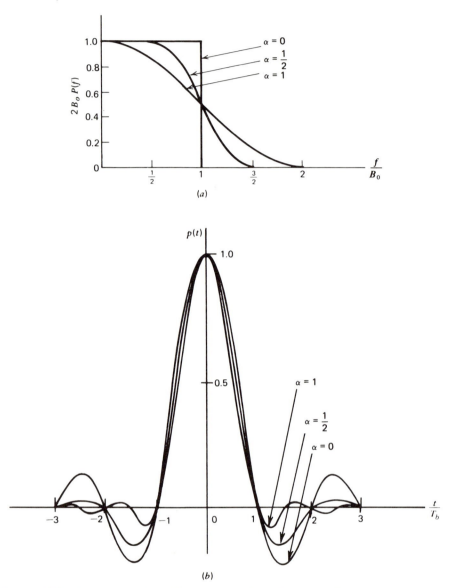

Figure 6.7 Responses for different rolloff factors. (a) Frequency response. (b) Time response. Note that $B_o = 1/2T_b$.

with an ideal low-pass filter (corresponding to $\alpha = 0$), and it is therefore easier to realize in practice. Also the function $P(f)$ exhibits odd symmetry about the cutoff frequency B_o of the ideal low-pass filter. The time response $p(t)$, that is, the inverse Fourier transform of $P(f)$, is defined by

$$p(t) = \text{sinc}(2B_ot) \frac{\cos(2\pi\alpha B_ot)}{1 - 16\alpha^2 B_o^2 t^2} \tag{6.36}$$

This function consists of the product of two factors: the factor $\mathrm{sinc}(2B_o t)$ associated with the ideal filter, and a second factor that decreases as $1/|t|^2$ for large $|t|$. The first factor ensures zero crossings of $p(t)$ at the desired sampling instants of time $t = iT_b$ with i an integer (positive and negative). The second factor reduces the tails of the pulse considerably below that obtained from the ideal low-pass filter, so that the transmission of binary waves using such pulses is relatively insensitive to sampling time errors. In fact, the amount of intersymbol interference resulting from this timing error decreases as the rolloff factor α is increased from zero to unity.

The time response $p(t)$ is shown plotted in Fig. 6.7b for $\alpha = 0, 0.5$, and 1. For the special case of $\alpha = 1$, the function $p(t)$ simplifies to

$$p(t) = \frac{\mathrm{sinc}(4B_o t)}{1 - 16B_o^2 t^2}$$

This time response exhibits two interesting properties:

1. At $t = \pm T_b/2 = \pm 1/4B_o$, we have $p(t) = 0.5$; that is, the pulse width measured at half amplitude is exactly equal to the bit duration T_b.
2. There are zero crossings at $t = \pm 3T_b/2, \pm 5T_b/2, \ldots$ in addition to the usual zero crossings at the sampling times $t = \pm T_b, \pm 2T_b, \ldots$.

These two properties are particularly useful in generating a timing signal from the received signal for the purpose of synchronization. However, this requires the use of a transmission bandwidth double that required for the ideal case corresponding to $\alpha = 0$.

EXAMPLE 1   BANDWIDTH REQUIREMENTS OF THE T1 CARRIER

In Section 5.8, we described the signal format for the T1 carrier system that is used to multiplex 24 independent voice inputs, based on an 8-bit PCM word. It was shown that the bit duration of the resulting time-division multiplexed signal (including a framing bit) is

$$T_b = 0.647 \ \mu s$$

In Section 6.1, we remarked that, for various reasons, the T1 carrier uses a bipolar signal format to represent the input binary data. Thus, assuming an ideal low-pass characteristic for the channel, it follows that the minimum (optimum) transmission bandwidth of the T1 system is

$$B_o = \frac{1}{2T_b} = 772 \text{ kHz}$$

However, a more realistic value for the necessary transmission bandwidth is obtained by using a cosine-rolloff characteristic with $\alpha = 1$. From Eq. (6.34), we see that the transmission bandwidth required to accommodate the raised cosine spectrum is given by

$$\begin{aligned} B &= 2B_o - f_1 \\ &= 2B_o - B_o(1 - \alpha) \\ &= B_o(1 + \alpha) \end{aligned} \qquad (6.37)$$

where, in the second line, we have made use of Eq. (6.35). Hence, putting $\alpha = 1$ in Eq. (6.37), we find that

$$B = 2B_o = \frac{1}{T_b} = 1.544 \text{ MHz}$$

6.5 CORRELATIVE CODING

Thus far we have treated intersymbol interference as an undesirable phenomenon that produces a degradation in system performance. Indeed, its very name connotes a nuisance effect. Nevertheless, by adding intersymbol interference to the transmitted signal in a controlled manner, it is possible to achieve a bit rate of $2B_o$ bits per second in a channel of bandwidth B_o hertz. Such schemes are called *correlative coding* or *partial-response* signaling schemes.* The design of these schemes is based on the premise that since intersymbol interference introduced into the transmitted signal is known, its effect can be compensated at the receiver. Thus correlative coding may be regarded as a practical means of achieving the theoretical maximum signaling rate of $2B_o$ bits per second in a bandwidth of B_o hertz, as postulated by Nyquist, using realizable and perturbation-tolerant filters.

(1) Duobinary Signaling

The basic idea of correlative coding will now be illustrated by considering the specific example of *duobinary signaling*, where "duo" implies doubling of the transmission capacity of a straight binary system.

Consider a binary input sequence $\{b_k\}$ consisting of uncorrelated binary digits each having duration T_b seconds, with symbol 1 represented by a pulse of amplitude $+1$ volt, and symbol 0 by a pulse of amplitude -1 volt. When this sequence is applied to a *duobinary encoder*, it is converted into a *three-level output*, namely, -2, 0, and $+2$ volts. To produce this transformation, we may use the scheme shown in Fig. 6.8. The binary sequence $\{b_k\}$ is first passed through a simple filter involving a single delay element. For every unit impulse applied to the input of this filter, we get two unit impulses spaced T_b seconds apart at the filter output. We may therefore express the digit c_k at the duobinary coder output as the sum of the present binary digit b_k and its previous value b_{k-1}, as shown by

$$c_k = b_k + b_{k-1} \tag{6.38}$$

One of the effects of the transformation described by Eq. 6.38 is to change the input sequence $\{b_k\}$ of uncorrelated binary digits into a sequence $\{c_k\}$ of correlated digits. This correlation between the adjacent transmitted levels may be

---

* Correlative and partial response are synonomous; both terms are used in the literature. The idea of correlative coding was originated by Lender (1963). Lender's work was generalized for binary data transmission by Kretzmer (1966). For further details on correlative techniques see Kabal and Pasupathy (1975), Pasupathy (1977), and Lender (1981, pp. 144–182).

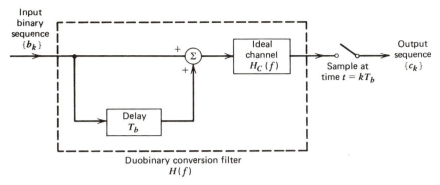

Figure 6.8   Duobinary signaling scheme.

viewed as introducing intersymbol interference into the transmitted signal in an artificial manner. However, this intersymbol interference is under the designer's control, which is the basis of correlative coding.

An ideal delay element, producing a delay of T_b seconds, has the transfer function $\exp(-j2\pi f\,T_b)$, so that the transfer function of the simple filter shown in Fig. 6.8 is $1 + \exp(-j2\pi f\,T_b)$. Hence, the overall transfer function of this filter connected in cascade with the ideal channel $H_C(f)$ is

$$
\begin{aligned}
H(f) &= H_C(f)[1 + \exp(-j2\pi f\,T_b)] \\
&= H_C(f)[\exp(j\pi f\,T_b) + \exp(-j\pi f\,T_b)]\exp(-j\pi f\,T_b) \qquad (6.39) \\
&= 2H_C(f)\cos(\pi f\,T_b)\exp(-j\pi f\,T_b),
\end{aligned}
$$

For an ideal channel of bandwidth $B_o = R_b/2$, we have

$$
H_C(f) = \begin{cases} 1 & |f| \le R_b/2 \\ 0 & \text{otherwise} \end{cases} \qquad (6.40)
$$

Thus the overall frequency response has the form of a half-cycle cosine function, as shown by

$$
H(f) = \begin{cases} 2\cos(\pi f\,T_b)\exp(-j\pi f\,T_b) & |f| \le R_b/2 \\ 0 & \text{otherwise} \end{cases} \qquad (6.41)
$$

for which the amplitude response and phase response are as shown in Fig. 6.9a and Fig. 6.9b, respectively. An advantage of this frequency response is that it can be easily approximated in practice.

The corresponding value of the impulse response consists of two sinc pulses, time-displaced by T_b seconds, as shown by (except for a scaling factor)

$$
\begin{aligned}
h(t) &= \frac{\sin(\pi t/T_b)}{\pi t/T_b} + \frac{\sin[\pi(t - T_b)/T_b]}{\pi(t - T_b)/T_b} \\
&= \frac{\sin(\pi t/T_b)}{\pi t/T_b} - \frac{\sin(\pi t/T_b)}{\pi(t - T_b)/T_b} \qquad (6.42) \\
&= \frac{T_b^2 \sin(\pi t/T_b)}{\pi t(T_b - t)}.
\end{aligned}
$$

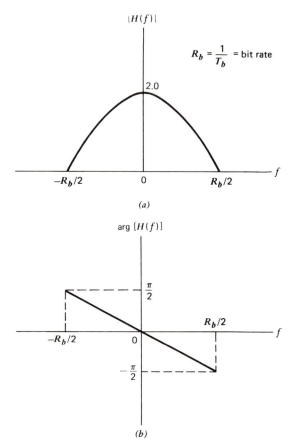

Figure 6.9 Frequency response of duobinary conversion filter. (a) Amplitude response. (b) Phase response.

which is shown plotted in Fig. 6.10. We see that the overall impulse response $h(t)$ has only *two* distinguishable values at the sampling instants.

The original data $\{b_k\}$ may be detected from the duobinary-coded sequence $\{c_k\}$ by subtracting the previous decoded binary digit from the currently received digit c_k in accordance with Eq. 6.38. Specifically, letting $\hat{b}_k$ represent the *estimate* of the original binary digit b_k as conceived by the receiver at time t equal to kT_b, we have

$$\hat{b}_k = c_k - \hat{b}_{k-1} \qquad (6.43)$$

It is apparent that if c_k is received without error and if also the previous estimate $\hat{b}_{k-1}$ at time $t = (k-1)T_b$ corresponds to a correct decision, then the current estimate $\hat{b}_k$ will be correct too. The technique of using a stored estimate of the previous symbol is called *decision feedback*.

We observe that the detection procedure just described is essentially an inverse of the operation of the simple filter at the transmitter. However, a drawback of this detection process is that once errors are made, they tend to

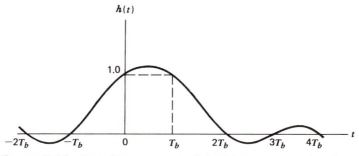

Figure 6.10   Impulse response of duobinary conversion filter.

propagate. This is due to the fact that a decision on the current binary digit b_k depends on the correctness of the decision made on the previous binary digit b_{k-1}.

A practical means of avoiding this error propagation is to use *precoding* before the duobinary coding, as shown in Fig. 6.11. The precoding operation performed on the input binary sequence $\{b_k\}$ converts it into another binary sequence $\{a_k\}$ defined by

$$a_k = b_k + a_{k-1} \qquad \text{modulo-2} \tag{6.44}$$

Module-2 addition is equivalent to the EXCLUSIVE-OR operation. (Modulo-2 addition is discussed in more detail in Appendix H.) An EXCLUSIVE-OR gate operates as follows. The output of an EXCLUSIVE-OR gate is a 1 if exactly one input is a 1; otherwise, the output is a 0. The resulting precoder output $\{a_k\}$ is next applied to the duobinary coder, thereby producing the sequence $\{c_k\}$ that is related to $\{a_k\}$ as follows:

$$c_k = a_k + a_{k-1} \tag{6.45}$$

Note that unlike the linear operation of duobinary coding, the precoding is a nonlinear operation.

We assume that symbol 1 at the precoder output in Fig. 6.11 is represented by +1 volt and symbol 0 by −1 volt. Therefore, from Eqs. 6.44 and 6.45, we find that

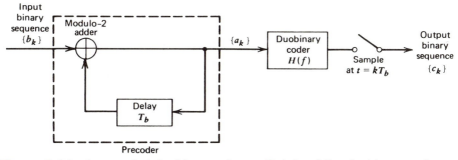

Figure 6.11   A precoded duobinary scheme. Details of the duobinary coder are given in Fig. 6.8.

Figure 6.12 Detector for recovering original binary sequence from the precoded duobinary coder output.

$$c_k = \begin{cases} \pm 2 \text{ volts, if } b_k \text{ is represented by symbol 0} \\ 0 \text{ volt, if } b_k \text{ is represented by symbol 1} \end{cases} \quad (6.46)$$

which is illustrated in Example 2. From Eq. 6.46 we deduce the following decision rule for detecting the original input binary sequence $\{b_k\}$ from $\{c_k\}$:

$$b_k = \begin{cases} \text{symbol 0} & \text{if } |c_k| > 1 \text{ volt} \\ \text{symbol 1} & \text{if } |c_k| < 1 \text{ volt} \end{cases} \quad (6.47)$$

According to Eq. 6.47, the detector consists of a rectifier, the output of which is compared to a threshold of 1 volt, and the original binary sequence $\{b_k\}$ is thereby detected. A block diagram of the detector is shown in Fig. 6.12. A useful feature of this detector is that no knowledge of any input sample other than the present one is required. Hence, error propagation cannot occur in the detector of Fig. 6.12.

EXAMPLE 2
Consider the input binary sequence 0010110. To proceed with the precoding of this sequence, which involves feeding the precoder output back to the input, we add an extra bit to the precoder output. This extra bit is chosen arbitrarily as a bit 1. Hence, using Eq. 6.44, we find that the sequence $\{a_k\}$ at the precoder output is as shown in row 2 of Table 6.3. We assume that symbol 1 is represented by $+1$ volt and symbol 0 by -1 volt. Accordingly, the precoder output has the amplitudes shown in row 3. Finally, using Eq. 6.38, we find that the duobinary coder output has the amplitudes given in row 4 of Table 6.3.

To detect the original binary sequence, we apply the decision rule of Eq. 6.47, and so obtain the sequence given in row 5 of Table 6.3. This shows that, in the absence of noise, the original binary sequence is detected correctly.

Table 6.3

Binary sequence $\{b_k\}$		0	0	1	0	1	1	0
Binary sequence $\{a_k\}$	1	1	1	0	0	1	0	0
Polar representation of pre-coder output, a_k (volts)	$+1$	$+1$	$+1$	-1	-1	$+1$	-1	-1
Duobinary coder output, c_k (volts)		2	2	0	-2	0	0	-2
Sequence obtained by applying decision rule of Eq. (6.47)		0	0	1	0	1	1	0

EXAMPLE 3
Consider the circuit shown in Fig. 6.13, which may be viewed as a differential encoder connected in cascade with a special form of correlative coder. The differential encoder

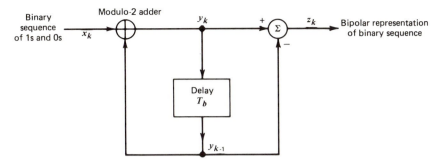

Figure 6.13 Circuit for generating bipolar format.

consists of a modulo-2 adder and 1-bit delay element. The correlative coder consists of a 1-bit delay element and summer. A single delay element is shown in Fig. 6.13, since it is common to both the differential encoder and the correlative coder.

The circuit of Fig. 6.13 has the property that it can convert the unipolar format for a binary sequence into the corresponding bipolar format. We illustrate this property by examining the input binary sequence 0110100011.

Let x_k denote the input to the differential encoder and y_k denote the resulting output. We may then characterize this part of the scheme in Fig. 6.13 by writing

$$y_k = x_k + y_{k-1} \qquad \text{modulo-2} \tag{6.48}$$

where y_{k-1} is the delayed version of y_k, and the addition is performed modulo-2. In Fig. 6.13, we note that the differential encoder output y_k is the same as the correlative coder input. Hence, we may express the correlative coder output as follows:

$$z_k = y_k - y_{k-1} \tag{6.49}$$

Thus, starting with an arbitrary initial bit (say, 1) for y_k, and using Eqs. 6.48 and 6.49, we may construct the sequences shown in Table 6.4. Examining the sequence $\{z_k\}$ at the correlative coder output, we see that it has three levels, with the 0s of the input binary sequence $\{x_k\}$ being left intact, but the 1s being made to alternate in polarity. In other words, the output of the circuit shown in Fig. 6.13 represents the bipolar representation of the original binary sequence.

Table 6.4 Illustrating Operation of the Circuit in Fig. 6.13

x_k		0	1	1	0	1	0	0	0	1	1
y_k	1	1	0	1	1	0	0	0	0	1	0
y_{k-1}		1	1	0	1	1	0	0	0	0	1
z_k		0	-1	1	0	-1	0	0	0	1	-1

(2) Modified Duobinary Technique

The *modified duobinary* technique involves a correlation span of two binary digits. This is achieved by subtracting input binary digits spaced $2T_b$ seconds apart, as indicated in the block diagram of Fig. 6.14. The output of the modified

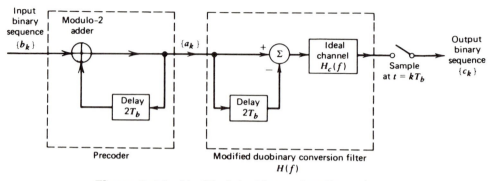

Figure 6.14 Modified duobinary signaling scheme.

duobinary conversion filter is related to the sequence $\{a_k\}$ at its input as follows:

$$c_k = a_k - a_{k-2} \tag{6.50}$$

Here, again, we find that a three-level signal is generated. If $a_k = \pm 1$ volt, as assumed previously, c_k takes on one of three values: 2, 0, and -2 volts.

The overall transfer function of the tapped-delay-line filter connected in cascade with the ideal channel, as in Fig. 6.14, is given by

$$H(f) = H_C(f)[1 - \exp(-j4\pi f\, T_b)]$$
$$= 2jH_C(f)\sin(2\pi f\, T_b)\exp(-j2\pi f\, T_b) \tag{6.51}$$

where $H_C(f)$ is as defined in Eq. 6.40. We, therefore, have an overall frequency response in the form of a half-cycle sine function, as shown by

$$H(f) = \begin{cases} 2j\,\sin(2\pi f\, T_b)\exp(-j2\pi f\, T_b) & |f| \le R_b/2 \\ 0 & \text{elsewhere} \end{cases} \tag{6.52}$$

The corresponding amplitude response and phase response of the modified duobinary-coder are shown in Fig. 6.15a and Fig. 6.15b, respectively. Note that the phase response in Fig. 6.15b does not include the constant 90° phase shift due to the multiplying factor j. A useful feature of the modified duobinary coder is the fact that its output has no dc component. This property is important since, in practice, many communication channels cannot transmit a dc component.

The impulse response of the modified duobinary coder consists of two sinc pulses that are time-displaced by $2T_b$ seconds, as shown by (except for a scaling factor)

$$\begin{aligned} h(t) &= \frac{\sin(\pi t/T_b)}{\pi t/T_b} - \frac{\sin[\pi(t - 2T_b)/T_b]}{\pi(t - 2T_b)/T_b} \\ &= \frac{\sin(\pi t/T_b)}{\pi t/T_b} - \frac{\sin(\pi t/T_b)}{\pi(t - 2T_b)/T_b} \\ &= \frac{2T_b^2\,\sin(\pi t/T_b)}{\pi t(2T_b - t)}. \end{aligned} \tag{6.53}$$

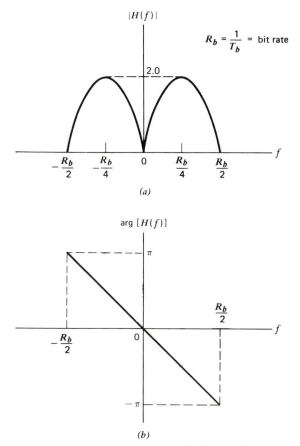

Figure 6.15 Frequency response of modified duobinary conversion filter. (a) Amplitude response. (b) Phase response.

This impulse response is plotted in Fig. 6.16, which shows that it has *three* distinguishable levels at the sampling instants.

In order to eliminate the possibility of error propagation in the modified duobinary system, we use a precoding procedure similar to that used for the duobinary case. Specifically, prior to the generation of the modified duobinary signal, a modulo-2 logical addition is used on signals $2T_b$ seconds apart, as shown by (see Fig. 6.14)

$$a_k = b_k + a_{k-2} \qquad \text{modulo-2} \qquad (6.54)$$

where $\{b_k\}$ is the input binary sequence and $\{a_k\}$ is the sequence at the precoder output. Note that modulo-2 addition and modulo-2 subtraction are the same. The sequence $\{a_k\}$ thus produced is then applied to the modified duobinary conversion filter.

In the case of Fig. 6.14, the output digit c_k equals 0, +2, or −2 volt. Also we find that b_k can be extracted from c_k by disregarding the polarity of c_k, as was

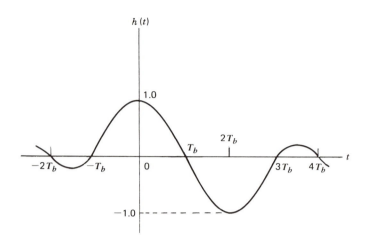

Figure 6.16 Impulse response of modified duobinary conversion filter.

done with the duobinary technique. Specifically, we may extract the original sequence $\{b_k\}$ at the receiver using the following decision rule:

$$b_k = \begin{cases} \text{symbol 0} & \text{if } |c_k| < 1 \text{ volt} \\ \text{symbol 1} & \text{if } |c_k| > 1 \text{ volt} \end{cases} \qquad (6.55)$$

(3)   Generalized Form of Correlative Coding

The duobinary and modified duobinary techniques have correlation spans of 1 binary digit and 2 binary digits, respectively. It is a straightforward matter to generalize these two techniques to other schemes, which are known collectively as *correlative coding schemes*. This generalization is shown in Fig. 6.17, where $H_C(f)$ is defined in Eq. 6.40. It involves the use of a tapped-delay-line filter with tap weights $w_0, w_1, \ldots, w_{N-1}$. Specifically, a correlative sample c_k is obtained from a superposition of N successive input sample values b_k, as shown by

$$c_k = \sum_{n=0}^{N-1} w_n b_{k-n} \qquad (6.56)$$

Thus, by choosing various combinations of integer values for the w_n, we obtain different forms of correlative coding schemes to suit individual applications. For example, in the duobinary case, we have

$$w_0 = +1$$
$$w_1 = +1$$

and $w_n = 0$ for $n \geq 2$. In the modified duobinary case, we have

$$w_0 = +1$$
$$w_1 = 0$$
$$w_2 = -1$$

and $w_n = 0$ for $n \geq 3$.

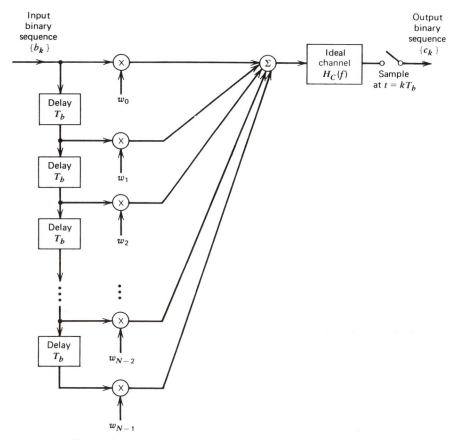

Tapped-delay-line filter

Figure 6.17   Generalized correlative coding scheme.

6.6   EYE PATTERN

One way to study intersymbol interference in a PCM or data transmission system experimentally is to apply the received wave to the vertical deflection plates of an oscilloscope and to apply a sawtooth wave at the transmitted symbol rate $R = 1/T$ to the horizontal deflection plates. The waveforms in successive symbol intervals are thereby translated into one interval on the oscilloscope display, as illustrated in Fig. 6.18 for the case of a binary wave for which $T = T_b$. The resulting display is called an *eye pattern* because of its resemblance to the human eye for binary waves. The interior region of the eye pattern is called the *eye opening*.

An eye pattern provides a great deal of information about the performance of the pertinent system, as described next (see Fig. 6.19):

1.   The width of the eye opening defines the time interval over which the received wave can be sampled without error from intersymbol interfer-

Binary
data   1   0   1   1   0   1

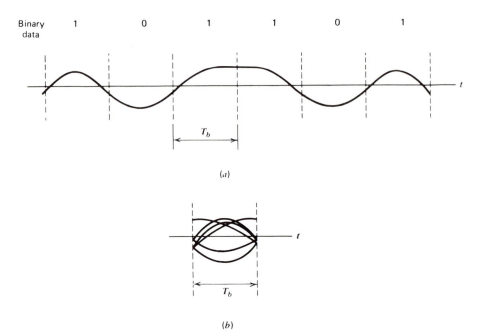

(*a*)

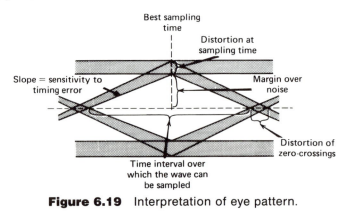

(*b*)

Figure 6.18 (*a*) Distorted binary wave. (*b*) Eye pattern.

ence. It is apparent that the preferred time for sampling is the instant of time at which the eye is open widest.

2. The sensitivity of the system to timing error is determined by the rate of closure of the eye as the sampling time is varied.

3. The height of the eye opening, at a specified sampling time, defines the margin over noise.

When the effect of intersymbol interference is severe, traces from the upper portion of the eye pattern cross traces from the lower portion, with the result that the eye is completely closed. In such a situation, it is impossible to avoid errors due to the combined presence of intersymbol interference and noise in the system, and a solution has to be found to correct for them.

Best sampling
time

Distortion at
sampling time

Slope = sensitivity to
timing error

Margin over
noise

Distortion of
zero-crossings

Time interval over
which the wave can
be sampled

Figure 6.19 Interpretation of eye pattern.

6.7   BASEBAND M-ARY PAM SYSTEMS

In the baseband binary PAM system of Fig. 6.5, the output of the pulse generator consists of binary pulses, that is, pulses with one of two possible amplitude
levels. On the other hand, in a baseband M-ary PAM system, the output of the
pulse generator takes on one of M possible amplitude levels, with M typically
an integer power of 2. Each amplitude level at the pulse generator output
corresponds to a distinct symbol, so that there are M distinct amplitude levels
to be transmitted. Indeed, the M-ary data format so produced is equivalent to
the extension of a binary data source to order $\log_2 M$. We may therefore view
M-ary data transmission as a generalization of the binary case.

Consider then an *M-ary PAM system* with a signal alphabet that contains M
equally likely and statistically independent symbols, with the symbol duration
denoted by T seconds. With $T = T_b \log_2 M$, we find that an M-ary PAM system is
able to transmit information at a rate that is $\log_2 M$ faster than the corresponding
binary PAM system for a fixed channel bandwidth. Equivalently, for a fixed bit
rate, an M-ary PAM system requires less channel bandwidth than a binary
PAM system. However, in order to realize the same average probability of
symbol error, an M-ary PAM system requires more transmitted power (see
Example 2 of Chapter 3). Accordingly, M-ary PAM is well-suited for the transmission of digital data over channels that offer a limited bandwidth and a high
signal-to-noise ratio.

Another matter of interest is that in the case of an M-ary system, the eye
pattern contains $(M - 1)$ eye openings stacked up vertically one upon the other,
where M is the number of discrete amplitude levels used to construct the
transmitted signal. In a strictly linear system with truly random data, all of
these eye openings would be identical. In practice, however, it is often possible
to discern asymmetries in the eye pattern, which are caused by nonlinearities in
the transmission channel.

6.8   ADAPTIVE EQUALIZATION FOR DATA TRANSMISSION

An efficient approach to *high-speed transmission* of digital data (e.g., computer
data) over a voice-grade telephone channel (which is characterized by a limited
bandwidth and high signal-to-noise ratio) involves the use of two basic signal
processing operations:

1. Discrete PAM by encoding the amplitudes of successive pulses in a peri-
    odic pulse train with a discrete set of possible amplitude levels.
2. A linear modulation scheme that offers bandwidth conservation to transmit
    the encoded pulse train over the telephone channel. (Spectrally efficient
    modulation schemes are considered in the next chapter).

At the receiving end of the system, the received wave is demodulated, and then
synchronously sampled and quantized. As a result of dispersion of the pulse
shape by the channel, we find that the number of detectable amplitude levels is
often limited by intersymbol interference rather than by additive noise. In

principle, if the channel is known precisely, it is virtually always possible to make the intersymbol interference (at the sampling instants) arbitrarily small by using a suitable pair of transmitting and receiving filters, so as to control the overall pulse shape in the manner described previously. The transmitting filter is placed directly before the modulator, whereas the receiving filter is placed directly after the demodulator. Thus, insofar as intersymbol interference is concerned, we may consider the data transmission as being essentially baseband.

However, in a *switched telephone network*,* we find that two factors contribute to pulse distortion on different link connections: (1) differences in the transmission characteristics of the individual links that may be switched together, and (2) differences in the number of links in a connection. The result is that the telephone channel is random in the sense of being one of an ensemble of possible channels. Consequently, the use of a fixed pair of transmitting and receiving filters designed on the basis of average channel characteristics may not adequately reduce intersymbol interference. To realize the full transmission capability of a telephone channel, there is need for *adaptive equalization*.† By equalization we mean the process of correcting channel-induced distortion. This process is said to be adaptive when it adjusts itself continuously during data transmission by operating on the input signal.

Among the philosophies for adaptive equalization of data transmission systems, we have *prechannel equalization* at the transmitter, and *postchannel equalization* at the receiver. Because the first approach requires a feedback channel, we consider only adaptive equalization at the receiving end of the system. This equalization can be achieved, prior to data transmission, by training the filter with the guidance of a suitable *training sequence* transmitted through the channel so as to adjust the filter parameters to optimum values. The typical telephone channel changes little during an average data call, so that precall equalization with a training sequence is sufficient in most cases encountered in practice. The equalizer is positioned after the receiving filter in the receiver.

The adaptive equalizer consists of a tapped-delay-line filter (with as many as 100 taps or more), whose coefficients are updated (starting from zero initial values) in accordance with the *least-mean square* (LMS) algorithm; the LMS algorithm was described in Section 3.15. The adjustments to the filter coefficients are made in a step-by-step fashion synchronously with the incoming data.‡

There are two modes of operating the adaptive equalizer, as shown in Fig. 6.20. During the *training period*, a known sequence is transmitted and a syn-

---

* For a description of switched-communication networks, see Section 10.2.

† For tutorial papers on adaptive equalization, see Qureshi (1982, 1985).

‡ In a *synchronous equalizer*, the delay-line taps are spaced at the reciprocal of the symbol rate. In the equalization of telephone channels, the equalizer is sometimes designed with the taps spaced closer than the reciprocal of the symbol rate. Such equalizers are called *fractionally spaced equalizers*. For the same total time span, a fractionally spaced equalizer can effectively compensate for more severe delay distortion than the conventional synchronous equalizer. For details of fractionally spaced equalizers, see Gitlin and Weinstein (1981).

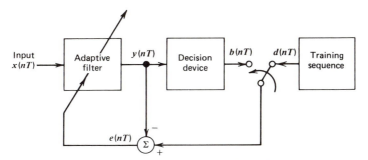

Figure 6.20   Illustrating the modes of operation of an adaptive equalizer.

chronized version of this signal is generated in the receiver where it is applied to the adaptive equalizer as the desired response. The training sequence may, for example, consist of a linear maximal-length or pseudo-noise (PN) sequence.* The length of this training sequence must be equal to or greater than that of the adaptive equalizer.

When the training period is completed, the adaptive equalizer is switched to its second mode of operation, the *decision-directed mode*. In this mode of operation, the error signal equals

$$e(nT) = b(nT) - y(nT)$$

where $y(nT)$ is the equalizer output and $b(nT)$ is the final (not necessarily) correct estimate of the transmitted symbol $b(nT)$. Now, in normal operation the decisions made by the receiver are correct with high probability. This means that the error estimates are correct most of the time, thereby permitting the adaptive equalizer to operate satisfactorily. Furthermore, an adaptive equalizer operating in a decision-directed mode is able to *track* relatively slow variations in channel characteristics.

The methods of implementing adaptive equalizers may be divided into three broad categories: *analog, hardwired digital,* and *programmable digital*, as described here:

1. The analog approach is primarily based on the use of *charge-coupled device* (CCD) technology. The basic circuit realization of the CCD is a row of *field-effect transistors* (FET) with drains and sources connected in series, and the drains capacitively coupled to the gates. The set of adjustable tap weights are stored in digital memory locations, and the multiplications of the analog sample values by the digitized tap weights take place in analog fashion. This approach has significant potential in applications where the symbol rate is too high for digital implementation.

2. In hardwired digital implementation of an adaptive equalizer, the equalizer input is first sampled and then quantized into a form suitable for storage in shift registers. The set of adjustable tap weights are also stored in shift registers. Logic circuits are used to perform the required digital arithmetic

* Pseudo-noise (PN) sequences are discussed in Chapters 8 and 9.

(e.g., multiply and accumulate). In this approach the circuitry is hard-wired for the sole purpose of performing equalization. Nonetheless, it is the most widely used method of implementing adaptive equalizers.

3. The use of a programmable digital processor in the form of a *microprocessor*, for example, offers flexibility in that the adaptive equalization is performed as a series of steps or instructions in the microprocessor. An important advantage of this approach is that the same hardware may be time-shared to perform a multiplicity of signal-processing functions such as filtering, modulation, and demodulation in a modem (*modulator-demo*dulator) used to transmit digital data over a telephone channel.

6.9   SUMMARY AND DISCUSSION

In this section, we summarize highlights of the baseband shaping problem for data transmission, and discuss related issues.

First, we may state that given an ideal low-pass channel of bandwidth B_o hertz, it is possible to transmit independent binary symbols through the channel at the maximum rate $R_b = 2B_o$ bits per second. Equivalently, given a bit rate $R_b = 1/T_b$, the bandwidth $B_o = 0.5R_b$ defines the minimum transmission bandwidth acceptable for distortionless transmission. The bandwidth B_o so defined is called the *Nyquist bandwidth*.*

For practical usefulness, however, the minimum bandwidth solution has to be modified. We do this by (1) permitting a channel bandwidth B in excess of the Nyquist bandwidth B_o, and (2) introducing a transition region shaped as one-half of a raised cosine. The width of the transition region is controlled by the rolloff factor α, defined as the *excess bandwidth* (i.e., the amount by which the channel bandwidth B exceeds the Nyquist bandwidth B_o) divided by the Nyquist bandwidth itself.

In the raised cosine solution, flexibility exists in the selection of the transmitting and receiving filters. This flexibility can be exploited to provide noise immunity. In particular, given a baseband channel of transfer function $H_C(f)$ and a message source of known waveform, we can optimize the transfer function $H_T(f)$ of the transmitting filter and the transfer function $H_R(f)$ of the receiving filter, so that the following three requirements are jointly satisfied:

1. Intersymbol interference is zero.
2. Probability of symbol error is minimized.
3. Constant power is transmitted.

This issue is discussed in Problem 6.4.7.

In the case of *correlative coding* or *partial response techniques*, we follow a different approach. Correlation is intentionally introduced between adjacent symbols with the result that data transmission through a channel limited to the Nyquist bandwidth becomes a practical reality. However, this remarkable performance is achieved at the expense of increased transmitted power. A popular

---

* The term "Nyquist bandwidth" is used in recognition of the pioneering work by Harry Nyquist on baseband shaping for telegraphy.

form of correlative coding is represented by duobinary signaling, which is combined with *precoding*. The latter operation is introduced for the purpose of preventing the occurrence of error propagation. The transfer function of the duobinary conversion filter is low-pass-like, defined by a *half-cosine* function. In the *modified duobinary signaling* format, the transfer function of the filter is band-pass-like, defined by a *full-sine* function; zero-frequency components of the input are therefore suppressed.

When the channel of interest offers a limited bandwidth and a high signal-to-noise ratio, the use of M-ary data transmission offers an attractive method for efficient utilization of the channel. Specifically, it enables the transmission of data through the channel at a rate that is $\log_2 M$ faster than the corresponding binary case. This improvement, however, is attained at the cost of an increase in transmitted power (needed to maintain an acceptable average probability of symbol error) and also an increase in system complexity.

Finally, it should be noted that although the various pulse-shaping techniques have been discussed in the context of baseband data-transmission systems (i.e., systems employing baseband channels such as coaxial cables and optical fibers), nevertheless, the theory is equally applicable to data transmission over voice-grade telephone channels and other band-pass channels that rely on the use of *linear modulation* of a sinusoidal carrier wave. (Linear modulation schemes for band-pass data transmission are discussed in the next chapter.) Moreover, when the channel is linear with unknown and time-varying characteristics, we may use an *adaptive equalizer* to correct for channel-induced variations. The *least-mean-square (LMS) algorithm* provides a simple and yet effective method for updating the equalizer coefficients in accordance with variations in channel characteristics.

PROBLEMS

P6.1   DISCRETE PAM SIGNALS

Problem 6.1.1   The unipolar, polar, and bipolar formats shown in Fig. 6.1 are all of the NRZ type. Construct the RZ versions of these formats for the binary sequence 0110100011.

Problem 6.1.2   Construct the NRZ bipolar format for the binary sequence 011010110.

Problem 6.1.3   Given the binary sequence 011010110, construct the polar octal (8-level) format of the NRZ type using (a) natural code, and (b) Gray code.

P6.2   POWER SPECTRA OF DISCRETE PAM SIGNALS

Problem 6.2.1   Show that, for the bipolar format, the autocorrelation function $R_A(n)$, that is, $E[A_k A_{k-n}]$ is zero for $n > 1$, where A_k is a random variable representing the kth bit of the input binary sequence. Assume statistically independent and equally likely message bits.

Problem 6.2.2   Consider a random binary sequence where bits are statistically independent and equally likely. Determine the power spectral density for the following two representations of the sequence:

(a)   RZ unipolar format.
(b)   RZ polar format.

Comment on your results.

Problem 6.2.3   Determine the power spectral density of the polar quaternary format of NRZ type, based on the natural code. Assume statistically independent and equally likely message bits.

P6.3   INTERSYMBOL INTERFERENCE

Problem 6.3.1   A periodic sequence, consisting of 1s and 0s, is applied to the (normalized) low-pass RC filter shown in Fig. P6.1. The filter output is sampled at midpoints of the signaling intervals, and then compared to a threshold of zero volts. If the threshold is exceeded, a decision is made in favor of symbol 1; otherwise, the decision is made in favor of symbol 0. Using this procedure, construct the resulting binary sequence for the following three values of bit duration:

(a) $T_b = 5$ s
(b) $T_b = 1$ s
(c) $T_b = 0.25$ s

Comment on your results.

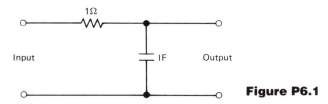

Figure P6.1

Problem 6.3.2   The sampled output of the receiving filter in Fig. 6.5 equals $y(t_i)$, and the subsequent decision proceeds as follows:

$$\text{if} \quad y(t_i) > 0 \quad \text{say binary symbol 1}$$

and

$$\text{if} \quad y(t_i) < 0 \quad \text{say binary symbol 0}$$

Show that the presence of intersymbol interference in the system can cause erroneous decisions.

P6.4   NYQUIST'S CRITERION FOR DISTORTIONLESS BASEBAND TRANSMISSION

Problem 6.4.1   The overall pulse shape $p(t)$ of a binary PAM system is defined by

$$p(t) = \operatorname{sinc}\!\left(\frac{t}{T_b}\right)$$

where T_b is the bit duration of the input binary data. The amplitude levels at the pulse generator output are $+1$ volt or -1 volt, depending on whether the binary symbol at the input is 1 or 0, respectively. Sketch the waveform at the output of the receiving filter in response to the input data 001101001.

Problem 6.4.2 Consider a baseband binary PAM system that is designed to have a raised cosine spectrum $P(f)$, as in Eq. 6.34. Show that the resulting pulse $p(t)$ is as defined in Eq. 6.36. How would this pulse be modified if the system was designed to have a linear phase response?

Problem 6.4.3 A computer puts out binary data at the rate of 56 kilobits per second. The computer output is transmitted using a baseband binary PAM system that is designed to have a raised cosine spectrum. Determine the transmission bandwidth required for each of the following rolloff factors:

(a) $\alpha = 0.25$
(b) $\alpha = 0.5$
(c) $\alpha = 0.75$
(d) $\alpha = 1.0$

Problem 6.4.4 An analog signal is sampled, quantized, and encoded into a binary PCM wave. The number of representation levels used is 128. A synchronizing pulse is added at the end of each code-word representing a sample of the analog signal. The resulting PCM wave is transmitted over a channel of bandwidth 12 kHz using a binary PAM system with a raised cosine spectrum. The rolloff factor is unity.

(a) Find the rate (in bits per second) at which information is transmitted through the channel.
(b) Find the rate at which the analog signal is sampled. What is the maximum possible value for the highest frequency component of the analog signal?

Problem 6.4.5 A binary PAM wave is to be transmitted over a low-pass channel with an absolute maximum bandwidth of 75 kHz. The bit duration is 10 microseconds. Find a raised cosine spectrum that satisfies these requirements.

Problem 6.4.6 An analog signal is sampled, quantized, and encoded into a binary PCM wave. The specifications of the PCM system include the following:

$$\text{Sampling rate} = 8 \text{ kHz}$$

$$\text{Number of representation levels} = 64$$

The PCM wave is transmitted over a baseband channel using binary PAM. Determine the minimum bandwidth required for transmitting the PCM wave, assumed to be of binary form.

Problem 6.4.7 In this problem we explore the issue of optimizing the transmitting and receiving filters of a binary PAM system for noise immunity.

Suppose that white noise of zero mean and power spectral density $N_0/2$ is added to the input of the receiving filter. Assume that the baseband response of the system is shaped to produce zero intersymbol interference.

(a) Show that the signal-to-noise ratio at the output of the receiving filter is given by

$$\frac{a^2}{\sigma_N^2} = \frac{2PT_b}{N_0} \left[\int_{-\infty}^{\infty} \frac{|P(f)|^2}{|H_C(f)|^2|H_R(f)|^2} \, df \int_{-\infty}^{\infty} |H_R(f)|^2 df \right]^{-1}$$

where

a = amplitude of sampled output of the receiving filter

σ_N^2 = noise variance at the output of the receiving filter

P = average transmitted power

T_b = bit duration

$P(f)$ = composite transfer function defined in Eq. 6.21

$H_C(f)$= transfer function of the channel

$H_R(f)$ = transfer function of the receiving filter

(b) Use Schwarz's inequality to minimize a^2/σ_N^2; Schwarz's inequality is discussed in Appendix G. Hence, show that the optimum receiving filter is defined by

$$|H_R(f)|_{\text{opt}}^2 = \frac{C_1|P(f)|}{|H_C(f)|}$$

where C_1 is a constant. Correspondingly, show that the optimum transmitting filter is defined by

$$|H_T(f)|_{\text{opt}}^2 = \frac{C_2|P(f)|}{|V(f)|^2|H_C(f)|}$$

where C_2 is another constant, and $V(f)$ is the Fourier transform of the message source waveform.

(c) Show that the maximum value of the signal-to-noise ratio is given by

$$\left(\frac{a^2}{\sigma_N^2}\right)_{\text{max}} = \frac{2PT_b}{N_0}\left[\int_{\infty}^{-\infty}\frac{|P(f)|}{|H_C(f)|}\,df\right]^{-2}$$

(d) Discuss the choice of a message source waveform for which the transmitting and receiving filters have essentially the same transfer function (except for scaling factors).

P6.5   CORRELATIVE CODING

Problem 6.5.1   The duobinary, ternary, and bipolar signaling techniques have one common feature: they all employ three amplitude levels. In what way does the duobinary technique differ from the other two?

Problem 6.5.2   The binary data 001101001 are applied to the input of a duobinary system.

(a) Construct the duobinary coder output and corresponding receiver output, without a precoder.
(b) Suppose that due to error during transmission, the level at the receiver input produced by the second digit is reduced to zero. Construct the new receiver output.

Problem 6.5.3   Repeat Problem 6.5.2, assuming the use of a precoder in the transmitter.

Problem 6.5.4

(a)   Find the frequency response and impulse response of the correlative coder part of the circuit shown in Fig. 6.13.

(b)   Show that the power spectral density of the signal $z(t)$ appearing at the output of the circuit in Fig. 6.13 is the same as that of the bipolar format given in Eq. 6.15.

Problem 6.5.5   The binary data 011100101 are applied to the input of a modified duobinary system.

(a)   Construct the modified duobinary coder output and corresponding receiver output, without a precoder.

(b)   Suppose that due to error during transmission, the level produced by the third digit is reduced to zero. Construct the new receiver output.

Problem 6.5.6   Repeat Problem 6.5.5 assuming the use of a precoder in the transmitter.

Problem 6.5.7   Using conventional analog filter design methods, it is difficult to approximate the frequency response of the modified duobinary system defined by Eq. 6.52. To get around this problem, we may use the arrangement shown in Fig. P.6.2. Justify the validity of this scheme.

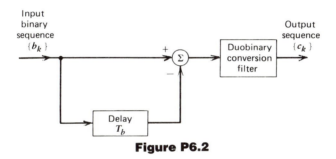

Figure P6.2

P6.6   EYE PATTERN

Problem 6.6.1   A binary wave using polar signaling is generated by representing symbol 1 by a pulse of amplitude $+1$ volt and symbol 0 by a pulse of amplitude -1 volt; in both cases the pulse duration equals the bit duration. This signal is applied to a low-pass RC filter with transfer function:

$$H(f) = \frac{1}{1 + jf/f_0}$$

Construct the eye pattern for the filter output for the following sequences:

(a)   Alternating 1s and 0s.

(b)   A long sequence of 1s followed by a long sequence of 0s.

(c)   A long sequence of 1s followed by a single 0 and then a long sequence of 1s.

Assume a bit rate of $2f_0$ bits per second.

Problem 6.6.2   The binary sequency 011010 is transmitted through a channel having a raised cosine characteristic with rolloff factor of unity. Assume the use of polar signaling, with symbols 1 and 0 represented by $+1$ and -1 volt, respectively.

(a) Construct, to scale, the received wave, and indicate the best sampling times for regeneration.
(b) Construct the eye pattern for this received wave and show that it is completely open.
(c) Determine the zero crossings of the received wave.

P6.7   BASEBAND M-ARY PAM SYSTEMS

Problem 6.7.1   Repeat Problems 6.4.3, given that each set of three successive binary digits in the computer output are coded into one of eight possible amplitude levels, and the resulting signal is transmitted by using an 8-level PAM system designed to have a raised cosine spectrum.

Problem 6.7.2   Repeat Problem 6.4.6, assuming that each pulse of the transmitted PCM wave is allowed to take on the following number of amplitude levels:

(a) $M = 4$
(b) $M = 8$

CHAPTER SEVEN

DIGITAL MODULATION TECHNIQUES

When it is required to transmit digital data over a band-pass channel, it is necessary to modulate the incoming data onto a carrier wave (usually sinusoidal) with fixed frequency limits imposed by the channel. The data may represent digital computer outputs or PCM waves generated by digitizing voice or video signals. The channel may be a telephone channel, microwave radio link, satellite channel, or an optical fiber. In any event, the modulation process involves switching or keying the amplitude, frequency, or phase of the carrier in accordance with the incoming data. Thus there are three basic modulation techniques for the transmission of digital data; they are known as amplitude-shift keying (ASK), frequency-shift keying (FSK), and phase-shift keying (PSK), which may be viewed as special cases of amplitude modulation, frequency modulation, and phase modulation, respectively.

This chapter is devoted to a study of digital modulation techniques: their noise performance, spectral properties, their merits and limitations, applications, and other related issues. For the noise analysis, we use the *signal-space approach* described in Chapter 3. We begin the chapter by presenting an overview of various modulation formats available to the designer of digital communication systems.

7.1 DIGITAL MODULATION FORMATS

Modulation is defined as *the process by which some characteristic of a carrier is varied in accordance with a modulating wave.** In digital communications, the modulating wave consists of *binary* data or an *M-ary* encoded version of it. For the carrier, it is customary to use a sinusoidal wave. With a sinusoidal carrier, the feature that is used by the modulator to distinguish one signal from another is a step change in the amplitude, frequency, or phase of the carrier. The result of this modulation process is *amplitude-shift keying (ASK)*, *frequency-shift keying (FSK)*, *or phase-shift keying (PSK)*, respectively, as illustrated in Fig. 7.1 for the special case of a source of binary data.

Ideally, PSK and FSK signals have a constant envelope, as shown in Fig. 7.1. This feature makes them impervious to amplitude nonlinearities, as en-

* This definition is taken from the *IEEE Standard Dictionary of Electrical and Electronics Terms* (Wiley-Interscience, 1972, p. 351).

273

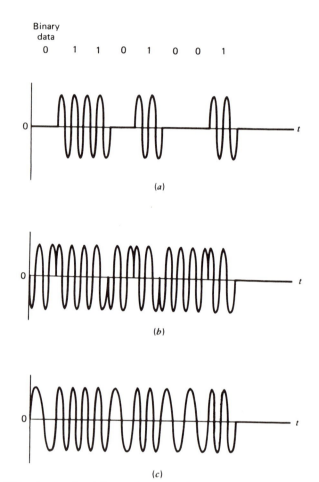

Binary
data

0   1   1   0   1   0   0   1

(a)

(b)

(c)

Figure 7.1 Waveforms for (a) ampliture-shift keying, (b) phase-shift keying, and (c) frequency-shift keying.

countered in microwave radio links and satellite channels. Accordingly, we find that, in practice, PSK and FSK signals are much more widely used than ASK signals.

In the more general case of M-ary signaling, the modulator produces one of an available set of $M = 2^m$ distinct signals in response to m bits of source data at a time. Clearly, binary modulation is a special case of M-ary modulation with $M = 2$.

In the waveforms shown in Fig. 7.1, a single feature of the carrier (i.e., amplitude, phase, or frequency) undergoes modulation. Sometimes, a *hybrid* form of modulation is used. For example, changes in both amplitude and phase of the carrier are combined to produce *amplitude-phase keying* (APK). The use of hybrid techniques opens up yet another format for digital modulation.

To perform demodulation at the receiver, we have the choice of *coherent* or *noncoherent detection.* In the ideal form of coherent detection, exact replicas of the possible arriving signals are available at the receiver. This means that the receiver has exact knowledge of the carrier wave's phase reference, in which case we say the receiver is *phase-locked* to the transmitter. Coherent detection is performed by cross-correlating the received signal with each one of the replicas, and then making a decision based on comparisons with preselected thresholds. In noncoherent detection, on the other hand, knowledge of the carrier wave's phase is not required. The complexity of the receiver is thereby reduced but at the expense of an inferior error performance, compared to a coherent system.

We thus see that there are a multitude of modulation/detection schemes available to the designer of a digital communication system required for data transmission over a band-pass channel. Each scheme offers *system trade-offs* of its own. The final choice made by the designer is determined by the way in which the available primary communication resources, *transmitted power* and *channel bandwidth,* are best exploited. In particular, the choice is made in favor of the scheme that attains as many of the following design goals as possible:

1. Maximum data rate.
2. Minimum probability of symbol error.
3. Minimum transmitted power.
4. Minimum channel bandwidth.
5. Maximum resistance to interfering signals.
6. Minimum circuit complexity.

Some of these goals pose conflicting requirements; for example, goals (1) and (2) are in conflict with goals (3) and (4). The best we can therefore do is to satisfy as many of these goals as possible.

7.2 COHERENT BINARY MODULATION TECHNIQUES

As mentioned previously, binary modulation has three basic forms: amplitude-shift keying (ASK), phase-shift keying (PSK), and frequency-shift keying (FSK). In this section, we present the noise analysis for the coherent detection of PSK and FSK signals, assuming an *additive white Gaussian noise (AWGN) model.* The analysis is based on the signal-space approach described in Chapter 3. The noise analysis for the coherent detection of ASK signals is presented as a problem to the reader (see Problem 7.2.9). It turns out that although the signal constellations for ASK and FSK are radically different, nevertheless, for large signal-to-noise ratios, they have the same probability of error for an AWGN channel. A *signal constellation* refers to a set of possible message points.

(1) Coherent Binary PSK

In a coherent binary PSK system, the pair of signals, $s_1(t)$ and $s_2(t)$, used to represent binary symbols 1 and 0, respectively, are defined by

$$s_1(t) = \sqrt{\frac{2E_b}{T_b}} \cos(2\pi f_c t) \tag{7.1}$$

$$s_2(t) = \sqrt{\frac{2E_b}{T_b}} \cos(2\pi f_c t + \pi) = -\sqrt{\frac{2E_b}{T_b}} \cos(2\pi f_c t) \tag{7.2}$$

where $0 \leq t < T_b$, and E_b is the *transmitted signal energy per bit*. In order to ensure that each transmitted bit contains an integral number of cycles of the carrier wave, the carrier frequency f_c is chosen equal to n_c/T_b for some fixed integer n_c. A pair of sinusoidal waves that differ only in a relative phase-shift of 180 degrees, as defined above, are referred to as *antipodal signals*.

From Eqs. 7.1 and 7.2, it is clear that there is only one basis function of unit energy, namely

$$\phi_1(t) = \sqrt{\frac{2}{T_b}} \cos(2\pi f_c t) \qquad 0 \leq t < T_b \tag{7.3}$$

Then we may expand the transmitted signals $s_1(t)$ and $s_2(t)$ in terms of $\phi_1(t)$ as follows

$$s_1(t) = \sqrt{E_b}\phi_1(t) \qquad 0 \leq t < T_b \tag{7.4}$$

and

$$s_2(t) = -\sqrt{E_b}\phi_1(t) \qquad 0 \leq t < T_b \tag{7.5}$$

A coherent binary PSK system is therefore characterized by having a signal space that is one-dimensional (i.e., $N = 1$), and with two message points (i.e., $M = 2$), as shown in Fig. 7.2. The coordinates of the message points equal

$$\begin{aligned} s_{11} &= \int_0^{T_b} s_1(t)\phi_1(t)dt \\ &= +\sqrt{E_b} \end{aligned} \tag{7.6}$$

and

$$\begin{aligned} s_{21} &= \int_0^{T_b} s_2(t)\phi_1(t)dt \\ &= -\sqrt{E_b} \end{aligned} \tag{7.7}$$

The message point corresponding to $s_1(t)$ is located at $s_{11} = +\sqrt{E_b}$, and the message point corresponding to $s_2(t)$ is located at $s_{21} = -\sqrt{E_b}$.

To realize a *rule for making a decision* in favor of symbol 1 or symbol 0, we apply Eq. 3.64. Specifically, we must partition the signal space of Fig. 7.2 into two regions:

1. The set of points closest to the message point at $+\sqrt{E_b}$.
2. The set of points closest to the message point at $-\sqrt{E_b}$.

This is accomplished by constructing the midpoint of the line joining these two message points, and then marking off the appropriate decision regions. In Fig.

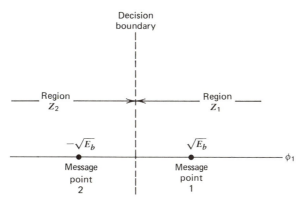

Figure 7.2 Signal space diagram for coherent binary PSK system.

7.2, these decision regions are marked Z_1 and Z_2, according to the message point around which they are constructed.

The decision rule is now simply to guess signal $s_1(t)$ or binary symbol 1 was transmitted if the received signal point falls in region Z_1, and guess signal $s_2(t)$ or binary symbol 0 was transmitted if the received signal point falls in region Z_2. Two kinds of erroneous decisions may, however, be made. Signal $s_2(t)$ is transmitted, but the noise is such that the received signal point falls inside region Z_1 and so the receiver decides in favor of signal $s_1(t)$. Alternatively, signal $s_1(t)$ is transmitted, but the noise is such that the received signal point falls inside region Z_2 and so the receiver decides in favor of signal $s_2(t)$.

To calculate the probability of making an error of the first kind, we note from Fig. 7.2 that the decision region associated with symbol 1 or signal $s_1(t)$ is described by

$$Z_1: \qquad 0 < x_1 < 1$$

where x_1 is the observation scalar:

$$x_1 = \int_0^{T_b} x(t)\phi_1(t)dt \tag{7.8}$$

where $x(t)$ is the *received signal*. From Eq. 3.49 we deduce that the likelihood function, when symbol 0 or signal $s_2(t)$ is transmitted, is defined by

$$
\begin{aligned}
f_{X_1}(x_1|0) &= \frac{1}{\sqrt{\pi N_0}} \exp\left[-\frac{1}{N_0}(x_1 - s_{21})^2 \right] \\
&= \frac{1}{\sqrt{\pi N_0}} \exp\left[-\frac{1}{N_0}(x_1 + \sqrt{E_b})^2 \right]
\end{aligned}
\tag{7.9}
$$

The conditional probability of the receiver deciding in favor of symbol 1, given that symbol 0 was transmitted, is therefore

$$
\begin{aligned}
P_e(0) &= \int_0^\infty f_{X_1}(x_1|0)dx_1 \\
&= \frac{1}{\sqrt{\pi N_0}} \int_0^\infty \exp\left[-\frac{1}{N_0}(x_1 + \sqrt{E_b})^2 \right] dx_1
\end{aligned}
\tag{7.10}
$$

Putting

$$z = \frac{1}{\sqrt{N_0}} (x_1 + \sqrt{E_b}) \qquad (7.11)$$

and changing the variable of integration from x_1 to z, we may rewrite Eq. 7.10 in the form

$$P_e(0) = \frac{1}{\sqrt{\pi}} \int_{\sqrt{E_b/N_0}}^{\infty} \exp(-z^2)dz$$

$$= \frac{1}{2} \operatorname{erfc}\left(\sqrt{\frac{E_b}{N_0}}\right) \qquad (7.12)$$

where $\operatorname{erfc}(\sqrt{E_b/N_0})$ is the complementary error function (see Appendix E).

Similarly, we may show that $P_e(1)$, the conditional probability of the receiver deciding in favor of symbol 0, given that symbol 1 was transmitted, also has the same value as in Eq. 7.12. Thus, averaging the conditional error probabilities $P_e(0)$ and $P_e(1)$, we find that the average probability of symbol error for coherent binary PSK equals

$$P_e = \frac{1}{2} \operatorname{erfc}\left(\sqrt{\frac{E_b}{N_0}}\right) \qquad (7.13)$$

It is of interest to note that, as a rule, whenever the observation space is partitioned in a symmetric manner (as in Fig. 7.2, for example), then the conditional symbol error probabilities and the average probability of symbol error will all have the same value.

To generate a binary PSK wave, we see from Eqs. 7.3–7.5 that we have to represent the input binary sequence in polar form with symbols 1 and 0 represented by constant amplitude levels of $+\sqrt{E_b}$ and $-\sqrt{E_b}$, respectively. This

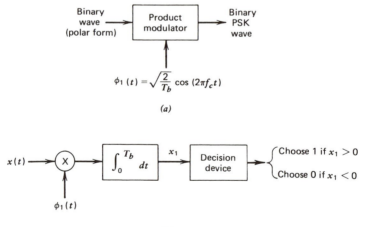

Figure 7.3 Block diagrams for (a) binary PSK transmitter, and (b) coherent binary PSK receiver.

binary wave and a sinusoidal carrier were $\phi_1(t)$ (whose frequency $f_c = n_c/T_b$ for some fixed integer n_c) are applied to a product modulator, as in Fig. 7.3a. The carrier and the timing pulses used to generate the binary wave are usually extracted from a common master clock. The desired PSK wave is obtained at the modulator output.

To detect the original binary sequence of 1s and 0s, we apply the noisy PSK wave $x(t)$ (at the channel output) to a correlator, which is also supplied with a locally generated coherent reference signal $\phi_1(t)$, as in Fig. 7.3b. The correlator output, x_1, is compared with a threshold of zero volts. If $x_1 > 0$, the receiver decides in favor of symbol 1. On the other hand, if $x_1 < 0$, it decides in favor of symbol 0.

(2) Coherent Binary FSK

In a binary FSK system, symbols 1 and 0 are distinguished from each other by transmitting one of two sinusoidal waves that differ in frequency by a fixed amount. A typical pair of sinusoidal waves is described by

$$s_i(t) = \begin{cases} \sqrt{\dfrac{2E_b}{T_b}} \cos(2\pi f_i t) & 0 \leq t \leq T_b \\ 0 & \text{elsewhere} \end{cases} \tag{7.14}$$

where $i = 1, 2$, and E_b is the transmitted signal energy per bit, and the transmitted frequency equals

$$f_i = \frac{n_c + i}{T_b} \qquad \text{for some fixed integer } n_c \text{ and } i = 1, 2 \tag{7.15}$$

Thus symbol 1 is represented by $s_1(t)$, and symbol 0 by $s_2(t)$.

From Eq. 7.14 we observe directly that the signals $s_1(t)$ and $s_2(t)$ are orthogonal, but not normalized to have unit energy. We therefore deduce that the most useful form for the set of orthonormal basis functions is

$$\phi_i(t) = \begin{cases} \sqrt{\dfrac{2}{T_b}} \cos(2\pi f_i t) & 0 \leq t \leq T_b \\ 0 & \text{elsewhere} \end{cases} \tag{7.16}$$

where $i = 1, 2$. Correspondingly, the coefficient s_{ij} for $i = 1, 2$, and $j = 1, 2$, is defined by

$$\begin{aligned} s_{ij} &= \int_0^{T_b} s_i(t)\phi_j(t)\,dt \\ &= \int_0^{T_b} \sqrt{\frac{2E_b}{T_b}} \cos(2\pi f_i t) \sqrt{\frac{2}{T_b}} \cos(2\pi f_j t)\,dt \\ &= \begin{cases} \sqrt{E_b} & i = j \\ 0 & i \neq j \end{cases} \end{aligned} \tag{7.17}$$

Thus a coherent binary FSK system is characterized by having a signal space that is two-dimensional (i.e., $N = 2$) with two message points (i.e., $M = 2$), as in Fig. 7.4. The two message points are defined by the signal vectors:

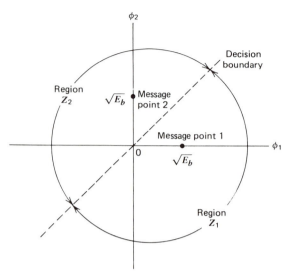

Figure 7.4 Signal space diagram for coherent binary FSK system.

$$\mathbf{s}_1 = \begin{bmatrix} \sqrt{E_b} \\ 0 \end{bmatrix} \tag{7.18}$$

and

$$\mathbf{s}_2 = \begin{bmatrix} 0 \\ \sqrt{E_b} \end{bmatrix} \tag{7.19}$$

Note that the distance between the two message points is equal to $\sqrt{2E_b}$.

The observation vector $\mathbf{x}$ has two elements, x_1 and x_2, that are defined by, respectively

$$x_1 = \int_0^{T_b} x(t)\phi_1(t)dt \tag{7.20}$$

and

$$x_2 = \int_0^{T_b} x(t)\phi_2(t)dt \tag{7.21}$$

where $x(t)$ is the received signal, the form of which depends on which symbol was transmitted. Given that symbol 1 was transmitted, $x(t)$ equals $s_1(t) + w(t)$, where $w(t)$ is the sample function of a white Gaussian noise process of zero mean and power spectral density $N_0/2$. If, on the other hand, symbol 0 was transmitted, $x(t)$ equals $s_2(t) + w(t)$.

Now, applying the decision rule of Eq. 3.64, we find that the observation space is partitioned into two decision regions, labeled as Z_1 and Z_2 in Fig. 7.4. Accordingly, the receiver decides in favor of symbol 1 if the received signal point represented by the observation vector $\mathbf{x}$ falls inside region Z_1. This occurs when $x_1 > x_2$. If, on the other hand, we have $x_1 < x_2$, the received signal point falls inside region Z_2, and the receiver decides in favor of symbol 0. The decision boundary, separating region Z_1 from region Z_2, is defined by $x_1 = x_2$.

Define a new Gaussian random variable L whose sample value l is equal to the difference between x_1 and x_2; thus

$$l = x_1 - x_2 \qquad (7.22)$$

The mean value of the random variable L depends on which binary symbol was transmitted. Given that symbol 1 was transmitted, the Gaussian random variables X_1 and X_2, whose sample values are denoted by x_1 and x_2, have mean values equal to $\sqrt{E_b}$ and zero, respectively. Correspondingly, the conditional mean of the random variable L, given that symbol 1 was transmitted, is given by

$$
\begin{aligned}
E[L|1] &= E[X_1|1] - E[X_2|1] \\
&= + \sqrt{E_b}
\end{aligned}
\qquad (7.23)
$$

On the other hand, given that symbol 0 was transmitted, the random variables X_1 and X_2 have mean values equal to zero and $\sqrt{E_b}$, respectively. Correspondingly, the conditional mean of the random variable L, given that symbol 0 was transmitted, is given by

$$
\begin{aligned}
E[L|0] &= E[X_1|0] - E[X_2|0] \\
&= - \sqrt{E_b}
\end{aligned}
\qquad (7.24)
$$

The variance of the random variable L is independent of which binary symbol was transmitted. Since the random variables X_1 and X_2 are statistically independent, each with a variance equal to $N_0/2$, it follows that

$$
\begin{aligned}
\text{Var}[L] &= \text{Var}[X_1] + \text{Var}[X_2] \\
&= N_0
\end{aligned}
\qquad (7.25)
$$

Suppose we know that symbol 0 was transmitted. Then the corresponding value of the conditional probability density function of the random variable L equals

$$f_L(l|0) = \frac{1}{\sqrt{2\pi N_0}} \exp\left[-\frac{(l + \sqrt{E_b})^2}{2N_0} \right] \qquad (7.26)$$

Since the condition $x_1 > x_2$, or, equivalently, $l > 0$, corresponds to the receiver making a decision in favor of symbol 1, we deduce that the conditional probability of error, given that symbol 0 was transmitted, is given by

$$
\begin{aligned}
P_e(0) &= P(l > 0 | \text{symbol 0 was sent}) \\
&= \int_0^\infty f_L(l|0)\,dl \\
&= \frac{1}{\sqrt{2\pi N_0}} \int_0^\infty \exp\left[-\frac{(l + \sqrt{E_b})^2}{2N_0} \right] dl
\end{aligned}
\qquad (7.27)
$$

Put

$$\frac{l + \sqrt{E_b}}{\sqrt{2N_0}} = z \qquad (7.28)$$

Then, changing the variable of integration from l to z, we may rewrite Eq. 7.27 as follows

$$P_e(0) = \frac{1}{\sqrt{\pi}} \int_{\sqrt{E_b/2N_0}}^{\infty} \exp(-z^2)dz$$

$$= \frac{1}{2} \text{erfc}\left(\sqrt{\frac{E_b}{2N_0}} \right) \tag{7.29}$$

Similarly, we may show that $P_e(1)$, the conditional probability of error, given that symbol 1 was transmitted, has the same value as in Eq. 7.29. Accordingly, averaging $P_e(0)$ and $P_e(1)$, we find that the average probability of symbol error for coherent binary FSK is

$$P_e = \frac{1}{2} \text{erfc}\left(\sqrt{\frac{E_b}{2N_0}} \right) \tag{7.30}$$

Comparing Eqs. 7.13 and 7.30, we see that in a coherent binary FSK system we have to double the *bit energy-to-noise density ratio*, E_b/N_0, in order to maintain the same average error rate as in a coherent binary PSK system. This result is in perfect accord with the signal space diagrams of Fig. 7.2 and 7.4, where we see that in a binary PSK system the distance between the two message points is equal to $2\sqrt{E_b}$, whereas in a binary FSK system the corresponding distance is $\sqrt{2E_b}$. This shows that, in an AWGN channel, the detection performance of equal energy binary signals depends only on the "distance" between the two pertinent message points in the signal space. In particular, the larger we make this distance, the smaller will the average probability of error be. This is intuitively appealing, since the larger the distance between the message points, the less will be the probability of mistaking one signal for the other.

To generate a binary FSK signal, we may use the scheme shown in Fig. 7.5a. The input binary sequence is represented in its on–off form, with symbol 1 represented by a constant amplitude of $\sqrt{E_b}$ volts and symbol 0 represented by zero volts. By using an *inverter* in the lower channel in Fig. 7.5a, we in effect make sure that when we have symbol 1 at the input, the oscillator with frequency f_1 in the upper channel is switched on while the oscillator with frequency f_2 in the lower channel is switched off, with the result that frequency f_1 is transmitted. Conversely, when we have symbol 0 at the input, the oscillator in the upper channel is switched off, and the oscillator in the lower channel is switched on, with the result that frequency f_2 is transmitted. The two frequencies f_1 and f_2 are chosen to equal integer multiples of the bit rate $1/T_b$, as in Eq. 7.15.

In the transmitter of Fig. 7.5a, we assume that the two oscillators are synchronized, so that their outputs satisfy the requirements of the two orthonormal basis functions $\phi_1(t)$ and $\phi_2(t)$, as in Eq. 7.17. Alternatively, we may use a single keyed (voltage-controlled) oscillator. In either case, the frequency of the modulated wave is shifted with a *continuous phase*, in accordance with the input binary wave. That is to say, phase continuity is always maintained,

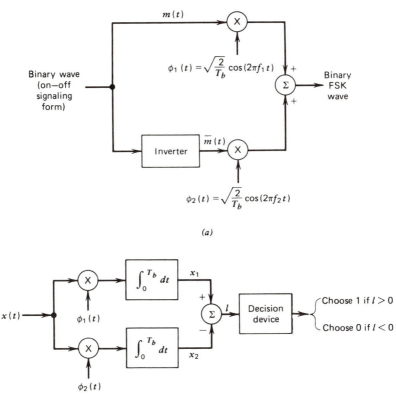

Figure 7.5 Block diagrams for (a) binary FSK transmitter, and (b) coherent binary FSK receiver.

including the inter-bit switching times. We refer to this form of digital modulation as *continuous-phase frequency-shift keying* (CPFSK).

In order to detect the original binary sequence given the noisy received wave $x(t)$, we may use the receiver shown in Fig. 7.5b. It consists of two correlators with a common input, which are supplied with locally generated coherent reference signals $\phi_1(t)$ and $\phi_2(t)$. The correlator outputs are then subtracted, one from the other, and the resulting difference, l, is compared with a threshold of zero volts. If $l > 0$, the receiver decides in favor of 1. On the other hand, if $l < 0$, it decides in favor of 0.

7.3 COHERENT QUADRATURE-MODULATION TECHNIQUES

The provision of reliable performance, exemplified by a very low probability of error, is one important goal in the design of a digital communication system. Another important goal is the efficient utilization of channel bandwidth. (More is said on this issue in Section 7.8.) In this section, we study two bandwidth-conserving modulation schemes for the transmission of binary data. They are both examples of the *quadrature-carrier multiplexing system,* which produces a modulated wave described as follows:

$$s(t) = s_I(t) \cos(2\pi f_c t) - s_Q(t) \sin(2\pi f_c t) \tag{7.31}$$

where $s_I(t)$ is the *in-phase component* of the modulated wave, and $s_Q(t)$ is the *quadrature component.* This terminology is in recognition of the associated *cosine* or *sine* versions of the carrier wave, which are in phase-quadrature with each other. Both $s_I(t)$ and $s_Q(t)$ are related to the input data stream in a way that is characteristic of the type of modulation used.

In the sequel, we first study a quadrature-carrier signaling technique known as quadriphase-shift keying, which is an extension of binary PSK. Next, we consider minimum shift keying, which is a special form of continuous-phase frequency-shift keying (CPFSK). In this latter scheme, the receiver carries out the coherent detection in two successive bit intervals.

(1) Quadriphase-shift Keying

As with binary PSK, this modulation scheme is characterized by the fact that the information carried by the transmitted wave is contained in the phase. In particular, in *quadriphase-shift keying* (QPSK), the phase of the carrier takes on one of four equally spaced values, such as $\pi/4$, $3\pi/4$, $5\pi/4$, and $7\pi/4$, as shown by

$$s_i(t) = \begin{cases} \sqrt{\dfrac{2E}{T}} \cos\left[2\pi f_c t + (2i-1)\dfrac{\pi}{4}\right] & 0 \le t \le T \\ 0 & \text{elsewhere} \end{cases} \tag{7.32}$$

where $i = 1, 2, 3, 4$, and E is the transmitted signal energy per symbol, T is the symbol duration, and the carrier frequency f_c equals n_c/T for some fixed integer n_c. Each possible value of the phase corresponds to a unique pair of bits called a *dibit.* Thus, for example, we may choose the foregoing set of phase values to represent the Gray encoded set of dibits: 10, 00, 01, and 11.

Using a well-known trigonometric identity, we may rewrite Eq. 7.32 in the equivalent form:

$$s_i(t) = \begin{cases} \sqrt{\dfrac{2E}{T}} \cos\left[(2i-1)\dfrac{\pi}{4}\right] \cos(2\pi f_c t) \\ \qquad - \sqrt{\dfrac{2E}{T}} \sin\left[(2i-1)\dfrac{\pi}{4}\right] \sin(2\pi f_c t) & 0 \le t \le T \\ 0 & \text{elsewhere} \end{cases} \tag{7.33}$$

where $i = 1, 2, 3, 4$. Based on this representation, we can make the following observations:

1. There are only two orthonormal basis functions, $\phi_1(t)$ and $\phi_2(t)$, contained in the expansion of $s_i(t)$. The appropriate forms for $\phi_1(t)$ and $\phi_2(t)$ are defined by

Table 7.1   Signal-Space Characterization of QPSK

Input dibit $0 \leqslant t \leqslant T$	Phase of QPSK signal (*radians*)	Coordinates of message points	
		s_{i1}	s_{i2}
10	$\pi/4$	$+\sqrt{E/2}$	$-\sqrt{E/2}$
00	$3\pi/4$	$-\sqrt{E/2}$	$-\sqrt{E/2}$
01	$5\pi/4$	$-\sqrt{E/2}$	$+\sqrt{E/2}$
11	$7\pi/4$	$+\sqrt{E/2}$	$+\sqrt{E/2}$

$$\phi_1(t) = \sqrt{\frac{2}{T}} \cos(2\pi f_c t) \qquad 0 \leqslant t \leqslant T \tag{7.34}$$

and

$$\phi_2(t) = \sqrt{\frac{2}{T}} \sin(2\pi f_c t) \qquad 0 \leqslant t \leqslant T \tag{7.35}$$

2.   There are four message points, and the associated signal vectors are defined by

$$\mathbf{s}_i = \begin{bmatrix} \sqrt{E} \cos\left((2i - 1)\dfrac{\pi}{4}\right) \\ -\sqrt{E} \sin\left((2i - 1)\dfrac{\pi}{4}\right) \end{bmatrix} \qquad i = 1, 2, 3, 4 \tag{7.36}$$

The elements of the signal vectors, namely, s_{i1} and s_{i2}, have their values summarized in Table 7.1. The first two columns of this table give the associated dibits and phase of the QPSK signal.

Accordingly, a QPSK signal is characterized by having a two-dimensional signal constellation (i.e., $N = 2$) and four message points (i.e., $M = 4$), as illustrated in Fig. 7.6.

EXAMPLE 1

Figure 7.7 illustrates the sequences and waveforms involved in the generation of a QPSK signal. The input binary sequence 01101000 is shown in Fig. 7.7a. This sequence is divided into two other sequences, consisting of odd- and even-numbered bits of the input sequence. These two sequences are shown in the top lines of Figs. 7.7b and 7.7c. The waveforms representing the in-phase and quadrature components of the QPSK signal are also shown in Figs. 7.7b and 7.7c, respectively. These two waveforms may individually be viewed as examples of a binary PSK signal. Adding them, we get the QPSK waveform shown in Fig. 7.7d.

To realize the decision rule for the detection of the transmitted data sequence, we partition the signal space into four regions, in accordance with Eq. 3.64, as described here:

1.   The set of points closest to the message point associated with signal vector $\mathbf{s}_1$.
2.   The set of points closest to the message point associated with signal vector $\mathbf{s}_2$.

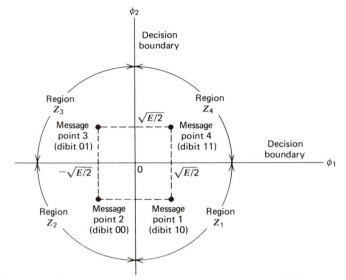
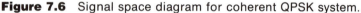

Figure 7.6 Signal space diagram for coherent QPSK system.

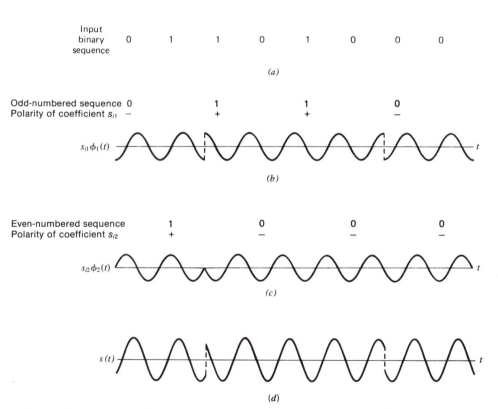

Figure 7.7 (a) Input binary sequence. (b) Odd-numbered bits of input sequence and associated binary PSK wave. (c) Even-numbered bits of input sequence and associated binary PSK wave. (d) QPSK waveform.

3. The set of points closest to the message point associated with signal vector s_3.

4. The set of points closest to the message point associated with signal vector s_4.

This is accomplished by constructing the perpendicular bisectors of the square formed by joining the four message points, and then marking off the appropriate regions. We thus find that the decision regions are quadrants whose vertices coincide with the origin. These regions are marked as Z_1, Z_2, Z_3, and Z_4, in Fig. 7.6, according to the message point about which they are constructed.

The received signal, $x(t)$, is defined by

$$x(t) = s_i(t) + w(t) \qquad \begin{matrix} 0 \leq t \leq T \\ i = 1, 2, 3, 4 \end{matrix} \qquad (7.37)$$

where $w(t)$ is the sample function of a white Gaussian noise process of zero mean and power spectral density $N_0/2$. The observation vector, $\mathbf{x}$, of a coherent QPSK receiver has two elements, x_1 and x_2, that are defined by

$$x_1 = \int_0^T x(t)\phi_1(t)dt$$

$$= \sqrt{E} \cos\left[(2i - 1) \frac{\pi}{4}\right] + w_1 \qquad (7.38)$$

and

$$x_2 = \int_0^T x(t)\phi_2(t)dt$$

$$= -\sqrt{E} \sin\left[(2i - 1) \frac{\pi}{4}\right] + w_2 \qquad (7.39)$$

where $i = 1, 2, 3, 4$.

Thus x_1 and x_2 are sample values of independent Gaussian random variables with mean values equal to $\sqrt{E} \cos[(2i - 1)\pi/4]$ and $-\sqrt{E} \sin[(2i - 1)\pi/4]$, respectively, and with a common variance equal to $N_0/2$.

The decision rule is now simply to guess $s_1(t)$ was transmitted if the received signal point associated with the observation vector $\mathbf{x}$ falls inside region Z_1, guess $s_2(t)$ was transmitted if the received signal point falls inside region Z_2, and so on. An erroneous decision will be made if, for example, signal $s_4(t)$ is transmitted but the noise $w(t)$ is such that the received signal point falls outside region Z_4.

We note that, owing to the symmetry of the decision regions, the probability of interpreting the received signal point correctly is the same regardless of which particular signal was actually transmitted. Suppose, for example, we know that signal $s_4(t)$ was transmitted. The receiver will then make a correct decision provided that the received signal point represented by the observation vector $\mathbf{x}$ lies inside region Z_4 of the signal space diagram in Fig. 7.6. Accordingly, for a correct decision when signal $s_4(t)$ is transmitted, the elements x_1 and x_2 of the observation vector $\mathbf{x}$ must be both positive, as illustrated in Fig. 7.8. This means that the *probability of a correct decision, P_c,* equals the conditional

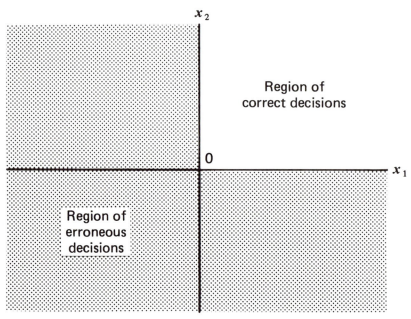

Figure 7.8   Illustrating the region of correct decisions and the region of erroneous decisions, given that signal $s_4(t)$ was transmitted.

probability of the joint event $x_1 > 0$ and $x_2 > 0$, given that signal $s_4(t)$ was transmitted. Since the random variables X_1 and X_2 (with sample values x_1 and x_2, respectively) are independent, P_c also equals the product of the conditional probabilities of the events $x_1 > 0$ and $x_2 > 0$, both given that signal $s_4(t)$ was transmitted. Furthermore, both X_1 and X_2 are Gaussian random variables with a conditional mean equal to $\sqrt{E/2}$ and a variance equal to $N_0/2$. Hence, we may write

$$P_c = \int_0^\infty \frac{1}{\sqrt{\pi N_0}} \exp\left[-\frac{(x_1 - \sqrt{E/2})^2}{N_0}\right] dx_1$$

$$\cdot \int_0^\infty \frac{1}{\sqrt{\pi N_0}} \exp\left[-\frac{(x_2 - \sqrt{E/2})^2}{N_0}\right] dx_2 \qquad (7.40)$$

where the first integral on the right side is the conditional probability of the event $x_1 > 0$ and the second integral is the conditional probability of the event $x_2 > 0$, both given that signal $s_4(t)$ was transmitted. Let

$$\frac{x_1 - \sqrt{E/2}}{\sqrt{N_0}} = \frac{x_2 - \sqrt{E/2}}{\sqrt{N_0}} = z \qquad (7.41)$$

Then, changing the variables of integration from x_1 and x_2 to z, we may rewrite Eq. 7.40 in the form

$$P_c = \left[\frac{1}{\sqrt{\pi}} \int_{-\sqrt{E/2N_0}}^\infty \exp(-z^2) dz\right]^2 \qquad (7.42)$$

However, from the definition of the complementary error function, we find that

$$\frac{1}{\sqrt{\pi}} \int_{-\sqrt{E/2N_0}}^{\infty} \exp(-z^2)\,dz = 1 - \tfrac{1}{2}\mathrm{erfc}\left(\sqrt{\frac{E}{2N_0}}\right) \tag{7.43}$$

Accordingly, we have

$$P_c = \left[1 - \tfrac{1}{2}\mathrm{erfc}\left(\sqrt{\frac{E}{2N_0}}\right)\right]^2$$

$$= 1 - \mathrm{erfc}\left(\sqrt{\frac{E}{2N_0}}\right) + \tfrac{1}{4}\mathrm{erfc}^2\left(\sqrt{\frac{E}{2N_0}}\right) \tag{7.44}$$

The average probability of symbol error for coherent QPSK is therefore

$$P_e = 1 - P_c$$

$$= \mathrm{erfc}\left(\sqrt{\frac{E}{2N_0}}\right) - \tfrac{1}{4}\mathrm{erfc}^2\left(\sqrt{\frac{E}{2N_0}}\right) \tag{7.45}$$

In the region where $(E/2N_0) \gg 1$, we may ignore the second term on the right side of Eq. 7.45, and so approximate the formula for the average probability of symbol error for coherent QPSK as

$$P_e \simeq \mathrm{erfc}\left(\sqrt{\frac{E}{2N_0}}\right) \tag{7.46}$$

In a QPSK system, we note that there are two bits per symbol. This means that the transmitted signal energy per symbol is twice the signal energy per bit; that is,

$$E = 2E_b \tag{7.47}$$

Thus, expressing the average probability of symbol error in terms of the ratio E_b/N_0, we may write

$$P_e \simeq \mathrm{erfc}\left(\sqrt{\frac{E_b}{N_0}}\right) \tag{7.48}$$

Consider next the generation and demodulation of QPSK. Figure 7.9a shows the block diagram of a typical QPSK transmitter. The input binary sequence $b(t)$ is represented in polar form, with symbols 1 and 0 represented by $+\sqrt{E_b}$ and $-\sqrt{E_b}$ volts, respectively. This binary wave is divided by means of a *demultiplexer* into two separate binary waves consisting of the odd- and even-numbered input bits. These two binary waves are denoted by $b_1(t)$ and $b_2(t)$. We note that in any signaling interval, the amplitudes of $b_1(t)$ and $b_2(t)$ equal s_{i1} and s_{i2}, respectively, depending on the particular dibit that is being transmitted. The two binary waves $b_1(t)$ and $b_2(t)$ are used to modulate a pair of quadrature carriers or orthonormal basis functions: $\phi_1(t)$ equal to $\sqrt{2/T}\cos(2\pi f_c t)$ and $\phi_2(t)$ equal to $\sqrt{2/T}\sin(2\pi f_c t)$. The result is a pair of binary PSK waves, which may be detected independently due to the orthogonality of $\phi_1(t)$ and $\phi_2(t)$. Finally,

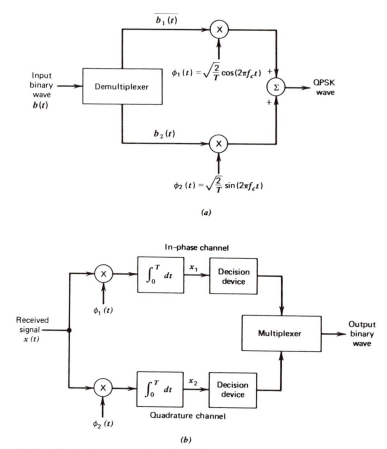

Figure 7.9 Block diagrams for (a) QPSK transmitter, and (b) QPSK receiver.

the two binary PSK waves are added to produce the desired QPSK wave. Note that the symbol duration, T, of a QPSK wave is twice as long as the bit duration, T_b, of the input binary wave. That is, for a given bit rate $1/T_b$, a QPSK wave requires half the transmission bandwidth of the corresponding binary PSK wave. Equivalently, for a given transmission bandwidth, a QPSK wave carries twice as many bits of information as the corresponding binary PSK wave.

The QPSK receiver consists of a pair of correlators with a common input and supplied with a locally generated pair of coherent reference signals $\phi_1(t)$ and $\phi_2(t)$, as in Fig. 7.9b. The correlator outputs, x_1 and x_2, are each compared with a threshold of zero volts. If $x_1 > 0$, a decision is made in favor of symbol 1 for the upper or in-phase channel output, but if $x_1 < 0$ a decision is made in favor of symbol 0. Similarly, if $x_2 > 0$, a decision is made in favor of symbol 1 for the lower or quadrature channel output, but if $x_2 < 0$, a decision is made in favor of symbol 0. Finally, these two binary sequences at the in-phase and quadrature channel outputs are combined in a *multiplexer* to reproduce the original binary sequence at the transmitter input with the minimum probability of symbol error.

(2) Minimum Shift Keying

In the coherent detection of binary FSK signals described in Section 7.2, the phase information contained in the received signal was not fully exploited, other than to provide for synchronization of the receiver to the transmitter. We now show that by proper utilization of the phase when performing detection, it is possible to improve the noise performance of the receiver significantly. This improvement is, however, achieved at the expense of increased receiver complexity.

Consider a continuous-phase frequency-shift keying (CPFSK) signal, which is defined for the interval $0 \leqslant t \leqslant T_b$, as follows:

$$s(t) = \begin{cases} \sqrt{\dfrac{2E_b}{T_b}} \cos[2\pi f_1 t + \theta(0)] & \text{for symbol 1} \\ \sqrt{\dfrac{2E_b}{T_b}} \cos[2\pi f_2 t + \theta(0)] & \text{for symbol 0} \end{cases} \tag{7.49}$$

where E_b is the transmitted signal energy per bit, and T_b is the bit duration. The phase $\theta(0)$, denoting the value of the phase at time $t = 0$, depends on the past history of the modulation process. The frequencies f_1 and f_2 are sent in response to binary symbols 1 and 0 appearing at the modulator input, respectively.

Another useful way of representing the CPFSK signal $s(t)$ is to express it in the conventional form of an *angle-modulated wave* as follows

$$s(t) = \sqrt{\frac{2E_b}{T_b}} \cos[2\pi f_c t + \theta(t)] \tag{7.50}$$

where $\theta(t)$ is the phase of $s(t)$. When the phase $\theta(t)$ is a continuous function of time, we find that the modulated wave $s(t)$ itself is also continuous at all times, including the inter-bit switching times.

The nominal carrier frequency f_c is chosen as the arithmetic mean of the two frequencies f_1 and f_2, as shown by

$$f_c = \frac{1}{2}(f_1 + f_2) \tag{7.51}$$

The phase $\theta(t)$ of a CPFSK signal increases or decreases linearly with time during each bit period of T_b seconds, as shown by

$$\theta(t) = \theta(0) \pm \frac{\pi h}{T_b} t \qquad 0 \leqslant t \leqslant T_b \tag{7.52}$$

where the plus sign corresponds to sending symbol 1, and the minus sign corresponds to sending symbol 0. The parameter h is defined by

$$h = T_b(f_1 - f_2) \tag{7.53}$$

We refer to h as the *deviation ratio*, measured with respect to the bit rate $1/T_b$. From Eq. 7.52, we find that at time $t = T_b$

$$\theta(T_b) - \theta(0) = \begin{cases} \pi h & \text{for symbol 1} \\ -\pi h & \text{for symbol 0} \end{cases} \tag{7.54}$$

That is to say, the sending of symbol 1 increases the phase of the CPFSK signal $s(t)$ by πh radians, whereas the sending of symbol 0 reduces it by an equal amount.

The variation of phase $\theta(t)$ with time t follows a path consisting of a sequence of straight lines, the slopes of which represent frequency changes. Figure 7.10 depicts possible paths starting from time $t = 0$. A plot like that shown in Fig. 7.10 is called a *phase tree*. The tree makes clear the transitions of phase across interval boundaries of the incoming sequence of data bits. Moreover, it is evident from Fig. 7.10 that the phase of the CPFSK signal is an odd or even multiple of πh radians at odd or even multiples of the bit duration T_b, respectively. Since all phase shifts are modulo-2π, the case of $h = 1/2$ is of special interest, because then the phase can take on only the two values $\pm\pi/2$ at odd multiples of T_b, and only the two values 0 and π at even multiples of T_b, as in Fig. 7.11. This second graph is called a *phase trellis,* since a ''trellis'' is a tree-like structure with remerging branches. Each path from left to right through the trellis of Fig. 7.11 corresponds to a specific binary sequence input. For example, the path shown in bold-face in Fig. 7.11 corresponds to the binary sequence 01101000 with $\theta(0) = 0$. Henceforth, we assume that $h = 1/2$.

Using a well-known trigonometric identity in Eq. 7.49, we may express the CPFSK signal $s(t)$ in terms of its in-phase and quadrature components as follows:

$$s(t) = \sqrt{\frac{2E_b}{T_b}}\,\cos[\theta(t)]\cos(2\pi f_c t) - \sqrt{\frac{2E_b}{T_b}}\,\sin[\theta(t)]\sin(2\pi f_c t) \qquad (7.55)$$

Consider first the in-phase component $\sqrt{2E_b/T_b}\cos[\theta(t)]$. With the deviation

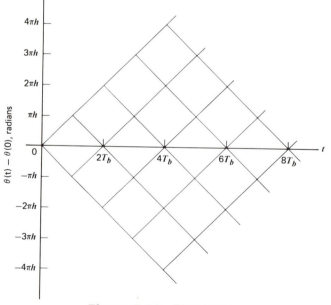

Figure 7.10 Phase tree.

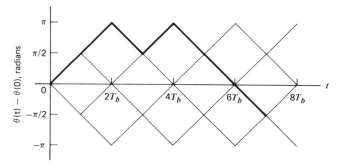

Figure 7.11 Phase trellis; bold-faced path represents the sequence 1101000.

ratio $h = 1/2$, we have from Eq. 7.52 that

$$\theta(t) = \theta(0) \pm \frac{\pi}{2T_b} t \qquad 0 \le t \le T_b \qquad (7.56)$$

where the plus sign corresponds to symbol 1 and the minus sign corresponds to symbol 0. A similar result holds for $\theta(t)$ in the interval $-T_b \le t \le 0$, except that the algebraic sign is not necessarily the same in both intervals. Since the phase $\theta(0)$ is 0 or π, depending on the past history of the modulation process, we find that, in the interval $-T_b \le t \le T_b$, the polarity of $\cos[\theta(t)]$ depends only on $\theta(0)$, regardless of the sequence of 1s and 0s transmitted before or after $t = 0$. Thus, for this time interval, the in-phase component, $s_I(t)$, consists of a *half-cosine pulse* defined as follows:

$$
\begin{aligned}
s_I(t) &= \sqrt{\frac{2E_b}{T_b}} \cos[\theta(t)] \\
&= \sqrt{\frac{2E_b}{T_b}} \cos[\theta(0)] \cos\left(\frac{\pi}{2T_b} t\right) \\
&= \pm \sqrt{\frac{2E_b}{T_b}} \cos\left(\frac{\pi}{2T_b} t\right) \qquad -T_b \le t \le T_b \qquad (7.57)
\end{aligned}
$$

where the plus sign corresponds to $\theta(0) = 0$ and the minus sign corresponds to $\theta(0) = \pi$. In a similar way, we may show that, in the interval $0 \le t \le 2T_b$, the quadrature component, $s_Q(t)$, consists of a *half-sine pulse*, whose polarity depends only on $\theta(T_b)$, as shown by

$$
\begin{aligned}
s_Q(t) &= \sqrt{\frac{2E_b}{T_b}} \sin[\theta(t)] \\
&= \sqrt{\frac{2E_b}{T_b}} \sin[\theta(T_b)] \sin\left(\frac{\pi}{2T_b} t\right) \\
&= \pm \sqrt{\frac{2E_b}{T_b}} \sin\left(\frac{\pi}{2T_b} t\right) \qquad 0 \le t \le 2T_b \qquad (7.58)
\end{aligned}
$$

where the plus sign corresponds to $\theta(T_b) = \pi/2$ and the minus sign corresponds to $\theta(T_b) = -\pi/2$.

With $h = 1/2$, we find from Eq. 7.53 that the frequency deviation (i.e., the difference between the two signaling frequencies f_1 and f_2) equals half the bit rate. This is the minimum frequency spacing that allows the two FSK signals representing symbols 1 and 0, as in Eq. 7.49, to be coherently orthogonal in the sense that they do not interfere with one another in the process of detection. It is for this reason, a CPFSK signal with a deviation ratio of one-half is referred to as *minimum-shift keying (MSK)*.* Since the frequency spacing is only half as much as the conventional spacing of $1/T_b$ that is used in the coherent detection of binary FSK signals (see Section 7.2), it is also referred to as *fast FSK*.†

From the foregoing discussion we see that since the phase states $\theta(0)$ and $\theta(T_b)$ can each assume one of two possible values, any one of four possibilities can arise, as described here:

1. The phase $\theta(0) = 0$ and $\theta(T_b) = \pi/2$, corresponding to the transmission of symbol 1.
2. The phase $\theta(0) = \pi$ and $\theta(T_b) = \pi/2$, corresponding to the transmission of symbol 0.
3. The phase $\theta(0) = \pi$ and $\theta(T_b) = -\pi/2$ (or, equivalently, $3\pi/2$, modulo 2π), corresponding to the transmission of symbol 1.
4. The phase $\theta(0) = 0$ and $\theta(T_b) = -\pi/2$, corresponding to the transmission of symbol 0.

This, in turn, means that the MSK signal itself may assume any one of four possible forms, depending on the values of $\theta(0)$ and $\theta(T_b)$.

From the expansion of Eq. 7.55, we deduce that in the case of an MSK signal the appropriate form for the orthonormal basis functions $\phi_1(t)$ and $\phi_2(t)$ is as follows:

$$\phi_1(t) = \sqrt{\frac{2}{T_b}} \cos\left(\frac{\pi}{2T_b} t\right) \cos(2\pi f_c t) \qquad -T_b \le t \le T_b \qquad (7.59)$$

and

$$\phi_2(t) = \sqrt{\frac{2}{T_b}} \sin\left(\frac{\pi}{2T_b} t\right) \sin(2\pi f_c t) \qquad 0 \le t \le 2T_b \qquad (7.60)$$

Note that both $\phi_1(t)$ and $\phi_2(t)$ are defined for a period equal to twice the bit duration. This is necessary, so as to ensure that they satisfy the condition of orthogonality.

Correspondingly, we may express the MSK signal in the form

$$s(t) = s_1\phi_1(t) + s_2\phi_2(t) \qquad 0 \le t \le T_b \qquad (7.61)$$

where the coefficients s_1 and s_2 are related to the phase states $\theta(0)$ and $\theta(T_b)$, respectively. To evaluate s_1 we integrate the product $s(t)\phi_1(t)$ between the limits $-T_b$ and T_b [in accordance with the observation interval $-T_b \le t \le T_b$ for

---

* The MSK signal was first described in Doelz and Heald (1961). For a tutorial review of MSK and comparison with QPSK, see Pasupathy (1979).

† See deBuda (1972).

the in-phase component of $s(t)$, as in Eq. 7.58]. We thus obtain

$$s_1 = \int_{-T_b}^{T_b} s(t)\phi_1(t)dt$$

$$= \sqrt{E_b}\cos[\theta(0)] \qquad -T_b \leqslant t \leqslant T_b \qquad (7.62)$$

Similarly, to evaluate s_2 we integrate the product $s(t)\phi_2(t)$ between the limits 0 and $2T_b$ [in accordance with the observation interval $0 \leqslant t \leqslant 2T_b$ for the quadrature component of $s(t)$, as in Eq. 7.59]. We therefore obtain

$$s_2 = \int_0^{2T_b} s(t)\phi_2(t)dt$$

$$= -\sqrt{E_b}\sin[\theta(T_b)] \qquad 0 \leqslant t \leqslant 2T_b \qquad (7.63)$$

Note that in Eqs. 7.62 and 7.63

1. Both integrals are evaluated for a time interval equal to twice the bit duration, for which $\phi_1(t)$ and $\phi_2(t)$ are orthogonal.
2. Both the lower and upper limits of the product integration used to evaluate the coefficient s_1 are shifted by T_b seconds with respect to those used to evaluate the coefficient s_2.
3. The time interval $0 \leqslant t \leqslant T_b$, for which the phase states $\theta(0)$ and $\theta(T_b)$ are defined, is common to both integrals.

Accordingly, the signal constellation for an MSK signal is two-dimensional (i.e., $N = 2$), with four message points (i.e., $M = 4$), as illustrated in Fig. 7.12. The coordinates of the message points are as follows: $(+\sqrt{E_b}, -\sqrt{E_b})$, $(-\sqrt{E_b}, -\sqrt{E_b})$, $(-\sqrt{E_b}, +\sqrt{E_b})$, and $(+\sqrt{E_b}, +\sqrt{E_b})$. The possible values of $\theta(0)$ and $\theta(T_b)$, corresponding to these four message points, are also included in Fig. 7.12.

Comparing Figs. 7.6 and 7.12, we see that the signal space diagrams for QPSK and MSK signals have an identical format. Note, however, that the coordinates of the message points for the QPSK signal in Fig. 7.6 are expressed in terms of the signal energy per symbol, E, whereas for the MSK signal in Fig. 7.12, they are expressed in terms of the signal energy per bit, E_b, with $E_b = E/2$. The basic difference between QPSK and MSK signals is in the choice of the orthonormal signals $\phi_1(t)$ and $\phi_2(t)$. For a QPSK signal, $\phi_1(t)$ and $\phi_2(t)$ are represented by a pair of quadrature carriers, as in Eqs. 7.34 and 7.35, whereas for an MSK signal, they are represented by a pair of sinusoidally modulated quadrature carriers, as in Eqs. 7.59 and 7.60.

Table 7.2 presents a summary of the values of $\theta(0)$ and $\theta(T_b)$, as well as the corresponding values of s_1 and s_2 that are calculated for the time intervals $-T_b \leqslant t \leqslant T_b$ and $0 \leqslant t \leqslant 2T_b$, respectively. The first column of this table indicates whether symbol 1 or symbol 0 was sent in the interval $0 \leqslant t \leqslant T_b$. Note that the coordinates of the message points, s_1 and s_2, have opposite signs when symbol 1 is sent in this interval, but the same sign when symbol 0 is sent. Accordingly, for a given input data sequence, we may use the entries of Table

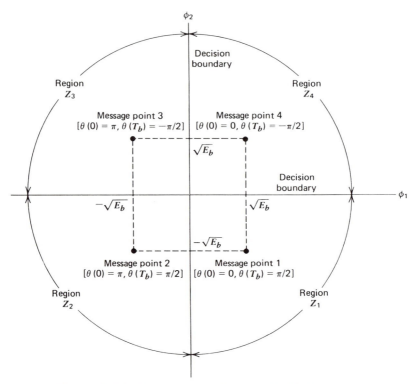

Figure 7.12 Signal space diagram for MSK system.

7.2 to derive, on a bit-by-bit basis, the two sequences of coefficients required to scale $\phi_1(t)$ and $\phi_2(t)$, and thereby determine the MSK signal $s(t)$.

EXAMPLE 2
Figure 7.13 shows the sequences and waveforms involved in the generation of an MSK signal for the binary sequence 1101000. The input binary sequence is shown Fig. 7.13a. Assuming that, at time $t = 0$, the phase $\theta(0)$ is zero, the sequence of phase states is as

Table 7.2 Signal-space Characterization of MSK

Transmitted binary symbol, $0 \leq t \leq T_b$	Phase states (*radians*) $\theta(0)$	$\theta(T_b)$	Coordinates of message points s_1	s_2
1	0	$+\pi/2$	$+\sqrt{E_b}$	$-\sqrt{E_b}$
0	π	$+\pi/2$	$-\sqrt{E_b}$	$-\sqrt{E_b}$
1	π	$-\pi/2$	$-\sqrt{E_b}$	$+\sqrt{E_b}$
0	0	$-\pi/2$	$+\sqrt{E_b}$	$+\sqrt{E_b}$

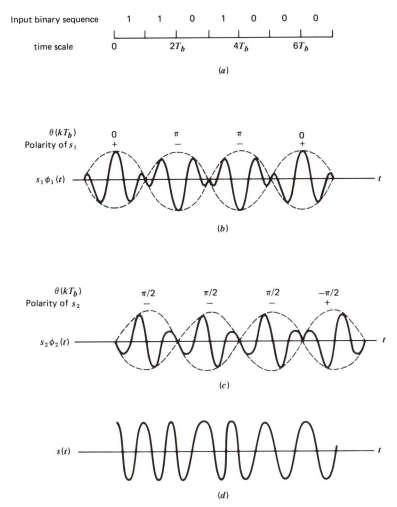

Figure 7.13 Sequence and waveforms for MSK signal. (a) Input binary sequence. (b) Scaled time function $s_1\phi_1(t)$. (c) Scaled time function $s_2\phi_2(t)$. (d) MSK signal s(t) *obtained by adding* $s_1\phi_1(t)$ *and* $s_2\phi_2(t)$ *on a bit-by-bit basis.*

shown in Fig. 7.11, modulo-2π. The polarities of the two sequences of factors used to scale the time functions $\phi_1(t)$ and $\phi_2(t)$ are shown in the top lines of Figs. 7.13b and 7.13c. Note that these two sequences are *offset* relative to each other by an interval equal to the bit duration T_b. The waveforms of the resulting in-phase and quadrature components are also shown in Figs. 7.13b and 7.13c. Adding these two modulated waveforms, we get the desired MSK signal shown in Fig. 7.13d.

In the case of an AWGN channel, the received signal is given by

$$x(t) = s(t) + w(t) \qquad (7.64)$$

where $s(t)$ is the transmitted MSK signal, and $w(t)$ is the sample function of a white Gaussian noise process of zero mean and power spectral density $N_0/2$. In order to decide whether symbol 1 or symbol 0 was transmitted in the interval $0 \le t \le T_b$, say, we have to establish a procedure for the use of $x(t)$ to detect the phase states $\theta(0)$ and $\theta(T_b)$. For the optimum detection of $\theta(0)$, we first determine the projection of the received signal $x(t)$ onto the reference signal $\phi_1(t)$, obtaining

$$x_1 = \int_{-T_b}^{T_b} x(t)\phi_1(t)dt$$
$$= s_1 + w_1 \qquad -T_b \le t \le T_b \tag{7.65}$$

where s_1 is as defined by Eq. 7.62, and w_1 is the sample value of a Gaussian random variable of zero mean and variance $N_0/2$. From the signal space diagram of Fig. 7.12, we observe that if $x_1 > 0$, the receiver chooses the estimate $\hat{\theta}(0) = 0$. On the other hand, if $x_1 < 0$, it chooses the estimate $\hat{\theta}(0) = \pi$.

Similarly, for the optimum detection of $\theta(T_b)$, we determine the projection of the received signal $x(t)$ onto the second reference signal $\phi_2(t)$, obtaining

$$x_2 = \int_0^{2T_b} x(t)\phi_2(t)dt$$
$$= s_2 + w_2 \qquad 0 \le t \le 2T_b \tag{7.66}$$

where s_2 is as defined by Eq. 7.63 and w_2 is the sample value of another independent Gaussian random variable of zero mean and variance $N_0/2$. Referring again to the signal space diagram of Fig. 7.12, we observe that if $x_2 > 0$, the receiver chooses the estimate $\hat{\theta}(T_b) = -\pi/2$. If, on the other hand, $x_2 < 0$, it chooses the estimate $\hat{\theta}(T_b) = \pi/2$.

To reconstruct the original binary sequence, we interleave the above two sets of phase decisions, as described next (see Table 7.2):

1. If we have the estimates $\hat{\theta}(0) = 0$ and $\hat{\theta}(T_b) = -\pi/2$, or alternatively if we have the estimates $\hat{\theta}(0) = \pi$ and $\hat{\theta}(T_b) = \pi/2$, the receiver makes a final decision in favor of symbol 0.
2. If we have the estimates $\hat{\theta}(0) = \pi$ and $\hat{\theta}(T_b) = -\pi/2$, or alternatively if we have the estimates $\hat{\theta}(0) = 0$ and $\hat{\theta}(T_b) = \pi/2$, the receiver makes a final decision in favor of symbol 1.

Earlier we remarked that the MSK and QPSK signals have similar signal space diagrams. It follows, therefore, that for the case of an AWGN channel, they will have the same formula for their average probability of symbol error. Accordingly, the average probability of symbol error for the MSK is given by

$$P_e = \text{erfc}\left(\sqrt{\frac{E_b}{N_0}}\right) - \tfrac{1}{4}\,\text{erfc}^2\left(\sqrt{\frac{E_b}{N_0}}\right) \tag{7.67}$$

Here again, we may ignore the second term on the right-hand side of Eq. 7.67 in the region where $E_b/N_0 \gg 1$, and so approximate the formula for the average probability of symbol error as

$$P_e \simeq \mathrm{erfc}\left(\sqrt{\frac{E_b}{N_0}}\right) \tag{7.68}$$

Thus, comparing Eqs. 7.13 and 7.68, we find that for high values of E_b/N_0, the average probability of error for an MSK system is approximately the same as that for a coherent binary PSK system.

Consider next the generation and demodulation of MSK. Figure 7.14 shows the block diagram of a typical MSK transmitter. The advantage of this method of generating MSK signals is that the signal coherence and deviation ratio are largely unaffected by variations in the input data rate. Two input sinusoidal waves, one of frequency $f_c = n_c/4T_b$ for some fixed integer n_c, and the other of frequency $1/4T_b$, are first applied to a product modulator. This produces two phase-coherent sine waves at frequencies f_1 and f_2, which are related to f_c and the bit rate $1/T_b$ by Eqs. 7.51 and 7.53 for $h = 1/2$. These two sinusoidal waves are separated from each other by two narrow-band filters, one centered at f_1

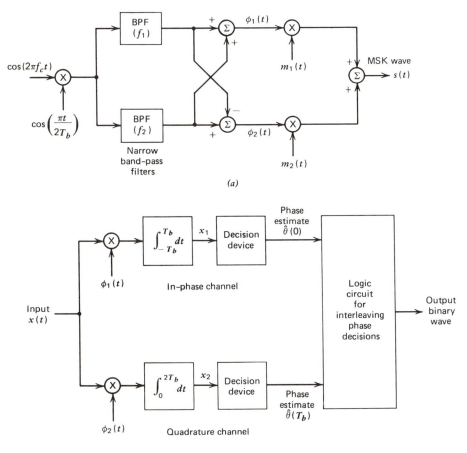

Figure 7.14   Block diagrams for (a) MSK transmitter, and (b) MSK receiver.

and the other at f_2. The resulting filter outputs are next summed to produce the pair of quadrature carriers or orthonormal basis functions $\phi_1(t)$ and $\phi_2(t)$. Finally, $\phi_1(t)$ and $\phi_2(t)$ are multiplied with two binary waves $m_1(t)$ and $m_2(t)$, both of which have a bit rate equal to $1/2T_b$. These two binary waves are extracted from the incoming binary wave $b(t)$ in the manner described in Example 2.

Figure 7.14*b* shows the block diagram of a typical MSK receiver. The received signal $x(t)$ is correlated with locally generated replicas of the coherent reference signals, $\phi_1(t)$ and $\phi_2(t)$. Note that in both cases the integration interval is $2T_b$ seconds, and that the integration in the quadrature channel is delayed by T_b seconds with respect to that in the in-phase channel. The resulting in-phase and quadrature channel correlator outputs, x_1 and x_2, are next compared with a threshold of zero volts, and estimates of the phase $\theta(0)$ and $\theta(T_b)$ are derived in the manner described previously. Finally, these phase decisions are interleaved so as to reconstruct the original input binary wave $b(t)$ with the minimum average probability of symbol error.

7.4 NONCOHERENT BiNARY MODULATION TECHNIQUES

Coherent detection exploits knowledge of the carrier wave's phase reference, thereby providing the optimum error performance attainable with a digital modulation format of interest. When, however, it is impractical to have knowledge of the carrier phase at the receiver, we resort to the use of *noncoherent detection*. In this section, we study the noise performance of *noncoherent binary FSK* signals, assuming an AWGN channel. In the case of phase-shift keying, we cannot have "noncoherent PSK," because noncoherent means doing without phase information. Nevertheless, there is a "pseudo PSK" technique called *differential phase-shift keying (DPSK)*, which may be viewed as the noncoherent form of PSK. In this section, we also study the noise performance of DPSK. The approach taken is to treat noncoherent binary FSK and DPSK signals as special cases of *noncoherent orthogonal modulation*.

For the noise analysis of noncoherent binary ASK signals, the reader is referred to Problem 7.4.6. As with the coherent case, it turns out that for large signal-to-noise ratios noncoherent binary ASK and FSK signals have the same average probability of error for an AWGN channel.

(1) Noncoherent Orthogonal Modulation

Consider a binary signaling scheme that involves the use of two orthogonal signals $s_1(t)$ and $s_2(t)$, which have equal energy. During the interval $0 \leq t \leq T$, one of these two signals is sent over an imperfect channel that shifts the carrier phase by an unknown amount. Let $g_1(t)$ and $g_2(t)$ denote the phase-shifted versions of $s_1(t)$ and $s_2(t)$, respectively. It is assumed that the signals $g_1(t)$ and $g_2(t)$ remain orthogonal and of equal energy, regardless of the unknown carrier phase. We refer to such a signaling scheme as *noncoherent orthogonal modulation*. Depending on how we define the orthogonal pair of signals $s_1(t)$ and $s_2(t)$, noncoherent binary FSK and DPSK may be treated as special cases of this modulation scheme.

The channel also introduces an additive white Gaussian noise $w(t)$ of zero mean and power spectral density $N_0/2$. We may thus express the received signal $x(t)$ as

$$x(t) = \begin{cases} g_1(t) + w(t) & 0 \leqslant t \leqslant T \\ g_2(t) + w(t) & 0 \leqslant t \leqslant T \end{cases} \tag{7.69}$$

The requirement is to use $x(t)$ to discriminate between $s_1(t)$ and $s_2(t)$, regardless of the carrier phase.

For this purpose, we employ a receiver as shown in Fig. 7.15a. The receiver consists of a pair of filters matched to the basis functions $\phi_1(t)$ and $\phi_2(t)$ that are scaled versions of the transmitted signals $s_1(t)$ and $s_2(t)$, respectively. Because the carrier phase is unknown, the receiver relies on amplitude as the only possible discriminant. Accordingly, the matched filter outputs are envelope detected, sampled, and then compared with each other. If the upper path in

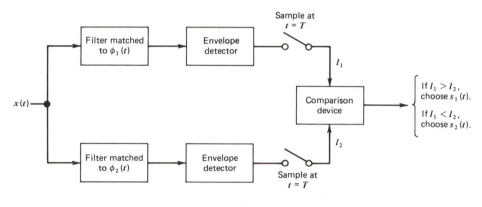

(a)

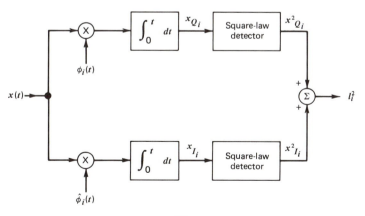

(b)

Figure 7.15 (a) Generalized binary receiver for noncoherent orthogonal modulation. (b) Quadrature receiver equivalent to either one of the two matched filters in part a; the index $i = 1, 2$.

Fig. 7.15*a* has an output amplitude l_1 greater than the output amplitude l_2 of the lower path, the receiver makes a decision in favor of $s_1(t)$. If the converse is true, it decides in favor of $s_2(t)$. When they are equal, the decision may be made by flipping a fair coin. In any event, a decision error occurs when the matched filter that rejects signal component of the received signal $x(t)$ has a larger output amplitude (due to noise alone) than the matched filter that passes it.

From the discussion presented in Section 3.9, we note that a noncoherent matched filter (constituting the upper or lower path in the receiver of Fig. 7.15*a*) may be viewed as being equivalent to a *quadrature receiver*. The quadrature receiver itself has two paths. One version of the quadrature receiver is shown in Fig. 7.15*b*. In the upper path, called the *in-phase path,* the received signal $x(t)$ is correlated against the basis function $\phi_i(t)$, representing a scaled version of the transmitted signal $s_1(t)$ or $s_2(t)$ with zero carrier phase. In the lower path, called the *quadrature path,* on the other hand, $x(t)$ is correlated against another basis function $\hat{\phi}_i(t)$, representing the version of $\phi_i(t)$ that results from shifting the carrier phase by $-90°$. Naturally, $\phi_i(t)$ and $\hat{\phi}_i(t)$ are orthogonal to each other.*

The average probability of error for the noncoherent receiver of Fig. 7.15*a* will now be calculated by making use of the equivalence depicted in Fig. 7.15*b*. In particular, we observe that since the carrier phase is unknown, noise at the output of each matched filter in Fig. 7.15*a* has *two degrees of freedom,* namely, in-phase and quadrature. Accordingly, the noncoherent receiver of Fig. 7.15*a* has a total of four noise parameters that are *statistically independent and identically distributed.* These four noise parameters have sample values denoted by x_{I1}, x_{Q1}, x_{I2}, and x_{Q2}; the first two account for degrees of freedom associated with the upper path of Fig. 7.15*a*, and the latter two account for degrees of freedom associated with the lower path.

The receiver of Fig. 7.15*a* has a *symmetric* structure. Hence, the probability of choosing $s_2(t)$, given that $s_1(t)$ was transmitted, is the same as the probability of choosing $s_1(t)$, given that $s_2(t)$ was transmitted. This means that the average probability of error may be obtained by transmitting $s_1(t)$ and calculating the probability of choosing $s_2(t)$, or vice versa.

---

* In the literature, the signal $\hat{\phi}_i(t)$ is referred to as the *Hilbert transform* of $\phi_i(t)$. To illustrate the nature of this relationship, let

$$\phi_i(t) = m(t)\cos(2\pi f_i t)$$

where $m(t)$ is a band-limited message signal. Typically, the carrier frequency f_i is greater than the highest frequency component of $m(t)$. Then, the Hilbert transform of $\phi_i(t)$ is defined by

$$\hat{\phi}_i(t) = m(t)\sin(2\pi f_i t)$$

Since

$$\cos\left(2\pi f_i - \frac{\pi}{2}\right) = \sin(2\pi f_i t),$$

we see that $\hat{\phi}_i(t)$ is indeed obtained from $\phi_i(t)$ by shifting the carrier $\cos(2\pi f_i t)$ by $-90°$. An important property of Hilbert transformation is that a signal $\phi_i(t)$ and its Hilbert transform $\hat{\phi}_i(t)$ are orthogonal to each other. For details of the Hilbert transform and its properties, see Haykin (1983, pp. 74–79).

Suppose that signal $s_1(t)$ is transmitted for the interval $0 \leqslant t \leqslant T$. An error occurs if the receiver noise $w(t)$ is such that the output l_2 of the lower path in Fig. 7.15a is greater than the output l_1 of the upper path. Then the receiver makes a decision in favor of $s_2(t)$ rather than $s_1(t)$. To calculate the probability of error so made, we must have the probability density function of the random variable L_2 (represented by sample value l_2). Since the filter in the lower path is matched to $s_2(t)$, and $s_2(t)$ is orthogonal to the transmitted signal $s_1(t)$, it follows that the output of this matched filter is due to *noise alone*. Let x_{I2} and x_{Q2} denote the in-phase and quadrature components of the matched filter output in the lower path of Fig. 7.15. Then, from the equivalence depicted in Fig. 7.15b, we see that (for $i = 2$)

$$l_2 = \sqrt{x_{I2}^2 + x_{Q2}^2} \tag{7.70}$$

Figure 7.16a shows a geometric interpretation of this relation. The channel noise $w(t)$ is both white (with power spectral density $N_0/2$) and Gaussian (with zero mean). Correspondingly, we find that the random variables X_{I2} and X_{Q2} (represented by sample values x_{I2} and x_{Q2}) are both Gaussian distributed with zero mean and variance $N_0/2$; see Section 3.4. Hence, we may write

$$f_{X_{I2}}(x_{I2}) = \frac{1}{\sqrt{\pi N_0}} \exp\left(- \frac{x_{I2}^2}{N_0}\right) \tag{7.71}$$

and

$$f_{X_{Q2}}(x_{Q2}) = \frac{1}{\sqrt{\pi N_0}} \exp\left(- \frac{x_{Q2}^2}{N_0}\right) \tag{7.72}$$

Next, we use a well-known result in probability theory, namely, the fact that the envelope of a Gaussian process is *Rayleigh distributed*. (For a discussion of the Rayleigh distribution, see Appendix D.) Specifically, for the situation at hand, we may state that the random variable L_2 (whose sample value l_2 is related to x_{I2} and x_{Q2} by Eq. 7.70) has the following probability density function:

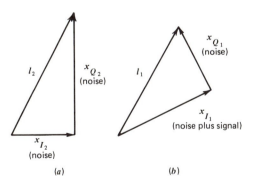

(a)   (b)

Figure 7.16 Geometric interpretations of the two path outputs l_1 and l_2 in the generalized noncoherent receiver.

$$f_{L2}(l_2) = \begin{cases} \dfrac{2l_2}{N_0} \exp\left(-\dfrac{l_2^2}{N_0}\right) & l_2 \geq 0 \\ 0 & \text{elsewhere} \end{cases} \tag{7.73}$$

Figure 7.17 shows a plot of this probability density function. The conditional probability that $l_2 > l_1$, given the sample value l_1, is defined by the shaded area in Fig. 7.17. Hence, we have

$$P(l_2 > l_1 | l_1) = \int_{l_1}^{\infty} f_{L_2}(l_2)\,dl_2 \tag{7.74}$$

Substituting Eq. 7.73 in Eq. 7.74, and integrating, we get

$$P(l_2 > l_1 | l_1) = \exp\left(-\dfrac{l_1^2}{N_0}\right) \tag{7.75}$$

Consider next the output amplitude l_1, pertaining to the upper path in Fig. 7.15a. Since the filter in this path is matched to $s_1(t)$, and it is assumed that $s_1(t)$ is transmitted, it follows that l_1 is due to *signal plus noise*. Let x_{I1} and x_{Q1} denote the components at the output of the matched filter (in the upper path of Fig. 7.15a) that are in phase and in quadrature to the received signal. Then, from the equivalence depicted in Fig. 7.15b, we see that (for $i = 1$)

$$l_1 = \sqrt{x_{I1}^2 + x_{Q1}^2} \tag{7.76}$$

Figure 7.16b presents a geometric interpretation of this relation. Since $\hat{s}_1(t)$ is orthogonal to $s_1(t)$, it is obvious that x_{I1} is due to signal plus noise, whereas x_{Q1} is due to noise alone. Specifically, the random variable X_{I1} represented by sample value x_{I1} is Gaussian distributed with mean $\sqrt{E}$ and variance $N_0/2$, where E is the signal energy per symbol. The random variable X_{Q1} represented by sample value x_{Q1} is Gaussian distributed with zero mean and variance $N_0/2$. Hence, we may express the probability density functions of these two independent random variables as follows:

$$f_{X_{I1}}(x_{I1}) = \dfrac{1}{\sqrt{\pi N_0}} \exp\left(-\dfrac{(x_{I1} - \sqrt{E})^2}{N_0}\right) \tag{7.77}$$

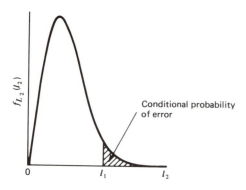

Conditional probability of error

Figure 7.17 Calculation of the conditional probability that $l_2 > l_1$, given l_1.

and

$$f_{X_{Q1}}(x_{Q1}) = \frac{1}{\sqrt{\pi N_0}} \exp\left(- \frac{x_{Q1}^2}{N_0}\right) \tag{7.78}$$

Since the two random variables X_{I1} and X_{Q1} are independent, their joint probability density function is simply the product of the probability density functions given in Eqs. 7.77 and 7.78.

To find the average probability of error, we have to average the conditional probability of error given in Eq. 7.75 over all possible values of l_1. Naturally, this calculation requires knowledge of the probability density function of random variable L_1 represented by sample value l_1. The standard method is now to combine Eqs. 7.77 and 7.78 to find the probability density function of L_1 due to signal plus noise. However, this leads to rather complicated calculations involving the use of Bessel functions.* This analytic difficulty may be circumvented by the following approach. Given x_{I1} and x_{Q1}, an error occurs when (in Fig. 7.15a) the lower path's output amplitude l_2 due to noise alone exceeds l_1 due to signal plus noise, where we have (from Eq. 7.76)

$$l_1^2 = x_{I1}^2 + x_{Q1}^2 \tag{7.79}$$

The probability of such an occurrence is obtained by substituting Eq. 7.79 in Eq. 7.75, as shown by

$$P(\text{error}|x_{I1}, x_{Q1}) = \exp\left(- \frac{x_{I1}^2 + x_{Q1}^2}{N_0}\right) \tag{7.80}$$

This is now a conditional probability of error, conditional on the output of the matched filter in the upper path taking on values X_{I1} and X_{Q1}. This conditional probability multiplied by the joint probability density function of X_{I1} and X_{Q1} is then the *error-density at given x_{I1} and x_{Q1}*. Since X_{I1} and X_{Q1} are statistically independent, their joint probability density function equals the product of their individual probability density functions. The resulting error-density is a complicated expression in x_{c1} and x_{s1}. However, the average probability of error (which is the issue of interest) may be obtained in a relatively simple manner. We first use Eqs. 7.77, 7.78, and 7.80 to evaluate the desired error-density as

$$P(\text{error}|x_{I1}, x_{Q1})f_{X_{I1}}(x_{I1})f_{X_{Q1}}(x_{Q1})$$
$$= \frac{1}{\pi N_0} \exp\left\{- \frac{1}{N_0} [x_{I1}^2 + x_{Q1}^2 + (x_{I1} - \sqrt{E})^2 + x_{Q1}^2]\right\} \tag{7.81}$$

Completing the square in the exponent of Eq. 7.81, we may rewrite the exponent as

$$x_{I1}^2 + x_{Q1}^2 + (x_{I1} - \sqrt{E})^2 + x_{Q1}^2 = 2\left(x_{I1} - \frac{\sqrt{E}}{2}\right)^2 + 2x_{Q1}^2 + \frac{E}{2} \tag{7.82}$$

---

* See, for example, Arthurs and Dym (1962) and Whalen (1971).

Next, we substitute Eq. 7.82 in Eq. 7.81, and integrate the error-density over all x_{I1} and x_{Q1}. We thus evaluate the average probability of error as

$$P_e = \int_{-\infty}^{\infty} \int_{-\infty}^{\infty} P(\text{error}|x_{I1},x_{Q1}) f_{X_{I1}}(x_{I1}) f_{X_{Q1}}(x_{Q1}) dx_{I1} dx_{Q1}$$

$$= \frac{1}{\pi N_0} \exp\left(-\frac{E}{2N_0}\right) \int_{-\infty}^{\infty} \exp\left[-\frac{2}{N_0}\left(x_{I1} - \frac{\sqrt{E}}{2}\right)^2\right] dx_{I1}$$

$$\cdot \int_{-\infty}^{\infty} \exp\left(-\frac{2x_{Q1}^2}{N_0}\right) dx_{Q1} \tag{7.83}$$

We now use the following identities:*

$$\int_{-\infty}^{\infty} \exp\left[-\frac{2}{N_0}\left(x_{I1} - \frac{\sqrt{E}}{2}\right)^2\right] dx_{I1} = \sqrt{\frac{N_0 \pi}{2}} \tag{7.84}$$

and

$$\int_{-\infty}^{\infty} \exp\left(-\frac{2x_{Q1}^2}{N_0}\right) dx_{Q1} = \sqrt{\frac{N_0 \pi}{2}} \tag{7.85}$$

Accordingly, Eq. 7.83 simplifies as follows:

$$P_e = \frac{1}{2} \exp\left(-\frac{E}{2N_0}\right) \tag{7.86}$$

This is the desired formula for the average probability of error for noncoherent orthogonal modulation.

(1) Noncoherent Binary FSK

In the binary FSK case, the transmitted signal is defined by

$$s_i(t) = \begin{cases} \sqrt{\dfrac{2E_b}{T_b}} \cos(2\pi f_i t) & 0 \le t \le T_b \\ 0 & \text{elsewhere} \end{cases} \tag{7.87}$$

where the carrier frequency f_i equals one of two possible values f_1 and f_2. The transmission of frequency f_1 represents symbol 1, and the transmission of frequency f_2 represents symbol 0. For the noncoherent detection of this frequency-modulated wave, the receiver consists of a pair of matched filters followed by envelope detectors, as in Fig. 7.18. The filter in the upper path of the receiver is matched to $\sqrt{2/T_b} \cos(2\pi f_1 t)$, and the filter in the lower path is matched to $\sqrt{2/T_b} \cos(2\pi f_2 t)$, and $0 \le t \le T_b$. The resulting envelope detector outputs are sampled at $t = T_b$, and their values are compared. The envelope samples of the upper and lower paths in Fig. 7.18 are shown as l_1 and l_2,

---

* Equation 7.84 is obtained by considering a Gaussian-distributed variable with mean $\sqrt{E}/2$ and variance $N_0/4$, and recognizing that the total area under the curve of a random variable's probability density function equals unity. Equation 7.85 follows as a special case of Eq. 7.84.

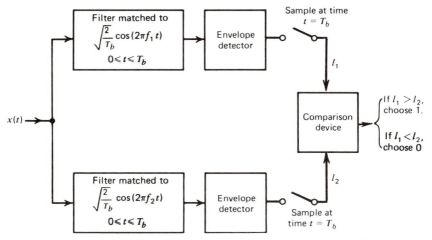

Figure 7.18 Noncoherent receiver for detection of binary FSK signals.

respectively; then, if $l_1 > l_2$, the receiver decides in favor of symbol 1, and if $l_1 < l_2$ it decides in favor of symbol 0.

The noncoherent binary FSK described is a special case of noncoherent orthogonal modulation with

$$T = T_b \tag{7.88}$$

and

$$E = E_b \tag{7.89}$$

where T_b is the bit duration and E_b is the signal energy per bit. Hence, substituting Eq. 7.89 in Eq. 7.86, we find that the average probability of error for noncoherent binary FSK is given by

$$P_e = \frac{1}{2} \exp\left(-\frac{E_b}{2N_0}\right) \tag{7.90}$$

(2) Differential Phase-shift Keying

We may view *differential phase-shift keying* as the noncoherent version of the PSK. It eliminates the need for a coherent reference signal at the receiver by combining two basic operations at the transmitter: (1) *differential encoding* of the input binary wave, and (2) *phase-shift keying*—hence, the name, *differential phase-shift keying* (DPSK). In effect, to send symbol 0 we phase advance the current signal waveform by 180°, and to send symbol 1 we leave the phase of the current signal waveform unchanged. The receiver is equipped with a *storage* capability, so that it can measure the *relative phase difference* between the waveforms received during two successive bit intervals. Provided that the unknown phase θ contained in the received wave varies slowly (that is, slow enough for it to be considered essentially constant over two bit intervals), the phase difference between waveforms received in two successive bit intervals will be independent of θ.

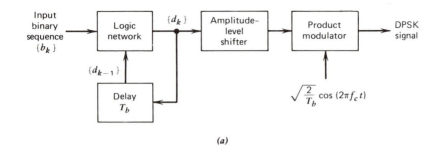

(a)

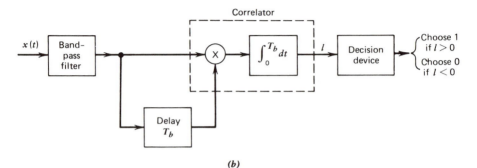

(b)

Figure 7.19 Block diagrams for (a) DPSK transmitter, and (b) DPSK receiver.

DPSK is another example of noncoherent orthogonal modulation, when it is considered over two bit intervals. Suppose the transmitted DPSK signal equals $\sqrt{E_b/2T_b}\ \cos(2\pi f_c t)$ for $0 \le t \le T_b$, where T_b is the bit duration and E_b is the signal energy per bit. Let $s_1(t)$ denote the transmitted DPSK signal for $0 \le t \le 2T_b$ for the case when we have binary symbol 1 at the transmitter input for the second part of this interval, namely, $T_b \le t \le 2T_b$. The transmission of 1 leaves the carrier phase unchanged, and so we define $s_1(t)$ as

$$s_1(t) = \begin{cases} \sqrt{\dfrac{E_b}{2T_b}}\ \cos(2\pi f_c t) & 0 \le t \le T_b \\[2ex] \sqrt{\dfrac{E_b}{2T_b}}\ \cos(2\pi f_c t) & T_b \le t \le 2T_b \end{cases} \tag{7.91}$$

Let $s_2(t)$ denote the transmitted DPSK signal for $0 \le t \le 2T_b$ for the case when we have binary symbol 0 at the transmitter input for $T_b \le t \le 2T_b$. The transmission of 0 advances the carrier phase by 180°, and so we define $s_2(t)$ as

$$s_2(t) = \begin{cases} \sqrt{\dfrac{E_b}{2T_b}}\ \cos(2\pi f_c t) & 0 \le t \le T_b \\[2ex] \sqrt{\dfrac{E_b}{2T_b}}\ \cos(2\pi f_c t + \pi) & T_b \le t \le 2T_b \end{cases} \tag{7.92}$$

We readily see from Eqs. 7.91 and 7.92 that $s_1(t)$ and $s_2(t)$ are indeed orthogonal over the two-bit interval $0 \leq t \leq 2T_b$. In other words, DPSK is a special case of noncoherent orthogonal modulation with

$$T = 2T_b \qquad (7.93)$$

and

$$E = 2E_b \qquad (7.94)$$

Hence, substituting Eq. 7.94 in Eq. 7.86, we find that the average probability of error for DPSK is given by

$$P_e = \frac{1}{2} \exp\left(-\frac{E_b}{N_0}\right) \qquad (7.95)$$

The next issue of interest is the generation and demodulation of DPSK. The differential encoding process at the transmitter input starts with an arbitrary first bit, serving as reference, and thereafter the differentially encoded sequence $\{d_k\}$ is generated by using the *logical* equation

$$d_k = d_{k-1}b_k + \bar{d}_{k-1}\bar{b}_k \qquad \text{modulo-2} \qquad (7.96)$$

where b_k is the input binary digit at time kT_b, and d_{k-1} is the previous value of the differentially encoded digit. The use of an overbar denotes logical *inversion*. Table 7.3 illustrates the logical operations involved in the use of Eq. 7.96, assuming that the reference bit added to the differentially encoded sequence $\{d_k\}$ is a 1. The differentially encoded sequence $\{d_k\}$ thus generated is used to phase-shift key a carrier with the phase angles 0 and π radians, as illustrated in the last row of Table 7.3.

The block diagram of a DPSK transmitter is shown in Fig. 7.19a. It consists, in part, of a logic network and a one-bit delay element interconnected so as to convert an input binary sequence $\{b_k\}$ into a differentially encoded sequence

Table 7.3 Illustrating the Generation of DPSK Signal

$\{b_k\}$	1	0	0	1	0	0	1	1
$\{\bar{b}_k\}$	0	1	1	0	1	1	0	0
$\{d_{k-1}\}$	1	1	0	1	1	0	1	1
$\{\bar{d}_{k-1}\}$	0	0	1	0	0	1	0	0
$\{b_k d_{k-1}\}$	1	0	0	1	0	0	1	1
$\{\bar{b}_k \bar{d}_{k-1}\}$	0	0	1	0	0	1	0	0
Differentially encoded sequence $\{d_k\}$	1	1	0	1	1	0	1	1	1
Transmitted phase (*radians*)	0	0	π	0	0	π	0	0	0

$\{d_k\}$ in accordance with Eq. 7.96. This sequence is amplitude-level shifted and then used to modulate a carrier wave of frequency f_c, thereby producing the desired DPSK wave.

A method of demodulating the DPSK signal is shown in Fig. 7.19b. At the receiver input, the received DPSK signal plus noise is passed through a band-pass filter centered at the carrier frequency f_c, so as to limit the noise power. The filter output and a delayed version of it, with the delay equal to the bit duration T_b, are applied to a correlator, as depicted in Fig. 7.19b. The resulting correlator output is proportional to the cosine of the difference between the carrier phase angles in the two correlator inputs. The correlator output is finally compared with a threshold of zero volts, and a decision is thereby made in favor of symbol 1 or symbol 0. If the correlator output is positive, the phase difference between the waveforms received during the pertinent pair of bit intervals lies inside the range $-\pi/2$ to $\pi/2$, and the receiver decides in favor of symbol 1. If, on the other hand, the correlator output is negative, the phase difference lies outside the range $-\pi/2$ to $\pi/2$, modulo-2π, and the receiver decides in favor of symbol 0.

7.5 COMPARISON OF BINARY AND QUATERNARY MODULATION TECHNIQUES

Two systems having an unequal number of symbols may be compared in a meaningful way only if they use the same amount of energy to transmit each bit of information. It is the total amount of energy needed to transmit the complete message that represents the cost of the transmission, not the amount of energy needed to transmit a particular symbol satisfactorily. Accordingly, in comparing the different data transmission systems considered before, we will use, as the basis of our comparison, the average probability of symbol error expressed as a function of the bit energy-to-noise density ratio E_b/N_0.

In Table 7.4,* we have summarized the expressions for the average probability of symbol error P_e for the coherent PSK, conventional coherent FSK with one-bit decoding, DPSK, noncoherent FSK, QPSK, and MSK, when operating over an AWGN channel. In Fig. 7.20† we have used these expressions to plot P_e as a function of E_b/N_0.

Based on the performance curves shown in Fig. 7.20, the summary of formulas given in Table 7.4, and the defining equations for the pertinent modulation formats, we can make the following statements.

1. The error rates for all the systems decrease monotonically with increasing values of E_b/N_0.

---

* In Table 7.4, we have also included the expression for the average probability of symbol error for the *coherent detection of differentially encoded binary PSK*. Differential encoding is used to resolve the phase ambiguity problem that arises in synchronizing PSK receivers. This issue is discussed in Section 7.12.

† The average probability of error for the coherent detection of differentially encoded binary PSK is practically the same as that of QPSK and MSK; hence, the use of the same curve for these three modulation formats in Fig. 7.20.

Table 7.4 Summary of Formulas for the Symbol Error Probability for Different Data Transmission Systems

	Error probability, P_e
Coherent binary signaling:	
(a) Coherent PSK	$\frac{1}{2}\operatorname{erfc}(\sqrt{E_b/N_0})$
(b) Coherent detection of differentially encoded PSK	$\operatorname{erfc}(\sqrt{E_b/N_0}) - \frac{1}{2}\operatorname{erfc}^2(\sqrt{E_b/N_0})$
(c) Coherent FSK	$\frac{1}{2}\operatorname{erfc}(\sqrt{E_b/2N_0})$
Noncoherent binary signaling:	
(a) DPSK	$\frac{1}{2}\exp(-E_b/N_0)$
(b) Noncoherent FSK	$\frac{1}{2}\exp(-E_b/2N_0)$
Coherent quadrature signaling:	
(a) QPSK (b) MSK	$\operatorname{erfc}(\sqrt{E_b/N_0}) - \frac{1}{4}\operatorname{erfc}^2(\sqrt{E_b/N_0})$

2. For any value of E_b/N_0, coherent PSK produces a smaller error rate than any of the other systems. Indeed, it may be shown that in the case of systems restricted to one-bit decoding, perturbed by additive white Gaussian noise, coherent PSK system is the optimum system for transmitting binary data in the sense that it achieves the minimum probability of symbol error for a given value of E_b/N_0.*

3. Coherent PSK and DPSK require an E_b/N_0 that is 3 dB less than the corresponding values for conventional coherent FSK and noncoherent FSK, respectively, to realize the same error rate.

4. At high values of E_b/N_0, DPSK and noncoherent FSK perform almost as well (to within about 1 dB) as coherent PSK and conventional coherent FSK, respectively, for the same bit rate and signal energy per bit.

5. In QPSK two orthogonal carriers $\sqrt{2/T}\cos(2\pi f_c t)$ and $\sqrt{2/T}\sin(2\pi f_c t)$ are used, where the carrier frequency f_c is an integral multiple of the symbol rate $1/T$, with the result that two independent bit streams can be transmitted and subsequently detected in the receiver. At high values of E_b/N_0, coherently detected binary PSK and QPSK have about the same error rate performance for the same value of E_b/N_0.

6. In MSK the two orthogonal carriers $\sqrt{2/T_b}\cos(2\pi f_c t)$ and $\sqrt{2/T_b}\sin(2\pi f_c t)$ are modulated by the two antipodal symbol shaping pulses $\cos(\pi t/2\,T_b)$ and $\sin(\pi t/2\,T_b)$, respectively, over $2T_b$ intervals, where T_b is the bit duration. Correspondingly, the receiver uses a coherent phase decoding process over two successive bit intervals to recover the original bit stream. We thus find that MSK has exactly the same error rate performance as QPSK.

* See Stein (1964).

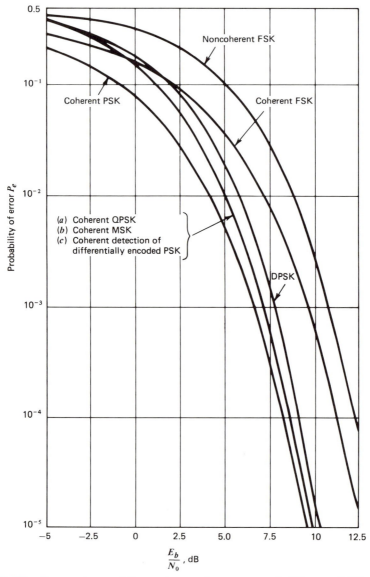

Figure 7.20 Comparison of the noise performances of different PSK and FSK schemes.

7. MSK scheme differs from the other signaling schemes in that its receiver has *memory*. In particular, MSK receiver makes decisions based on observations over two successive bit intervals. Thus, although the transmitted signal has a binary format represented by the transmission of two distinct frequencies, the presence of memory in the receiver makes it assume a form that has in-phase and quadrature paths as in QPSK.

7.6   M-ARY MODULATION TECHNIQUES

In an *M-ary signaling scheme*, we may send one of M possible signals, $s_1(t)$, $s_2(t)$, . . ., $s_M(t)$, during each signaling interval of duration T. For almost all applications, the number of possible signals $M = 2^n$, where n is an integer. The symbol duration $T = nT_b$, where T_b is the bit duration. These signals are generated by changing the amplitude, phase, or frequency of a carrier in M discrete steps. Thus, we have M-ary ASK, M-ary PSK, and M-ary FSK digital modulation schemes. The QPSK system considered in Section 7.3 is an example of M-ary PSK with $M = 4$.

Another way of generating M-ary signals is to combine different methods of modulation into a hybrid form. For example, we may combine discrete changes in both the amplitude and phase of a carrier to produce *M-ary amplitude-phase keying* (APK). A special form of this hybrid modulation, called *M-ary* QAM, has some attractive properties.

M-ary signaling schemes are preferred over binary signaling schemes for transmitting digital information over band-pass channels when the requirement is to conserve bandwidth at the expense of increased power. In practice, we rarely find a communication channel that has the exact bandwidth required for transmitting the output of an information source by means of binary signaling schemes. Thus, when the bandwidth of the channel is less than the required value, we may use M-ary signaling schemes so as to utilize the channel efficiently.

To illustrate the bandwidth-conservation capability of M-ary signaling schemes, consider the transmission of information consisting of a binary sequence with bit duration T_b. If we were to transmit this information by means of binary PSK, for example, we require a bandwidth inversely proportional to T_b. However, if we take blocks of n bits and use an M-ary PSK scheme with $M = 2^n$ and symbol duration $T = nT_b$, the bandwidth required is inversely proportional to $1/nT_b$. This shows that the use of M-ary PSK enables a reduction in transmission bandwidth by the factor $n = \log_2 M$ over binary PSK.

In this section we consider three different M-ary signaling schemes. They are M-ary PSK, M-ary QAM, and M-ary FSK, each of which offers virtues of its own.

(1)   M-ary PSK

In M-ary PSK, the phase of the carrier takes on one of M possible values, namely, $\theta_i = 2i\pi/M$, where $i = 0, 1, . . ., M - 1$. Accordingly, during each signaling interval of duration T, one of the M possible signals

$$s_i(t) = \sqrt{\frac{2E}{T}} \cos\left(2\pi f_c t + \frac{2\pi i}{M}\right) \qquad i = 0, 1, ..., M - 1 \qquad (7.97)$$

is sent, where E is the signal energy per symbol. The carrier frequency $f_c = n_c/T$ for some fixed integer n_c.

Each $s_i(t)$ may be expanded in terms of two basis functions $\phi_1(t)$ and $\phi_2(t)$ defined as

$$\phi_1(t) = \sqrt{\frac{2}{T}} \cos(2\pi f_c t) \qquad 0 \le t \le T \tag{7.98}$$

$$\phi_2(t) = \sqrt{\frac{2}{T}} \sin(2\pi f_c t) \qquad 0 \le t \le T \tag{7.99}$$

Both $\phi_1(t)$ and $\phi_2(t)$ have unit energy. The signal constellation of M-ary PSK is therefore two-dimensional. The M message points are equally spaced on a circle of radius $\sqrt{E}$ and center at the origin, as illustrated in Fig. 7.21 for *octaphase-shift-keying* (i.e., $M = 8$). This figure also includes the corresponding decision boundaries indicated by dashed lines.

The optimum receiver for coherent M-ary PSK (assuming perfect synchronization with the transmitter) is shown in block diagram form in Fig. 7.22. It includes a pair of correlators with reference signals in phase quadrature. The two correlator outputs, denoted as x_I and x_Q, are fed into a *phase discriminator* that first computes the phase estimate

$$\hat{\theta} = \tan^{-1}\left(\frac{x_Q}{x_I}\right) \tag{7.100}$$

The phase discriminator then selects from the set $\{s_i(t), i = 0, \ldots, M - 1\}$ that particular signal whose phase is closest to the estimate $\hat{\theta}$.

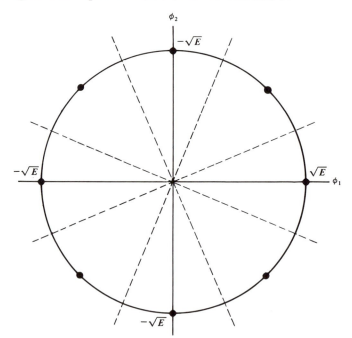

Figure 7.21 Signal constellation for octaphase-shift-keying (i.e., $M = 8$). The decision boundaries are shown as dashed lines.

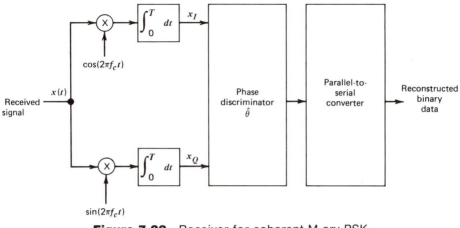

Figure 7.22   Receiver for coherent M-ary PSK.

In the presence of noise, the decision-making process in the phase discrimi-nator is based on the noisy inputs

$$x_I = \sqrt{E} \cos\left(\frac{2\pi i}{M}\right) + w_I \qquad i = 0,1,..., M - 1 \qquad (7.101)$$

and

$$x_Q = -\sqrt{E} \sin\left(\frac{2\pi i}{M}\right) + w_Q \qquad i = 0,1,..., M - 1 \qquad (7.102)$$

where w_I and w_Q are samples of two independent Gaussian random variables W_I and W_Q whose mean is zero and common variance equals

$$\sigma^2 = \frac{N_0}{2}$$

In Fig. 7.21, we see that the message points exhibit circular symmetry. Moreover, both random variables W_I and W_Q have a symmetric probability density function. The implication of these symmetries is that in an M-ary PSK system, the average probability of symbol error, P_e, is independent of the par-ticular signal $s_i(t)$ that is transmitted. We may therefore simplify the calculation of P_e by setting $\theta_i = 0$, which corresponds to the message point whose coordi-nates along the $\phi_1(t)$- and $\phi_2(t)$-axes are $+\sqrt{E}$ and 0, respectively. The decision region pertaining to this message point [i.e., the signal $s_0(t)$] is bounded by the threshold $\hat{\theta} = -\pi/M$ below the $\phi_1(t)$-axis and the threshold $\theta = +\pi/M$ above the $\phi_1(t)$-axis. The probability of correct reception is therefore

$$P_c = \int_{-\pi/M}^{\pi/M} f_\Theta(\hat{\theta}) \, d\hat{\theta} \qquad (7.103)$$

where $f_\theta(\hat{\theta})$ is the probability density function of the random variable Θ whose sample value equals the phase discriminator output $\hat{\theta}$ produced in response to a received signal that consists of the signal $s_0(t)$ plus AWGN. That is,

$$\hat{\theta} = \tan^{-1}\left(\frac{W_Q}{\sqrt{E} + W_I}\right) \qquad (7.104)$$

The phase $\hat{\theta}$ is recognized to be the same as the phase of a sine wave plus narrow-band noise (see appendix D). As such, the probability density function $f_\Theta(\hat{\theta})$ has a known value. Specifically, for $-\pi \le \hat{\theta} \le \pi$, we may write

$$f_\Theta(\hat{\theta}) = \frac{1}{2\pi} \exp\left(-\frac{E}{N_0}\right)$$
$$+ \sqrt{\frac{E}{\pi N_0}} \cos\hat{\theta} \exp\left(-\frac{E}{N_0}\sin^2\hat{\theta}\right)\left[1 - \frac{1}{2}\mathrm{erfc}\left(\frac{E}{N_0}\cos\hat{\theta}\right)\right] \quad (7.105)$$

The probability density function $f_\Theta(\hat{\theta})$ is shown plotted versus θ in Fig. 7.23* for various values of E/N_0. We see that it approaches an impulse-like appearance about θ as E/N_0 assumes high values.

A decision error is made if the angle $\hat{\theta}$ falls outside $-(\pi/M) \le \hat{\theta} \le (\pi/M)$. The probability of symbol error is therefore

$$P_e = 1 - P_c$$
$$= 1 - \int_{-\pi/M}^{\pi/M} f_\Theta(\hat{\theta}) \, d\hat{\theta} \qquad (7.106)$$

In general, the integral in Eq. 7.106 does not reduce to a simple form, except for $M = 2$ and $M = 4$. Hence, for $M > 4$, it must be evaluated by using numerical integration.†

However, for large M and high values of E/N_0, we may derive an approximate formula for P_e. For high values of E/N_0 and for $|\hat{\theta}| < \pi/2$, we may use the approximation (see the upper bound of Eq. E8 in Appendix E)

$$\mathrm{erfc}\left(-\frac{E}{N_0}\cos\hat{\theta}\right) \simeq \sqrt{\frac{N_0}{\pi E}}\frac{1}{\cos\hat{\theta}}\exp\left(-\frac{E}{N_0}\cos^2\hat{\theta}\right) \qquad (7.107)$$

Hence, using this approximation for the complementary error function in Eq. 7.105 and simplifying terms, we finally get

$$f_\Theta(\hat{\theta}) \simeq \sqrt{\frac{E}{\pi N_0}}\cos\hat{\theta}\exp\left(-\frac{E}{N_0}\sin^2\hat{\theta}\right) \qquad |\hat{\theta}| < \frac{\pi}{2} \qquad (7.108)$$

Thus, substituting Eq. 7.108 in Eq. 7.106, we get

$$P_e \simeq 1 - \sqrt{\frac{E}{\pi N_0}}\int_{-\pi/M}^{\pi/M}\cos\hat{\theta}\exp\left(-\frac{E}{N_0}\sin^2\hat{\theta}\right) d\hat{\theta} \qquad (7.109)$$

Changing the variable of integration from $\hat{\theta}$ to

$$z = \sqrt{\frac{E}{N_0}}\sin\hat{\theta}$$

---

* This figure is taken from Thomas (1969).
† For a table of values of P_e for varying E/N_0 and for $M = 2, 4, 8, 16, 32, 64$, see Lindsey and Simon (1973, pp. 232–233).

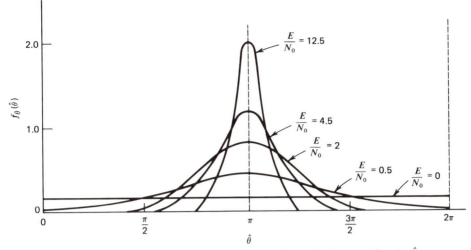

Figure 7.23 Probability density function of phase estimate $\hat{\theta}$.

we may rewrite Eq. 7.109 as

$$P_e \simeq 1 - \frac{2}{\sqrt{\pi}} \int_0^{\sqrt{E/N_0}\sin(\pi/M)} \exp(-z^2)dz$$

$$= \operatorname{erfc}\left(\sqrt{\frac{E}{N_0}} \sin\left(\frac{\pi}{M}\right) \right) \tag{7.110}$$

This is the desired approximate formula for the probability of symbol error that results from the use of coherent M-ary PSK for $M \geq 4$. The approximation becomes extremely tight, for fixed M, as E/N_0 is increased.

Coherent M-ary PSK requires exact knowledge of the carrier frequency and phase for the receiver to be accurately synchronized to the transmitter. When carrier recovery at the receiver is impractical, we may use differential encoding based on the phase difference between successive symbols at the cost of some degradation in performance. If the incoming data are encoded by a phase shift rather than by absolute phase, the receiver performs detection by comparing the phase of one symbol with that of the previous symbol, and the need for a coherent reference is thereby eliminated. This procedure is the same as that described for binary DPSK in Section 7.3. The exact calculation of probability of symbol error for the differential detection of differential M-ary PSK (commonly referred to as *M-ary* DPSK) is much too complicated for $M > 2$. However, for large values of E/N_0 and $M \geq 4$, the probability of symbol error is approximately given by[*]

$$P_e \simeq \operatorname{erfc}\left(\sqrt{\frac{2E}{N_0}} \sin\left(\frac{\pi}{2M}\right) \right) \qquad M \geq 4 \tag{7.111}$$

Comparing the approximate formulas of Eqs. 7.110 and 7.111, we see that for $M \geq 4$ an M-ary DPSK system attains the same probability of symbol error as

[*] See Lindsey and Simon (1973, p. 248).

the corresponding M-ary PSK system provided that the transmitted energy per symbol is increased by the following factor:

$$k(M) = \frac{\sin^2\left(\frac{\pi}{M}\right)}{2 \sin^2\left(\frac{\pi}{2M}\right)} \qquad M \geqslant 4 \tag{7.112}$$

For example, $k(4) = 1.7$. That is, differential QPSK (which is noncoherent) is approximately 2.3 dB poorer in performance than coherent QPSK.

(2) M-ary QAM

In an M-ary PSK system, in-phase and quadrature components of the modulated signal are interrelated in such a way that the envelope is constrained to remain constant. This constraint manifests itself in a circular constellation for the message points. However, if this constraint is removed, and the in-phase and quadrature components are thereby permitted to be independent, we get a new modulation scheme called *M-ary quadrature amplitude modulation* (QAM). In this modulation scheme, the carrier experiences amplitude as well as phase modulation.

The signal constellation for M-ary QAM consists of a *square lattice* of message points, as illustrated in Fig. 7.24 for $M = 16$. The corresponding signal constellations for the in-phase and quadrature components of the amplitude-phase modulated wave are shown in Figs. 7.25a and 7.25b, respectively. The basic format of the signal constellations shown in the latter figures is recognized to be that of a *polar L-ary ASK signal* with $L = 4$. Thus, in general, an M-ary QAM scheme enables the transmission of $M = L^2$ independent symbols over the same channel bandwidth as that required for one polar L-ary ASK scheme.

The general form of M-ary QAM is defined by the transmitted signal

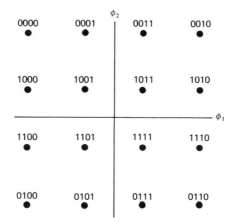

Figure 7.24 Signal-constellation of M-ary QAM for $M = 16$. (The message points are identified with 4-bit Gray codes for later discussion.)

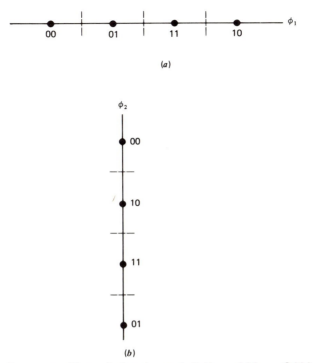

(a)

(b)

Figure 7.25   Decomposition of signal constellation of M-ary QAM (for $M = 16$) into two signal-space diagrams for (a) in-phase comment $\phi_1(t)$, and (b) quadrature comment $\phi_2(t)$. (The message points are identified by 2-bit Gray codes for later discussion.)

$$s_i(t) = \sqrt{\frac{2E_0}{T}}\, a_i \cos(2\pi f_c t)$$

$$+ \sqrt{\frac{2E_0}{T}}\, b_i \sin(2\pi f_c t) \qquad 0 \le t \le T \qquad (7.113)$$

where E_0 is the energy of the signal with the lowest amplitude, and a_i and b_i are a pair of independent integers chosen in accordance with the location of the pertinent message point. The signal $s_i(t)$ consists of two phase-quadrature carriers, each of which is modulated by a set of discrete amplitudes; hence, the name "quadrature amplitude modulation."

The signal $s_i(t)$ can be expanded in terms of a pair of basis functions:

$$\phi_1(t) = \sqrt{\frac{2}{T}} \cos(2\pi f_c t) \qquad 0 \le t \le T \qquad (7.114)$$

and

$$\phi_2(t) = \sqrt{\frac{2}{T}} \sin(2\pi f_c t) \qquad 0 \le t \le T \qquad (7.115)$$

The coordinates of the ith message point are $a_i\sqrt{E}$ and $b_i\sqrt{E_0}$, where (a_i,b_i) is an element of the L-by-L matrix:

$$\{a_i,b_i\} = \begin{bmatrix} (-L+1,L-1) & (-L+3,L-1) & \dots & (L-1,L-1) \\ (-L+1,L-3) & (-L+3,L-3) & \dots & (L-1,L-3) \\ \vdots & \vdots & & \vdots \\ (-L+1,-L+1) & (-L+3,-L+1) & \dots & (L-1,-L+1) \end{bmatrix}$$

(7.116)

where

$$L = \sqrt{M} \qquad (7.117)$$

For example, for the 16-QAM whose signal constellation is depicted in Fig. 7.24, where $L = 4$, we have the matrix

$$\{a_i,b_i\} = \begin{bmatrix} (-3,3) & (-1,3) & (1,3) & (3,3) \\ (-3,1) & (-1,1) & (1,1) & (3,1) \\ (-3,-1) & (-1,-1) & (1,-1) & (3,-1) \\ (-3,-3) & (-1,-3) & (1,-3) & (3,-3) \end{bmatrix}$$

(7.118)

To calculate the probability of symbol error for M-ary QAM, we proceed as follows:

1. Since the in-phase and quadrature components of M-ary QAM are independent, the probability of correct detection for such a scheme may be written as

$$P_c = (1 - P'_e)^2 \qquad (7.119)$$

where P'_e is the probability of symbol error for either component.

2. The signal constellation for the in-phase or quadrature component has a geometry similar to that for discrete pulse-amplitude modulation (PAM) with a corresponding number of amplitude levels. We may therefore adapt the formula of Eq. 3.75 to fit the terminology of the signal constellations shown in Fig. 7.25, and so we write

$$P'_e = \left(1 - \frac{1}{L}\right) \text{erfc}\left(\sqrt{\frac{E_0}{N_0}}\right) \qquad (7.120)$$

where L is the square root of M.

3. The probability of symbol error for M-ary QAM is given by

$$\begin{aligned} P_e &= 1 - P_c \\ &= 1 - (1 - P'_e)^2 \\ &\approx 2P'_e \end{aligned} \qquad (7.121)$$

where it is assumed that P'_e is small compared to unity. Hence, using Eqs. 7.117 and 7.120 in Eq. 7.121, we find that the probability of symbol error for M-ary QAM is (for all practical purposes) given by

$$P_e \approx 2\left(1 - \frac{1}{\sqrt{M}}\right) \text{erfc}\left(\sqrt{\frac{E_0}{N_0}}\right) \qquad (7.122)$$

The transmitted energy in M-ary QAM is variable in that its instantaneous value depends on the particular symbol transmitted. It is logical to express P_e in terms of the *average* value of the transmitted energy rather than E_0. Assuming that the L amplitude levels of the in-phase or quadrature component are equally likely, we have

$$E_{av} = 2 \left[\frac{2E_0}{L} \sum_{i=1}^{L/2} (2i - 1)^2 \right] \tag{7.123}$$

where the multiplying factor of 2 accounts for the equal contributions made by the in-phase and quadrature components. The limits of the summation take account of the symmetric nature of the pertinent amplitude levels around zero. Summing the series in Eq. 7.123, we get

$$E_{av} = \frac{2(L^2 - 1)E_0}{3}$$

$$= \frac{2(M - 1)E_0}{3} \tag{7.124}$$

Accordingly, we may rewrite Eq. 7.122 in terms of E_{av} as

$$P_e \simeq 2 \left(1 - \frac{1}{\sqrt{M}} \right) \text{erfc} \left(\sqrt{\frac{3E_{av}}{2(M - 1)N_0}} \right) \tag{7.125}$$

which is the desired result.

The case of $M = 4$ is of special interest. The signal constellation for this value of M is shown in Fig. 7.26, which is recognized to be the same as that for QPSK. Indeed, putting $M = 4$ in Eq. 7.27 and noting from Fig. 7.26 that E_{av} equals E, where E is the energy per symbol, we find that the resulting formula for the probability of symbol error becomes identical to that in Eq. 7.46, and so it should.

Figure 7.27*a* shows the block diagram of an M-ary QAM transmitter. The

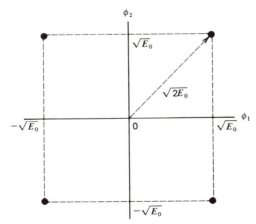

Figure 7.26 Signal constellation for the special case of M-ary QAM for $M = 4$.

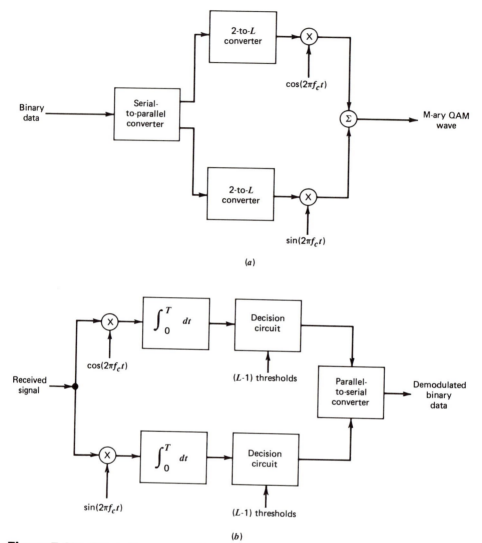

Figure 7.27   Block diagrams of M-ary QAM system. (a) Transmitter. (b) Receiver.

serial-to-parallel converter accepts a binary sequence at a bit rate $R_b = 1/T_b$ and produces two parallel binary sequences whose bit rates are $R_b/2$ each. The *2-to-L level converters*, where $L = \sqrt{M}$, generate polar L-level signals in response to the respective in-phase and quadrature channel inputs. Quadrature-carrier multiplexing of the two polar L-level signals so generated produces the desired M-ary QAM signal. Figure 7.27b shows the block diagram of the corresponding receiver. Decoding of each baseband channel is accomplished at the output of the pertinent decision circuit, which is designed to compare the L-level signals against $L - 1$ decision thresholds. The two binary sequences so detected are then combined in the parallel-to-serial converter to reproduce the original binary sequence.

(3) M-ary FSK
In an M-ary FSK scheme, the transmitted signals are defined by

$$s_i(t) = \sqrt{\frac{2E}{T}} \cos\left[\frac{\pi}{T}(n_c + i)t\right] \qquad 0 \le t \le T \tag{7.126}$$

where $i = 1, 2, \ldots , M$, and the carrier frequence $f_c = n_c/2T$ for some fixed integer n_c. The transmitted signals are of equal duration T and have equal energy E. Since the individual signal frequencies are separated by $1/2T$ hertz, the signals in Eq. 7.126 are orthogonal; that is

$$\int_0^T s_i(t)\, s_j(t)dt = 0 \qquad i \ne j \tag{7.127}$$

For coherent M-ary FSK, the optimum receiver consists of a bank of M correlators or matched filters, with the signals in Eq. 7.126 providing the pertinent references. At the sampling times $t = kT$, the receiver makes decisions based on the largest matched filter output.

An upper bound for the probability of symbol error may be obtained by applying the *union bound* of Section 3.6. The resulting bound is given by (see Problem 7.6.5)

$$P_e \le \frac{1}{2}(M - 1)\, \text{erfc}\left(\sqrt{\frac{E}{2N_0}}\right) \tag{7.128}$$

For fixed M, this bound becomes increasingly tight as E/N_0 is increased. Indeed, it becomes a good approximation to P_e for values of $P_e \le 10^{-3}$. Moreover, for $M = 2$ (i.e., binary FSK), the bound of Eq. 7.128 becomes an equality.

Coherent detection of M-ary FSK requires the use of exact phase references, the provision for which at the receiver can be costly and difficult to maintain. We may avoid the need for such a provision by using noncoherent detection, which results in a slightly inferior performance. In a noncoherent receiver, the individual matched filters are followed by envelope detectors that destroy the phase information.

The probability of symbol error for the noncoherent detection of M-ary FSK is given by*

$$P_e = \sum_{k=1}^{M-1} \frac{(-1)^{k+1}}{k+1}\binom{M-1}{k}\exp\left(-\frac{kE}{(k+1)N_0}\right) \tag{7.129}$$

where $\binom{M-1}{k}$ is a binomial coefficient, that is,

$$\binom{M-1}{k} = \frac{(M-1)!}{(M-1-k)!k!} \tag{7.130}$$

The leading term of the series in Eq. 7.129 provides an upper bound on the

---

* See Lindsey and Simon (1973, p. 489).

probability of symbol error for the noncoherent detection of M-ary FSK:

$$P_e \leqslant \frac{M-1}{2} \exp\left(-\frac{E}{2N_0}\right) \tag{7.131}$$

For fixed M, this bound becomes increasingly close to the actual value of P_e as E/N_0 is increased. Indeed, for $M = 2$ (i.e., binary FSK), the bound of Eq. 7.131 becomes an equality.

(4) Comparison of M-ary Digital Modulation Techniques

We conclude this section on M-ary digital modulation techniques by presenting some notes on the comparative performances and merits of the three M-ary modulation schemes considered.

In Table 7.5 we have summarized typical values of power-bandwidth requirements for coherent binary and M-ary PSK schemes, assuming an average probability of symbol error equal to 10^{-4} and that the systems operate in identical noise environments. This table shows that, among the family of M-ary PSK signals, QPSK (corresponding to $M = 4$) offers the best trade-off between power and bandwidth requirements. It is for this reason that we find QPSK is widely used in practice. For $M > 8$, power requirements become excessive; accordingly, M-ary PSK schemes with $M > 8$ are not as widely used in practice. Also, coherent M-ary PSK schemes require considerably more complex equipment than coherent binary PSK schemes for signal generation or detection, especially when $M > 8$.

Basically, M-ary PSK and M-ary QAM have similar spectral and bandwidth characteristics. For $M > 4$, however, the two schemes have different signal constellations. For M-ary PSK the signal constellation is circular, whereas for M-ary QAM it is rectangular. Moreover, a comparison of these two constellations reveals that the distance between the message points of M-ary PSK is smaller than the distance between the message points of M-ary QAM, for a fixed peak transmitted power. This basic difference between the two schemes is illustrated in Fig. 7.28 for $M = 16$. Accordingly, in an AWGN channel, M-ary QAM outperforms the corresponding M-ary PSK in error performance for $M > 4$. However, the superior performance of M-ary QAM can be realized only if the channel is free from nonlinearities.

Table 7.5[a] Comparison of Power-Bandwidth Requirements for M-ary PSK with Binary PSK. Probability of Symbol Error = 10^{-4}

Value of M	$\dfrac{\text{(Bandwidth)}_{\text{M-ary}}}{\text{(Bandwidth)}_{\text{Binary}}}$	$\dfrac{\text{(Average power)}_{\text{M-ary}}}{\text{(Average power)}_{\text{Binary}}}$
4	0.5	0.34 dB
8	0.333	3.91 dB
16	0.25	8.52 dB
32	0.2	13.52 dB

[a] See Shanmugam (1979, p. 424).

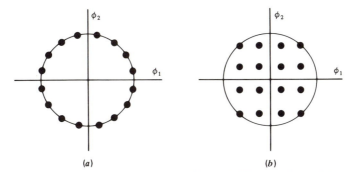

Figure 7.28 Signal constellations. (a) M-ary QPSK, and (b) M-ary QAM for $M = 16$.

As for M-ary FSK, we find that for a fixed probability of error, increasing M results in a reduced power requirement. However, this reduction in transmitted power is achieved at the cost of increased channel bandwidth. More will be said on this issue in Section 7.9.

7.7 POWER SPECTRA

The description given in Eq. 7.31 for a band-pass signal $s(t)$ contains the definitions of ASK, PSK, and FSK signals, depending on the way in which the in-phase component $s_I(t)$ and the quadrature component $s_Q(t)$ are defined. The analysis of such signals may be simplified by using *complex notation*.* In particular, we may express $s(t)$ in the form

$$s(t) = s_I(t)\cos(2\pi f_c t) - s_Q(t)\sin(2\pi f_c t)$$
$$= \text{Re}[\tilde{s}(t)\exp(j2\pi f_c t)] \qquad (7.132)$$

where $\text{Re}[\cdot]$ is the real part of the expression contained inside the square brackets. We also have

$$\tilde{s}(t) = s_I(t) + js_Q(t) \qquad (7.133)$$

and

$$\exp(j2\pi f_c t) = \cos(2\pi f_c t) + j\sin(2\pi f_c t) \qquad (7.134)$$

The signal $\tilde{s}(t)$ is called the *complex envelope* of the band-pass signal $s(t)$. The components $s_I(t)$ and $s_Q(t)$ and therefore $\tilde{s}(t)$ are all low-pass signals. They are uniquely defined in terms of the band-pass signal $s(t)$ and the carrier frequency f_c provided that the half-bandwidth of $s(t)$ is less than the carrier frequency f_c.

Let $S_B(f)$ denote the power spectral density of the complex envelope $\tilde{s}(t)$. We refer to $S_B(f)$ as the *baseband power spectral density*. The power spectral

* For a treatment of the complex representation of band-pass signals, see Appendix C.

density, $S_S(f)$, of the original band-pass signal $s(t)$ is a frequency-shifted version of $S_B(f)$, except for a scaling factor, as shown by

$$S_S(f) = \tfrac{1}{4}[S_B(f - f_c) + S_B(f + f_c)] \tag{7.135}$$

It is therefore sufficient to evaluate the baseband power spectral density $S_B(f)$. Since $\tilde{s}(t)$ is a low-pass signal, the calculation of $S_B(f)$ should be simpler than the calculation of $S_S(f)$. In the sequel, we calculate the baseband power spectral density for binary PSK, binary FSK, QPSK, and MSK signals.

(1) Power Spectra of Binary PSK and FSK Signals

Consider first the case of a binary PSK wave. From the modulator of Fig. 7.3a, we see that the complex envelope of a binary PSK wave consists of an in-phase component only. Furthermore, depending on whether we have a symbol 1 or a symbol 0 at the modulator input during the signaling interval $0 \leq t \leq T_b$, we find that this in-phase component equals $+g(t)$ or $-g(t)$, respectively, where $g(t)$ is the *symbol shaping function* defined by

$$g(t) = \begin{cases} \sqrt{\dfrac{2E_b}{T_b}} & 0 \leq t \leq T_b \\ 0 & \text{elsewhere} \end{cases} \tag{7.136}$$

We assume that the input binary wave is random, with symbols 1 and 0 equally likely and the symbols transmitted during the different time slots being statistically independent. In Appendix D, it is shown that the power spectral density of a random binary wave so described is equal to energy spectral density of the symbol shaping function divided by the symbol duration. The energy spectral density of a Fourier transformable signal $g(t)$ is defined as the squared magnitude of the signal's Fourier transform. Hence, the baseband power spectral density of a binary PSK wave equals

$$\begin{aligned} S_B(f) &= \frac{2E_b \sin^2(\pi T_b f)}{(\pi T_b f)^2} \\ &= 2E_b \operatorname{sinc}^2(T_b f) \end{aligned} \tag{7.137}$$

This power spectrum falls off as the inverse square of frequency.

Consider next the case of a binary FSK wave, for which the two transmitted frequencies f_1 and f_2 differ by an amount equal to the bit rate $1/T_b$, and their arithmetic mean equals the nominal carrier frequency f_c. We also assume that phase continuity is always maintained, including inter-bit switching times. We may thus express the binary FSK signal as follows*:

$$s(t) = \sqrt{\frac{2E_b}{T_b}} \cos\left(2\pi f_c t \pm \frac{\pi t}{T_b}\right) \qquad 0 \leq t \leq T_b$$

* This special form of binary FSK signal is sometimes referred to as *Sunde's* FSK; see Sunde (1959).

and using a well-known trigonometric identity, we get

$$s(t) = \sqrt{\frac{2E_b}{T_b}} \cos\left(\pm \frac{\pi t}{T_b}\right) \cos(2\pi f_c t) - \sqrt{\frac{2E_b}{T_b}} \sin\left(\pm \frac{\pi t}{T_b}\right) \sin(2\pi f_c t)$$

$$= \sqrt{\frac{2E_b}{T_b}} \cos\left(\frac{\pi t}{T_b}\right) \cos(2\pi f_c t) \mp \sqrt{\frac{2E_b}{T_b}} \sin\left(\frac{\pi t}{T_b}\right) \sin(2\pi f_c t) \qquad (7.138)$$

In the last line of Eq. 7.138, the plus sign corresponds to transmitting symbol 0, and the minus sign corresponds to transmitting symbol 1. As before, we assume that the symbols 1 and 0 in the random binary wave at the modulator input are equally likely, and that the symbols transmitted in adjacent time slots are statistically independent. Then, based on the representation of Eq. 7.138, we may make the following observations pertaining to the in-phase and quadrature components of a binary FSK signal with continuous phase:

1. The in-phase component is completely independent of the input binary wave. It equals $\sqrt{2E_b/T_b} \cos(\pi t/T_b)$ for all values of time t. The power spectral density of this component therefore consists of two delta functions, weighted by the factor $E_b/2T_b$, and occurring at $f = \pm 1/2T_b$.
2. The quadrature component is directly related to the input binary wave. During the signaling interval $0 \leq t \leq T_b$, it equals $+g(t)$ when we have symbol 1, and $-g(t)$ when we have symbol 0. The symbol shaping function $g(t)$ is defined by

$$g(t) = \begin{cases} \sqrt{\dfrac{2E_b}{T_b}} \sin\left(\dfrac{\pi t}{T_b}\right) & 0 \leq t \leq T_b \\ 0 & \text{elsewhere} \end{cases} \qquad (7.139)$$

The energy spectral density of this symbol shaping function equals

$$\Psi_g(f) = \frac{8E_b T_b \cos^2(\pi T_b f)}{\pi^2 (4T_b^2 f^2 - 1)^2} \qquad (7.140)$$

The power spectral density of the quadrature component equals $\Psi_g(f)/T_b$. It is also apparent that the in-phase and quadrature components of the binary FSK wave are independent of each other. Accordingly, the baseband power spectral density of the binary FSK wave (whose two tones are separated by the bit rate $1/T_b$) equals the sum of the power spectral densities of these two components, as shown by

$$S_B(f) = \frac{E_b}{2T_b} \left[\delta\left(f - \frac{1}{2T_b}\right) + \delta\left(f + \frac{1}{2T_b}\right) \right] + \frac{8E_b \cos^2(\pi T_b f)}{\pi^2 (4T_b^2 f^2 - 1)^2} \qquad (7.141)$$

Substituting Eq. 7.141 in Eq. 7.135, we find that the power spectrum of the binary FSK signal contains two discrete frequency components located at $(f_c + 1/2T_b) = f_1$ and $(f_c - 1/2T_b) = f_2$, with their average powers adding up to one-half the total power of the binary FSK signal. The presence of these two discrete frequency components provides a means of synchronizing the receiver with the transmitter.

Note also that the baseband power spectral density of a binary FSK signal with continuous phase ultimately falls off as the inverse fourth power of frequency. This is readily established by taking the limit in Eq. 7.141 as f approaches infinity. If, however, the FSK signal exhibits phase discontinuity at the inter-bit switching instants (this arises when the two oscillators supplying frequencies f_1 and f_2 operate independently of each other), the power spectral density ultimately falls off as the inverse square of frequency (see Problem 7.7.1). Accordingly, an FSK signal with continuous phase does not produce as much interference outside the signal band of interest as an FSK signal with discontinuous phase.

In Fig. 7.29 we have plotted the baseband power spectra of Eqs. 7.137 and 7.141. In both cases, $S_B(f)$ is shown normalized with respect to $2E_b$, and the frequency is normalized with respect to the bit rate $R_b = 1/T_b$. The difference in the rates of falloff of these spectra can be explained on the basis of the pulse shape $g(t)$. The smoother the pulse, the faster is the drop of spectral tails to zero. Thus, with binary FSK (with continuous phase) having a smoother pulse shape, it has lower sidelobes than binary PSK.

(2) Power Spectra of QPSK and MSK Signals

Consider first the case of QPSK. We assume that the binary wave at the modulator input is random, with symbols 1 and 0 being equally likely, and with the symbols transmitted during adjacent time slots being statistically independent. We may make the following observations pertaining to the in-phase and quadrature components of a QPSK signal:

1. Depending on the dibit sent during the signaling interval $-T_b \leq t \leq T_b$, the in-phase component equals $+g(t)$ or $-g(t)$, and similarly for the quadrature component. The $g(t)$ denotes the symbol shaping function, defined by

$$g(t) = \begin{cases} \sqrt{\dfrac{E}{T}} & 0 \leq t \leq T \\ 0 & \text{elsewhere} \end{cases} \tag{7.142}$$

    Hence, the in-phase and quadrature components have a common power spectral density, namely, $E \operatorname{sinc}^2(Tf)$.

2. The in-phase and quadrature components are statistically independent. Accordingly, the baseband power spectral density of the QPSK signal equals the sum of the individual power spectral densities of the in-phase and quadrature components, and so we may write

$$\begin{aligned} S_B(f) &= 2E \operatorname{sinc}^2(Tf) \\ &= 4E_b \operatorname{sinc}^2(2T_b f) \end{aligned} \tag{7.143}$$

Consider next the MSK signal. Here again we assume that the input binary wave is random, with symbols 1 and 0 equally likely, and the symbols transmitted during different time slots being statistically independent. In this case, we may make the following observations:

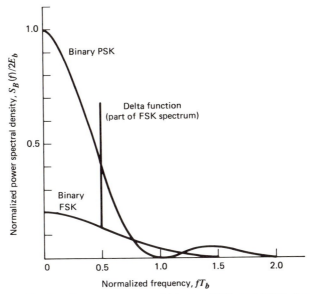

Figure 7.29   Power spectra of binary PSK and FSK signals.

1. Depending on the value of phase state $\theta(0)$, the in-phase component equals $+g(t)$ or $-g(t)$, where

$$g(t) = \begin{cases} \sqrt{\dfrac{2E_b}{T_b}} \cos\left(\dfrac{\pi t}{2T_b}\right) & -T_b \le t \le T_b \\ 0 & \text{elsewhere} \end{cases} \qquad (7.144)$$

The energy spectral density of this symbol-shaping function equals

$$\Psi_g(f) = \frac{32E_bT_b}{\pi^2} \left[\frac{\cos(2\pi T_b f)}{16T_b^2 f^2 - 1}\right]^2 \qquad (7.145)$$

Hence, the power spectral density of the in-phase component equals $\Psi_g(f)/2T_b$.

2. Depending on the value of the phase state $\theta(T_b)$, the quadrature component equals $+g(t)$ or $-g(t)$, where

$$g(t) = \begin{cases} \sqrt{\dfrac{2E_b}{T_b}} \sin\left(\dfrac{\pi t}{2T_b}\right) & 0 \le t \le 2T_b \\ 0 & \text{elsewhere} \end{cases} \qquad (7.146)$$

The energy spectral density of this second symbol shaping function is also given by Eq. 7.145. Hence, the in-phase and quadrature components have the same power spectral density.

3. As with the QPSK signal, the in-phase and quadrature component of the MSK signal are also statistically independent. Hence, the baseband power spectral density of the MSK signal is given by

$$S_B(f) = 2\left[\frac{\Psi_g(f)}{2T_b}\right]$$

$$= \frac{32E_b}{\pi^2}\left[\frac{\cos(2\pi T_b f)}{16T_b^2 f^2 - 1}\right]^2 \tag{7.147}$$

The baseband power spectra of Eqs. 7.143 and 7.147 for the QPSK and MSK signals, respectively, are shown plotted in Fig. 7.30. The power spectral density is normalized with respect to $4E_b$, and the frequency is normalized with respect to the bit rate $1/T_b$. Note that for $f \gg 1/T_b$, the baseband power spectral density of the MSK signal falls off as the inverse fourth power of frequency, whereas in the case of the QPSK signal, it falls off as the inverse square of frequency. Accordingly, MSK does not produce as much interference outside the signal band of interest as QPSK. This is a desirable characteristic of MSK, especially when it operates with a bandwidth limitation.

(3) Power Spectra of M-ary Signals

Binary PSK and QPSK are special cases of M-ary PSK signals. The symbol duration of M-ary PSK is defined by

$$T = T_b \log_2 M \tag{7.148}$$

where T_b is the bit duration. Proceeding in a manner similar to that described for QPSK signal, we may show that the baseband power spectral density of M-ary PSK signal is given by

$$S_B(f) = 2E \ \mathrm{sinc}^2(Tf)$$
$$= 2E_b \log_2 M \ \mathrm{sinc}^2(T_b f \log_2 M) \tag{7.149}$$

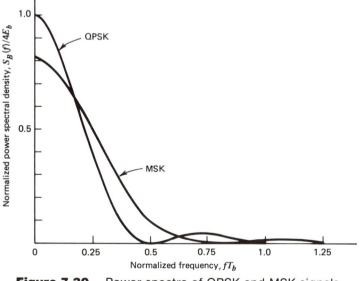

Figure 7.30 Power spectra of QPSK and MSK signals.

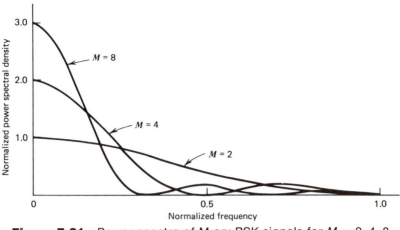

Figure 7.31 Power spectra of M-ary PSK signals for $M = 2, 4, 8$.

In Fig. 7.31, we show the normalized power spectral density $S_B(f)/2E_b$ plotted versus the normalized frequency fT_b for three different values of M, namely, $M = 2, 4, 8$.

The spectral analysis of M-ary FSK signals is much more complicated than that of M-ary PSK signals. A case of particular interest occurs when the frequencies assigned to the multilevels make the frequency spacing uniform and the frequency deviation $k = 0.5$. That is, the M signal frequencies are separated by $1/2T$, where T is the symbol duration. For $k = 0.5$, the baseband power spectral density of M-ary FSK signals is defined by*

$$S_B(f) = 4E_b \left[\frac{1}{2M} \sum_{i=1}^{M} \left(\frac{\sin\gamma_i}{\gamma_i} \right)^2 + \frac{1}{M^2} \sum_{i=1}^{M} \sum_{j=1}^{M} \cos(\gamma_i + \gamma_j) \left(\frac{\sin\gamma_i}{\gamma_i} \right)^2 \left(\frac{\sin\gamma_j}{\gamma_j} \right)^2 \right]$$

(7.150)

where

$$\gamma_i = \left(fT_b - \frac{\alpha_i}{4} \right) \pi$$

$$\alpha_i = 2i - (M + 1) \qquad i = 1, 2, \ldots, M$$

Equation 7.150 is shown plotted in Fig. 7.32 for $M = 2, 4, 8$.

7.8 BANDWIDTH EFFICIENCY

Channel bandwidth and *transmitted power* constitute two primary "communication resources," the efficient utilization of which provides the motivation for the search for *spectrally efficient* schemes. The primary objective of spectrally

---

* Anderson and Salz (1965) present a detailed analysis of the spectra of M-ary FSK for an arbitrary value of frequency deviation k. Equation 7.150 is a special case of the formula derived in this paper for $k = 0.5$.

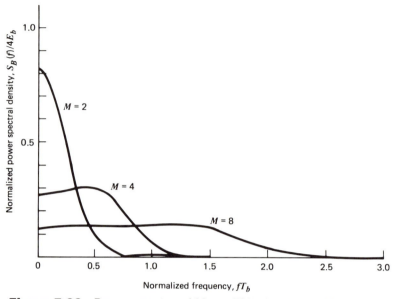

Figure 7.32 Power spectra of M-ary FSK signals for M = 2, 4, 8.

efficient modulation is to maximize the *bandwidth efficiency*, defined as the ratio of data rate to channel bandwidth; it is measured in units of bits per second per hertz. A secondary objective is to achieve this bandwidth efficiency at a minimum practical expenditure of average signal power or, equivalently, in a channel perturbed by additive white Gaussian noise, a minimum practical expenditure of average signal-to-noise ratio. Bandwidth efficiency is also referred to as *spectral efficiency*.

With the data rate denoted by R_b and the channel bandwidth by B, we may express the bandwidth efficiency, ρ, as

$$\rho = \frac{R_b}{B} \qquad \text{bits/s/Hz} \qquad (7.151)$$

The data rate R_b is well defined. Unfortunately, however, there is no universally satisfying definition for the bandwidth B. This means that the bandwidth efficiency of a digital modulation scheme depends on the particular definition adopted for the bandwidth of the modulated signal. In the sequel, we consider the evaluation of the bandwidth efficiency of M-ary PSK signals, followed by that of M-ary FSK signals.

(1) Bandwidth Efficiency of M-ary PSK Signals

The power spectra of M-ary PSK signals possess a *main lobe* bounded by well-defined *spectral nulls* (i.e., frequencies at which the power spectral density is zero). Accordingly, the spectral width of the main lobe provides a simple and popular measure for the bandwidth of M-ary PSK signals. This definition is referred to as the *null-to-null bandwidth*. With the null-to-null bandwidth encompassing the main lobe of the power spectrum of an M-ary signal, we find that it contains most of the signal power. This is readily seen by looking at the power spectral plots of Fig. 7.31.

The channel bandwidth required to pass M-ary PSK signals (more precisely, the main spectral lobe of M-ary PSK signals) is given by

$$B = \frac{2}{T} \tag{7.152}$$

where T is the symbol duration. But the symbol duration T is related to the bit duration T_b by Eq. 7.148. Moreover, the bit rate $R_b = 1/T_b$. Hence, we may redefine the channel bandwidth of Eq. 7.152 in terms of the bit rate R_b as

$$B = \frac{2R_b}{\log_2 M} \tag{7.153}$$

Equivalently, we may express the bandwidth efficiency of M-ary PSK signals as

$$\rho = \frac{R_b}{B}$$
$$= \frac{\log_2 M}{2} \tag{7.154}$$

Table 7.6 gives the values of ρ calculated from Eq. 7.154 for varying M.

Table 7.6   Bandwidth Efficiency of M-ary PSK Signals

M	2	4	8	16	32	64
ρ (bits/s/Hz)	0.5	1	1.5	2	2.5	3

(2)   Bandwidth Efficiency of M-ary FSK Signals

Consider next an M-ary FSK signal that consists of an *orthogonal* set of M frequency-shifted signals. When the orthogonal signals are detected coherently, the adjacent signals need only be separated from each other by a frequency difference $1/2T$ so as to maintain orthogonality (see Problem 7.2.7). Hence, we may define the channel bandwidth required to transmit M-ary FSK signals as

$$B = \frac{M}{2T} \tag{7.155}$$

For multilevels with frequency assignments that make the frequency spacing uniform and equal to $1/2T$, the bandwidth B of Eq. 7.155 contains a large fraction of the signal power. This is readily confirmed by looking at the baseband power spectral plots shown in Fig. 7.32. Hence, substituting Eq. 7.152 in Eq. 7.155 and using $R_b = 1/T_b$, we may redefine the channel bandwidth B for

M-ary FSK signals as

$$B = \frac{R_b M}{2 \log_2 M} \tag{7.156}$$

Hence, the bandwidth efficiency of M-ary FSK signals is given by

$$\rho = \frac{R_b}{B}$$

$$= \frac{2 \log_2 M}{M} \tag{7.157}$$

Table 7.7 gives the values of ρ calculated from Eq. 7.157 for varying M.

Comparing Tables 7.6 and 7.7, we see that increasing the number of levels M tends to increase the bandwidth efficiency of M-ary PSK signals, but tends to decrease the bandwidth efficiency of M-ary FSK signals. In other words, M-ary PSK signals are spectrally efficient, whereas M-ary FSK signals are spectrally inefficient.

Table 7.7 Bandwidth Efficiency of M-ary FSK Signals

M	2	4	8	16	32	64
ρ (bits/s/Hz)	1	1	0.75	0.5	0.3125	0.1875

7.9 M-ARY MODULATION FORMATS VIEWED IN THE LIGHT OF THE CHANNEL CAPACITY THEOREM

It is informative to compare the bandwidth–power exchange capabilities of M-ary PSK and M-ary FSK signals in the light of Shannon's *channel capacity theorem*. The yardstick for this comparison is provided by the *ideal system* for error-free transmission. By the ideal system, we mean one that follows the channel capacity theorem (see Section 2.9). Consider first a coherent M-ary PSK system that employs a *nonorthogonal* set of M phase-shifted signals for the transmission of binary data. Each signal in the set represents a K-bit word, where $K = \log_2 M$. Using the definition of null-to-null bandwidth, we may express the bandwidth efficiency of M-ary PSK as in Eq. 7.154, which is reproduced here for convenience:

$$\frac{R_b}{B} = \frac{\log_2 M}{2}$$

In Fig. 7.33, we show the operating points for different numbers of phase levels $M = 2^K$, where $K = 1, 2, 3, 4, 5, 6$. Each point corresponds to an average probability of symbol error $P_e = 10^{-5}$. In the figure we have also included the capacity curve for the ideal system. We observe from Fig. 7.33 that as M is increased, the bandwidth efficiency is improved, but the value of E_b/N_0 required for error-free transmission moves away from the Shannon limit. This

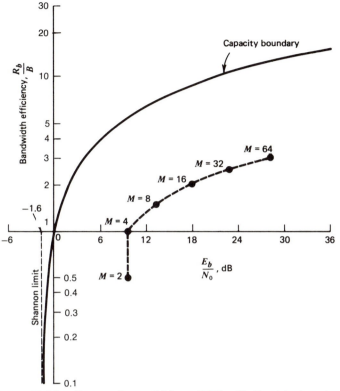

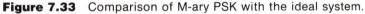

Figure 7.33   Comparison of M-ary PSK with the ideal system.

behavior of M-ary PSK is also displayed by the curves shown in the error-rate diagram of Fig. 7.34.

Consider next a coherent M-ary FSK system that uses an *orthogonal* set of M frequency-shifted signals for the transmission of binary data, with the separation between adjacent signal frequencies set at $1/2T$, where T is the symbol period. As with the M-ary PSK, each signal in the set represents a K-bit word, where $K = \log_2 M$. The bandwidth efficiency of this second set is defined in Eq. 7.157, which is reproduced here for convenience:

$$\frac{R_b}{B} = \frac{2 \log_2 M}{M}$$

In Fig. 7.35, we show the operating points for different numbers of frequency levels $M = 2^K$, where $K = 1, 2, 3, 4, 5, 6$ for an average probability of symbol error $P_e = 10^{-5}$. In the figure we have also included the capacity curve for the ideal system. We see that increasing M in (orthogonal) M-ary FSK has the opposite effect to that in (nonorthogonal) M-ary PSK. In particular, as M is increased, which is equivalent to increased bandwidth requirement, the operating point moves closer to the Shannon limit. This effect is also exemplified in the error-rate diagram of Fig. 7.36, where the average probability of symbol error, P_e, is shown plotted versus E_b/N_0 for the same set of values of M used in

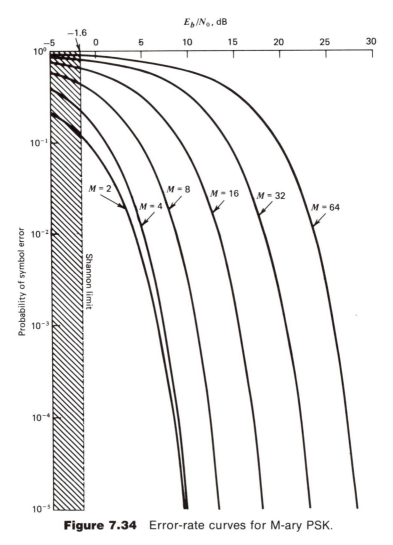

Figure 7.34 Error-rate curves for M-ary PSK.

Fig. 7.35. We clearly see that as M is increased, the system approaches the Shannon limit. However, the price paid for this performance is a large channel bandwidth that approaches infinity in the limit.

7.10 EFFECT OF INTERSYMBOL INTERFERENCE

From the study of digital modulation schemes presented in the preceding sections, we see that the error-performance analysis of these schemes in the presence of additive white Gaussian noise is well-understood for both coherent and noncoherent reception. In practice, however, we find that because bandwidth occupancy of the channel is a major factor in the design of these systems, there is indeed a second source of interference, namely, intersymbol interference (ISI), which must be accounted for in error rate calculations. As explained in

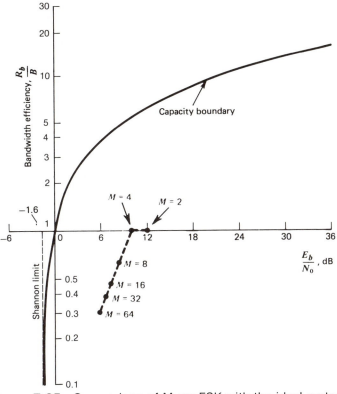

Figure 7.35 Comparison of M-ary FSK with the ideal system.

Section 6.2, intersymbol interference is generated by the use of band-limiting filters at the transmitter output, in the transmission medium, and at the receiver input, or combinations thereof. When ISI is present, we find that (for the case of coherent reception) the correlation receiver or the matched filter receiver is no longer optimum, with the result that there is degradation in the actual error rate of the receiver. This also applies to noncoherent receivers.

Owing to the linear nature of the detection process in coherent PSK systems, we find that the effect of ISI on the performance of these systems has been treated in great detail in the literature. In particular, numerical methods have been developed for calculating the average probability of symbol error in coherent M-ary PSK systems in the combined presence of additive white Gaussian noise and intersymbol interference.* On the other hand, in the case of noncoherent reception (as, for example, DPSK), the detection process is inherently nonlinear. Accordingly, we find that under these conditions the performance analysis of noncoherent receivers is very difficult.†

---

* See Shimbo, Fang, and Celebiler (1973) and Prabhu (1973).

† For a rigorous method for calculating the error rate of M-ary DPSK in the combined presence of additive white Gaussian noise and intersymbol interference, see Prabhu and Salz (1981).

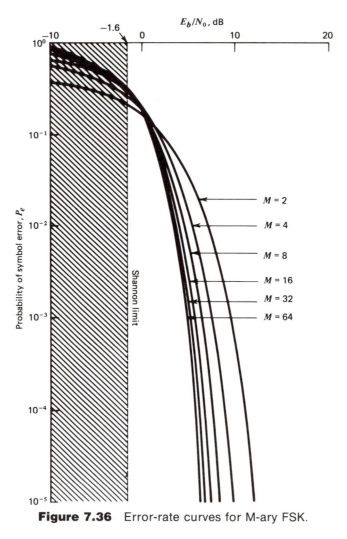

Figure 7.36 Error-rate curves for M-ary FSK.

(1) Digital Computer Simulation

When explicit performance analysis of a band-pass data transmission system defies a satisfactory solution, the use of *digital computer simulation* provides the only alternative approach to actual hardware evaluation. The speed and flexibility usually associated with digital computers are compelling reasons for adopting this approach.

In order to simulate a given band-pass data-transmission system, we first develop a baseband (low-pass) equivalent model for the system.* To do this we proceed as follows:

---

* For a discussion of the complex baseband representation of band-pass signals and systems, see Appendix C.

1. The transmitted signal $s(t)$ is represented by its *complex* (low-pass) *envelope* $\tilde{s}(t)$. These two signals are related by Eq. 7.132, which is reproduced here for completeness:

$$s(t) = \text{Re}[\tilde{s}(t)\exp(j2\pi f_c t)]$$

where f_c is the carrier frequency, and $\text{Re}[\cdot]$ denotes the real part of the quantity in question; it is assumed that f_c is larger than one-half the bandwidth of the transmitted band-pass signal $s(t)$. The complex envelope $\tilde{s}(t)$ equals

$$\tilde{s}(t) = s_I(t) + js_Q(t)$$

where $s_I(t)$ and $s_Q(t)$ are the in-phase and quadrature components of the transmitted signal $s(t)$, respectively. The characteristics of the input data waveform and the particular signaling method used to transmit the data are completely contained in the description of the complex envelope $\tilde{s}(t)$. Furthermore, by using the complex envelope $\tilde{s}(t)$, we eliminate the need for simulating the high-frequency carrier component.

2. To simulate the effect of noise at the front end of the receiver, we add to $s(t)$ the complex envelope

$$\tilde{w}(t) = w_I(t) + jw_Q(t)$$

where the in-phase component $w_I(t)$ and the quadrature component $w_Q(t)$ are modeled as sample functions of independent, zero-mean, white Gaussian noise processes, and with each one having a power spectral density of $N_0/2$ watts per hertz. Thus the received signal is represented by the complex envelope

$$\tilde{x}(t) = \tilde{s}(t) + \tilde{w}(t)$$
$$= [s_I(t) + w_I(t)] + j[s_Q(t) + w_Q(t)]$$

3. Correspondingly, the matched filter (or correlation) receiver and any band-limiting filter in the system are replaced by their respective complex equivalent low-pass versions. Here, we use the fact that the actual impulse response, $h(t)$, of a band-pass filter and the *complex (loss-pass) impulse response*, $\tilde{h}(t)$, of its baseband equivalent are related by

$$h(t) = 2\text{Re}[\tilde{h}(t)\exp(j2\pi f_c t)]$$

It is assumed that f_c is larger than one-half of the system bandwidth. Let $\tilde{h}(t)$ be expressed in the form

$$\tilde{h}(t) = h_I(t) + jh_Q(t)$$

where $h_I(t)$ is the in-phase component of the complex impulse response, and $h_Q(t)$ is the quadrature component.

4. The complex envelope of the band-pass system's output $\tilde{y}(t)$ is computed by convolving $\tilde{x}(t)$ with $\tilde{h}(t)$, as shown by

$$\tilde{y}(t) = \tilde{h}(t) \star \tilde{x}(t)$$

where $\star$ denotes convolution. Expressing $\tilde{y}(t)$ in terms of its in-phase component $y_I(t)$ and quadrature component $y_Q(t)$, we may also write

$$y_I(t) = h_I(t) \star x_I(t) - h_Q(t) \star x_Q(t) \tag{7.158}$$

and

$$y_Q(t) = h_Q(t) \star x_I(t) + h_I(t) \star x_Q(t) \tag{7.159}$$

The block diagram of Fig. 7.37 depicts the relationships described in Eqs. 7.158 and 7.159 between the in-phase and quadrature components of the input and output signals of the system.

When the block diagram of Fig. 7.37 is simulated on a digital computer, the various time functions are naturally represented in their sampled forms in accordance with the sampling theorem. Moreover, the *discrete Fourier transform (DFT)* is used to replace the time-domain operation of convolution by an equivalent operation that involves the multiplication of DFTs. In this context, it is the normal practice to use an efficient procedure known as the *fast Fourier transform (FFT) algorithm* for computing the DFT. For details of the DFT and the FFT algorithm, the reader is referred to Appendix A.

In carrying out the simulation, it is customary to assume that the symbols of the alphabet used in the particular system under study are equally likely, and that the symbols transmitted in adjacent time slots are statistically independent. One way of accomplishing this requirement is to use linear maximal-length or pseudo-noise sequences of sufficient length.* Suppose that in a particular simulation run a total of N such symbols are transmitted and, say, L of them are misinterpreted by the receiver. Then, with N assumed to be large enough, the average probability of error will (almost always) be approximately equal to

$$P_e \simeq \frac{L}{N}$$

and the approximation becomes better as N approaches infinity. This suggests that in order to measure (with some degree of confidence) an average probability of symbol error as low as 10^{-5}, for example, the number N will have to be at least as large as 10^7; that is, 100 times the reciprocal of the error rate. This results in a standard deviation or root mean-square measurement error of no more than 10 percent.

In the case of linear systems, we may avoid the use of such long simulation runs on a computer (which can be quite expensive) by applying an *indirect* procedure to evaluate the effects of the transmitted signal and noise separately. We illustrate the procedure by considering a quadrature signaling system. Using the baseband equivalent model of the receiver, we first compute the amplitude of the correlator output (or decision device input) in the in-phase channel

---

* Maximal-length sequences are pseudo-noise-like in their statistical characteristics; hence the alternative name "pseudo-noise." These sequences are considered in Chapters 8 and 9.

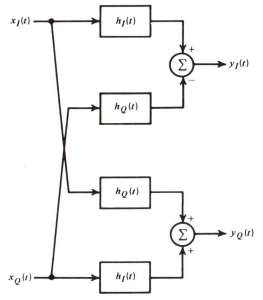

Figure 7.37 Block diagram illustrating the relationships between the in-phase and quadrature components of the response of a band-pass filter and those of the input signal.

of the receiver in response to the in-phase component, $s_I(t)$, of the transmitted signal. This computation is done for each transmitted symbol. Let $a_{I,n}$ denote the value of this amplitude for symbol m_n, say. The subscript I refers to the in-phase channel. Next, we compute the variance of the noise at the same point in the receiver, in response to the in-phase noise component, $w_I(t)$. Let σ_I^2 denote this variance. Thus, for symbol m_n we compute the conditional probability of error $P_{eI}(n)$ by using the formula

$$P_{eI}(n) = \frac{1}{2}\,\text{erfc}\left(\frac{\gamma_{i,n}}{2}\right) \qquad n = 1, 2, \ldots, N$$

where $\gamma_{i,n} = a_{I,n}^2/\sigma_I^2$, and N is the total number of symbols transmitted.

The foregoing computation is next repeated for the quadrature channel of the receiver. Let $P_{eQ}(n)$ denote this probability of error, conditional on the transmission of symbol m_n. The subscript Q refers to the quadrature channel. Since the in-phase and quadrature channels are independent, the corresponding value of the conditional probability of symbol error for the complete receiver equals

$$P_e(n) = P_{eI}(n) + P_{eQ}(n) - P_{eI}(n)P_{eQ}(n) \qquad n = 1, 2, \ldots, N$$

By averaging this result over the number of transmitted symbols, N, we get the average probability of symbol error for the complete receiver as

$$P_e \simeq \frac{1}{N}\sum_{n=1}^{N} P_e(n)$$

In the case of a binary signaling system, we have only an in-phase channel, so that $P_{eQ}(n)$ is zero and the average probability of symbol error reduces to

$$P_e \simeq \frac{1}{N} \sum_{n=1}^{N} P_{eI}(n)$$

By using the indirect procedure described, the length of a simulation run is reduced by several orders of magnitude compared to the direct procedure. The only requirement is that the pseudo-noise sequence used to generate the in-phase and quadrature components of the transmitted signal should have a length that is in excess of 100, say, and that all symbols of the alphabet for the system under study occur with approximately equal frequency.

7.11 BIT VERSUS SYMBOL ERROR PROBABILITIES

Thus far, the only figure of merit we have used to assess the noise performance of digital modulation schemes has been the average probability of symbol error. This figure of merit is the natural choice when messages of length $m = \log_2 M$ are transmitted, such as alphanumeric symbols or PCM code words. However, when the requirement is to transmit binary data, it is often more meaningful to use another figure of merit called the *probability of bit error* or *bit error rate* (BER). Although there are no unique relationships between these two figures of merit, it is fortunate that such relationships can be derived for two cases of practical interest, as shown here:

CASE 1

In the first case, the mapping from binary to M-ary symbols is performed in such a way that the two binary M-tuples corresponding to any pair of adjacent symbols in the M-ary modulation scheme differ in only one bit position. This mapping constraint is satisfied by using a *Gray code*.

Consider, for example, the Gray-encoded version of M-ary PSK, which is shown illustrated in Fig. 7.38 for $M = 4$ and $M = 8$. When the probability of symbol error P_e is acceptably small, we find that the probability of mistaking one symbol for either of the two "nearest" (in-phase) symbols is much greater

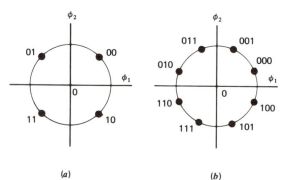

(a)   (b)

Figure 7.38 Signal constellations of M-ary PSK for (a) $M = 4$, and (b) $M = 8$ with the message points identified by 2-bit and 3-bit Gray codes, respectively.

than any other kind of symbol error. Moreover, given a symbol error, the most probable number of bit errors is one, subject to the aforementioned mapping constraint. Since there are $\log_2 M$ bits per symbol, it follows that the bit error rate is related to the probability of symbol error by the simple formula

$$\text{BER} = \frac{P_e}{\log_2 M} \qquad M \geqslant 2 \tag{7.160}$$

Note that the use of Eq. 7.160 for $M = 4$, in conjunction with Eq. 7.48 shows that QPSK has the same bit error rate as binary PSK (see Eq. 7.13).

For another example under Case 1, we may consider the Gray-encoded version of M-ary QAM. This form of modulation is illustrated in Fig. 7.24 for $M = 16$. The corresponding Gray-encoded versions of the in-phase and quadrature components are shown in Figs. 7.25a and 7.25b. It is evident from Fig. 7.25 that there is a change in only one bit position as we move from any symbol to an adjacent one horizontally or vertically. Here again, when the probability of symbol error is acceptably small, we find that the probability of mistaking a symbol for an adjacent one horizontally or vertically is much greater than any other kind of symbol error. Accordingly, as in M-ary PSK, the bit error rate for M-ary QAM is related to the probability of symbol error P_e by Eq. 7.160.

CASE 2

In the second case, we assume that all symbol errors are equally likely and occur with probability

$$\frac{P_e}{M - 1} = \frac{P_e}{2^K - 1}$$

where P_e is the average probability of symbol error, and $K = \log_2 M$. Since there are $\binom{K}{k}$ ways in which k bits out of k may be in error, it follows that the average number of bit errors per k-bit symbol is given by

$$\sum_{k=1}^{K} k\binom{K}{k} \frac{P_e}{2^K - 1} = \frac{2^{K-1}}{2^K - 1} K P_e \tag{7.161}$$

where

$$\binom{K}{k} = \frac{K!}{(K - k)! k!} \tag{7.162}$$

The bit error rate is obtained by dividing the result of Eq. 7.161 by K, as shown by

$$\text{BER} = \left(\frac{2^{K-1}}{2^K - 1}\right) P_e \tag{7.163}$$

or equivalently

$$\text{BER} = \left(\frac{\frac{M}{2}}{M - 1}\right) P_e \tag{7.164}$$

Note that for large M, the bit error rate approaches the limiting value of $P_e/2$.

The formula of Eq. 7.164, relating the bit error rate to the probability of symbol error, applies to M-ary FSK in which the M orthogonal symbols are equiprobable.

7.12 SYNCHRONIZATION

The coherent reception of a digitally modulated signal requires that the receiver be synchronous to the transmitter. We say that two sequences of events (representing a transmitter and a receiver) are *synchronous* relative to each other when the events in one sequence and the corresponding events in the other occur simultaneously. The process of making a situation synchronous, and maintaining it in this condition, is called *synchronization*.

From the discussion presented on the operation of digital modulation techniques, we recognize the need for two basic modes of synchronization:

1. When coherent detection is used, knowledge of both the frequency and phase of the carrier is necessary. The estimation of carrier phase and frequency is called *carrier recovery* or *carrier synchronization*.

2. To perform demodulation, the receiver has to know the instants of time at which the modulation can change its state. That is, it has to know the starting and finishing times of the individual symbols, so that it may determine when to sample and when to quench the product integrators. The estimation of these times is called *clock recovery* or *symbol synchronization*.

These two modes of synchronization can be coincident with each other, or they can occur sequentially one after the other. Naturally, in a noncoherent system, carrier synchronization is of no concern.

Both modes of the synchronization problem may be formulated in statistical terms as maximum-likelihood estimation problems.* In the sequel, we present a qualitative discussion of the synchronization problem, and describe circuits for carrier and clock recovery.

(1) Carrier Synchronization

The most straightforward method of carrier synchronization is to modulate the data-bearing signal onto a carrier in such a way that the power spectrum of the modulated signal contains a discrete component at the carrier frequency. Then a narrow-band *phase-locked loop* (PLL) can be used to track this component, thereby providing the desired reference signal at the receiver; a phase-locked loop consists of a voltage-controlled oscillator (VCO), a loop filter, and a multiplier that are connected together in the form of a negative feedback system. The disadvantage of such an approach is that since the residual component

---

* For a statistical analysis of the synchronization problems, and descriptions of carrier and clock recovery circuits, see Stiffler (1971), Lindsey (1972), and Lindsey and Simon (1973, Chapters 2 and 9).

does not convey any information other than the frequency and phase of the carrier, its transmission represents a waste of power.

Accordingly, modulation techniques that conserve power are always of interest in practice. In particular, the modulation employed is often of such a form that, in the absence of a dc component in the power spectrum of the data-bearing signal, the receiver requires the use of a *suppressed carrier-tracking loop* for providing a coherent secondary carrier (subcarrier) reference. For example, Fig. 7.39 shows the block diagram of a carrier-recovery circuit for M-ary PSK. This circuit is called the *Mth-power loop*. For the special case of $M = 2$, the circuit is called a *squaring loop*. However, when the squaring loop or its generalization is used for carrier recovery, we encounter a *phase ambiguity* problem. Consider, for example, the simple case of binary PSK. Since a squaring loop contains a squaring device at its input end, it is clear tht changing the sign of the input signal leaves the sign of the recovered carrier unaltered. In other words, the squaring loop exhibits a 180° phase ambiguity. Correspondingly, the generalization of the squaring loop for M-ary PSK exhibits M phase ambiguities in the interval $(0, 2\pi)$.

Another method for carrier recovery involves the use of a *Costas loop*. Figure 7.40 shows a Costas loop for binary PSK. The loop consists of two paths, one referred to as *in-phase* and the other referred to as *quadrature,* that are coupled together via a common voltage-controlled oscillator (VCO) to form a negative feedback system. When synchronization is attained, the demodulated data waveform appears at the output of the in-phase path, and the corre-

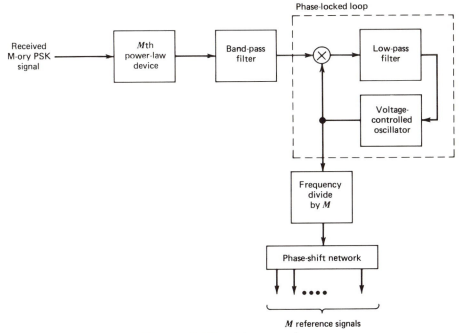

Figure 7.39 *Mth* power loop.

In-phase path

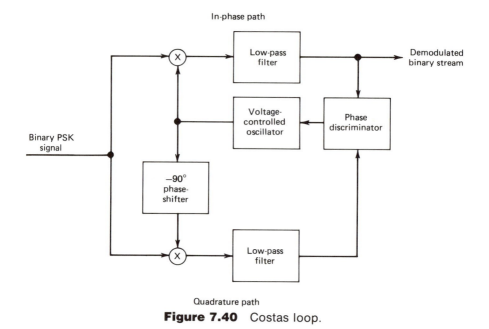

Quadrature path

Figure 7.40 Costas loop.

sponding output of the quadrature path is zero under ideal conditions. Analysis of the Costas loop shows that it too exhibits the same phase ambiguity problem as the squaring loop. Moreover, the Costas loop is equivalent to the squaring loop in terms of noise performance, provided that the two low-pass filters in the two paths of the Costas loop are the low-pass (baseband) equivalent of the band-pass filter in the squaring loop. The Costas loop may be generalized for M-ary PSK, in which case it also exhibits M phase ambiguities in the interval $(0, 2\pi)$. However, compared to the Mth order loop, the Mth order Costas loop has a practical disadvantage in that the amount of circuitry needed for its implementation becomes prohibitive for large M.

One method of resolving the phase ambiguity problem is to exploit differential encoding. Specifically, the incoming data sequence is differentially encoded before modulation at the transmitter, and differentially decoded after detection at the receiver, resulting in a small degradation in noise performance. This method is called the *coherent detection of differentially encoded M-ary PSK*.* As such, it is different from M-ary DPSK. For the special case of coherent detection of differentially encoded PSK, the average probability of symbol error is given by

$$P_e = \text{erfc}\left(\sqrt{\frac{E_b}{N_0}}\right) - \frac{1}{2}\,\text{erfc}^2\left(\sqrt{\frac{E_b}{N_0}}\right) \qquad (7.165)$$

In the region where $(E_b/N_0) \gg 1$, the second term on the right side of Eq. 7.165

---

* For the noise analysis of coherent detection of differentially encoded M-ary PSK, see Lindsey and Simon (1973, pp. 242–246).

has a negligible effect; hence, this modulation scheme has an average probability of symbol error practically the same as that for coherent QPSK or MSK. (The formula of 7.165 is included in the summary presented in Table 7.4.) For the coherent detection of differentially encoded QPSK, the average probability of symbol error is given by

$$P_e = 2\,\text{erfc}\left(\sqrt{\frac{E_b}{N_0}}\right) - 2\,\text{erfc}^2\left(\sqrt{\frac{E_b}{N_0}}\right) + \text{erfc}^3\left(\sqrt{\frac{E_b}{N_0}}\right) - \frac{1}{4}\,\text{erfc}^4\left(\sqrt{\frac{E_b}{N_0}}\right) \quad (7.166)$$

For large E_b/N_0, this average probability of symbol error is approximately twice that of coherent QPSK.

(2)   Symbol Synchronization

As mentioned previously, clock recovery (i.e., symbol synchronization) can be processed alongside carrier recovery. Alternatively, clock recovery is accomplished first, followed by carrier recovery; sometimes, the reverse procedure is followed. The choice of a particular approach or the other is determined by the application of interest.

In one approach, the symbol synchronization problem is solved by transmitting a clock along with the data-bearing signal, in multiplexed form. Then, at the receiver, the clock is extracted by appropriate filtering of the modulated waveforms. Such an approach minimizes the time required for carrier/clock recovery. However, a disadvantage of the approach is that a fraction of the transmitted power is allocated to the transmission of the clock.

In another approach, a good method is, first, to use a noncoherent detector to extract the clock. Here, use is made of the fact that clock timing is usually much more stable than carrier phase. Then, the carrier is recovered by processing the noncoherent detector output in each clocked interval.

In yet another approach, when clock recovery follows carrier recovery, the clock is extracted by processing demodulated (not necessarily detected) baseband waveforms, thereby avoiding any wastage of transmitted power. In the sequel, we use heuristic arguments to develop a symbol synchronizer that satisfies this requirement.

Consider first a rectangular pulse defined by (see Fig. 7.41a)

$$g(t) = \begin{cases} a & 0 \le t \le T \\ 0 & \text{otherwise} \end{cases}$$

The output of a filter matched to the pulse $g(t)$ is shown in Fig. 7.41b. We observe that the matched filter output attains its peak value at time $t = T$, and that it is symmetric about this point. Clearly, the proper time to sample the matched filter output is at $t = T$. Suppose, however, that the matched filter output is sampled *early* at $t = T - \Delta_0 T$ or *late* at $t = T + \Delta_0 T$. Then, the absolute values of the two samples so obtained will (on the average in the presence of additive noise) be equal, and smaller than the peak value at $t = T$. Moreover, the *error signal*, the difference between the absolute values of the two samples, is zero; and the proper sampling time is the midpoint between $t = T - \Delta_0 T$ and

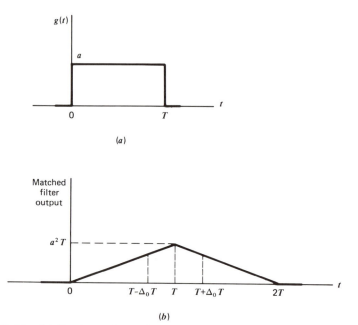

Figure 7.41 (a) Rectangular pulse, $g(t)$. (b) Output of filter matched to $g(t)$.

$t = T + \Delta_0 T$. This special condition may be viewed as an *equilibrium point*, in that if we deviate from it the error signal becomes nonzero.

Figure 7.42 shows the block diagram of a symbol synchronizer, which exploits this notion. For obvious reasons, it is called an *early–late gate symbol synchronizer of the absolute value type*. In Fig. 7.42 correlators are used in place of equivalent matched filters. Both correlators integrate over a full symbol interval T, with one starting $\Delta_0 T$ early relative to the transition time estimate and the other starting $\Delta_0 T$ late. An error signal, $e(kT)$, is generated by taking the difference between the absolute values of the two correlator outputs. For a given offset λT between the actual transition times $\{t(kT)\}$ and their local estimates $\{\hat{t}(kT)\}$, the error signal is zero when no transition occurs at $t(kT)$; otherwise, it is linearly proportional to λ, irrespective of the polarity of λ. This polarity independence is the result of taking absolute values of the correlator outputs before evaluating the difference between them. The error signal is low-pass filtered and then applied to a voltage-controlled oscillator that controls (through a symbol waveform generator) the charging and discharging instants of the correlators. The closed loop is designed to be narrowband relative to the symbol rate $1/T$. The instantaneous frequency of the local clock is advanced or retarded in an iterative manner until the equilibrium point is reached, and symbol synchronization is thereby established.

7.13 APPLICATIONS

In this section of the chapter, we describe applications of digital modulation techniques in (1) modems for data transmission over voice-grade telephone

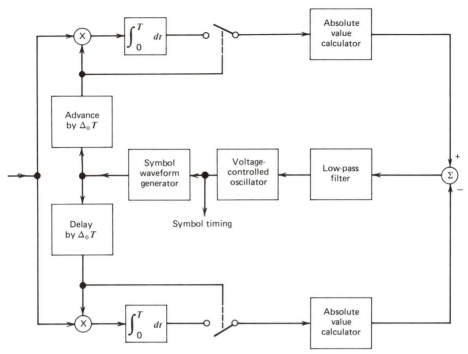

Figure 7.42 Early–late gate type of symbol synchronizer.

channels, (2) digital radio, and (3) digital communications by satellite. These three applications are considered in turn.

(1) Voice-grade Modems

The telephone network developed for voice communications is designed to pass frequencies in the range 300–3400 Hz. The widely spread composition of the network makes it an effective medium for digital communications. However, to transmit data over a telephone channel that is analog in nature, it is necessary to use a data transmitter to *mo*dulate a voice-frequency carrier and a data receiver to *dem*odulate this signal. A data transceiver is therefore commonly referred to as a *modem*.

The type of modulation selected for use in a modem depends on the application of interest.* When simplicity and economy are more important than bandwidth efficiency, binary FSK with noncoherent detection is the usual choice. Typically, the frequency shift in hertz is set at one-half to three-quarters of the maximum bit rate, and the bandwidth in hertz is approximately twice the bit rate. For example, in a binary FSK modem operating at 1200 b/s, the two frequencies are commonly selected as 1300 to 2100 Hz.

---

* Voice-band modems are discussed in Williams (1987, pp. 367–369), Pickholtz (1985, Chapter 3), Smith (1985, pp. 290–300), and Davey (1972).

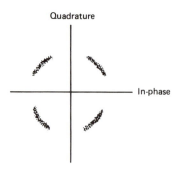

Quadrature

In-phase

Figure 7.43 Illustrating the effect of phase jitter on the signal constellation of a QPSK system.

However, FSK modems are inefficient in their use of bandwidth, with the result that their operation over voice-grade telephone channels is limited to 1800 b/s. Typically, at rates above 1200 b/s, other modulation schemes are used. In particular, four-phase DPSK (with carrier at 1300 Hz) has established itself as an international standard for modems operating at 2400 b/s. Another international modem standard is an eight-phase DPSK modem (with carrier at 1800 Hz) operating at 4800 b/s.

A shortcoming of M-ary PSK and DPSK schemes is their susceptibility to *phase jitter* experienced in telephone channels. Figure 7.43 illustrates the effect of phase jitter on the received signal constellations of QPSK. Without such an impairment, there would only be four possible points in the signal constellation, as in Fig. 7.6. However, over a long stream of received bits, the presence of phase jitter results in the *scatter diagram* depicted in Fig. 7.43. Accordingly, phase jitter imposes a practical limit on the number of phases that can be employed or, equivalently, the number of signal points that can be placed around a circle (corresponding to a prescribed signal energy per bit). Indeed, it is because of phase jitter that the use of DPSK is limited to 4800 b/s.

Given that we cannot expand the circle owing to power constraints, a solution to the phase jitter problem is to use M-ary QAM. The signal points are thereby placed on concentric circles, such that any pair of the signal points is phase-separated enough to withstand the effect of phase jitter. Moreover, to guard against other impairments such as noise and intersymbol interference, the signal points are also separated sufficiently from one another. Figure 7.44 shows the signal constellation of a typical 16-QAM modem. This modem permits the transmission of four bits at a time at a modulation (signaling) rate of 2400 bauds (i.e., a bit rate of 9600 b/s). However, the modem requires accurate carrier and clock recovery circuits. The modem is also vulnerable to channel imperfections, especially amplitude and group-delay distortions. To compensate for these distortions, the use of adaptive equalization is required (the subject of adaptive equalization was discussed in Section 6.8).

(2) Digital Radio

In *digital radio*, information originating from a source is transmitted to its destination by means of digital modulation techniques over an appropriate

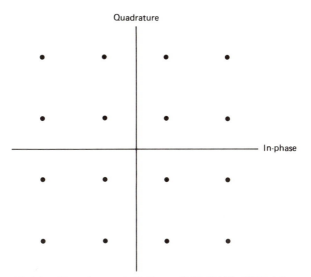

Figure 7.44 Signal constellation of 16-QAM, 9600 b/s modem.

number of microwave radio links, with each link offering a *line-of-sight* propagation path. Such a link is illustrated in Fig. 7.45. Antennas are placed on towers whose heights are large enough to provide a direct, unobstructed path between the transmitter and the receiver.

In radio systems, rigid transmission bandwidths are normally specified so as to establish well-defined, noninterleaving channels. Efficient utilization of these channels usually requires the use of multilevel signaling techniques, which makes it possible to achieve high bit rates despite bandwidth restrictions. This requirement arises because digital radios have to provide essentially the same bandwidth efficiency as analog FM radios that they are in many instances replacing. Consider, for example, the 4-GHz band for which the allocated bandwidth is typically 20 MHz. An analog microwave radio, based on frequency modulation of frequency-division multiplexed (FDM) single-sideband signals, carries 1500 voice channels over this bandwidth. In a digital radio, on the other hand, 64 kb/s PCM is frequently used for each voice channel. Accord-

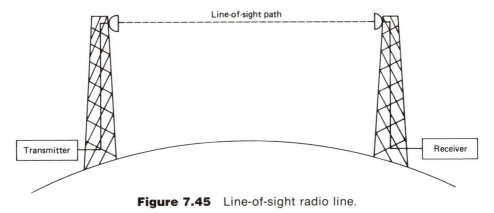

Figure 7.45 Line-of-sight radio line.

ingly, the digital radio must achieve a bandwidth efficiency of 4.8 b/s/Hz if it is to accommodate 1500 voice channels as does its analog counterpart. Such a bandwidth efficiency can be realized only by the use of high-level modulation formats.

From the inception of digital radio, linear modulation schemes have been employed almost exclusively. In particular, the use of M-ary QAM has received much attention. At present, there are 64-QAM radio systems in the field, and prototype 256-QAM systems have been demonstrated.* However, as the number of amplitude levels increases, so does the sensitivity of the modulated wave to equipment imperfections and propagation impairments. These issues are briefly considered in the sequel.

First of all, the successful application of high-level modulation schemes requires the use of highly accurate carrier and clock recovery circuits. For example, a 64-QAM radio suffers approximately 1 dB degradation in error performance for an rms carrier phase error of 1°. In addition, high-level modulation schemes are sensitive to linear distortions (i.e., amplitude and group-delay distortions) resulting from the use of filters in the transmission path. It is therefore of critical importance that all filters in both the transmitter and the receiver be equalized, so that their amplitude responses are essentially symmetric and their group-delay responses are essentially flat inside the bandwidth occupied by the modulated signal. Moreover, high-level modulation schemes are very sensitive to nonlinear distortion, and this sensitivity increases with the number of amplitude levels used. The primary source of nonlinear distortion in a digital radio is the power amplifier at the transmitter output. It is therefore necessary to operate the power amplifier at a large *backoff* from saturation or else provide some means of linearizing the amplifier.

Another issue that has to be considered in the design of digital radio systems is that of propagation impairments. Most of the time, a line-of-sight microwave radio link behaves as a wide-band, low-noise channel capable of providing highly reliable, high-speed data transmission. The channel, however, suffers from anomalous propagation conditions that arise from natural phenomena, which can cause the error performance of a digital radio to be severely degraded. Such anomalies manifest themselves by causing the transmitted signal to propagate along several paths, each of different electrical length. This phenomenon is called *multipath fading*.† Figure 7.46 illustrates the geometry of multipath fading. Three rays are included in the figure: a *direct* ray, a *refracted* ray associated with the atmosphere, and a *reflected* ray associated with the

---

* For an overview of developments in digital radio and a discussion of issues relating to equipment imperfections and propagation impairments, see Taylor and Hartman (1986). For a discussion and evaluation of various modulation formats used in digital radio, see Bellamy (1982, pp. 272–320), and Chamberlain et al. (1987). The last reference describes the technical requirements and design of a 256-QAM modem.

† For a tutorial review of statistical modeling of the multipath phenomenon, see Rummler, Coutts, and Liniger (1986). For a statistical analysis of a three-ray model for multipath fading, see Shafi (1987).

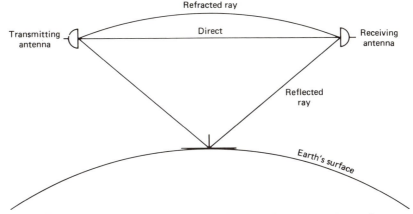

Figure 7.46   Illustrating direct and indirect paths in a radio environment.

terrain. Figure 7.47 shows a plot of the amplitude response of the channel. When the direct ray is out of phase with the indirect rays, deep notches appear in the amplitude response. The time delay τ between the direct and indirect rays determines the frequency separation between the notches, and their relative amplitudes determine the notch depth. In digital radio, this form of distortion leads to intersymbol interference, which in turn causes the error performance of the system to be severely degraded. The degradation is directly proportional to the transmitted bit rate, since increasing the bit rate corresponds to a reduction in bit duration and therefore greater vulnerability to intersymbol interference. To combat the deleterious effects of multipath fading in digital radio, diversity techniques are widely used either by themselves or in combination with adaptive equalizers.

In digital radio, diversity techniques are frequency-, space-, or angle-based. Each of these techniques may be used to provide a second version of the transmitted signal for processing at the receiver. *Frequency diversity* is implemented by switching operation of the system to an alternate channel when multipath fading causes outage in a working channel. In *space diversity*, two

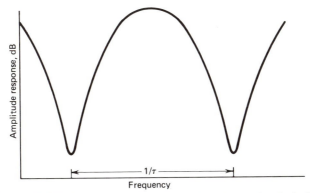

Figure 7.47   Frequency response due to multipath fading.

vertically spaced antennas are placed on a radio tower, with the separation between them chosen large enough to ensure independent multipath fadings. The receiver combines the two signals adaptively or else switches between them. *Angle diversity* employs two receiving antennas mounted side by side with different angles of elevation, or dual-beam antennas with angular offset between the two beams. The offset is designed to provide a difference in the amplitudes of received rays. It appears that the use of angle diversity offers equivalent or better performance than space diversity with lower implementation costs.*

(3) Digital Communications by Satellite

The use of *multiple access* provides a method for exploiting the broadcast capability of a satellite channel. A particular type of this method, known as *time-division multiple access* (TDMA), is well suited for digital communications.† In TDMA, a number of ground stations are enabled to access a satellite by having their individual transmissions reach the satellite in nonoverlapping time slots. The traveling-wave tube, constituting the RF power amplifier at the output of the satellite transponder, is thereby permitted to operate at or near saturation. Such a feature, which is essentially unique to TDMA, helps to maximize the downlink carrier-to-noise ratio. Moreover, since only one modulated carrier is present in the nonlinear transponder at any one time, the generation of intermodulation products is avoided.

Figure 7.48 illustrates basics of a TDMA network, in which transmissions are organized into *frames*. A frame contains N *bursts*. To compensate for variations in satellite range, a *guard time* is inserted between successive bursts to protect the system against overlap. One burst per frame is used as a *reference*. The remaining $N - 1$ bursts are allocated to ground stations on the basis of one burst per station. Thus, each station transmits once per frame. Typically, a burst consists of an initial portion called the *preamble*, which is followed by a *message* portion; in some systems a *postamble* is also included. The preamble consists of a part for carrier recovery, a part for symbol-timing recovery, a unique word for burst synchronization, a station identification code, and some housekeeping symbols. Two functionally different components may therefore be identified in each frame: revenue-producing component represented by message portions of the bursts, and system overhead represented by guard times, the reference burst, preambles, and postambles (if included).

Two important points emerge from this brief discussion of the TDMA network:

---

* The use of angle diversity on line-of-sight microwave links is of recent origin; for details, see Lin, Giger, and Alley (1987), and Balaban, Sweedyk, and Axeling (1987).

† There are two other types of multiple access, namely, *frequency-division multiple access* (FDMA) and *code-division multiple access* (CDMA). The former is used for analog communications. The latter is discussed in Chapter 10. For discussions of the TDMA network and system timing, see Spilker (1977, pp. 265–294), Bhargava et al. (1981, pp. 233–244), and Pratt and Bostian (1986, pp. 235–251).

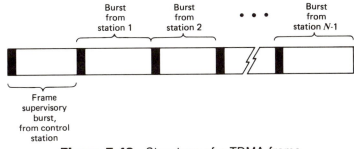

Figure 7.48   Structure of a TDMA frame.

1. Power efficiency in a satellite transponder is maximized by permitting the traveling-wave tube to operate at or near saturation.
2. The transmissions contain independent provisions for carrier and clock recovery to occur simultaneously, thereby keeping overhead due to recovery time to a minimum.

Accordingly, only a limited set of digital modulation techniques is desirable for satellite communications. In particular, point 1 constrains the modulation format to have a constant envelope, thereby excluding ASK and hybrid modulation techniques that involve ASK. Point 2 makes it feasible to employ coherent detection. We therefore find that in digital communications by satellite, primary interest is in the use of coherent binary PSK, coherent M-ary PSK, and coherent MSK. The modulation format in active use is QPSK because of its bandwidth-saving capability and manageable system complexity.

There is one other aspect of a TDMA network that deserves to be considered. At an earth-station terminal, parallel voice channels are PCM-encoded at a clock rate synchronous with the TDMA frame rate. In addition, there are nonsynchronous parallel input streams, for which asynchronous multiplexers are provided. In any case, the incoming bit streams that require transmission are continuous in time. On the other hand, the modulator output consists of a periodic burst of RF energy. Hence, the ground-station terminal must contain a data *buffer* that stores the data bits from one frame to the next. The total storage required is determined by the number of input bit streams, their bit rates, and the frame rate.

7.14   SUMMARY AND DISCUSSION

In this chapter, we covered various *modulation techniques* and related issues for the transmission of digital data over band-pass channels. Much of the mathematical analysis presented herein has been concerned with the performance evaluation of these techniques, with emphasis on the calculation of the *average probability of symbol error* committed by the receiver in the presence of additive noise. To carry out the analysis, we used *maximum likelihood detection* interpreted in terms of signal constellations and based on the following assumptions:

1. The signal waveforms, representing the source symbols, are transmitted with *equal probability*.
2. The transmitter is *average-power limited*.
3. The received signal consists of the transmitted signal and *additive white Gaussian noise* of zero mean and power spectral density $N_0/2$.
4. *For coherent detection*, the receiver is synchronized in time and phase-locked to the transmitter. Synchronization in time (i.e., *symbol synchronization*) ensures proper timing of the matched filters or correlators in relation to the incoming stream of information-bearing symbols. Phase locking (i.e., *carrier synchronization*) ensures the provision of replicas of the signal waveforms required for correlation. For *noncoherent detection*, there is no need for phase locking.

Based on these assumptions, we presented a detailed analysis of two classes of digital modulation techniques, namely, *phase-shift keying* (PSK) and *frequency-shift keying* (FSK). Coherent and noncoherent as well as binary and M-ary versions of these techniques were considered. *Quadriphase-shift keying* (QPSK), which is a special form of M-ary PSK, and *minimum-shift keying* (MSK), which is a special form of continuous phase frequency-shift keying, were given detailed attention. We also analyzed a special form of hybrid amplitude-phase keying known as *M-ary quadrature amplitude modulation* (QAM). The analysis of the coherent and noncoherent versions of *amplitude shift keying* (ASK) was presented as problems at the end of the chapter. For noncoherent ASK, it is assumed that the signal-to-noise ratio is high so as to simplify mathematical analysis of the problem.

In Section 7.5, we presented a summary of characteristics of binary and quaternary versions of PSK and FSK. Below, we highlight some additional characteristics of digital modulation schemes, with emphasis on M-ary schemes:

(a) Coherent versus Noncoherent
Of the three basic digital modulation techniques (i.e., ASK, PSK, and FSK), the performance of PSK is the most sensitive to lack of coherence. In particular, the lack of coherence in M-ary PSK results in a 3-dB degradation in signal-to-noise ratio for large M. The degradation is reduced for small M. Indeed, for $M = 2$, the degradation approaches zero when the signal-to-noise ratio is high.

On the other hand, ASK and FSK do not suffer as much degradation in performance due to incoherence. Specifically, when the average probability of symbol error is low enough to be of interest (say, $P_e \leq 10^{-5}$), the degradation is on the order of a decibel or less.

(b) Bandwidth and Power Considerations
M-ary FSK utilizes channel bandwidth inefficiently, since only a fraction of the bandwidth is occupied by the transmission of one of the M orthogonal tones during any symbol period. Specifically, beyond $M = 4$, the bandwidth efficiency of M-ary FSK decreases steadily with increasing M. Consequently, the

use of M-ary FSK is limited to situations where bandwidth conservation is not of primary concern. In fact, when the bandwidth is unconstrained, coherent M-ary FSK (in the limiting case as M approaches infinity) requires minimum power for error-free transmission over an additive white Gaussian noise channel, provided that (in accordance with the Shannon limit) the bit rate R_b is less than $(P/N_0)\log_2 e$, where P is the average power of the transmitted signal and $N_0/2$ is the noise power spectral density.

On the other hand, the M-ary versions of PSK and ASK provide a more efficient utilization of channel bandwidth than their FSK counterparts. Indeed, for these modulation schemes, the bandwidth efficiency increases monotonically with increasing number of levels M.

The channel utilization may be improved further by the hybrid use of amplitude and phase modulation, for a prescribed probability of error P_e. A popular form of this hybrid scheme is *M-ary QAM*. Unlike M-ary PSK, the in-phase and quadrature components of M-ary QAM are permitted to be independent. The application of M-ary QAM requires the channel to be linear. By contrast, M-ary PSK is resistant to nonlinearities in the channel, since the envelope of the transmitted signal is constrained to remain constant.

PROBLEMS

P7.2   COHERENT BINARY MODULATION TECHNIQUES

Problem 7.2.1   Binary data are transmitted over a microwave link at the rate of 10^6 bits per second and the power spectral density of the noise at the receiver input is 10^{-10} watts per hertz. Find the average carrier power required to maintain an average probability of error $P_e \leqslant 10^{-4}$ for coherent binary FSK. What is the required channel bandwidth?

Problem 7.2.2   The signal component of a coherent PSK system is defined by

$$s(t) = A_c k \sin(2\pi f_c t) \pm A_c \sqrt{1 - k^2} \cos(2\pi f_c t)$$

where $0 \leqslant t \leqslant T_b$, and the plus sign corresponds to symbol 1 and the minus sign to symbol 0. The first term represents a carrier component included for the purpose of synchronizing the receiver to the transmitter.

(a)   Show that, in the presence of additive white Gaussian noise of zero mean and power spectral density $N_0/2$, the average probability of error is

$$P_e = \frac{1}{2} \operatorname{erfc}\left(\sqrt{\frac{E_b}{N_0} (1 - k^2)} \right)$$

where

$$E_b = \frac{1}{2} A_c^2 T_b$$

(b)   Suppose that 10 percent of the transmitted signal power is allocated to the carrier component. Determine the E_b/N_0 required to realize a probability of error equal to 10^{-4}.

(c)   Compare this value of E_b/N_0 with that required for a conventional binary PSK system with the same probability of error.

Problem 7.2.3 A binary PSK signal is applied to a correlator supplied with a phase reference that differs from the exact carrier phase by ϕ radians. Determine the effect of the phase error ϕ on the average probability of error of the system.

Problem 7.2.4 Consider a phase-locked loop consisting of a multiplier, loop filter, and voltage-controlled oscillator (VCO). Let the signal applied to the multiplier input be a PSK signal defined by

$$s(t) = A_c \cos[2\pi f_c t + k_p m(t)]$$

where k_p is the phase sensitivity, and the data signal $m(t)$ takes on the value $+1$ volt for binary symbol 1 and -1 volt for binary symbol 0. The VCO output is

$$r(t) = A_c \sin[2\pi f_c t + \theta(t)]$$

(a) Evaluate the loop filter output, assuming that this filter removes only modulated components with carrier frequency $2f_c$.
(b) Show that this output is proportional to the data signal $m(t)$ when the loop is phase-locked, that is, $\theta(t) = 0$.

Problem 7.2.5 The signal vectors s_1 and s_2 are used to represent binary symbols 1 and 0, respectively, in a coherent binary FSK system. The receiver decides in favor of symbol 1 when

$$(\mathbf{x}, s_1) > (\mathbf{x}, s_2)$$

where $(\mathbf{x}, s_1)$ is the inner product of the observation vector $\mathbf{x}$ and the signal vector s_1, and similarly for $(\mathbf{x}, s_2)$. Show that this decision rule is equivalent to the condition $x_1 > x_2$, where x_1 and x_2 are the elements of the observation vector $\mathbf{x}$ (see Eqs. 7.20 and 7.21). Assume that the signal vectors s_1 and s_2 have equal energy.

Problem 7.2.6 An FSK system transmits binary data at the rate of 2.5×10^6 bits per second. During the course of transmission, white Gaussian noise of zero mean and power spectral density 10^{-20} watts per hertz is added to the signal. In the absence of noise, the amplitude of the received sinusoidal wave for digit 1 or 0 is 1 microvolt. Determine the average probability of symbol error, assuming coherent detection.

Problem 7.2.7

(a) In a coherent FSK system, the signals $s_1(t)$ and $s_2(t)$ representing symbols 1 and 0, respectively, are defined by

$$s_1(t), \; s_2(t) = A_c \cos\left[2\pi \left(f_c \pm \frac{\Delta f}{2}\right)t \right] \qquad 0 \leq t \leq T_b$$

Assuming that $f_c > \Delta f$, show that the correlation coefficient of the signals $s_1(t)$ and $s_2(t)$ is approximately given by

$$\frac{\int_0^{T_b} s_1(t)s_2(t)dt}{\int_0^{T_b} s_1^2(t)dt} \simeq \text{sinc}(2 \, \Delta f T_b)$$

(b) What is the minimum value of frequency shift Δf for which the signals $s_1(t)$ and $s_2(t)$ are orthogonal?
(c) What is the value of Δf that minimizes the average probability of symbol error?
(d) For the value of Δf obtained in part (c), determine the increase in E_b/N_0 required so that this coherent FSK system has the same noise performance as a coherent binary PSK system.
(e) Compare the result in part (d) with Fig. 7.20 at $P_e = 10^{-4}$, and discuss.

Problem 7.2.8 Set up a block diagram for the generation of a binary FSK signal $s(t)$ with continuous phase by using the representation

$$s(t) = \sqrt{\frac{2E_b}{T_b}} \cos\left(\frac{\pi t}{T_b}\right) \cos(2\pi f_c t) \mp \sqrt{\frac{2E_b}{T_b}} \sin\left(\frac{\pi t}{T_b}\right) \sin(2\pi f_c t)$$

Problem 7.2.9 In the on–off keying version of an ASK system, symbol 1 is represented by transmitting a sinusoidal carrier of amplitude $\sqrt{2E_b/T_b}$, where E_b is the bit energy and T_b is the bit duration. Symbol 0 is represented by switching off the carrier. Assume that symbols 1 and 0 occur with equal probability. Show that for the coherent detection of this ASK signal in an AWGN channel, the probability of error is given by

$$P_e = \frac{1}{2} \operatorname{erfc}\left(\sqrt{\frac{E_{av}}{2N_0}}\right)$$

where $N_0/2$ is the noise spectral density and E_{av} is the average signal energy.

Problem 7.2.10 Find the union bound for (a) coherent binary PSK, and (b) coherent binary FSK. Compare your results with the actual values of probability of error for these two schemes. (The union bound is discussed in Section 3.6.)

P7.3 COHERENT QUADRATURE-CARRIER MODULATION TECHNIQUES

Problem 7.3.1 Consider the QPSK signal described by the signal constellation shown in Fig. P7.1.

(a) Sketch the waveforms for the in-phase and quadrature components of this QPSK signal produced by the input binary sequence 1100100010.
(b) Sketch the waveform of the QPSK signal for the binary sequence in part (a), assuming that the carrier frequency is an integral multiple of the symbol rate $1/T$.

Problem 7.3.2

(a) Sketch the waveforms of the in-phase and quadrature components of the MSK signal in response to the input binary sequence 1100100010.
(b) Sketch the MSK waveform itself for the binary sequence specified in part (a).

Problem 7.3.3 In MSK, the initial state $\theta(0)$ may equal 0 or π radians. Demonstrate the following properties of MSK:

(a) The transmission of dibit 10 or 01 leaves the state of MSK unchanged.
(b) The transmission of dibit 00 or 11 changes the state of MSK from a phase value of 0 to π radians, or vice versa.

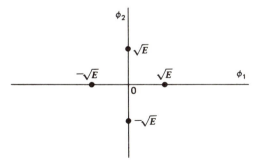

Figure P7.1

Problem 7.3.4 Show that the basis functions $\phi_1(t)$ and $\phi_2(t)$, defined in Eqs. 7.59 and 7.60 for the MSK scheme, are orthogonal to each other. Also, show that they have unit energy.

Problem 7.3.5 Establish the validity of Eq. 7.58 for the MSK scheme.

Problem 7.3.6 Repeat the calculations of Problem 7.2.1 by using the coherent MSK scheme.

Problem 7.3.7 In a special form of quadriphase-shift keying known as the *offset* QPSK, the in-phase data stream is delayed relative to the quadrature data stream by half a symbol period $T/2$.

(a) What is the average probability of symbol error for an offset QPSK system?
(b) What is the power spectral density of an offset QPSK signal produced by a random binary sequence in which symbols 1 and 0 (represented by ± 1 volt) are equally likely, and the symbols in different time slots are statistically independent and identically distributed?
(c) What are the similarities between offset QPSK and MSK, and what features distinguish them?

Problem 7.3.8 The values of E_b/N_0 required to realize an average probability of symbol error $P_e = 10^{-4}$ using coherent binary PSK and coherent FSK (conventional) systems are equal to 7.2 and 13.5, respectively. Using the approximation (see Appendix E)

$$\text{erfc}(u) \simeq \frac{1}{\sqrt{\pi u}} \exp(-u^2),$$

determine the separation in the values of E_b/N_0 for $P_e = 10^{-4}$ for the following pairs of modulation formats:

(a) Coherent binary PSK and QPSK.
(b) Coherent binary (conventional) FSK and MSK.

Problem 7.3.9 Find the union bound for coherent QPSK, and compare your result with the actual value of the probability of symbol error. (The union bound is discussed in Section 3.6.)

P7.4 NONCOHERENT BINARY MODULATION TECHNIQUES

Problem 7.4.1 Repeat the calculations of Problem 7.2.1 for noncoherent binary FSK.

Problem 7.4.2 There are two ways of detecting an MSK signal. One way is to use a coherent receiver to take full account of the phase information content of the MSK signal. Another way is to use a noncoherent receiver and disregard the phase information. The second method offers the advantage of simplicity of implementation at the expense of a degraded noise performance. By how many decibels do we have to increase the bit energy-to-noise density ratio, E_b/N_0, in the second case so as to realize an average probability of symbol error equal to 10^{-5} in both cases?

Problem 7.4.3 Repeat the calculations of Problem 7.3.8 for the following pairs of modulation formats:

(a) Coherent binary (conventional) FSK and noncoherent binary FSK.
(b) Coherent binary PSK and DPSK.

Problem 7.4.4 Repeat the calculations of Problem 7.2.1 for DPSK.

Problem 7.4.5 The binary sequence 1100100010 is applied to the DPSK transmitter of Fig. 7.19a.
(a) Sketch the resulting waveform at the transmitter output.
(b) Applying this waveform to the DPSK receiver of Fig. 7.19b, show that, in the absence of noise, the original binary sequence is reconstructed at the receiver output.

Problem 7.4.6 In the on–off keying version of an ASK system, symbol 1 is represented by transmitting a sinusoidal carrier of amplitude $\sqrt{2E_b/T_b}$, where E_b is the bit energy and T_b is the bit duration. Symbol 0 is represented by switching off the carrier. Assume that symbols 1 and 0 occur with equal probability.

For an AWGN channel, determine the average probability of error P_e for noncoherent detection of this ASK signal. You may assume a large value of bit energy-to-noise density ratio. Let E_{av} be the average signal energy. Hence, show that

$$P_e \simeq \frac{1}{2} \exp\left(-\frac{E_{av}}{2N_0}\right)$$

NOTE When x is large, the modified Bessel function of the first kind of zero order may be approximated as follows

$$I_0(x) \simeq \frac{\exp(x)}{\sqrt{2\pi x}}$$

P7.6 M-ARY MODULATION TECHNIQUES

Problem 7.6.1 In an M-ary scheme, the duration of each signal is fixed independently of M, the number of signals in the set. Using qualitative arguments based on signal constellations, justify the validity of the following statements:

1. Increasing M tends to increase the average probability of symbol error, whereas increasing the energy content of each transmitted signal reduces the average probability of symbol error.
2. Increasing M produces the most degradation in the ASK performance, somewhat less degradation in PSK, and comparatively little improvement in FSK.

Problem 7.6.2 Consider an M-ary signaling scheme that uses a fixed signaling rate $1/T$, independently of the number of levels M. Show that

$$10 \log_{10}\left(\frac{E}{N_0}\right) = 10 \log_{10}\left(\frac{E_b}{N_0}\right) + 10 \log_{10}(\log_2 M)$$

where E_b is the signal energy per bit for $M = 2$, and E is the signal energy per symbol for any M. As usual, $N_0/2$ is the noise power spectral density.

Problem 7.6.3 Consider the signal constellations for quadriphase-shift keying (QPSK) and octaphase-shift keying (OPSK) shown in Fig. P7.2. Both constellations have the same radius.

(a) Show that for high signal-to-noise ratio, the asymptotic difference in E_b/N_0 between these two signal constellations may be expressed as $10 \log_{10}(3d_0^2/2d_4^2)$, where the distances d_0 and d_4 are shown in Fig. P7.2.
(b) Show that this asymptotic difference represents a loss of 3.57 dB.

Problem 7.6.4 A 16-level QAM system transmits information at a rate equal to that

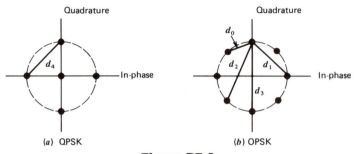

Figure P7.2

achievable by a combination of two QPSK systems and an eight-level PSK system. Identify the signal constellations for these three PSK systems.

Problem 7.6.5 Using the union bounding technique, show that the upper bound on the average probability of symbol error for M-ary FSK is defined by

$$P_e \leq \frac{1}{2}(M-1)\,\text{erfc}\left(\sqrt{\frac{E}{2N_0}}\right)$$

where M is the number of levels, and E/N_0 is the symbol energy-to-noise power spectral density ratio. For details of the union bound, see Section 3.6.

P7.7 POWER SPECTRA

Problem 7.7.1 A binary FSK signal with discontinuous phase is defined by

$$s(t) = \begin{cases} \sqrt{\dfrac{2E_b}{T_b}}\cos\left[2\pi\left(f_c + \dfrac{\Delta f}{2}\right)t + \theta_1\right] & \text{for symbol 1} \\[3em] \sqrt{\dfrac{2E_b}{T_b}}\cos\left[2\pi\left(f_c - \dfrac{\Delta f}{2}\right)t + \theta_2\right] & \text{for symbol 0} \end{cases}$$

where E_b is the signal energy per bit, T_b is the bit duration, and θ_1 and θ_2 are sample values of uniformly distributed random variables over the interval 0 to 2π. In effect, the two oscillators supplying the transmitted frequencies $f_c \pm \Delta f/2$ operate independently of each other. Assume that $f_c \gg \Delta f$.

(a) Evaluate the power spectral density of the FSK signal.
(b) Show that for frequencies far removed from the carrier frequency f_c, the power spectral density falls off as the inverse square of frequency.

Problem 7.7.2 A binary FSK signal has its two frequencies separated by one-half of the bit rate of the modulating binary wave. The requirement is to use an ordinary binary FSK receiver (operating on a bit-by-bit basis) to recover the binary wave. What form of preprocessing of the received signal is required to make this possible? Justify your answer.

P7.8 BANDWIDTH EFFICIENCY

Problem 7.8.1 The *equivalent noise bandwidth* of a band-pass signal is defined as the value of bandwidth that satisfies the relation

$4BS(f_c) = P$

where

$2B$ = noise equivalent bandwidth centered at the mid-band frequency f_c
$S(f_c)$ = maximum value of the power spectral density of the signal at $f = f_c$
P = average power of the signal

Show that the equivalent noise bandwidths of binary PSK, QPSK, and MSK normalized with respect to the data rate (measured in b/s) are as follows:

Type of Modulation	Noise Bandwidth/Bit Rate
Binary PSK	1.0
QPSK	0.5
MSK	0.62

Using this definition of bandwidth, calculate the bandwidth efficiency of binary PSK, QPSK, and MSK signals.

Problem 7.8.2 Calculate the bandwidth efficiency of an M-ary QAM scheme.

P7.9 M-ARY MODULATION FORMATS VIEWED IN THE LIGHT OF THE CHANNEL CAPACITY THEOREM

Problem 7.9.1 The loci of points for M-ary PSK and M-ary FSK shown in Figs. 7.33 and 7.35 are plotted for an average probability of symbol error $P_e = 10^{-5}$. How would these loci move relative to the capacity boundary as P_e is progressively reduced? Justify your answer.

Problem 7.9.2 In a coherent binary FSK system, symbols 0 and 1 are transmitted with equal probability. The system parameters are as follows:

Average transmitted power = 1 W
Noise power spectral density = 10^{-5} W/Hz
Transmitted bit rate = 10^{-4} b/s

Viewing the system as a binary symmetric channel, calculate the channel capacity C.

Problem 7.9.3 Consider an M-ary DPSK system operating over an AWGN channel. The values of E_b/N_0, required for a probability of symbol error $P_e = 2 \times 10^{-5}$, for varying M are as follows:

M	2	4	8	16	32
(E_b/N_0), dB	10	12	16	20.9	26

(a) Plot the bandwidth efficiency versus E_b/N_0 for the given values of M.
(b) Include a plot of the capacity boundary. Hence, comment on the effect of increasing M in the context of the Shannon limit.

Problem 7.9.4 Replot the bandwidth-efficiency diagram of Fig. 7.33 for 16-QAM, assuming an average probability of symbol error $P_e = 10^{-5}$. Hence, show that it approaches the capacity boundary more closely than does M-ary PSK.

P7.10 EFFECT OF INTERSYMBOL INTERFERENCE

Problem 7.10.1 Let P_{eI} and P_{eQ} denote the probabilities of symbol error for the in-phase and quadrature channels of a band-pass system. Show that the average probability of symbol error for the overall system is given by

$$P_e = P_{eI} + P_{eQ} - P_{eI}P_{eQ}$$

P7.11 SYNCHRONIZATION

Problem 7.11.1 The Costas loop of Fig. 7.40 for binary PSK signals has two paths, an in-phase and a quadrature one. How many paths does an Mth-order Costas loop have? Justify your answer.

CHAPTER EIGHT

ERROR-CONTROL CODING

In Chapter 7, we described modulation techniques for transmitting digital information over *noisy, band-pass channels*. Basically, *modulation* is a signal-processing operation that is used to provide for the *efficient* transmission of digital information over the channel. In this chapter, we study another important signal-processing operation, namely, *channel coding*, which is used to provide for the *reliable* transmission of digital information over the channel. In particular, we present a survey of *error-control coding* techniques that rely on the systematic addition of *redundant* symbols to the transmitted information so as to facilitate two basic objectives at the receiver: *error detection* and *error correction*. We also consider a type of code called a *trellis code* that combines coding with modulation.

We begin the chapter with some preliminary considerations that include a brief discussion of the role of coding in the reliable transmission of digital information over a noisy channel.

8.1 RATIONALE FOR CODING, AND TYPES OF CODES

The task facing the designer of a digital communication system is that of providing a cost-effective system for transmitting information from a sender (at one end of the system) at a rate and a level of reliability that are acceptable to a user (at the other end). The two key system parameters available to the designer are transmitted signal power and channel bandwidth. These two parameters, together with the power spectral density of receiver noise, determine the signal energy per bit-to-noise power density ratio, E_b/N_0. In Chapter 7, we showed that this ratio uniquely determines the bit error rate for a particular modulation scheme. Practical considerations usually place a limit on the value that we can assign to E_b/N_0. Accordingly, in practice, we often arrive at a modulation scheme and find that it is not possible to provide acceptable data quality (i.e., low enough error performance). For a fixed E_b/N_0, the only practical option available for changing data quality from problematic to acceptable is to use *coding*.

Another practical motivation for the use of coding is to reduce the required E_b/N_0 for a fixed bit error rate. This reduction in E_b/N_0 may, in turn, be exploited to reduce the required transmitted power or reduce the hardware costs by requiring a smaller antenna size.

Error control for data integrity may be exercised by means of *forward error correction*. Figure 8.1*a* shows the model of a digital communication system using such an approach. The discrete source generates information in the form of binary symbols. The *channel encoder* accepts message bits and adds *redundancy* according to a prescribed rule, thereby producing encoded data at a higher bit rate. The *channel decoder* exploits the redundancy to decide which

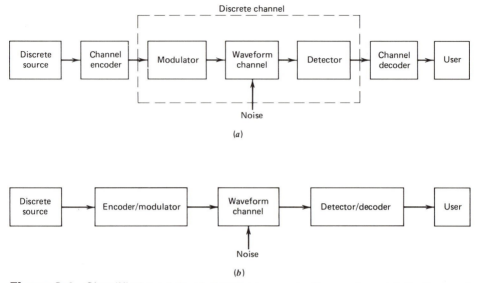

Figure 8.1 Simplified models of digital communication system. (*a*) Coding and modulation performed separately. (*b*) Coding and modulation combined.

message bit was actually transmitted. The combined goal of the channel encoder and decoder is to minimize the effect of channel noise. That is, the number of errors between the channel encoder input (derived from the source) and the channel decoder output (delivered to the user) is minimized.

The addition of redundancy in the coded messages implies the need for increased transmission bandwidth. Moreover, the use of coding adds *complexity* to the system, especially for the implementation of decoding operations in the receiver. Thus, the design trade-offs in the use of error-control coding to achieve acceptable error performance must include considerations of bandwidth and system complexity.

There are many different error-correcting codes (with roots in diverse mathematical disciplines) that we can use. Historically, these codes have been classified into *block codes* and *convolutional codes*. The distinguishing feature for the classification is the presence or absence of *memory* in the encoders for the two codes.

To generate an (n, k) block code, the channel encoder accepts information in successive k-bit *blocks*; for each block, it adds $n - k$ redundant bits that are algebraically related to the k message bits, thereby producing an overall encoded block of n bits, where $n > k$. The n-bit block is called a *code word*, and n is called the *block length* of the code. The channel encoder produces bits at the rate $R_0 = (n/k)R_s$, where R_s is the bit rate of the information generated by the source. The dimensionless ratio $r = k/n$ is called the *code rate*, where $0 < r < 1$. The bit rate R_0, coming out of the encoder, is called the *channel data rate*. Thus, the code rate is a dimensionless ratio, whereas the data rate produced by the source and the channel data rate are both measured in bits per second.

In a convolutional code, the encoding operation may be viewed as the *dis-*

crete-time convolution of the input sequence with the impulse response of the encoder. The duration of the impulse response equals the memory of the encoder. Accordingly, the encoder for a convolutional code operates on the incoming message sequence, using a "sliding window" equal in duration to its own memory. This, in turn, means that in a convolutional code, unlike a block code, the channel encoder accepts message bits as a continuous sequence, and thereby generates a continuous sequence of encoded bits at a higher rate.

Another way of classifying codes is as *linear* or *nonlinear*. A linear code distinguishes itself from a nonlinear code by the property that any two code words of a linear code can be added in modulo-2 arithmetic* to produce a third code word in the code. This property has far-reaching implications that become apparent later. The codes used in practical applications are almost always linear codes.

In the model depicted in Fig. 8.1*a*, the operations of coding and modulation are performed separately. When, however, bandwidth efficiency is of major concern, the most effective method of catering to forward error correction coding is to combine it with modulation as a single function, as shown in Fig. 8.1*b*. In such an approach, coding is redefined as the process of imposing certain patterns on the transmitted signal. The issue of combined modulation-coding is considered in Section 8.9.

Much of the material presented in this chapter relates to coding techniques suitable for forward error correction (FEC). There is, however, another major approach known as *automatic request for retransmission (ARQ)*, which is also widely used for solving the error-control problem. The philosophy of ARQ is quite different from that of FEC. Specifically, ARQ utilizes redundancy for the sole purpose of error detection. Upon detection, the receiver requests a repeat transmission, which necessitates the use of a return path (feedback channel). A description of ARQ is presented in Section 8.10.

8.2 DISCRETE MEMORYLESS CHANNELS

Returning to the model of Fig. 8.1*a*, the waveform channel is said to be memoryless if the detector output in a given interval depends only on the signal transmitted in that interval, and not on any previous transmission. Under this condition, we may model the combination of the modulator, the waveform channel, and the detector as a *discrete memoryless channel*. Such a channel is completely described by the set of transition probabilities $p(j|i)$, where i denotes a modulator input symbol, j denotes a demodulator output symbol, and $p(j|i)$ denotes the probability of receiving symbol j, given that symbol i was sent. (Discrete memoryless channels were described previously at some length in Section 2.4.)

The simplest discrete memoryless channel results from the use of binary input and binary output symbols. When binary coding is used, the modulator

---

* For details of modulo-2 arithmetic, see Appendix H. The arithmetic operations used throughout this chapter are performed modulo-2. Note, however, there are linear codes over different alphabets.

has only the binary symbols 0 and 1 as inputs. Likewise, the decoder has only binary inputs if binary quantization of the demodulator output is used, that is, a *hard decision* is made on the demodulator output as to which symbol was actually transmitted. In this situation, we have a *binary symmetric channel* (BSC) with a *channel transition distribution* as shown in Fig. 8.2. The binary symmetric channel, when derived from an additive white Gaussian noise (AWGN) channel, is completely described by the *transition probability p*. The majority of coded digital communication systems employ binary coding with hard-decision decoding, due to the simplicity of implementation offered by such an approach.

The use of hard decisions prior to decoding causes an irreversible loss of information in the receiver. To cure this problem, *soft-decision* coding is used. This is achieved by including a multilevel quantizer at the demodulator output, as illustrated in Fig. 8.3a for the case of binary PSK signals. The input–output characteristic of the quantizer is shown in Fig. 8.3b. As before, the modulator has only the binary symbols 0 and 1 as inputs. However, the demodulator output now has an alphabet with Q symbols. Assuming the use of the quantizer as characterized in Fig. 8.3b, we have $Q = 8$. Such a channel is called a *binary input Q-ary output discrete memoryless channel*. The corresponding channel transition distribution is shown in Fig. 8.3c. The form of this distribution, and consequently the decoder performance, depends on the location of the representation levels of the quantizer, which in turn depends on the signal level and noise variance. Accordingly, the demodulator must incorporate automatic gain control, if an effective multilevel quantizer is to be realized. Moreover, the use of soft decisions complicates the implementation of the decoder. Nevertheless, soft-decision decoding offers significant improvement in performance over hard-decision decoding.

(1) The Channel Coding Theorem Revisited

In Chapter 2, we established the concept of *channel capacity* which, for a discrete memoryless channel, represents the maximum amount of information transferred per channel use. The *channel coding theorem* states that if a discrete memoryless channel has capacity C and a source generates information at a rate less than C, then there exists a coding technique such that the output of the source may be transmitted over the channel with an arbitrarily low proba-

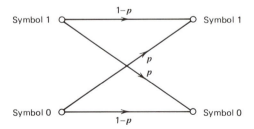

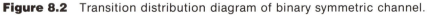

Figure 8.2 Transition distribution diagram of binary symmetric channel.

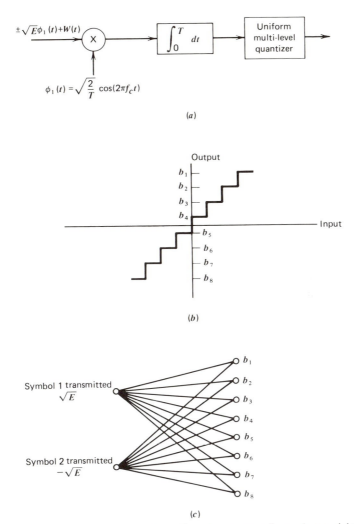

Figure 8.3 Binary input Q-ary output discrete memoryless channel (a) receiver, (b) transfer characteristic of multilevel quantizer, (c) channel transition distribution. Parts (b) and (c) are illustrated for 8 levels.

bility of symbol error. For the special case of a binary symmetric channel, the theorem tells us that if the code rate r is less than the channel capacity C, then it is possible to find a code that achieves error-free transmission over the channel. Conversely, it is not possible to find such a code if the code rate r is greater than the channel capacity C.

The channel coding theorem thus specifies the channel capacity C as a *fundamental limit* on the rate at which the transmission of reliable (error-free) messages can take place over a discrete memoryless channel. The issue that matters is not the signal-to-noise ratio, so long as it is large enough, but how the channel input is encoded.

The most unsatisfactory feature of the channel coding theorem, however, is its nonconstructive nature. The theorem only asserts the *existence of good codes*. But, the theorem does not tell us how to find them. We are still faced with the task of finding a good code that ensures reliable transmission of information over the channel. The error-control coding techniques presented in the remainder of this chapter provide different methods of achieving this important system requirement. We consider block codes first, followed by convolutional codes, and then trellis codes.

8.3 LINEAR BLOCK CODES

Consider an (n, k) linear block code in which the first portion of k bits is always identical to the message sequence to be transmitted. The $n - k$ bits in the second portion are computed from the message bits in accordance with a prescribed encoding rule that determines the mathematical structure of the code. Accordingly, these $n - k$ bits are referred to as *generalized parity check bits* or simply *parity bits*. Block codes in which the message bits are transmitted in unaltered form are called *systematic codes*. For applications requiring *both* error detection and error correction, the use of systematic block codes simplifies implementation of the decoder.

Let $m_0, m_1, \ldots, m_{k-1}$ constitute a block of k arbitrary message bits. Thus we have 2^k distinct message blocks. Let this sequence of message bits be applied to a linear block encoder, producing an n-bit code word whose elements are denoted by $x_0, x_1, \ldots, x_{n-1}$. Let $b_0, b_1, \ldots, b_{n-k-1}$ denote the $(n - k)$ parity bits in the code word. For the code to possess a systematic structure, a code word is divided into two parts, one of which is occupied by the message bits and the other by the parity bits. Clearly, we have the option of sending the message bits of a code word before the parity bits, or vice versa. The former option is illustrated in Fig. 8.4, and its use is assumed in the sequel.

According to the representation of Fig. 8.4, the $(n - k)$ left-most bits of a code word are identical to the corresponding parity bits, and the k right-most bits of the code word are identical to the corresponding message bits. We may therefore write

$$x_i = \begin{cases} b_i & i = 0, 1, \ldots, n - k - 1 \\ m_{i+k-n} & i = n - k, n - k + 1, \ldots, n - 1 \end{cases} \qquad (8.4)$$

The $(n - k)$ parity bits are *linear sums* of the k message bits, as shown by the generalized relation

$$b_i = p_{i0}m_0 + p_{i1}m_1 + \cdots + p_{i,k-1}m_{k-1} \qquad i = 0, 1, \ldots, n - k - 1 \quad (8.5)$$

where the coefficients are defined as follows:

$b_0, b_1, \ldots, b_{n-k-1}$	$m_0, m_1, \ldots, m_{k-1}$
Parity bits	Message bits

Figure 8.4 Structure of code word.

$$p_{ij} = \begin{cases} 1 & \text{if } b_i \text{ depends on } m_j \\ 0 & \text{otherwise} \end{cases} \tag{8.6}$$

These coefficients are chosen in such a way that the $(n - k)$ equations represented in Eq. 8.5 are *linearly independent;* that is, no equation in the set can be expressed as a linear combination of the remaining ones.

The system of Eqs. 8.4 and 8.5 defines the mathematical structure of the (n, k) linear block code. This system of equations may be rewritten in a compact form using matrix notation. To proceed with this reformulation, we define the 1-by-k *message vector* $\mathbf{m}$, the 1-by-$(n - k)$ *parity vector* $\mathbf{b}$, and the 1-by-n code vector $\mathbf{x}$ as follows, respectively:

$$\mathbf{m} = [m_0, m_1, \ldots, m_{k-1}] \tag{8.7}$$

$$\mathbf{b} = [b_0, b_1, \ldots, b_{n-k-1}] \tag{8.8}$$

and

$$\mathbf{x} = [x_0, x_1, \ldots, x_{n-1}] \tag{8.9}$$

Note that all three vectors are *row vectors.** Accordingly, we may rewrite the set of simultaneous equations in the compact matrix form:

$$\mathbf{b} = \mathbf{mP} \tag{8.10}$$

where $\mathbf{P}$ is the k-by-$(n - k)$ *coefficient matrix* defined by

$$\mathbf{P} = \begin{bmatrix} p_{00} & p_{10} & \cdots & p_{n-k-1,0} \\ p_{01} & p_{11} & \cdots & p_{n-k-1,1} \\ \vdots & \vdots & & \vdots \\ p_{0,k-1} & p_{1,k-1} & \cdots & p_{n-k-1,k-1} \end{bmatrix} \tag{8.11}$$

From the definitions given in Eqs. 8.7–8.9, we see that $\mathbf{x}$ may be expressed as a partitioned row vector in terms of the vectors $\mathbf{m}$ and $\mathbf{b}$ as follows:

$$\mathbf{x} = [\mathbf{b} \vdots \mathbf{m}] \tag{8.12}$$

Hence, substituting Eq. 8.10 in Eq. 8.12 and factoring out the common message vector $\mathbf{m}$, we get

$$\mathbf{x} = \mathbf{m}[\mathbf{P} \vdots \mathbf{I}_k] \tag{8.13}$$

where $\mathbf{I}_k$ is the k-by-k *identity matrix:*

$$\mathbf{I}_k = \begin{bmatrix} 1 & 0 & \cdots & 0 \\ 0 & 1 & \cdots & 0 \\ \vdots & \vdots & & \vdots \\ 0 & 0 & \cdots & 1 \end{bmatrix} \tag{8.14}$$

---

* The use of row vectors is adopted in this chapter for the sake of being consistent with the notation commonly used in the coding literature. This notation is different from that used in matrix algebra, where the vectors are commonly written as column vectors.

Define the *k-by-n generator matrix**

$$\mathbf{G} = [\mathbf{P} \vdots \mathbf{I}_k] \tag{8.15}$$

Then, we may simplify Eq. 8.13 as

$$\mathbf{x} = \mathbf{mG} \tag{8.16}$$

The full set of code words, referred to simply as *the code*, is generated in accordance with Eq. 8.16 by letting the message vector **m** range through the set of all 2^k binary *k*-tuples (1-by-*k* vectors). Moreover, the sum of any two code words is another code word. This basic property of linear block codes is called *closure*. To prove its validity, consider a pair of code vectors $\mathbf{x}_i$ and $\mathbf{x}_j$ corresponding to a pair of message vectors $\mathbf{m}_i$ and $\mathbf{m}_j$, respectively. Using Eq. 8.16, we may express the sum of $\mathbf{x}_i$ and $\mathbf{x}_j$ as

$$\mathbf{x}_i + \mathbf{x}_j = \mathbf{m}_i\mathbf{G} + \mathbf{m}_j\mathbf{G}$$
$$= (\mathbf{m}_i + \mathbf{m}_j)\mathbf{G}$$

The sum of $\mathbf{m}_i$ and $\mathbf{m}_j$ represents a new message vector, modulo-2. Correspondingly, the sum of $\mathbf{x}_i$ and $\mathbf{x}_j$ represents a new code vector, modulo-2. In other words, the sum of any two code words equals another code word in the code.

There is another way of expressing the relationship between the message bits and parity-check bits of a linear block code. Let **H** denote an $(n - k)$-by-*n* matrix, defined as

$$\mathbf{H} = [\mathbf{I}_{n-k} \vdots \mathbf{P}^T] \tag{8.17}$$

where $\mathbf{P}^T$ is an $(n - k)$-by-*k* matrix, representing the transpose of the coefficient matrix **P**, and $\mathbf{I}_{n-k}$ is the $(n - k)$-by-$(n - k)$ identity matrix. Accordingly, we may perform the following multiplication of partitioned matrices:

$$\mathbf{HG}^T = [\mathbf{I}_{n-k} \vdots \mathbf{P}^T] \left[\frac{\mathbf{P}^T}{\mathbf{I}_k}\right]$$
$$= \mathbf{P}^T + \mathbf{P}^T$$

where we have used the fact that multiplication of a rectangular matrix by an identity matrix of compatible dimensions leaves the matrix unchanged. In modulo-2 arithmetic, we have $\mathbf{P}^T + \mathbf{P}^T = \mathbf{0}$, where **0** denotes an $(n - k)$-by-*k* matrix that has zeros for all of its elements. Hence,

$$\mathbf{HG}^T = \mathbf{0} \tag{8.18}$$

Equivalently, we have $\mathbf{GH}^T = \mathbf{0}$. Postmultiplying both sides of Eq. 8.16 by $\mathbf{H}^T$, the transpose of **H**, and then using Eq. 8.18, we get

$$\mathbf{xH}^T = \mathbf{mGH}^T$$
$$= \mathbf{0} \tag{8.19}$$

---

* The generator matrix **G** of Eq. 8.15 is said to be in the *echelon canonical form* in that its *k* rows are linearly independent. That is, it is not possible to express any row of the matrix **G** as a linear combination of the remaining rows.

The matrix **H** is called the *parity-check matrix* of the code, and the equations specified by Eq. 8.19 are called *parity-check equations*.

EXAMPLE 1 REPETITION CODES

*Repetition codes** represent the simplest type of linear block codes. In particular, a single message bit is encoded into a block of identical n bits, producing an $(n, 1)$ block code. Such a code allows provision for a variable amount of redundancy. There are only two code words in the code: an all-zero code word and an all-one code word.

Consider, for example, the case of a repetition code with $k = 1$ and $n = 5$. In this case, we have four parity bits that are the same as the message bit. Hence, the identity matrix $\mathbf{I}_k = 1$, and the coefficient matrix **P** consists of a 1-by-4 vector that has 1 for all of its elements. Correspondingly, the generator matrix equals a row vector of all 1s, as shown by

$$\mathbf{G} = [1 \quad 1 \quad 1 \quad 1 \vdots 1]$$

The transpose of the coefficient matrix **P**, namely, matrix $\mathbf{P}^T$, consists of a 4-by-1 vector that has 1 for all of its elements. The identity matrix $\mathbf{I}_{n-k}$ consists of a 4-by-4 matrix. Hence, the parity-check matrix equals

$$\mathbf{H} = \begin{bmatrix} 1 & 0 & 0 & 0 & \vdots & 1 \\ 0 & 1 & 0 & 0 & \vdots & 1 \\ 0 & 0 & 1 & 0 & \vdots & 1 \\ 0 & 0 & 0 & 1 & \vdots & 1 \end{bmatrix}$$

Since the message vector consists of a single binary symbol, 0 or 1, it follows from Eq. 8.16 that there are only two code words: 00000 and 11111 in the (5, 1) repetition code, as expected. Note also that $\mathbf{HG}^T = \mathbf{0}$, modulo-2, in accordance with Eq. 8.18.

(1) Syndrome Decoding

The generator matrix **G** is used in the encoding operation at the transmitter. On the other hand, the parity-check matrix **H** is used in the decoding operation at the receiver. In the context of the latter operation, let **y** denote the 1-by-n *received vector* that results from sending the code vector **x** over a noisy channel. We express the vector **y** as the sum of the original code vector **x** and a vector **e**, as shown by

$$\mathbf{y} = \mathbf{x} + \mathbf{e} \tag{8.20}$$

The vector **e** is called the *error vector* or *error pattern*. The ith element of **e** equals 0 if the corresponding element of **y** is the same as that of **x**. On the other hand, the ith element of **e** equals 1 if the corresponding element of **y** is different from that of **x**, in which case an error is said to have occurred in the ith location. That is, for $i = 1, 2, \ldots, n$, we have

---

* It is noteworthy that the use of a repetition code may be viewed as analogous to that of *space diversity*. In a repetition code, we provide for error control by repeating each message bit two or more times. In space diversity, we improve reliability by having two or more receivers located at different points.

$$e_i = \begin{cases} 1 & \text{if an error has occurred in the } i\text{th location} \\ 0 & \text{otherwise} \end{cases} \qquad (8.21)$$

The receiver has the task of decoding the code vector $\mathbf{x}$ from the received vector $\mathbf{y}$. The algorithm commonly used to perform this decoding operation starts with the computation of a 1-by-$(n - k)$ vector called the *error-syndrome vector* or simply the *syndrome*.* The importance of the syndrome lies in the fact that it depends only upon the error pattern.

Given a 1-by-n received vector $\mathbf{y}$, the corresponding syndrome is formally defined as

$$\mathbf{s} = \mathbf{y}\mathbf{H}^T \qquad (8.22)$$

Accordingly, the syndrome has the following important properties:

PROPERTY 1
The syndrome depends only on the error pattern, and not on the transmitted code word.

To prove this property, we first use Eqs. 8.20 and 8.22 and then Eq. 8.19, obtaining

$$\begin{aligned} \mathbf{s} &= (\mathbf{x} + \mathbf{e})\mathbf{H}^T \\ &= \mathbf{x}\mathbf{H}^T + \mathbf{e}\mathbf{H}^T \\ &= \mathbf{e}\mathbf{H}^T \end{aligned} \qquad (8.23)$$

Hence, the parity-check matrix $\mathbf{H}$ of a code permits us to compute the syndrome $\mathbf{s}$, which depends only upon the error pattern $\mathbf{e}$. Property 1 is basic to the decoding of linear block codes.

PROPERTY 2
All error patterns that differ at most by a code word have the same syndrome.

For k message bits, there are 2^k distinct code vectors denoted as $\mathbf{x}_i$, $i = 0, 1, \ldots, 2^k - 1$. Correspondingly, for any error pattern $\mathbf{e}$, we define the 2^k distinct vectors $\mathbf{e}_i$ as

$$\mathbf{e}_i = \mathbf{e} + \mathbf{x}_i \qquad i = 0, 1, \ldots, 2^k - 1 \qquad (8.24)$$

The set of vectors $\{\mathbf{e}_i, i = 0, 1, \ldots, 2^k - 1\}$ so defined is called a *coset* of the code. In other words, a coset has exactly 2^k elements that differ at most by a code vector. Thus, an (n, k) linear block code has 2^{n-k} possible cosets. In any event, multiplying both sides of Eq. 8.24 by the matrix $\mathbf{H}^T$, we get

$$\begin{aligned} \mathbf{e}_i\mathbf{H}^T &= \mathbf{e}\mathbf{H}^T + \mathbf{x}_i\mathbf{H}^T \\ &= \mathbf{e}\mathbf{H}^T \end{aligned} \qquad (8.25)$$

* In medicine, the term "syndrome" is used to describe a pattern of symptoms that aids in the diagnosis of a "disease." In coding, the error pattern plays the role of the "disease," and a parity-check failure that of a "symptom." This use of "syndrome" was coined by Hagelbarger (1959).

which is independent of the index i. Accordingly, we may state that each coset of the code is characterized by a unique syndrome.

Property 2 shows that the syndrome **s** provides some information about the error pattern **e**, but not enough for the decoder to compute the exact value of the transmitted code vector. Nevertheless, knowledge of the syndrome **s** reduces the search for the error pattern **e** from 2^n to 2^k possibilities, namely, the elements that constitute the coset corresponding to **s**. The decoder then has the task of making the best selection from that coset.

PROPERTY 3
*The syndrome **s** is the sum of those columns of the matrix **H** corresponding to the error locations.*

To prove this property, let the parity-check matrix **H** be expanded in terms of its columns as follows:

$$\mathbf{H} = [\mathbf{h}_1, \mathbf{h}_2, \ldots, \mathbf{h}_n] \tag{8.26}$$

Then, substituting Eq. 8.26 in Eq. 8.23, we may express the syndrome as

$$\mathbf{s} = \mathbf{e}\mathbf{H}^T$$

$$= [e_1, e_2, \ldots, e_n] \begin{bmatrix} \mathbf{h}_1^T \\ \mathbf{h}_2^T \\ \vdots \\ \mathbf{h}_n^T \end{bmatrix}$$

$$= \sum_{i=1}^{n} e_i \mathbf{h}_i^T \tag{8.27}$$

where $\mathbf{h}_i^T$ is the ith row of the matrix $\mathbf{H}^T$, and e_i is the ith element of the error pattern **e**. The elements of **e** are defined in Eq. 8.21. Hence, substituting this definition in Eq. 8.27, we find that the syndrome **s** equals the sum of those rows of the matrix $\mathbf{H}^T$ that correspond to the error locations in **e**.

PROPERTY 4
With syndrome decoding, an (n, k) linear block code can correct up to t errors per code word, provided that n and k satisfy the Hamming bound

$$2^{n-k} \geq \sum_{i=0}^{t} \binom{n}{i} \tag{8.28}$$

where $\binom{n}{i}$ is a binomial coefficient, namely,

$$\binom{n}{i} = \frac{n!}{(n-i)!\, i!} \tag{8.29}$$

We prove Eq. 8.28 by examining the relationship between the syndrome and error patterns, and making use of combinatorial analysis. There are a total of 2^{n-k} syndromes, including the all-zero syndrome. Each syndrome corresponds to a specific error pattern. In general, we find that for an n-bit code word, there are $\binom{n}{i}$ *multiple-error patterns,* where i is the number of error locations in the n-by-1 error pattern **e**. Accordingly, the total number of all possible error patterns equals the sum of $\binom{n}{i}$ for $i = 0, 1, \ldots, t$, where t is the maximum number of error locations in **e**. If therefore an (n, k) linear block code is to be able to correct up to t errors, the total number of syndromes must *not* be less than the total number of all possible error patterns. In other words, the block length n and the number of message bits k must satisfy the Hamming bound of Eq. 8.28. A binary code for which the Hamming bound is satisfied with the equality sign is called a *perfect code.*

We stated earlier that, in general, the syndrome **s** does not uniquely specify the actual error pattern **e**. Rather, it identifies the coset that the error pattern belongs to. The *most likely* error pattern within the coset characterized by the syndrome **s** is the one with the largest probability. This assumes that the channel noise is additive. Hence, the decoding algorithm may proceed as follows:

1. For the received vector **y**, compute the syndrome $\mathbf{s} = \mathbf{y}\mathbf{H}^T$.
2. Within the coset characterized by the syndrome **s**, choose the error pattern with the largest probability. Call it $\mathbf{e}_o$.
3. For output, compute the code vector

$$\hat{\mathbf{x}} = \mathbf{y} + \mathbf{e}_o$$

(Note that in modulo-2 arithmetic, subtraction is the same as addition.)

(2) Minimum Distance Considerations

Consider a pair of code vectors **x** and **y** that have the same number of elements. The *Hamming distance* $d(\mathbf{x}, \mathbf{y})$ between such a pair of code vectors is defined as the number of locations in which their respective elements differ.

The *Hamming weight* $w(\mathbf{x})$ of a code vector **x** is defined as the number of nonzero elements in the code vector. Equivalently, we may state that the Hamming weight of a code vector is the distance between the code vector and an all-zero code vector.

The *minimum distance* $d_{\min}$ of a linear block code is defined as the smallest Hamming distance between any pair of code vectors in the code. That is, the minimum distance is the same as the smallest Hamming weight of the difference between any pair of code vectors. From the closure property of linear block codes, the sum (or difference) of two code vectors is another code vector. Accordingly, we may state that *the minimum distance of a linear block code is the smallest Hamming weight of the nonzero code vectors in the code.*

The minimum distance $d_{\min}$ is related to the structure of the parity-check matrix **H** of the code in a fundamental way. From Eq. 8.19 we know that a linear block code is defined by the set of all code vectors for which $\mathbf{x}\mathbf{H}^T = \mathbf{0}$, where $\mathbf{H}^T$ is the transpose of the parity-check matrix **H**. Let the matrix **H** be expressed in terms of its columns as in Eq. 8.26. Then, for a code vector **x** to satisfy the condition $\mathbf{x}\mathbf{H}^T = \mathbf{0}$, the vector **x** must have 1s in such positions that the corresponding rows of $\mathbf{H}^T$ sum to the zero vector **0**. However, by definition, the number of 1s in a code vector is the Hamming weight of the code vector. Moreover, the smallest Hamming weight of the nonzero code vectors in a linear block code equals the minimum distance of the code. Hence, *the minimum distance of a linear block code is defined by the minimum number of rows of the matrix $\mathbf{H}^T$ that sum to* **0**.

The minimum distance of a linear block code, d_{min}, is an important parameter of the code. Specifically, it determines the error-correcting capability of the code. Suppose an (n, k) linear block code is required to detect and correct all error patterns (over a binary symmetric channel), whose Hamming weight is less than or equal to t. That is, if a code vector $\mathbf{x}_i$ in the code is transmitted and the received vector is $\mathbf{y} = \mathbf{x}_i + \mathbf{e}$, we require that the decoder output $\hat{\mathbf{x}} = \mathbf{x}_i$, subject to the condition that the error pattern $\mathbf{e}$ has a Hamming weight $w(\mathbf{e}) \leq t$. We assume that the 2^k code vectors in the code are transmitted with equal probability. The best strategy for the decoder then is to pick the code vector closest to the received vector $\mathbf{y}$, that is, the one for which the Hamming distance $d(\mathbf{x}_i, \mathbf{y})$ is smallest. With such a strategy, the decoder will be able to detect and correct all error patterns of Hamming weight $w(\mathbf{e}) \leq t$, provided that the minimum distance of the code is equal to or greater than $2t + 1$. We may demonstrate the validity of this requirement by adopting a geometric interpretation of the problem. In particular, the 1-by-n code vectors and the 1-by-n received vector are represented as points in an n-dimensional space. Suppose we construct two spheres, each of radius t, around the points that represent code vectors $\mathbf{x}_i$ and $\mathbf{x}_j$. Let these two spheres be disjoint, as depicted in Fig. 8.5a. For this condition to be satisfied, we require that $d(\mathbf{x}_i, \mathbf{x}_j) \geq 2t + 1$. If then the code vector $\mathbf{x}_i$ is transmitted and the Hamming distance $d(\mathbf{x}_i, \mathbf{y}) \leq t$, it is clear that the decoder will pick $\mathbf{x}_i$ as it is the code vector closest to the received vector $\mathbf{y}$. If, on the other hand, the Hamming distance $d(\mathbf{x}_i, \mathbf{x}_j) \leq 2t$, the two spheres around $\mathbf{x}_i$ and $\mathbf{x}_j$ intersect, as depicted in Fig. 8.5b. Here we see that if $\mathbf{x}_i$ is transmitted, there exists a received vector $\mathbf{y}$ such that the Hamming distance $d(\mathbf{x}_i, \mathbf{y}) \leq t$, and yet $\mathbf{y}$ is as close to $\mathbf{x}_j$ as it is to $\mathbf{x}_i$. Clearly, there is now the possibility of the decoder picking the vector $\mathbf{x}_j$, which is wrong. We thus conclude that *an (n, k) linear block code has the power to correct all error patterns of weight t or less if, and only if*

$$d(\mathbf{x}_i, \mathbf{x}_j) \leq 2t + 1 \qquad \text{for all } \mathbf{x}_i \text{ and } \mathbf{x}_j$$

By definition, however, the smallest distance between any pair of code vectors in a code is the minimum distance of the code, d_{min}. We may therefore state

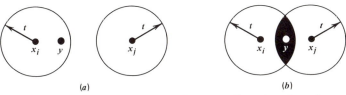

Figure 8.5 (a) Hamming distance $d(\mathbf{x}_i, \mathbf{x}_j) \geq 2t + 1$. (b) Hamming distance $d(\mathbf{x}_i, \mathbf{x}_j) < 2t$.

that *an (n, k) linear block code of minimum distance d_{min} can correct up to t errors, if and only if*

$$t \leq \left\lfloor \frac{1}{2}(d_{min} - 1) \right\rfloor \qquad (8.30)$$

where $\lfloor \cdot \rfloor$ *denotes the largest integer no greater than the enclosed number.* Equation 8.30 gives the error-correcting capability of a linear block code a quantitative meaning.

EXAMPLE 2 HAMMING CODES

Consider a family of (n, k) linear block codes that have the following parameters:

Block length: $n = 2^m - 1$

Number of message bits: $k = 2^m - m - 1$

Number of parity bits: $n - k = m$

where $m \geq 3$. These are the so-called *Hamming codes.**

Consider, for example, the (7,4) Hamming code with $n = 7$ and $k = 4$, corresponding to $m = 3$. The generator matrix of the code must have a structure that conforms to Eq. 8.15. The following matrix represents an appropriate generator matrix for the (7,4) Hamming code:

$$\mathbf{G} = \begin{bmatrix} 1 & 1 & 0 & | & 1 & 0 & 0 & 0 \\ 0 & 1 & 1 & | & 0 & 1 & 0 & 0 \\ 1 & 1 & 1 & | & 0 & 0 & 1 & 0 \\ 1 & 0 & 1 & | & 0 & 0 & 0 & 1 \end{bmatrix} \qquad (8.31)$$
$$\underbrace{}_{\mathbf{P}}\quad\underbrace{}_{\mathbf{I}_k}$$

The corresponding parity-check matrix is given by

$$\mathbf{H} = \begin{bmatrix} 1 & 0 & 0 & | & 1 & 0 & 1 & 1 \\ 0 & 1 & 0 & | & 1 & 1 & 1 & 0 \\ 0 & 0 & 1 & | & 0 & 1 & 1 & 1 \end{bmatrix} \qquad (8.32)$$
$$\underbrace{}_{\mathbf{I}_{n-k}}\quad\underbrace{}_{\mathbf{P}^T}$$

With $k = 4$, there are $2^k = 16$ distinct message words, which are listed in Table 8.1.

Table 8.1   Code Words of a (7,4) Hamming Code

Message Word	Code Word	Weight of Code Word	Message Word	Code Word	Weight of Code Word
0 0 0 0	0 0 0 0 0 0 0	0	1 0 0 0	1 1 0 1 0 0 0	3
0 0 0 1	1 0 1 0 0 0 1	3	1 0 0 1	0 1 1 1 0 0 1	4
0 0 1 0	1 1 1 0 0 1 0	4	1 0 1 0	0 0 1 1 0 1 0	3
0 0 1 1	0 1 0 0 0 1 1	3	1 0 1 1	1 0 0 1 0 1 1	4
0 1 0 0	0 1 1 0 1 0 0	3	1 1 0 0	1 0 1 1 1 0 0	4
0 1 0 1	1 1 0 0 1 0 1	4	1 1 0 1	0 0 0 1 1 0 1	3
0 1 1 0	1 0 0 0 1 1 0	3	1 1 1 0	0 1 0 1 1 1 0	4
0 1 1 1	0 0 1 0 1 1 1	4	1 1 1 1	1 1 1 1 1 1 1	7

---

* Hamming codes are named after their inventor Hamming (1950).

Table 8.2 Decoding
Table for the (7,4)
Hamming Code Defined in
Table 8.1

Syndrome	Error Pattern
0 0 0	0 0 0 0 0 0 0
1 0 0	1 0 0 0 0 0 0
0 1 0	0 1 0 0 0 0 0
0 0 1	0 0 1 0 0 0 0
1 1 0	0 0 0 1 0 0 0
0 1 1	0 0 0 0 1 0 0
1 1 1	0 0 0 0 0 1 0
1 0 1	0 0 0 0 0 0 1

For a given message word, the corresponding code word is obtained by using Eq. 8.16. Thus, the application of this equation results in the 16 code words listed in Table 8.1.

In Table 8.1 we have also listed the Hamming weights of the individual 16 code words in the (7,4) Hamming code. Since the smallest of the Hamming weights for the nonzero code words is 3, it follows that the minimum distance of the code is 3. Indeed, Hamming codes have the property that the minimum distance $d_{min} = 3$, independent of the value assigned to m.

To illustrate the relation between the minimum distance d_{min} and the structure of the parity-check matrix $\mathbf{H}$, consider the code word 0110100. In the matrix multiplication defined by Eq. 8.19, the nonzero elements of this code word "sift" out the second, third, and fifth columns of the matrix $\mathbf{H}$ yielding

$$\begin{bmatrix} 0 \\ 1 \\ 0 \end{bmatrix} + \begin{bmatrix} 0 \\ 0 \\ 1 \end{bmatrix} + \begin{bmatrix} 0 \\ 1 \\ 1 \end{bmatrix} = \begin{bmatrix} 0 \\ 0 \\ 0 \end{bmatrix}$$

We may perform similar calculations for the remaining 14 nonzero code words. We thus find that the smallest number of columns in $\mathbf{H}$ that sums to zero is 3, confirming the earlier statement that $d_{min} = 3$.

An important property of Hamming codes is that they satisfy the condition of Eq. 8.28 with the equality sign, assuming that $t = 1$. This means that Hamming codes are *single-error correcting binary perfect codes*. The single-error correcting capability of Hamming codes is also confirmed by the use of Eq. 8.30, which is satisfied for $d_{min} = 3$ and $t = 1$ with the equal sign.

Assuming single-error patterns, we may readily develop the entries presented in Table 8.2 by applying Property 3. In the right-hand column of the table, we show seven distinct single-error patterns. In the left-hand column, we show the corresponding syndromes. Each single-error pattern is associated with a unique syndrome that corresponds to the particular column of the matrix $\mathbf{H}$, in accordance with Property 3. The zero syndrome signifies no transmission errors.

8.4 CYCLIC CODES

Cyclic codes form a subclass of linear block codes. Indeed, many of the important linear block codes discovered to date are either cyclic codes or closely related to cyclic codes. An advantage of cyclic codes over most other types of codes is that they are easy to encode. Furthermore, cyclic codes possess a well-defined mathematical structure, which has led to the development of very efficient decoding schemes for them.

A binary code is said to be a *cyclic code* if it exhibits two fundamental properties:

1. *Linearity property: The sum of two code words is also a code word.*
2. *Cyclic property: Any cyclic shift of a code word is also a code word.*

Property 1 restates the fact that a cyclic code is a linear block code (i.e., it can be described as a parity-check code). To restate Property 2 in mathematical terms, let the n-tuple $(x_0, x_1, \ldots, x_{n-1})$ denote a code word of an (n, k) linear block code. The code is a cyclic code if the n-tuples

$$(x_{n-1}, x_0, \ldots, x_{n-2}),$$

$$(x_{n-2}, x_{n-1}, \ldots, x_{n-3}),$$

$$\vdots$$

$$(x_1, x_2, \ldots, x_0)$$

are all code words.

This formulation of the cyclic property suggests that we may treat the elements of a code word of length n as the coefficients of a polynomial of degree $(n - 1)$. That is, the code word with elements $x_0, x_1, \ldots, x_{n-1}$ may be represented in the form of a *code word polynomial* as follows:

$$x(D) = x_0 + x_1D + \cdots + x_{n-1}D^{n-1} \qquad (8.33)$$

where D is an arbitrary real variable. Naturally, for binary codes, the coefficients are 1s or 0s. Each power of D in the polynomial $x(D)$ represents a one-bit *cyclic shift* in time. Hence, multiplication of the polynomial $x(D)$ by D may be viewed as a cyclic shift or rotation to the right, subject to the constraint $D^n = 1$. Application of the constraint $D^n = 1$ achieves two objectives. First, it restores the polynomial $Dx(D)$ to order $n - 1$. Second, the right-most bit or coefficient x_{n-1} is fed back at the right. This special form of polynomial multiplication is referred to as *multiplication modulo $D^n - 1$*. For a single cyclic shift, we may thus write

$$Dx(D) \bmod(D^n - 1) = x_{n-1} + x_0D + \cdots + x_{n-2}D^{n-1} \qquad (8.34)$$

where mod is the abbreviation for "modulo." The code word polynomial in Eq. 8.34 is a polynomial representation of the code word $(x_{n-1}, x_0, \ldots, x_{n-2})$. For two cyclic shifts, we may write

$$D^2x(D) \bmod(D^n - 1) = x_{n-2} + X_{n-1}D + \cdots + x_{n-3}D^{n-1}$$

which is a polynomial representation of the code word $(x_{n-2}, x_{n-1}, \ldots, x_{n-3})$, and so on. In general, we may describe the cyclic property in polynomial notation by stating that if $x(D)$ is a code word polynomial, then the polynomial $D^ix(D) \bmod(D^n - 1)$ is also a code word polynomial for any cyclic shift i.

(1)   Generator Polynomial

An (n, k) cyclic code is specified by the complete set of code word polynomials of degree $(n - 1)$ or less, which contains a polynomial of minimum degree $(n - k)$ as a factor. This special factor, denoted by $g(D)$, is called the *generator polynomial* of the code. The degree of $g(D)$ is equal to the number of parity bits in the code. The generator polynomial $g(D)$ is equivalent to the generator matrix **G** as a description of the code.

The generator polynomial $g(D)$ has the following properties:

PROPERTY 1

The generator polynomial of an (n, k) cyclic code is unique in that it is the only code word polynomial of minimum degree $(n - k)$.

We justify this property by contradiction. Suppose there is another code word polynomial of the same degree as the generator polynomial $g(D)$. If this were possible, then we may add the two polynomials and thereby get a code word polynomial of degree less than $(n - k)$. This is impossible, since $(n - k)$ is the minimum degree. Hence, $g(D)$ is unique.

As a corollary to Property 1, we may expand the generator polynomial of an (n, k) cyclic code as follows:

$$g(D) = 1 + \sum_{i=1}^{n-k-1} g_i D^i + D^{n-k} \qquad (8.35)$$

where g_i is a coefficient equal to 0 or 1. That is, $g(D)$ has two terms with coefficient 1 separated by $(n - k - 1)$ terms with undetermined coefficients.

PROPERTY 2

Any multiple of the generator polynomial $g(D)$ is a code word polynomial, as shown by

$$x(D) = a(D)g(D) \bmod(D^n - 1) \qquad (8.36)$$

where $a(D)$ is a polynomial in D.

Let the polynomial $a(D)$ be expressed as

$$a(D) = a_0 + a_1 D + \cdots + a_{k-1} D^{k-1}$$

where $a_0, a_1, \ldots, a_{k-1}$ are equal to 0 or 1. We express the result of multiplying $g(D)$ by $a(D)$ in the expanded form

$$a(D)g(D) = a_0 g(D) + a_1 D g(D) + \cdots + a_{k-1} D^{k-1} g(D)$$

From the cyclic property, $D^n g(D)$ is a code word polynomial. Moreover, from the linearity property, a linear combination of code word polynomials is also a code word polynomial. It follows therefore that any multiple of $g(D)$, modulo $(D^n - 1)$, is a code word polynomial as stated in Eq. 8.36. Conversely, we may state that a binary polynomial of order $n - 1$ or less is a code word polynomial if and only if it is a multiple of $g(D)$.

Suppose we are given the generator polynomial $g(D)$, and the requirement is to encode the message sequence $(m_0, m_1, \ldots, m_{k-1})$ into an (n, k) *systematic cyclic* code. That is, the message bits are transmitted in unaltered form, as shown by the following structure for a code word (see Fig. 8.4):

$$\underbrace{(b_0, b_1, \ldots, b_{n-k-1},}_{n-k \text{ parity bits}} \underbrace{m_0, m_1, \ldots, m_{k-1})}_{k \text{ message bits}}$$

Let the *message polynomial* be defined by

$$m(D) = m_0 + m_1 D + \cdots + m_{k-1} D^{k-1} \tag{8.37}$$

However, according to the structure specified for a code word, the leading message bit, m_{k-1}, is the coefficient of D^{n-1} in the corresponding polynomial representation. To accommodate this requirement, we multiply the message polynomial $m(D)$ by D^{n-k}. We may thus write

$$D^{n-k}m(D) = m_0 D^{n-k} + m_1 D^{n-k+1} + \cdots + m_{k-1} D^{n-1} \tag{8.38}$$

Let the polynomial $D^{n-k}m(D)$ be divided by the generator polynomial $g(D)$, resulting in a quotient $a(D)$ and remainder $b(D)$. That is, we have

$$\frac{D^{n-k}m(D)}{g(D)} = a(D) + \frac{b(D)}{g(D)} \tag{8.39}$$

or, equivalently,

$$D^{n-k}m(D) = a(D)g(D) + b(D) \tag{8.40}$$

where, in general,

$$a(D) = a_0 + a_1 D + \cdots + a_{k-1} D^{k-1} \tag{8.41}$$

and

$$b(D) = b_0 + b_1 D + \cdots + b_{n-k-1} D^{n-k-1} \tag{8.42}$$

In modulo-2 arithmetic, $b(D) = -b(D)$. We may therefore rearrange Eq. 8.40 as

$$b(D) + D^{n-k}m(D) = a(D)g(D) \tag{8.43}$$

This polynomial is a multiple of the generator polynomial $g(D)$. Therefore, in accordance with Property 2, it represents a code word polynomial of the (n, k) cyclic code generated by $g(D)$. We may thus write

$$x(D) = b(D) + D^{n-k}m(D) \tag{8.44}$$

The degree of the remainder $b(D)$ is always less than the degree of the divisor, which is $n - k$. The term $D^{n-k}m(D)$ contains only terms whose exponents are equal to or greater than $n - k$. Accordingly, adding these two terms as in Eq. 8.44 will not modify any of the terms in either expression. It is also of interest to note that the coefficients of the remainder $b(D)$ are simply the $(n - k)$ parity bits. It was in anticipation of this result that we used the representation shown in Eq. 8.42 to denote the remainder $b(D)$.

In summary, there are three steps involved in the encoding procedure for an (n, k) cyclic code assured of a systematic structure:

1. Multiply the message polynomial $m(D)$ by D^{n-k}.
2. Divide $D^{n-k}m(D)$ by the generator polynomial $g(D)$, obtaining the remainder $b(D)$.
3. Add $b(D)$ to $D^{n-k}m(D)$, obtaining the code word polynomial $x(D)$.

(2) Parity-check Polynomial

An (n, k) cyclic code is uniquely specified by its generator polynomial $g(D)$ of order $(n - k)$. Such a code is also uniquely specified by another polynomial of degree k, which is called the *parity-check polynomial*. Earlier we remarked that the generator polynomial $g(D)$ is an equivalent representation of the generator matrix **G**. Correspondingly, the parity-check polynomial, denoted by $h(D)$, is an equivalent representation of the parity-check matrix **H**. We thus find that the matrix relation $\mathbf{HG}^T = \mathbf{0}$ presented in Eq. 8.18 for linear block codes corresponds to the relationship

$$h(D)g(D) \bmod(D^n - 1) = 0 \tag{8.45}$$

From Property 2, any multiple of the generator polynomial $g(D)$ of an (n, k) cyclic code is a code word polynomial. Accordingly, we find from Eq. 8.45 that any code word polynomial $x(D)$ in the code satisfies the following fundamental relation:

$$h(D)x(D) \bmod(D^n - 1) = 0 \tag{8.46}$$

In modulo-2 arithmetic, $1 - D^n$ has the same value as $1 + D^n$. Accordingly, we deduce from Eq. 8.45 that the generator and parity-check polynominals of a linear cyclic code have another property, stated as follows:

PROPERTY 3

The generator polynomial $g(D)$ and the parity-check polynomial $h(D)$ are factors of the polynomial $1 + D^n$, as shown by

$$h(D)g(D) = 1 + D^n \tag{8.47}$$

This property provides the basis for selecting the generator or parity-check polynomial of a cyclic code. In particular, we may state that if $g(D)$ is a polynomial of degree $(n - k)$, and it is also a factor of $1 + D^n$, then $g(D)$ is the generator polynomial of an (n, k) cyclic code. Equivalently, we may state that if $h(D)$ is a polynomial of degree k, and it is also a factor of $1 + D^n$, then $h(D)$ is the parity-check polynomial of an (n, k) cyclic code.

Any factor of $1 + D^n$ with degree $(n - k)$, the number of parity bits, can be used as a generator polynomial. For large values of n, the polynomial $1 + D^n$ may have many factors of degree $n - k$. Some of these polynomial factors generate good cyclic codes, while some of them generate bad cyclic codes. The issue of how to select generator polynomials that produce good cyclic codes is very difficult to resolve. Indeed, coding theorists have expended much effort in the search for good cyclic codes.

EXAMPLE 3 HAMMING CODES REVISITED

To illustrate issues relating to the polynomial representation of cyclic codes, we consider the generation of a (7,4) cyclic code. With the block length $n = 7$, we start by factorizing $1 + D^7$ into three *irreducible polynomials:*

$$1 + D^7 = (1 + D)(1 + D^2 + D^3)(1 + D + D^3)$$

By an irreducible polynomial we mean a polynomial that cannot be factored using only polynomials with coefficients from the binary field. An irreducible polynomial of degree m is said to be *primitive* if the smallest positive integer n for which the polynomial divides $1 + D^n$ is $n = 2^m - 1$. For the example at hand, only two polynomials, namely, $(1 + D^2 + D^3)$ and $(1 + D + D^3)$ are primitive. Let us take

$$g(D) = 1 + D + D^3$$

as the generator polynomial, whose degree equals the number of parity bits. This means that the parity-check polynomial is given by

$$h(D) = (1 + D)(1 + D^2 + D^3)$$
$$= 1 + D + D^2 + D^4$$

whose degree equals the number of message bits $k = 4$.

Next, we illustrate the procedure for the construction of a code word by using this generator polynomial to encode the message sequence 1001. The corresponding message polynomial is given by

$$m(D) = 1 + D^3$$

Hence, multiplying $m(D)$ by $D^{n-k} = D^3$, we get

$$D^{n-k}m(D) = D^3 + D^6$$

The second step is to divide $D^{n-k}m(D)$ by $g(D)$, the details of which are given below:

$$
\require{enclose}
\begin{array}{r}
D^3 + D \\
D^3 + D + 1 \enclose{longdiv}{} \\
\end{array}
$$

```
                      D³ + D
         ┌──────────────────────────────
D³ + D + 1│ D⁶              + D³
          │ D⁶    + D⁴ + D³
          ├──────────────────────────────
                   D⁴
                   D⁴     + D² + D
          ├──────────────────────────────
                          D² + D
```

Note that in this long division we have treated subtraction the same as addition, since we are operating in modulo-2 arithmetic. Accordingly, we have

$$\frac{D^3 + D^6}{1 + D + D^3} = D + D^3 + \frac{D + D^2}{1 + D + D^3}$$

That is, the quotient $a(D)$ and remainder $b(D)$ are as follows, respectively:

$$a(D) = D + D^3$$

and

$$b(D) = D + D^2$$

Hence, from Eq. 8.44 we find that the desired code word polynomial is

$$x(D) = b(D) + D^{n-k}m(D)$$
$$= D + D^2 + D^3 + D^6$$

The code word is therefore 0111001. The four right-most bits, 1001, are the specified message bits. The three left-most bits, 011, are the parity-check bits. The code word thus generated is exactly the same as the corresponding one shown in Table 8.1 for a (7,4) Hamming code.

We may generalize this result by stating that *any cyclic code generated by a primitive polynomial is a Hamming code of minimum distance 3.*

In the remainder of the example, we show that the generator polynomial $g(D)$ and the parity-check polynomial $h(D)$ uniquely specify the generator matrix **G** and the parity-check matrix **H**, respectively.

To construct the 4-by-7 generator matrix **G**, we start with four polynomials represented by $g(D)$ and three cyclic-shifted versions of it, as shown by

$$g(D) = 1 + D + D^3$$

$$Dg(D) = D + D^2 + D^4$$

$$D^2g(D) = D^2 + D^3 + D^5$$

$$D^3g(D) = D^3 + D^4 + D^6$$

From Property 1, the polynomials $g(D)$, $Dg(D)$, $D^2g(D)$, and $D^3g(D)$ represent code word polynomials in the (7,4) Hamming code. If the coefficients of these polynomials are used as the elements of the rows of a 4-by-7 matrix, we get the following generator matrix:

$$\mathbf{G} = \begin{bmatrix} 1 & 1 & 0 & 1 & 0 & 0 & 0 \\ 0 & 1 & 1 & 0 & 1 & 0 & 0 \\ 0 & 0 & 1 & 1 & 0 & 1 & 0 \\ 0 & 0 & 0 & 1 & 1 & 0 & 1 \end{bmatrix}$$

Clearly, the generator matrix **G** so constructed is not in systematic form. We can transform it into a systematic form by adding the first row to the third row, and adding the sum of the first two rows to the fourth row. These manipulations result in the new generator matrix

$$\mathbf{G} = \begin{bmatrix} 1 & 1 & 0 & 1 & 0 & 0 & 0 \\ 0 & 1 & 1 & 0 & 1 & 0 & 0 \\ 1 & 1 & 1 & 0 & 0 & 1 & 0 \\ 1 & 0 & 1 & 0 & 0 & 0 & 1 \end{bmatrix}$$

This generator matrix is exactly the same as that in Example 2 (see Eq. 8.31).

We next show how to construct the 3-by-7 parity-check matrix **H** from the parity-check polynomial $h(D)$. To do this, however, we first take the *reciprocal* of $h(D)$, defined as $D^k h(D^{-1})$. The polynomial $D^k h(D^{-1})$ is also a factor of $1 + D^n$. For the problem at hand, we form three polynomials represented by $D^4 h(D^{-1})$ and two cyclic-shifted versions of it, as shown by

$$D^4 h(D^{-1}) = 1 + D^2 + D^3 + D^4$$

$$D^5 h(D^{-1}) = D + D^3 + D^4 + D^5$$

$$D^6 h(D^{-1}) = D^2 + D^4 + D^5 + D^6$$

If the coefficients of these three polynomials are used as the elements of the rows of a 3-by-7 matrix, we get the following parity-check matrix:

$$\mathbf{H} = \begin{bmatrix} 1 & 0 & 1 & 1 & 1 & 0 & 0 \\ 0 & 1 & 0 & 1 & 1 & 1 & 0 \\ 0 & 0 & 1 & 0 & 1 & 1 & 1 \end{bmatrix}$$

Here again we see that the matrix $\mathbf{H}$ is not in systematic form. To transform it into a systematic form, we add the third row to the first row obtaining

$$\mathbf{H} = \begin{bmatrix} 1 & 0 & 0 & 1 & 0 & 1 & 1 \\ 0 & 1 & 0 & 1 & 1 & 1 & 0 \\ 0 & 0 & 1 & 0 & 1 & 1 & 1 \end{bmatrix}$$

This parity-check matrix is exactly the same as that of Example 2 (see Eq. 8.32).

(3) Encoder for Cyclic Codes

Earlier we showed that the encoding procedure for an (n, k) cyclic code in systematic form involves three steps: (1) multiplication of the message polynomial $m(D)$ by D^{n-k}, (2) division of $D^{n-k}m(D)$ by the generator polynomial $g(D)$ to obtain the remainder $b(D)$, and (3) addition of $b(D)$ to $D^{n-k}m(D)$ to form the desired code word polynomial. These three steps can be implemented by means of the encoder shown in Fig. 8.6, consisting of a *linear feedback shift register* with $(n - k)$ stages.

The boxes in Fig. 8.6 represent *flip-flops* or *delay elements*. The flip-flop is a device that resides in one of two possible states denoted by 0 and 1. An *external clock* (not shown in Fig. 8.6) controls the operation of all the flip-flops. Every time the clock ticks, the contents of the flip-flops (initially set in the state 0) are shifted out in the direction of the arrows. In addition to the flip-flops, the encoder of Fig. 8.6 includes a second set of logic elements, namely, *adders*, which compute the modulo-2 sums of their respective inputs. Finally, the *multipliers* multiply their respective inputs by the associated coefficients. In particular, if the coefficient $g_i = 1$, the multiplier is just a direct "connection." If, on the other hand, the coefficient $g_i = 0$, the multiplier is "no connection."

The operation of the encoder shown in Fig. 8.6 proceeds as follows:

1. The gate is switched on. Hence, the k message bits are shifted into the channel. As soon as the k message bits have entered the shift register, the

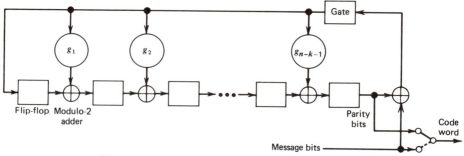

Figure 8.6 Encoder for an (n,k) cyclic code.

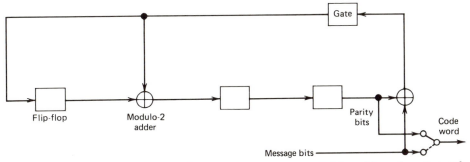

Figure 8.7   Encoder for the (7,4) cyclic code generated by $g(d) = 1 + D + D^3$.

resulting $(n - k)$ bits in the register form the parity bits [recall that the parity bits are the same as the coefficients of the remainder $b(D)$].

2. The gate is switched off, thereby breaking the feedback connections.
3. The contents of the shift register are shifted out into the channel.

EXAMPLE 4   ENCODER FOR THE (7,4) CYCLIC HAMMING CODE

Figure 8.7 shows the encoder for the (7,4) cyclic Hamming code generated by the polynomial $g(D) = 1 + D + D^3$. To illustrate the operation of this encoder, consider the message sequence (1001). The contents of the shift register are modified by the incoming message bits as in Table 8.3. After four shifts, the contents of the shift register, and therefore the parity bits, are (011). Accordingly, appending these parity bits to the message bits (1001), we get the code word (0111001). This result is exactly the same as that determined in Example 3.

Table 8.3   Contents of the Shift
Register in the Encoder of Fig. 8.7 for
Message Sequence (1001)

Shift	Input	Register Contents
		0 0 0 (initial state)
1	1	1 1 0
2	0	0 1 1
3	0	1 1 1
4	1	0 1 1

(4)   Calculation of the Syndrome

Suppose the code word $(x_0, x_1, \ldots, x_{n-1})$ is transmitted over a noisy channel, resulting in the received word $(y_0, y_1, \ldots, y_{n-1})$. From Section 8.3, we recall that the first step in the decoding of a linear block code is to calculate the syndrome for the received word. If the syndrome is zero, there are no transmission errors in the received word. If, on the other hand, the syndrome is nonzero, then the received word contains transmission errors that require correction.

In the case of a cyclic code in systematic form, the syndrome can be calculated easily. Let the received word be represented by a polynomial of degree

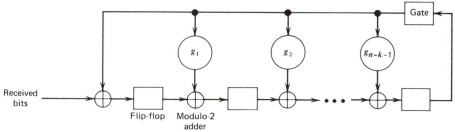

Figure 8.8 Syndrome calculator.

$n - 1$ or less, as shown by

$$y(D) = y_0 + y_1 D + \cdots + y_{n-1}D^{n-1} \tag{8.46}$$

Let $a(D)$ denote the quotient and $s(D)$ denote the remainder, which are the results of dividing $y(D)$ by the generator polynomial $g(D)$. We may therefore express $y(D)$ as follows

$$y(D) = a(D)g(D) + s(D) \tag{8.47}$$

The remainder $s(D)$ is a polynomial of degree $n - k - 1$ or less. It is called the *syndrome polynomial* in that its coefficients make up the $(n - k)$-by-1 syndrome **s**. When the syndrome polynomial $s(D)$ is nonzero, the presence of transmission errors in the received word is detected.

Figure 8.8 shows a *syndrome calculator* that is identical to the encoder of Fig. 8.6 except for the fact that the received bits are fed into the $(n - k)$ stages of the feedback shift register from the left. As soon as all the received bits have been shifted into the shift register, its contents define the desired syndrome **s**. Once we know **s**, we can determine the corresponding error pattern **e** and thereby make the appropriate correction, as described in Section 8.3.

EXAMPLE 5 SYNDROME CALCULATOR FOR THE (7,4) CYCLIC HAMMING CODE

For the (7,4) Hamming code generated by the polynomial $g(D) = 1 + D + D^3$, the syndrome calculator of Fig. 8.8 simplifies into the form shown in Fig. 8.9.

Let the transmitted code word be (0111001) and the received word be (0110001); that is, the middle bit is in error. As the received bits are fed into the shift register, initially set to zero, its contents are modified as in Table 8.4. At the end of the seventh shift, the

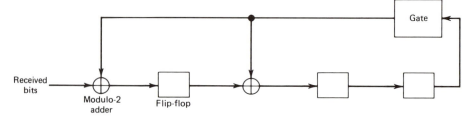

Figure 8.9 Syndrome calculator for the (7,4) cyclic code generated by the polynomial $g(D) = 1 + D + D^3$.

Table 8.4 Contents of the Syndrome Calculator in Fig. 8.6 for the Received Word 0110001

Shift	Input Bit	Contents of Shift Register
		000 (initial state)
1	1	100
2	0	010
3	0	001
4	0	110
5	1	111
6	1	001
7	0	110

syndrome is identified from the contents of the shift register as 110. Since the syndrome is nonzero, the received word is in error. Moreover, from Table 8.2, we see that the error pattern corresponding to this syndrome is 0001000. This indicates that the error is in the middle bit of the received word, which is indeed the case.

(5) Some Important Cyclic Codes

We conclude the discussion of cyclic codes by presenting the characteristics of five important classes of cyclic codes.*

(a) Cyclic Redundancy Check Codes

Cyclic codes are extremely well-suited for *error detection*. We say this for two reasons. First, they can be designed to detect many combinations of likely errors. Second, the implementation of both encoding and error-detecting circuits is practical. It is for these reasons that virtually all error-detecting codes used in practice are of the cyclic-code type. A cyclic code used for error-detection is referred to as *cyclic redundancy check (CRC) code*.

We define a *CRC error burst* of length B in an n-bit received word as a contiguous sequence of B bits in which the first and last bits and any number of intermediate bits are received in error. Such an error burst may also include an end-shifted version of a contiguous sequence. In any event, binary (n, k) CRC codes are capable of detecting the following error patterns:

1. All CRC error bursts of length $n - k$ or less.
2. A fraction of CRC error bursts of length equal to $n - k + 1$; the fraction equals $1 - 2^{-(n-k-1)}$.
3. A fraction of CRC error bursts of length greater than $n - k + 1$; the fraction equals $1 - 2^{-(n-k)}$.
4. All combinations of $d_{min} - 1$ (or less) errors.

---

* For more details on the codes listed in this subsection, see Odenwalder (1985), Michelson and Levesque (1985), Lin and Costello (1983), Clark and Cain (1981), McEliece (1977), MacWilliams and Sloane (1977), Peterson and Weldon (1972), and Berlekamp (1968).

5. All error patterns with an odd number of errors if the generator polynomial $g(D)$ for the code has an even number of nonzero coefficients.

Table 8.5 presents the generator polynomials of three CRC codes that have become international standards. All three contain $1 + D$ as a prime factor. The CRC-12 code is used when the character length is 6 bits. The other two are used for 8-bit characters.

Table 8.5 CRC Codes

Code	Generator Polynomial, $g(D)$
CRC-12 code	$1 + D + D^2 + D^3 + D^{11} + D^{12}$
CRC-16 code	$1 + D^2 + D^{15} + D^{16}$
CRC-CCITT[a] code	$1 + D^5 + D^{12} + D^{16}$

[a] CCITT is an abbreviation for "Comité Consultaitif International Téléphonique et Télégraphique," a Geneva-based organization made up of telephone companies from all over the world.

(b) Maximum-length Codes

For any positive integer $m \geqslant 3$, there exists a *maximum-length code* with the following parameters:

Block length: $n = 2^m - 1$
Number of message bits: $k = m$
Minimum distance: $d_{\min} = 2^{m-1}$

Maximum-length codes are generated by polynomials of the form

$$g(D) = \frac{1 + D^n}{h(D)} \tag{8.50}$$

where $h(D)$ is any primitive polynomial of degree m. Earlier we stated that any cyclic code generated by a primitive polynomial is a Hamming code of minimum distance 3 (see Example 3). It follows therefore that maximum-length codes are the *dual* of Hamming codes.

Maximum-length codes are also referred to as *pseudo-noise (PN) codes;* the name "pseudo-noise" is derived from the fact that these codes have correlation and spectral characteristics that resemble those of a white noise sequence. More will be said on them in Chapter 9.

(c) Golay Codes

The (23, 12) *Golay code* is a very special binary code that is capable of correcting any combination of three or fewer random errors in a block of 23 bits. The code has minimum distance of 7. Moreover, it is a perfect code in that it satisfies the Hamming bound of Eq. 8.30 for $t = 3$ with the equality sign, as shown by the number-theoretic fact:

$$1 + \binom{23}{1} + \binom{23}{2} + \binom{23}{3} = 2048 = 2^{11}$$

Indeed, the (23, 12) Golay code is the only known three-error correcting binary perfect cyclic code.

The (23, 12) Golay code is generated either by the polynomial

$$g_1(D) = 1 + D^2 + D^4 + D^5 + D^6 + D^{10} + D^{11} \qquad (8.51)$$

or by the polynomial

$$g_2(D) = 1 + D + D^5 + D^6 + D^7 + D^9 + D^{11} \qquad (8.52)$$

Both $g_1(D)$ and $g_2(D)$ are factors of $1 + D^{23}$. In particular,

$$1 + D^{23} = (1 + D)g_1(D)g_2(D) \qquad (8.52)$$

Unfortunately, the Golay code does not generalize to other combinations of code parameters n and k.

(d) Bose–Chaudhuri–Hocquenqhem (BCH) Codes

One of the most important and powerful classes of linear block codes are *BCH codes,* which are cyclic codes with a wide variety of parameters. The most common BCH codes are characterized as follows. Specifically, for any positive integers m (equal to or greater than 3) and t [less than $(2^m - 1)/2$] there exists a binary BCH code with the following parameters:*

Block length:	$n = 2^m - 1$
Number of message bits:	$k \geq n - mt$
Minimum distance:	$d_{min} \geq 2t + 1$

Each BCH code is a *t-error correcting code* in that it can detect and correct up to t random errors per code word. The Hamming single-error corecting codes can be described as BCH codes.

The BCH codes offer flexibility in the choice of code parameters, namely, block length and code rate. Furthermore, at block lengths of a few hundred or less, the BCH codes are among the best known codes of the same block length and code rate.

In Table 8.6, we present the code parameters and generator polynomials for binary block BCH codes of length up to $2^5 - 1$.

(e) Reed–Solomon Codes

The *Reed–Solomon codes*† are an important subclass of *nonbinary* BCH codes; they are often abbreviated as RS codes. The encoder for an RS code differs from a binary encoder in that it operates on multiple bits rather than individual bits. Specifically, the encoder for an RS(n,k) code on m-bit symbols groups the incoming binary data stream into blocks, each km bits long. Each block is treated as k symbols, with each symbol having m bits. The encoding algorithm expands a block of k symbols to n symbols by adding $n - k$ redundant symbols. When m is an integer power of two, the m-bit symbols are called

---

* This type of BCH codes is called *primitive BCH codes*. The details of their algebraic development are beyond the scope of this book. For a description of BCH codes and their decoding algorithms, see Lin and Costello (1983, pp. 141–183), and MacWilliams and Sloane (1977, pp. 257–293).

† The Reed–Solomon codes are named in honor of their inventors Reed and Solomon (1960); for details of Reed–Solomon codes, see MacWilliams and Sloane (1977, pp. 294–306).

Table 8.6[a]   Binary BCH Codes of Length up to $2^5 - 1$

n	k	t				Generator Polynomial					
7	4	1								1	011
15	11	1								10	011
15	7	2							111	010	001
15	5	3						10	100	110	111
31	26	1								100	101
31	21	2						11	101	101	001
31	16	3				1	000	111	110	101	111
31	11	5			101	100	010	011	011	010	101
31	6	7	11	001	011	011	110	101	000	100	111

Notations: n = block length

k = number of message bits

t = maximum number of detectable errors

The high-order coefficients of the generator polynomial $g(D)$ are at the left.

Example: Suppose we wish to construct the generator polynomial for (15,7) BCH code. From the table, we have (111 010 001) for the coefficients of the generator polynomial. Hence,

$$g(D) = D^8 + D^7 + D^6 + D^4 + 1$$

[a] For a table of binary BCH codes of length $2^{10} - 1$, see Lin and Costello (1983, pp. 583–598).

bytes. A popular value of m is 8; indeed, 8-bit RS codes are extremely powerful.

A t-error-correcting RS code has the following parameters:

Block length:	$n = 2^m - 1$ symbols
Message size:	k symbols
Parity-check size:	$n - k = 2t$ symbols
Minimum distance:	$d_{min} = 2t + 1$ symbols

The block length of the RS code is one less than the size of a code symbol, and the minimum distance is one greater than the number of parity symbols. Indeed, the minimum distance is always equal to the design distance of the code. It is easy to show that no (n,k) linear block code can have minimum distance greater than $n - k + 1$. An (n,k) linear block code for which the minimum distance equals $n - k + 1$ is called a *maximum-distance separable* code. Therefore, every RS code is a maximum-distance separable code. Accordingly, the RS codes make highly efficient use of redundancy, and block lengths and symbol sizes can be adjusted readily to accommodate a wide range of message sizes. Moreover, the RS codes provide a wide range of code rates that can be chosen to optimize performance. Finally, efficient decoding techniques are available for use with RS codes.

EXAMPLE 6

Consider a single-error-correcting RS code with a 2-bit byte (symbol). That is, $t = 1$ and $m = 2$. Denoting the four possible symbols as 0, 1, 2, and 3, we can write their binary representations as follows:

$$0:00$$
$$1:01$$
$$2:10$$
$$3:11$$

The code has the following parameters:

$$n = 2^2 - 1 = 3 \text{ bytes} = 6 \text{ bits}$$

$$n - k = 2 \text{ bytes} = 4 \text{ bits}$$

$$\text{Code rate}: r = \frac{k}{n} = \frac{1}{3}$$

It can correct any *in-phase burst* (i.e., spanning a symbol) of length 2.

8.5 CONVOLUTIONAL CODES

In block coding, the encoder accepts a k-bit message block and generates an n-bit code word. Thus, code words are produced on a block-by-block basis. Clearly, provision must be made in the encoder to buffer an entire message block before generating the associated code word. There are applications, however, where the message bits come in *serially* rather than in large blocks, in which case the use of a buffer may be undesirable. In such situations, the use of *convolutional coding** may be the preferred method. A convolutional encoder operates on the incoming message sequence continuously in a serial manner.

The encoder of a binary convolutional code with rate $1/n$, measured in bits per symbol, may be viewed as a *finite-state machine* that consists of an M-stage shift register with prescribed connections to n modulo-2 adders, and a multiplexer that serializes the outputs of the adders. An L-bit message sequence produces a coded output sequence of length $n(L + M)$ bits. The *code rate* is therefore given by

$$2(5+3) = 16 \cdot$$

$$r = \frac{L}{n(L + M)} \quad \text{bits/symbol} \tag{8.54}$$

Typically, we have $L \gg M$. Hence, the code rate simplifies as

$$r \simeq \frac{1}{n} \quad \text{bits/symbol} \tag{8.55}$$

The *constraint length*† of a convolutional code, expressed in terms of mes-

---

* Convolutional codes were first introduced by Elias (1955) as an alternative to block codes.

† In the literature, the constraint length of a convolutional code is also expressed in terms of coded symbols, as the number of output code symbols that are influenced by a single message bit. According to this definition, the constraint length equals nK, where n is the number of modulo-2 adders.

sage bits, is defined as the number of shifts over which a single message bit can influence the encoder output. In an encoder with an M-stage shift register, the *memory* of the encoder equals M message bits, and $K = M + 1$ shifts are required for a message bit to enter the shift register and finally come out. Hence, the constraint length of the encoder is K.

Figure 8.10a shows a convolutional encoder with $n = 2$ and $K = 3$. Hence, the code rate of this encoder is 1/2. The encoder of Fig. 8.10a operates on the incoming message sequence, one bit at a time.

We may generate a binary convolutional code with rate k/n by using k sepa-

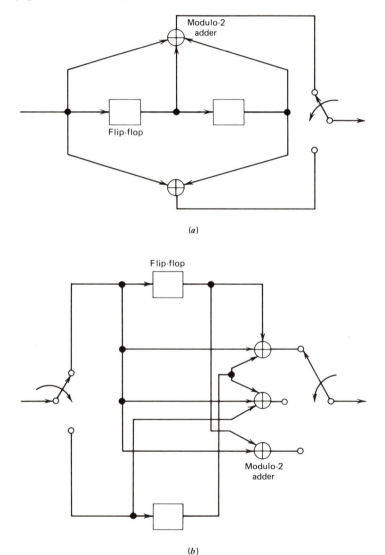

(a)

(b)

Figure 8.10 (a) Constraint length $-$ 3, rate $- \frac{1}{2}$ convolutional encoder. (b) Constraint length $-$ 2, rate $- \frac{2}{3}$ convolutional encoder.

rate shift registers with prescribed connections to n modulo-2 adders, an input multiplexer and an output multiplexer. An example of such an encoder is shown in Fig. 8.10b, where $k = 2$, $n = 3$, and the two shift registers have $K = 2$ each. The code rate is 2/3. In this second example, the encoder processes the incoming message sequence two bits at a time.

The convolutional codes generated by the encoders of Fig. 8.10 are *nonsystematic* codes. Unlike block coding, the use of nonsystematic codes is ordinarily preferred over systematic codes in convolutional coding.

In the sequel, we will use the encoder of Fig. 8.10a to illustrate the theory of convolutional encoders, using a time-domain approach first and then a transform-domain approach. The encoder of Fig. 8.10b is left as an exercise to the reader (see Problems 8.5.4 and 8.5.11).

(1) Time-domain Approach

The time-domain behavior of a binary convolutional encoder with code rate $1/n$ may be defined in terms of a set of n *impulse responses*. The simple encoder of Fig. 8.10a has a code rate of 1/2. We need two impulse responses to characterize its behavior in the time domain. Let the sequence $(g_0^{(1)}, g_1^{(1)}, \ldots, g_M^{(1)})$ denote the impulse response of the "input-top adder–output path" of the encoder, and the sequence $(g_0^{(2)}, g_1^{(2)}, \ldots, g_M^{(2)})$ denote the impulse response of the "input-bottom adder–output path." These two impulse responses are obtained by determining the two output sequences of the encoder that are produced in response to the input sequence $(1, 0, 0, \ldots)$. The impulse responses so defined are called the *generator sequences* of the code.

As the name implies, a convolutional encoder operates by performing convolutions on the incoming message sequence. Let $(m_0, m_1, m_2, \ldots)$ denote the message sequence that enters the encoder of Fig. 8.10a, one bit at a time (starting with m_0). The encoder generates the two output sequences, denoted by $\{x_i^{(1)}\}$ and $\{x_i^{(2)}\}$, by convolving the message sequence with the impulse responses of the input-top adder–output and input-bottom adder-output paths, respectively. Thus, the top output sequence is defined by the *convolution sum*:

$$x_i^{(1)} = \sum_{\ell=0}^{M} g_\ell^{(1)} m_{i-\ell} \qquad i = 0, 1, 2, \ldots \tag{8.56}$$

where $m_{i-\ell} = 0$ for all $\ell > i$. Likewise, the bottom output sequence of the encoder is described by

$$x_i^{(2)} = \sum_{\ell=0}^{M} g_\ell^{(2)} m_{i-\ell} \qquad i = 0, 1, 2, \ldots \tag{8.57}$$

After convolution, the two sequences $\{x_i^{(1)}\}$ and $\{x_i^{(2)}\}$ are combined by the multiplexer to produce the encoder output sequence $\{x_i\}$, as shown by

$$\{x_i\} = \{x_0^{(1)}, x_0^{(2)}, x_1^{(1)}, x_1^{(2)}, x_2^{(1)}, x_2^{(2)}, \ldots\} \tag{8.58}$$

This sequence is then fed into the discrete channel input.

EXAMPLE 7

The convolutional encoder of Fig. 8.10a has the following two generator sequences, each of length 3 (the same as the constraint length $K = 3$):

1. **Input-top adder–output path:**

$$(g_0^{(1)}, g_1^{(1)}, g_2^{(1)}) = (1,1,1)$$

2. **Input-bottom adder–output path:**

$$(g_0^{(2)}, g_1^{(2)}, g_2^{(2)}) = (1,0,1)$$

Note that the generator sequences can be determined directly from the circuit diagram of the encoder. In particular, the impulse response of either input–output path of the encoder is the same as the corresponding sequence of connections from the shift register to the pertinent adder, with a 1 representing a "connection" and a 0 representing "no connection."

Let the incoming message sequence be as follows:

$$(m_0, m_1, m_2, m_3, m_4) = (10011)$$

Then, the use of Eq. 8.56 yields the following values for the elements that constitute the top output sequence:

$$
\begin{aligned}
x_0^{(1)} &= g_0^{(1)} m_0 \\
&= 1 \times 1 = 1 \\
x_1^{(1)} &= g_0^{(1)} m_1 + g_1^{(1)} m_0 \\
&= 1 \times 0 + 1 \times 1 = 1 \\
x_2^{(1)} &= g_0^{(1)} m_2 + g_1^{(1)} m_1 + g_2^{(1)} m_0 \\
&= 1 \times 0 + 1 \times 0 + 1 \times 1 = 1 \\
x_3^{(1)} &= g_0^{(1)} m_3 + g_1^{(1)} m_2 + g_2^{(1)} m_1 \\
&= 1 \times 1 + 1 \times 0 + 1 \times 0 = 1 \\
x_4^{(1)} &= g_0^{(1)} m_4 + g_1^{(1)} m_3 + g_2^{(1)} m_2 \\
&= 1 \times 1 + 1 \times 1 + 1 \times 0 = 0 \\
x_5^{(1)} &= g_1^{(1)} m_4 + g_2^{(1)} m_3 \\
&= 1 \times 1 + 1 \times 1 = 0 \\
x_6^{(1)} &= g_2^{(1)} m_4 \\
&= 1 \times 1 = 1
\end{aligned}
$$

Hence, the top output is (1111001).

Consider next the evaluation of the bottom output sequence. The use of Eq. 8.57 yields the following values for its elements:

$$
\begin{aligned}
x_0^{(2)} &= g_0^{(2)} m_0 \\
&= 1 \times 1 = 1 \\
x_1^{(2)} &= g_0^{(2)} m_1 + g_1^{(1)} m_0 \\
&= 1 \times 0 + 0 \times 1 = 0 \\
x_2^{(2)} &= g_0^{(2)} m_2 + g_1^{(2)} m_1 + g_2^{(2)} m_0 \\
&= 1 \times 0 + 0 \times 0 + 1 \times 1 = 1
\end{aligned}
$$

$$x_3^{(2)} = g_0^{(2)} m_3 + g_1^{(2)} m_2 + g_2^{(2)} m_1$$
$$= 1 \times 1 + 0 \times 0 + 1 \times 0 = 1$$
$$x_4^{(2)} = g_0^{(2)} m_4 + g_1^{(2)} m_3 + g_2^{(1)} m_2$$
$$= 1 \times 1 + 0 \times 1 + 1 \times 0 = 1$$
$$x_5^{(2)} = g_1^{(2)} m_4 + g_2^{(2)} m_3$$
$$= 0 \times 1 + 1 \times 1 = 1$$
$$x_6^{(2)} = g_2^{(2)} m_4$$
$$= 1 \times 1 = 1$$

Hence, the bottom output sequence is (1011111). Finally, multiplexing the two output sequences, we get the encoded sequence

$$\{x_i\} = (11, 10, 11, 11, 01, 01, 11)$$

Note that the message sequence of length $L = 5$ bits produces an output coded sequence of length $n (L + K - 1) = 14$ bits. Note also that for the shift register to be restored to its zero-initial state, a terminating sequence of $K - 1 = 2$ zeros is appended to the last input bit of the message sequence (to appreciate this effect, examine the calculations of $x_6^{(1)}$ for the upper path and $x_6^{(2)}$ for the lower path). The terminating sequence of $K - 1$ zeros is called the *tail of the message*.

(2)   Transform-domain Approach

From the study of linear filter theory, we know that the convolution integral, which describes the linear filtering operation in the time domain, is replaced by the multiplication of Fourier transforms in the frequency domain. Since a convolutional encoder is a linear time-invariant finite-state machine, we may simplify computation of the adder outputs by applying an appropriate transformation. Let the impulse response of each path in the encoder be replaced by a polynomial whose coefficients are represented by the respective elements of the impulse response. Thus, for the input-top adder-output path of the encoder, we define the polynomial

$$g^{(1)}(D) = g_0^{(1)} + g_1^{(1)}D + \cdots + g_M^{(1)}D^M \qquad (8.59)$$

where $g_0^{(1)}, g_1^{(1)}, \ldots, g_M^{(1)}$ are the elements of the impulse response of the path. The variable D denotes a *unit-delay operator*, with the power of D defining the number of time units by which the associated bit in the impulse response is delayed with respect to the first bit, $g_0^{(1)}$. Similarly, for the input-bottom adder-output path, we may define the corresponding polynomial:

$$g^{(2)}(D) = g_0^{(2)} + g_1^{(2)}D + \cdots + g_M^{(2)}D^M \qquad (8.60)$$

where $g_0^{(2)}, g_1^{(2)}, \ldots, g_M^{(2)}$ are the elements of the impulse response of the second path. The polynomials $g^{(1)}(D)$ and $g^{(2)}(D)$ are called the *generator polynomials* of the code. Note that the coefficients of $g^{(1)}(D)$ and $g^{(2)}(D)$ can also be written directly by inspection of the circuit diagram of the encoder. In particular, the coefficient of D^i is a 1 if there is a ''connection'' from the ith stage of the shift register to the input of the adder of interest, and a 0 if there is ''no connection," where $i = 0, 1, \ldots, K - 1$. Note also that $i = 0$ corresponds to the left-most stage and $i = K - 1$ corresponds to the right-most stage.

Consider next the message sequence $\{m_0, m_1, m_2, \ldots, m_{L-1}\}$, for which we define the *message polynomial*:

$$m(D) = m_0 + m_1 D + m_2 D^2 + \cdots + m_{L-1} D^{L-1} \tag{8.61}$$

where L is the length of the message sequence. Accordingly, the convolution sums of Eqs. 8.56 and 8.57 are replaced by polynomial multiplications, as shown by the following two input–output relations, respectively:

$$x^{(1)}(D) = g^{(1)}(D)\, m(D) \tag{8.62}$$

and

$$x^{(2)}(D) = g^{(2)}(D)\, m(D) \tag{8.63}$$

Having determined the two *output polynomials* $x^{(1)}(D)$ and $x^{(2)}(D)$, we can obtain the corresponding output sequences by simply reading off their individual coefficients, as illustrated in the following example.

EXAMPLE 8

The impulse response of the input-top adder–output path of the convolutional encoder in Fig. 8.10a is (1,1,1). Hence, the corresponding generator polynomial is given by

$$g^{(1)}(D) = 1 + D + D^2$$

The impulse response of the input-bottom adder–output path of the encoder is (101). Hence, the corresponding generator polynomial is given by

$$g^{(2)}(D) = 1 + D^2$$

For the message sequence (10011), we have the polynomial representation

$$m(D) = 1 + D^3 + D^4$$

Hence, the use of Eq. 8.62 yields the top output polynomial (in modulo-2 arithmetic)

$$x^{(1)}(D) = (1 + D + D^2)(1 + D^3 + D^4)$$
$$= 1 + D + D^2 + D^3 + D^6$$

From this we deduce that the top output sequence is (1111001), which is exactly the same as the result obtained in Example 7. Next, the use of Eq. 8.63 yields the bottom output polynomial

$$x^{(2)}(D) = (1 + D^2)(1 + D^3 + D^4)$$
$$= 1 + D^2 + D^3 + D^4 + D^5 + D^6$$

The bottom output sequence is therefore (1011111). Here again, this result is the same as that obtained in Example 7. Note, however, that the computational effort involved in determining the output sequences of the encoder is much less than the time-domain approach.

(3) Code Tree, Trellis, and State Diagram

Traditionally, the structural properties of a convolutional encoder are portrayed in graphical form by using any one of three equivalent diagrams: code tree, trellis, and state diagram. We will use the convolutional encoder of Fig. 8.10a as a running example to demonstrate the insights that each one of these three diagrams can provide.

We begin the discussion with the *code tree* of Fig. 8.11. Each branch of the tree represents an input symbol, with the corresponding pair of output binary symbols indicated on the branch. The convention used to distinguish the input

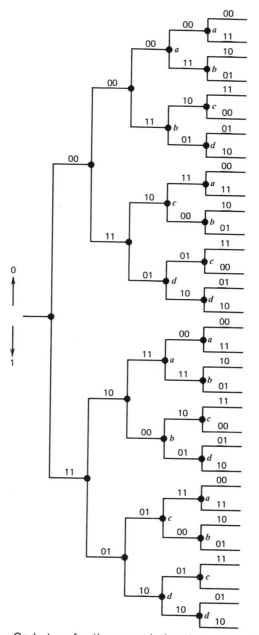

Figure 8.11   Code tree for the convolutional encoder of Fig. 8.10a.

binary symbols 0 and 1 is as follows. An input 0 specifies the upper branch of a bifurcation, while input 1 specifies the lower branch. A specific *path* in the tree is traced from left to right in accordance with the input (message) sequence. The corresponding coded symbols on the branches of that path constitute the sequence supplied by the encoder to the discrete channel input. Consider, for example, the message sequence (10011) applied to the input of the encoder of Fig. 8.10a. Following the procedure just described, we find that the corresponding encoded sequence is (11,10,11,11,01), which agrees with the first 5 pairs of bits of the encoded sequence $\{x_i\}$ derived in Example 7.

From the diagram of Fig. 8.11, we observe that the tree becomes *repetitive* after the first three branches. Indeed, beyond the third branch, the two nodes labeled **a** are identical, and so are all the other node pairs that are identically labeled. We may establish this repetitive property of the tree by examining the associated encoder of Fig. 8.10a. The encoder has memory $M = K - 1 = 2$ message bits. Hence, when the third message bit enters the encoder, the first message bit is shifted out of the register. Consequently, after the third branch, the message sequences (100 $m_3 m_4$ . . .) and (000 $m_3 m_4$ . . .) generate the same code symbols, and the pair of nodes labeled **a** may be joined together. The same reasoning applies to other nodes.

Accordingly, we may collapse the code tree of Fig. 8.11 into the new form shown in Fig. 8.12, called a *trellis*. It is so called since a trellis is a tree-like structure with remerging branches. The convention used in Fig. 8.12 to distinguish between input symbols 0 and 1 is as follows. A code branch produced by an input 0 is drawn as a solid line, while a code branch produced by an input 1 is drawn as a dashed line. As before, each input (message) sequence corresponds to a specific path through the trellis. For example, we readily see from Fig. 8.12 that the message sequence (10011) produces the encoded output sequence (11,10,11,11,01), which agrees with our previous result.

A trellis is more instructive than a tree in that it brings out explicitly the fact that the associated convolutional encoder is a finite-state machine. We define the *state* of a convolutional encoder of rate $1/n$ as the most recent $(K - 1)$ message bits moved into the encoder's shift register. At time j, the portion of the message sequence containing the most recent K bits is written as $(m_{j-K+1},$

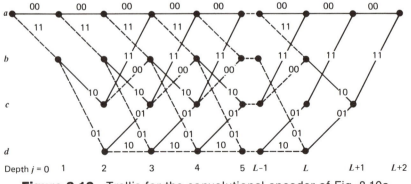

Figure 8.12 Trellis for the convolutional encoder of Fig. 8.10a.

$m_{j-K+2}, \ldots, m_{j-1}, m_j)$, where m_j is the *current* bit. The $(K-1)$-bit state of the encoder at time j is written simply as $(m_{j-1}, \ldots, m_{j-K+2}, m_{j-K+1})$. In the case of the simple convolutional encoder of Fig. 8.10a, we have $(K-1) = 2$. Hence, the state of this encoder can assume any one of four possible values, as described in Table 8.7. The trellis contains $(L + K)$ *levels*, where L is the length of the incoming message sequence, and K is the constraint length of the code. The levels of the trellis are labeled as $j = 0, 1, \ldots, L + K - 1$ in Fig. 8.12 for $K = 3$. Level j is also referred to as *depth j*; both terms are used interchangeably. The first $(K - 1)$ levels correspond to the encoder's departure from the initial state **a**, and the last $(K - 1)$ levels correspond to the encoder's return to the state **a**. Clearly, not all the states can be reached in these two portions of the trellis. However, in the central portion of the trellis, for which the level j lies in the range $K - 1 \leqslant j \leqslant L$, all the states of the encoder are reachable. Note also that the central portion of the trellis exhibits a fixed periodic structure.

Consider next a portion of the trellis corresponding to times j and $j + 1$. We assume that $j \geqslant 2$, so that it is possible for the current state of the encoder to be **a, b, c,** or **d**. For convenience of presentation, we have reproduced this portion of the trellis in Fig. 8.13. The left nodes represent the four possible current states of the encoder, while the right nodes represent the next states. Clearly,

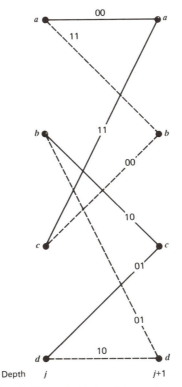

Figure 8.13 A portion of the central part of the trellis for the encoder of Fig. 8.10a.

Table 8.7 State Table for
the Convolutional Encoder
of Fig. 8.10a

State	Binary description
a	00
b	10
c	01
d	11

we may coalesce the left and right nodes. By so doing, we obtain the *state diagram* of the encoder, shown in Fig. 8.14. The nodes of the figure represent the four possible states of the encoder, with each node having two incoming branches and two outgoing branches. A transition from one state to another in response to input 0 is represented by a solid branch, while a transition in response to input 1 is represented by a dashed branch. The binary label on each branch represents the encoder's output as it moves from one state to another. Suppose, for example, the current state of the encoder is (01). The application of input 1 to the encoder of Fig. 8.10a results in the state (10) and the encoded output (00). Accordingly, with the help of this state diagram, we may readily determine the output of the encoder of Fig. 8.10a for any incoming message sequence. We simply start at state **a** and walk through the state diagram in

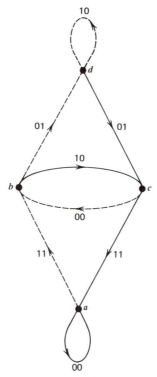

Figure 8.14 State diagram of the convolutional encoder of Fig. 8.10a.

accordance with the message sequence. We follow a solid branch if the input is a "0," and a dashed branch if it is a "1." As each branch is traversed, we output the corresponding binary label on the branch. Consider, for example, the message sequence (10011). For this input we follow the path **abcabd**, and therefore output the sequence (11,10,11,11,01), which agrees exactly with our earlier result. Thus, the input-output relation of a convolutional encoder is completely described by its state diagram.

8.6 MAXIMUM-LIKELIHOOD DECODING OF CONVOLUTIONAL CODES

In this section, we develop the *Viterbi algorithm** for the decoding of convolutional codes. The development proceeds in two stages. First, we show that for a binary symmetric channel, the maximum-likelihood decoder reduces to a minimum Hamming distance decoder. Next, we use the trellis representation of a convolutional code to establish the basic concepts involved in the formulation of the Viterbi algorithm. We refer to the decoding procedure as an "algorithm" because it has historically been implemented in software form on a computer or microprocessor. More recently, however, Viterbi decoders have been implemented in VLSI form.

Let **m** denote a *message vector*, and **x** denote the corresponding *code vector* applied by the encoder to the input of a discrete memoryless channel. Let **y** denote the *received vector*, which may differ from the transmitted code vector due to channel noise. Given the received vector **y**, the decoder is required to make an *estimate* $\hat{\mathbf{m}}$ of the message vector. Since there is a one-to-one correspondence between the message vector **m** and the code vector **x**, the decoder may equivalently produce an estimate $\hat{\mathbf{x}}$ of the code vector. We may then put $\hat{\mathbf{m}} = \mathbf{m}$ if and only if $\hat{\mathbf{x}} = \mathbf{x}$. Otherwise, a *decoding error* is committed in the receiver. The *decoding rule* for choosing the estimate $\hat{\mathbf{x}}$, given the received vector **y**, is said to be optimum when the *probability of decoding error* is minimized. From the material presented in Chapter 3, we may state that for equiprobable messages the probability of decoding error is minimized if the estimate $\hat{\mathbf{x}}$ is chosen to maximize the *log-likelihood function*. Let $p(\mathbf{y}|\mathbf{x})$ denote the conditional probability of receiving **y**, given that **x** was sent. The log-likelihood function equals $\ln p(\mathbf{y}|\mathbf{x})$. The *maximum likelihood decoder* or decision rule is described as follows:

Choose the estimate $\hat{x}$ if $\ln p(y|x)$ is maximum. (8.64)

---

* In 1967, Viterbi proposed a decoding algorithm for convolutional codes (Viterbi, 1967). Later, it was shown that the algorithm was equivalent to a technique known as "dynamic programming" in operations research (Omura, 1969). Notwithstanding this revelation, the algorithm has become known as the *Viterbi algorithm*. The algorithm was recognized by Forney to be a maximum-likelihood decoder (Forney, 1973). For introductory treatment of the Viterbi algorithm, see Hayes (1975), Forney (1973), McEliece (1977, pp. 206–213), Clark and Cain (1981, pp. 231–238), Michele-son and Levesque (1985, pp. 299–336), and Bhargava et al. (1981, pp. 371–402). The following two references present advanced mathematical treatments of the subject: Viterbi and Omura (1979, pp. 235–261) and Lin and Costello (1983, pp. 315–349).

Consider now the special case of a binary symmetric channel. In this case, both the transmitted code vector **x** and the received vector **y** represent binary sequences of length N, say. Naturally, these two sequences may differ from each other in some locations because of channel noise. Let x_i and y_i denote the ith elements of **x** and **y**, respectively. We then have

$$p(\mathbf{y}|\mathbf{x}) = \prod_{i=1}^{N} p(y_i|x_i) \qquad (8.65)$$

Correspondingly, the log-likelihood function equals

$$\ln p(\mathbf{y}|\mathbf{x}) = \sum_{i=1}^{N} \ln p(y_i|x_i) \qquad (8.66)$$

Let

$$p(y_i|x_i) = \begin{cases} p & \text{if } y_i \neq x_i \\ 1 - p & \text{if } y_i = x_i \end{cases}$$

Suppose also that the received vector **y** differs from the transmitted code vector **x** in exactly d positions. The number d is referred to as the *Hamming distance* between vector **y** and **x** (a formal definition of the Hamming distance and discussion of related issues were presented in Section 8.3). Then, we may rewrite the log-likelihood function in Eq. 8.66 as

$$\ln p(\mathbf{y}|\mathbf{x}) = d \ln p + (N - d) \ln(1 - p)$$

$$= d \ln\left(\frac{p}{1 - p}\right) + N \ln(1 - p) \qquad (8.67)$$

In general, the probability of an error occurring is low, such that we may assume $p < 1/2$. We also recognize that $N \ln(1 - p)$ is a constant for all **x**. Accordingly, we may state the maximum-likelihood decoding rule for the binary symmetric channel as follows:

Choose the estimate $\hat{\mathbf{x}}$ that minimizes the Hamming distance between the two vectors **y** *and* **x**. $\qquad (8.68)$

That is, for the binary symmetric channel, the maximum-likelihood decoder reduces to a *minimum distance decoder*. In such a decoder, the received vector **y** is compared with each possible transmitted code vector **x**, and the particular one closest to **y** is chosen as the correct transmitted code vector. The term "closest" is used in the sense of minimum number of differing binary symbols (i.e., Hamming distance) between the code vectors under investigation.

(1) The Viterbi Algorithm

The equivalence between maximum likelihood decoding and minimum distance decoding for a binary symmetric channel implies that we may decode a convolutional code by choosing a path in the code tree whose coded sequence differs from the received sequence in the fewest number of places. Since a code tree is equivalent to a trellis, we may equally limit our choice to the possible paths in the trellis representation of the code. The reason for preferring the trellis over the tree is that the number of nodes at any level of the trellis does not continue to grow as the number of incoming message bits increases; rather, it remains constant at 2^{K-1}, where K is the constraint length of the code.

Consider then the trellis diagram of Fig. 8.12 for a convolutional code with rate $r = 1/2$ and constraint length $K = 3$. We observe that at level $j = 3$, there are two paths entering any of the four nodes in the trellis. Moreover, those two paths will be identical onward from that point. Clearly, a minimum distance decoder may make a decision at that point as to which of those two paths to retain, without any loss of performance. A similar decision may be made at level $j = 4$, and so on.

This sequence of decisions is exactly what the Viterbi algorithm does as it walks through the trellis. The algorithm operates by computing a "metric" for every possible path in the trellis. The metric for a particular path is defined as the Hamming distance between the coded sequence represented by that path and the received sequence.* Thus, for each node (state) in the trellis of Fig. 8.12, the algorithm compares the two paths entering the node. The path with the lower metric is retained, and the other path is discarded. This computation is repeated for every level j of the trellis in the range $M \le j \le L$, where $M = K - 1$ is the encoder's memory and L is the length of the incoming message sequence. The paths that are retained by the algorithm are called *survivors*. For a convolutional code of constraint length $K = 3$, no more than $2^{K-1} = 4$ survivor paths and their metrics will ever be stored. This relatively small list of paths is always guaranteed to contain the maximum-likelihood choice.

A difficulty that may arise in the application of the Viterbi algorithm is the possibility that when the paths entering a state are compared, their metrics are found to be identical. In such a situation, we may make the choice by flipping a fair coin.

In summary, the Viterbi algorithm is a maximum-likelihood decoder, which is optimum for a white Gaussian noise channel. For an L-bit message sequence, and an encoder of memory M, the algorithm proceeds as follows (assuming that the encoder is initially in the all-zero state at $j = 0$):

Step 1 Starting at level (i.e., time unit) $j = M$, compute the metric for the single path entering each state of the encoder. Store the path (survivor) and its metric for each state.

Step 2 Increment the level j by 1. Compute the metric for all the paths entering each state by adding the metric of the incoming branches to the metric of the connecting survivor from the previous time unit. For each state, identify the path with the lowest metric as the survivor of step 2. Store the survivor and its metric.

Step 3 If level $j < L + M$, repeat step 2. Otherwise, stop.

---

* The definition given here for a metric is the negative of that in Chapter 3. There the metric was defined as the logarithm of the likelihood function.

Note that the memory is $M = K - 1$, where K is the constraint length of the code.

The great advantage of the Viterbi algorithm is that (for a constraint length K, code rate $r = k/n$, convolutional code) the number of operations performed in decoding L bits is $L2^{n(K-1)}$, which is linear in L. However, the number of operations performed per decoded bit is an exponential function of the constraint length K. This exponential dependence on K limits the utilization of the Viterbi algorithm as a practical decoding technique to relatively short constraint-length codes (typically, in the range of 7 to 11).

EXAMPLE 9

Suppose that the encoder of Fig. 8.10a generates an all-zero sequence that is sent over a binary symmetric channel, and that the received sequence is (0100010000 . . .). There are two errors in the received sequence due to noise in the channel: one in the second location and the other in the sixth location. We wish to show that this double-error pattern is correctable through the application of the Viterbi decoding algorithm.

In Fig. 8.15a, we show the result of applying step 1 of the algorithm for level $j = 2$. There are four paths (survivors), one for each of the four states of the encoder. The figure also includes the metric of each path.

In the left side of Fig. 8.15b, we show the two paths entering each of the four states at level $j = 3$, together with their individual metrics. In the right side of this figure, we show the four survivors that result from applying step 2 of the algorithm for $j = 3$. In Figs. 8.15c and 8.15d, we show the corresponding results obtained from application of step 2 of the algorithm for level $j = 4$ and $j = 5$, respectively.

Examining the four survivors in Fig. 8.15d, we see that the all-zero path has the smallest metric. This clearly shows that the all-zero sequence is the maximum likelihood choice of the Viterbi decoding algorithm, which agrees exactly with the transmitted sequence.

EXAMPLE 10

Suppose next that the received sequence is (1100010000 . . .), which contains three errors compared to the transmitted all-zero sequences.

In Fig. 8.16a, we show the paths (survivors) that result from applying Step 1 of the Viterbi decoding algorithm for $j = 2$.

In Figs. 8.16b and 8.16c, we show the survivors that result from applying step 2 of the algorithm for $j = 3$ and $j = 4$, respectively.

We see that in this example the correct path has been eliminated by level $j = 3$. Clearly, a triple-error pattern is uncorrectable by the Viterbi algorithm when applied to a convolutional code of rate 1/2 and constraint length $K = 3$.

The explanation for why this convolutional code can correct up to two errors in the received sequence (as in Example 9), but fails when there are three errors (as in Example 10), will be presented in the next section.

8.7 DISTANCE PROPERTIES OF CONVOLUTIONAL CODES

The performance of a convolutional code depends not only on the decoding algorithm used but also on the distance properties of the code. In this context, the most important single measure of a convolutional code's ability to combat

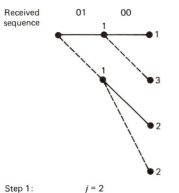

Step 1: $j = 2$ (a)

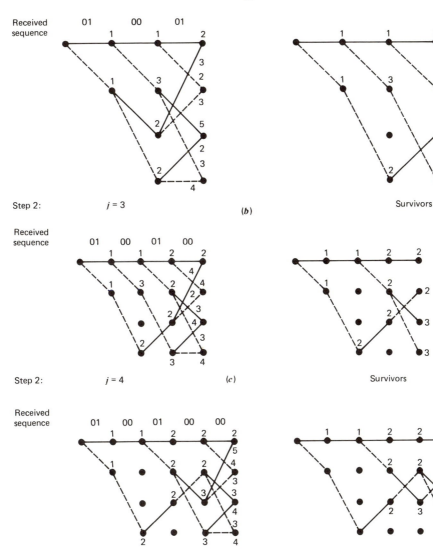

Figure 8.15   Illustrating steps in the Viterbi algorithm.

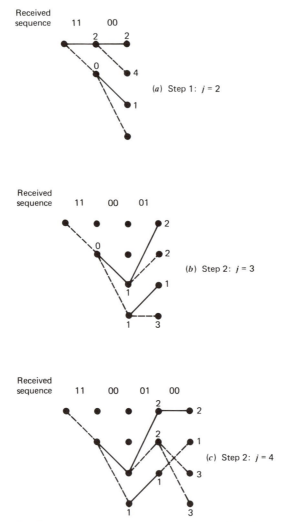

Figure 8.16 Illustrating breakdown of the Viterbi algorithm.

channel noise is the free distance, denoted by d_{free}. The *free distance* of a convolutional code is defined as the minimum Hamming distance between any two code words in the code.

The calculation of the free distance d_{free} is closely linked with that of the *generating function* of a convolutional code, which may be viewed as the transfer function of the encoder with respect to state transitions. Moreover, the generating function provides useful information about decoding error performance.

The state diagram of a convolutional code provides an effective tool for determining the generating function of the code. We illustrate the procedure by considering the encoder of Fig. 8.10a with rate $r = 1/2$ and constraint length $K = 3$. The state diagram of the encoder is shown in Fig. 8.14. We begin by modifying this state diagram into the form shown in Fig. 8.17, where the all-zero state **a** has been split up into an *initial* state $\mathbf{a}_0$ and a *final* state $\mathbf{a}_1$, and the self-loop at **a** has been removed. The other major change we have made in Fig. 8.17 is in the way the branches are labeled. The exponent of D on a branch describes the Hamming weight of the encoder output corresponding to that branch. The exponent of I describes the Hamming weight of the corresponding input; hence, for input 0 we have $I^0 = 1$ and for input 1 we have $I^1 = I$. The exponent of L is always equal to one, corresponding to the fact that the length of each branch is one. For example, the path $\mathbf{a}_0\mathbf{bcbdca}_1$ has the label $D^7L^6I^3$. This means that the Hamming weight of the corresponding encoder output sequence is 7, the path length is 6, and the Hamming weight of the encoder input sequence is 3.

We define a *fundamental path* as a path beginning at the initial state $\mathbf{a}_0$ and ending at the final state $\mathbf{a}_1$. Let $T_{d,l,i}$ denote the number of paths from $\mathbf{a}_0$ to $\mathbf{a}_1$ with label $D^dL^lI^i$. We may then define the *complete path enumerator* for the encoder as the generating function

$$T(D, L, I) = \sum_{d=1}^{\infty} \sum_{l=1}^{\infty} \sum_{i=1}^{\infty} T_{d,l,i}D^dL^lI^i \tag{8.70}$$

For the encoder at hand, we may calculate the generating function $T(D, L, I)$ by viewing the modified state diagram of Fig. 8.17 as a *signal-flow graph* with a single input and single output. Correspondingly, the generating function $T(D, L, I)$ may be viewed as the gain of the signal-flow graph. Thus, treating the nodes as summing junctions and the branch labels as gains, and assuming a unity input, we may write

$$\zeta_b = D^2LI + LI\zeta_c$$
$$\zeta_c = DL\zeta_b + DL\zeta_d$$
$$\zeta_d = DLI\zeta_b + DLI\zeta_d$$
$$T(D, L, I) = D^2L\zeta_c \tag{8.71}$$

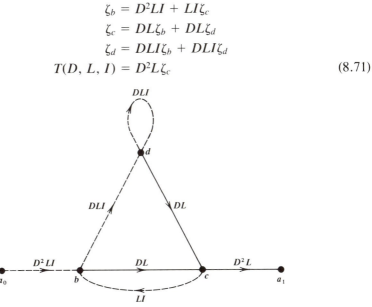

Figure 8.17 Modified state diagram.

where ζ_b, ζ_c, and ζ_d are dummy variables for the partial sums at the intermediate nodes. The four equations in Eq. 8.71 represent a set of algebraic state equations that may be solved to yield the result

$$T(D, L, I) = \frac{D^5 L^3 I}{1 - DLI(1 + L)} \tag{8.72}$$

Using the series expansion of $(1 - DLI(1 + L))^{-1}$, we may rewrite the expression for the generating function in the form of a power series as follows:

$$T(D, L, I) = D^5 L^3 I + D^6 L^4 I^2 (1 + L) + D^7 L^5 I^3 (1 + L)^2 + \cdots \tag{8.73}$$

Accordingly, we may make the following observations in the context of the trellis diagram for the encoder of Fig. 8.10*a*:

1. There are no fundamental paths at distance 0, 1, 2, or 3 from the all-zero path.
2. There is a single fundamental path at distance 5 from the all-zero path. It diverges from the latter path three branches back. Moreover, it differs from the all-zero path in the single input bit.
3. There are two fundamental paths at distance 6 from the all-zero path. One path diverges from the all-zero path four branches back, and the other five branches back. Both paths differ from the all-zero path in two input bits.
4. There are four fundamental paths at distance 7 from the all-zero path. One path diverges from the all-zero path five branches back, two other paths diverge from the all-zero path six branches back, and the fourth path diverges seven branches back. They all differ from the all-zero path in three input bits.

The reader is invited to check the validity of these statements by working through the trellis diagram of Fig. 8.12.

In a general convolutional code, the smallest Hamming distance of any fundamental path from the all-zero path equals the free distance of the code. Thus, for the convolutional code whose generating function is described by Eq. 8.72, the free distance $d_{\text{free}} = 5$. This implies that up to two errors in the received sequence are correctable, for two or fewer transmission errors will cause the received sequence to be at most at a Hamming distance of 2 from the transmitted sequence but at least at a Hamming distance of 3 from any other code sequence in the code. In other words, in spite of the presence of any pair of transmission errors, the received sequence remains closer to the transmitted sequence than any other possible code sequence. However, this statement is no longer true if there are three or more (closely spaced) transmission errors in the received sequence. These observations confirm the results reported earlier in Examples 9 and 10.

(1) Bound on the Bit Error Rate

Earlier we stated that the generating function of a convolutional code provides useful information about the decoding error performance of the code. The most useful measure of error performance is the *bit error rate* (i.e., *probability of bit error*). The generating function is useful in that it helps us evaluate a *bound* on the bit error rate for a given decoding scheme.

We confine our attention to the subclass of binary-input, output-symmetric channels that includes the binary-input AWGN channel, and its symmetrically quantized form, namely, the binary input Q-ary output symmetric discrete memoryless channel. Suppose then a convolutional code, with rate $r = k/n$, is used over such a channel, and the length of the incoming message sequence is large. We assume that a maximum likelihood decoding algorithm (i.e., the Viterbi algorithm) is used in the receiver. The bit error rate, defined as the number of bit errors per decoded bit, is bounded by*

$$(\text{BER}) \leq \frac{1}{k} \frac{\partial T(D, I)}{\partial I} \bigg|_{I=1, D=Z} \tag{8.74}$$

where the parameter Z is a function of the channel transition probabilities. Note that in Eq. 8.74 we have set L equal to one, since we are not interested in path lengths. Hence, given the generating function of a convolutional code, we may use Eq. 8.74 to calculate a bound on the bit error rate incurred by maximum-likelihood decoding of the code.

From the discussion presented earlier, we recognize that the generating function of a convolutional code, with free distance d_{free}, may be expressed as follows (for $L = 1$):

$$T(D, I) = \sum_{d=d_{\text{free}}}^{\infty} \sum_{i=1}^{\infty} T_{d,i} D^d I^i \tag{8.75}$$

Hence, differentiating $T(D, I)$ with respect to I, and then setting $I = 1$ and $D = Z$, we get

$$\frac{\partial T(D, I)}{\partial I} \bigg|_{I=1, D=Z} = \sum_{d=d_{\text{free}}}^{\infty} \sum_{i=1}^{\infty} i T_{d,i} Z^d$$

$$= \sum_{d=d_{\text{free}}}^{\infty} A_d Z^d \tag{8.76}$$

where

$$A_d = \sum_{i=1}^{\infty} i T_{d,i} \tag{8.77}$$

Accordingly, the use of Eq. 8.76 in Eq. 8.77 shows that the bit error rate is bounded by

$$(\text{BER}) \leq \frac{1}{k} \sum_{d=d_{\text{free}}}^{\infty} A_d Z^d \tag{8.78}$$

---

* For a proof of Eq. 8.74, see Viterbi and Omura (1979, pp. 242–252) and Lin and Costello (1983, pp. 322–329).

Note that the parameter k refers to the number of inputs employed in the encoder. Thus, for a rate-$1/n$ convolutional code, the encoder has a single input, and so $k = 1$.

(2) Asymptotic Coding Gain

It is instructive to evaluate the bound of Eq. 8.78 for two special channels: the binary symmetric channel, and the binary-input AWGN channel, assuming the use of *binary* PSK *with coherent detection*. This is done here:

1. *Binary symmetric channel.*   For this channel, the parameter Z is defined by

$$Z = 2\sqrt{p(1 - p)} \qquad (8.79)$$

The transition probability p is itself defined by (see Eq. 7.13 for the probability of error for binary PSK)

$$p = \frac{1}{2} \text{erfc}\left(\sqrt{\frac{E}{N_0}}\right) \qquad (8.80)$$

where E is the *energy per code symbol*, and $N_0/2$ is the noise power spectral density. We may approximate the complementary error function as follows (see Eq. E9 in Appendix E):

$$\text{erfc}\left(\sqrt{\frac{E}{N_0}}\right) \simeq \frac{1}{\sqrt{\pi}} \exp\left(-\frac{E}{N_0}\right) \qquad (8.81)$$

Correspondingly, we may approximate the transition probability p as

$$p \simeq 0.282 \exp\left(-\frac{E}{N_0}\right) \qquad (8.82)$$

For $p \ll 1$, corresponding to large values of E/N_0, we therefore find from Eq. 8.79 that the parameter Z is approximately given by

$$Z \simeq 1.06 \exp\left(-\frac{E}{2N_0}\right) \qquad (8.83)$$

For a convolutional code of rate r, we transmit r code symbols per message bit. Hence, the symbol energy E is related to the bit energy E_b as follows:

$$E = rE_b \qquad (8.84)$$

Thus, we may rewrite Eq. 8.83 in terms of E_b/N_0 as

$$Z \simeq 1.06 \exp\left(-\frac{rE_b}{2N_0}\right) \qquad (8.85)$$

Substituting this value for the parameter Z in Eq. 8.78, we get

$$\text{BER} \leqslant \frac{1}{k} \sum_{d=d_{\text{free}}}^{\infty} (1.06)^d A_d \exp\left(-\frac{drE_b}{2N_0}\right) \qquad (8.86)$$

For large values of E_b/N_0, the summation term in Eq. 8.86 is dominated by the first exponential term. Accordingly, we may further approximate the bound on the bit error rate (with coding) as

$$\text{BER} \simeq \frac{1}{k} (1.06)^{d_{\text{free}}} A_{d_{\text{free}}} \exp\left(-\frac{d_{\text{free}} r E_b}{2N_0}\right) \qquad (8.87)$$

For the convolutional encoder of Fig. 8.10a, $k = 1$ and $A_{d_{\text{free}}} = 1$. On the other hand, if no coding is used, the transition probability p reduces to the bit error rate for binary PSK. Thus, for large E_b/N_0, the bit error rate (without coding) is approximately given by

$$\text{BER} \simeq 0.282 \exp\left(-\frac{E_b}{N_0}\right) \qquad (8.88)$$

Comparing Eq. 8.87 to Eq. 8.86, we see that for a fixed E_b/N_0, the (negative) exponent with coding is larger than the exponent without coding by a factor equal to $d_{\text{free}} r/2$. For large values of E_b/N_0, the exponential term dominates the expression for the probability of error. Therefore, as a figure of merit for measuring the improvement in error performance made by the use of coding with hard-decision decoding, we may define the *asymptotic coding gain* (in decibels) as follows:

$$G_a = 10 \log_{10}\left(\frac{d_{\text{free}} r}{2}\right), \text{ dB} \qquad (8.89)$$

2. *Binary input AWGN channel.* Consider next the case of a memoryless binary input AWGN channel with no output quantization [i.e., the output amplitude lies in the interval $(-\infty, \infty)$]. For this channel, the parameter Z is defined as (see Problem 8.2.3)

$$Z = \exp\left(-\frac{r E_b}{N_0}\right) \qquad (8.90)$$

Thus, proceeding in a similar fashion to that described for the binary symmetric channel, we find that for large E_b/N_0 the bit error rate for the binary input AWGN channel (with coding) is approximately bounded by

$$\text{BER} \simeq \frac{1}{k} (1.06)^{d_{\text{free}}} A_{d_{\text{free}}} \exp\left(-\frac{d_{\text{free}} r E_b}{N_0}\right) \qquad (8.91)$$

Accordingly, comparing the exponent in Eq. 8.91 (with coding) with that in Eq. 8.88 (without coding), we deduce that the asymptotic coding gain for the case of a binary input AWGN channel is defined by

$$G_a = 10 \log_{10}(d_{\text{free}} r), \text{ dB} \qquad (8.92)$$

From Eqs. 8.92 and 8.89 we see that the asymptotic coding gain for the binary-input AWGN channel is greater than that for the binary symmetric channel by 3 dB. In other words, for large E_b/N_0, the transmitter for a binary symmetric channel must generate an additional 3 dB of signal energy (or power) over that for a binary-input AWGN channel, if we arc to achieve the same error performance. Clearly, then, there is an advantage to be gained by permitting an unquantized demodulator output instead of making hard decisions. This improvement in performance, however, is attained at the cost of increased decoder complexity due to the requirement for accepting analog inputs.

The asymptotic coding gain for a binary-input AWGN channel is approximated to within a small fraction of dB by a binary unit Q-ary output discrete memoryless channel with the number of quantizing levels $Q = 8$ (see Problems 8.7.4). This means that we may avoid the need for an analog decoder by using a soft-decision decoder that performs finite output quantization (typically, $Q = 8$), and yet realize a performance close to the optimum.

In summary, the error-correcting capability and the asymptotic coding gain of a convolutional code are determined by the free distance d_{free} of the code. These properties confirm d_{free} as the most important measure of a convolutional code. That is, the larger d_{free} is, the better the code is. Indeed, much effort has been expended on the issue of finding good convolutional codes with a large free distance.

8.8 SEQUENTIAL DECODING OF CONVOLUTIONAL CODES

In Section 8.6, we established that the Viterbi decoding algorithm for a convolutional code is maximum likelihood. As such, the error performance of the algorithm over a discrete memoryless channel is optimum. We also remarked that the exponential dependence of the computational complexity of the algorithm on the constraint length K limits the practical utility of the algorithm to relatively short constraint-length codes. Unfortunately, the imposition of such a restriction on K also limits the error-correcting capability of the code, since the larger K is, the larger the free distance of the code is likely to be. To resolve this dilemma, we need a decoding algorithm that avoids computing the likelihood, or metric, of every path in the trellis, thereby reducing computational complexity and allowing the constraint length K to take on very large values. Indeed, there is a suboptimum class of such algorithms known as *sequential decoding algorithms*.* The complexity of a sequential decoder is essentially independent of the constraint length K, so that very large values of K can be employed. Accordingly, although sequential decoding algorithms are not quite as good as maximum-likelihood decoding algorithms for a fixed code, they make up for that defect by the practical feasibility of using very large K.

Sequential decoding is an intuitive trial-and-error technique for searching out the correct path in a code tree. During the course of this search, the decoder

---

* Sequential decoding was first introduced by Wozencraft (1957), long before the advent of the Viterbi decoding algorithm. The subject is treated at an introductory level in the following references: Gallager (1968, pp. 263–286), McEliece (1977, pp. 219–228), Clark and Cain (1981, pp. 297–328), Michelson and Levesque (1985, pp. 337–371), and Bhargava et al. (1981, pp. 402–421). For advanced mathematical treatments of the subject, see Lin and Costello (1983, pp. 350–387) and Viterbi and Omura (1979, pp. 349–381).

moves forward or backward in the code tree, one node at a time. The decision whether to move forward or backward is determined by the manner in which the *metric* of the algorithm varies along the path followed by the decoder.

(1) Fano Metric

Consider a constraint length K, code rate $r = k/n$, convolutional code over a discrete memoryless channel. Let c_{ij} denote the ith bit of the binary label on the jth branch of the associated code tree. Let y_{ij} denote the corresponding bit of the received sequence at the channel output. For a binary input-Q-ary output channel, the *bit metric* associated with this bit is defined by

$$\gamma_{ij} = \log_2\left(\frac{p(y_{ij}|c_{ij})}{p(y_{ij})}\right) - r \tag{8.93}$$

where $p(y_{ij}|c_{ij})$ is a transition probability of the channel, and $p(y_{ij})$ is the nominal probability of the channel output y_{ij}.

The binary label on each branch of the code tree has n bits. Hence, we may use the bit metric γ_{ij} to define the *branch metric* associated with the jth branch as follows:

$$\gamma_j = \sum_{i=1}^{n} \gamma_{ij}$$

$$= \sum_{i=1}^{n} \left[\log_2\left(\frac{p(y_{ij}|c_{ij})}{p(y_{ij})}\right) - r\right] \tag{8.94}$$

As in the Viterbi algorithm, the branch metric γ_j is an integer-valued metric (see Example 11). Note also that γ_j varies from one branch of the code tree to another, depending on the number of agreements between the n bits that constitute the binary label of branch j and the corresponding n bits of the received sequence.

Suppose the sequential decoder has followed a path with l branches, starting from the origin of the code tree. We may then define the associated *path metric* as

$$\Gamma(l) = \sum_{j=1}^{l} \gamma_j$$

$$= \sum_{j=1}^{l} \sum_{i=1}^{n} \left[\log_2\left(\frac{p(y_{ij}|c_{ij})}{p(y_{ij})}\right) - r\right] \qquad l = 1, 2, \ldots, \tag{8.95}$$

This metric is called the *Fano metric.*[*]

---

[*] The Fano metric is so named in honor of its originator, Fano (1963), who used heuristic arguments for its adoption. Later on, Massey (1972) presented analytic justification that this metric is optimum for sequential decoding. It is optimum only in the sense that the path of least Fano metric is the most likely path looked at so far.

Rewriting the path metric for a path of $(l + 1)$ branches, we have

$$\Gamma(l + 1) = \sum_{j=1}^{l+1} \gamma_j$$

$$= \sum_{j=1}^{l} \gamma_j + \gamma_{l+1} \qquad (8.96)$$

The first term, $\sum_{j=1}^{l} \gamma_j$, is recognized as the path metric for a path of l branches. The second term, γ_{l+1}, represents the branch metric of the *last* branch in the path of $(l + 1)$ branches. We may therefore reformulate Eq. 8.96 as

$$\Gamma(l + 1) = \Gamma(l) + \gamma_{l+1} \qquad (8.97)$$

This *recursive equation* provides the basis of an efficient method for computing the path metric. In particular, as we modify a hypothesized path through the tree by adding or subtracting a branch, we simply update the path metric by adding or subtracting the corresponding branch metric.

EXAMPLE 11

Consider a convolutional code with rate $r = 1/2$, used over a binary symmetric channel with transition probability $p = 0.04$. That is,

$$p(1|0) = p(0|1) = p = 0.04$$
$$p(0|0) = p(1|1) = 1 - p = 0.96$$

Each branch in the code tree has a binary label consisting of 2 bits. Hence, the branch metric γ can assume any one of three possible values, as outlined here:

1. The two bits of the binary label on the particular branch are in perfect agreement with the corresponding two bits of the received sequence. Hence, with $n = 2$ and $r = 1/2$, the use of Eq. 8.94 yields

$$\gamma_j = 2\log_2 p(0|0) + 2(1 - 1/2)$$
$$= 2\log_2(1 - p) + 1$$
$$= 0.88 \qquad j = 1, 2, \ldots$$

2. There is agreement in only one bit between the 2-bit binary label on the particular branch and the corresponding two bits of the received sequence. Hence, the branch metric equals

$$\gamma_j = \log_2[p(1|0)p(0|0)] + 2(1 - 1/2)$$
$$= \log_2[p(1 - p)] + 1$$
$$= -3.7 \qquad j = 1, 2, \ldots$$

3. There is no agreement between the 2-bit binary label on the particular branch and the corresponding two bits of the received sequence. Hence,

$$\gamma_j = 2\log_2 p(1|0) + 2(1 - 1/2)$$
$$= 2\log_2 p + 1$$
$$= -8.29 \qquad j = 1, 2, \ldots$$

Dividing by 0.88, and then rounding off these results to the nearest integer, we get 1,

−4, and −9 as the three possible values for the branch metric. Note that the branch metric γ_j assumes its only positive value when there is complete agreement between the binary label of a chosen branch and the corresponding bits of the received sequence. It assumes its most negative value when there is no agreement at all.

(2) Fano Algorithm

Several algorithms have been devised for the sequential decoding of convolutional codes. From a practical viewpoint, the *Fano algorithm** is probably the most important of these algorithms. A useful feature of the Fano algorithm is that it uses very little storage.

The code tree is basic to the understanding of the Fano algorithm. At every stage of the algorithm, the decoder is located at some node in the code tree. From this node, the decoder looks forward (i.e., deeper) into the tree. In so doing, if the decoder sees a particular node that it favors, it moves forward to that node. If not, the decoder takes a backward step and then attempts to move forward along another branch in the tree. The decision whether or not the decoder favors a node is made by comparing the path metric at that node with a *running threshold* maintained by the decoder. The running threshold, T_l, is defined as an integer multiple of the *threshold spacing* Δ, which is a design parameter. The path metric $\Gamma(l)$ is computed in accordance with the recursive relation of Eq. 8.97.

It is important that the threshold spacing Δ be carefully selected. When Δ is too small, the decoder frequently backtracks even when it is on the correct path. On the other hand, when Δ is too large, incorrect turns are not identified quickly. Accordingly, a compromise choice of Δ is required to minimize computation.

In the Fano algorithm, the decoder moves forward or backward through the tree, always one node at a time. To illustrate how these moves are made, consider Fig. 8.18, which shows part of a tree for a rate $1/n$ convolutional code. Suppose that the decoder has traced a path up to node P at depth l in the tree. Let $\Gamma_P(l)$ denote the path metric at node P, and T_l denote the current running threshold. First, the decoder tries to make a forward move. In particular, it

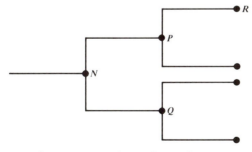

Figure 8.18 Part of a code tree.

---

* The Fano algorithm (Fano, 1963) was devised as a refinement of the first sequential decoding algorithm invented by Wozencraft (1957).

looks forward (at depth $l + 1$) to the "best" of the two succeeding nodes, that is, the node with the largest metric. Let R be the best node, viewed from node P. If the path metric $\Gamma_R(l + 1)$ at node R satisfies the condition $\Gamma_R(l + 1) \geqslant T_l$, the decoder *moves forward* to node R. If the node R is being visited for the first time, the threshold is *tightened*. By this we mean that the threshold T_l is increased by the largest possible multiple of the threshold spacing Δ, such that the new running threshold does not exceed $\Gamma_R(l + 1)$. If the node R has been examined previously, the threshold-tightening step is bypassed. Then the decoder looks forward again to the best succeeding node.

However, if $\Gamma_R(l + 1) < T_l$, the decoder *looks backward* to the preceeding node N at depth $l - 1$. Let the path metric at node N be $\Gamma_N(l - 1)$. If $\Gamma_N(l - 1) \geqslant T_l$, the decoder *moves back* to node N. If this backward move was made from the "worst" node viewed from node N, the decoder looks back again to the node preceeding node N. If not, the decoder looks forward to the "next best" node Q at depth l (*looks sideways*). Let $\Gamma_Q(l)$ denote the path metric at node Q. If $\Gamma_Q(l) \geqslant T_l$, the decoder accepts node Q (*moves sideways*). Then, the decoder attempts to move forward again from node Q.

If it turns out that both $\Gamma_R(l + 1)$ and $\Gamma_N(l - 1)$ are less than the current running threshold T_l at node N, the decoder can move neither forward nor backward. In such an event, the threshold is *loosened*. That is, T_l is reduced by the threshold spacing Δ. Then, the decoder looks forward again at node R with a lower threshold.

The *initial conditions* at the beginning of the Fano algorithm are set as follows. The *initial node* in the code tree is taken to be the *origin,* corresponding to an all-zero state. We take the *initial metric* $\Gamma(0)$ to be zero, and by convention $\Gamma(-1) = -\infty$. The *initial threshold* T_0 is set at zero.

The significance of putting $\Gamma(-1) = -\infty$ is that whenever the decoder looks backward from the origin of the tree, the threshold is always reduced by the spacing Δ. This, in turn, allows the algorithm to continue its forward search for a correct path.

The details of the Fano algorithm are presented in the form of a flowchart in Fig. 8.19. Two special nodes feature in this flowchart: the "best" node and the "worst" node. Viewed from a particular node in the tree, we formally define the best node as that node one branch deeper into the tree that has the largest path metric. The worst node is the node with the smallest path metric. Another point that should be made is that ties in metric values are resolved by the flip of a fair coin.

In addition to the computer requirements for executing the Fano algorithm, the sequential decoder contains a *buffer* for storing the (quantized) received sequence, and a *replica* of the encoder. For any hypothesized message sequence, the decoder can effectively trace the corresponding path in the code tree by running the sequence through the replica encoder, and the need for storing the entire tree is thereby eliminated.

Data is usually organized in *frames,* wherein the correct state is always known at the frame boundaries. This is done because the Fano algorithm and other sequential decoding algorithms suffer from a major problem called *buffer*

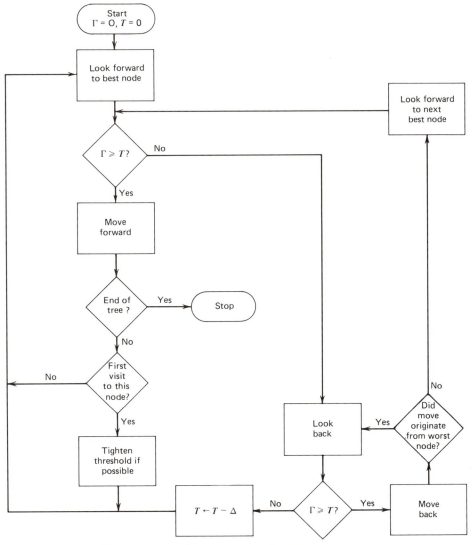

Figure 8.19 Flow chart for the Fano algorithm.

overflow, which occurs when the number of computations per frame is exceeded. The source of the problem is that the number of computations required for the algorithm to advance one node more deeply into the code tree is a highly ill-behaved random variable. In the event of buffer overflow, the particular sequence of received bits constituting a frame cannot be successfully decoded in the time allowed, and it is considered to be an erased frame. The relative frequency with which this event occurs is called the *probability of frame erasure*. Typically, sequentially decoded convolutional codes have very large constraint lengths, such that the probability of decoding error (due to channel noise) is negligible compared to the probability of frame erasure.

EXAMPLE 12

To illustrate the Fano algorithm, consider the code tree of Fig. 8.20, which corresponds to a rate -1/2, constraint length-7 convolutional code. The number of message bits is $L = 2$. The figure also includes the value of the path metric at each node of the tree. These values have been calculated for the all-zero transmitted code sequence (00,00,00,00) and the received sequence (10,00,01,00). There are two errors in the received sequence in locations 1 and 6. The threshold spacing is set at $\Delta = 4$.

In Table 8.8, the behavior of the Fano algorithm is summarized. For each step of the algorithm, the table presents the node at which the decoder resides, and the corresponding values of the path metric Γ and the running threshold T. The decoder starts at the origin, node A of the code tree, where the initial conditions are $\Gamma = 0$ and $T = 0$. These conditions are represented as step 1 in Table 8.8. At step 2, the decoder looks forward into the tree and sees nodes B and C. At both nodes, the path metric is -4, which is less than the running threshold $T = 0$. Hence, the decoder remains at node A, with the threshold reduced to $T = -4$. At step 3, the decoder may choose to move forward to node B or C. In Table 8.8, it is assumed that the choice is made in favor of the lower path; hence, the move to node C. The table shows that at step 3 through step 9, both inclusive, the decoder goes forward, then backward, forward, and then backward again, ending with the threshold loosened down to $T = -8$. Next, the decoder moves forward, backward, and then forward again, stopping at the terminal node L at step 20. Hence, it outputs the correct sequence (00,00,00,00). Note that at step 10, the decoder again faces the problem of choosing between two forward nodes at the same path metric. In Table 8.8, the decision is made in favor of the lower path; hence, the forward move to node G.

The reader is encouraged to check the validity of each step in Table 8.8.

Table 8.8 Illustrating Application of the Fano Algorithm

Step	Node	Metric, Γ	Threshold, T
1	A	0	0
2	A	0	−4
3	C	−4	−4
4	A	0	−4
5	B	−4	−4
6	D	−3	−4
7	B	−4	−4
8	A	0	−4
9	A	0	−8
10	C	−4	−8
11	G	−8	−8
12	K	−7	−8
13	G	−8	−8
14	C	−4	−8
15	A	0	−8
16	B	−4	−8
17	D	−3	−8
18	H	−7	−8
19	L	−6	−8
20	−STOP−		

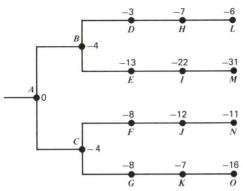

Figure 8.20   Tree for rate-$\frac{1}{2}$, constraint length-7, convolutional code.

8.9   TRELLIS CODES

In the traditional approach to channel coding described in the preceeding sections of the chapter, encoding is performed separately from modulation in the transmitter; likewise for decoding and detection in the receiver. Moreover, error control is provided by transmitting additional redundant bits in the code, which has the effect of lowering the information bit rate per channel bandwidth. That is, bandwidth efficiency is traded for increased power efficiency.

In order to attain a more effective utilization of the available bandwidth and power, coding and modulation have to be treated as a single entity. To deal with this new situation, we redefine coding as *the process of imposing certain patterns on the transmitted signal*. Indeed, this definition includes the traditional idea of parity coding.

The *trellis codes** for band-limited channels result from combining convolutional coding with modulation. The combination is itself referred to as *trellis coded modulation* (TCM). This form of signaling has two basic features:

1.   The number of signal points in the constellation used is larger than what is required for the modulation format of interest with the same data rate; the additional points allow redundancy for forward error-control coding without sacrificing bandwidth.

2.   Convolutional coding is used to introduce a certain dependency between

---

* There are two types of trellis codes for band-limited channels. The first type combines convolutional coding with multilevel signaling. This type of trellis codes first appeared in the work of Ungerboeck; their historical evolution is described in Ungerboeck (1982). The second type of trellis codes involves the use of continuous-phase frequency-shift keying (CPFSK) with specified modulation indices or frequency shifts (Anderson and deBuda (1976); Anderson and Taylor, 1978). The inspiration for the latter approach was derived from earlier work by deBuda on fast frequency-shift keying (deBuda, 1972). For a comprehensive survey of trellis codes of the first type, see Ungerboeck (1987), Forney et al. (1984), and Calderbank and Sloane (1987); the paper by Calderbank and Sloane presents an important generalization of trellis coding. For a detailed treatment of the second type of trellis codes, see Anderson et al. (1986). It is noteworthy that intersymbol interference may also be modeled as a trellis code, and so trellis decoders may be employed as equalizers (Forney, 1973).

successive signal points, such that only certain *patterns* or *sequences of signal points* are permitted.

The permissible sequence of signals may be modeled as a trellis structure; hence, the name "trellis codes."

In the presence of AWGN, maximum likelihood decoding of trellis codes consists of finding that path through the trellis with *minimum squared Euclidean distance* to the received sequence. Thus, in the design of trellis codes, the emphasis is on optimizing the Euclidean distance between code vectors (or, equivalently, code words) rather than optimizing the Hamming distance of an error-correcting code. The reason for this approach is that, except for conventional coding with binary PSK and QPSK, optimizing the Hamming distance is not the same as optimizing the squared Euclidean distance. Accordingly, in the sequel, the Euclidean distance is adopted as the distance measure of interest. Moreover, while a more general treatment is possible, the discussion is (by choice) confined to the case of *two-dimensional constellations of signal points*. The implication of such a choice is to restrict the development of trellis codes to multilevel (amplitude and/or phase) modulation schemes such as M-ary PSK and M-ary QAM.

The approach used to design this type of a trellis code involves partitioning a constellation of interest successively into 2, 4, 8, . . . subsets with increasing minimum Euclidean distance between their respective signal points. Such a mapping rule is called *mapping by set partitioning*.

In Fig. 8.21, we illustrate the partitioning procedure by considering a circular constellation that corresponds to *octal phase-shift keying* (OPSK) or 8-PSK.

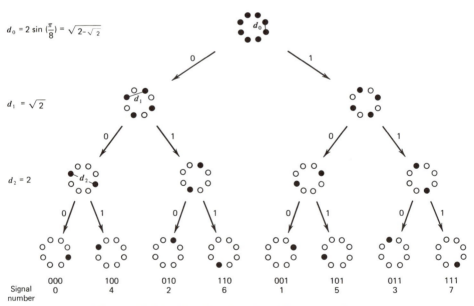

Figure 8.21 Partitioning of 8-PSK constellation.

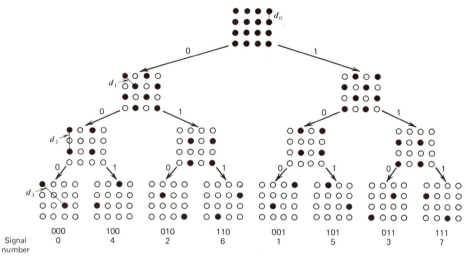

Figure 8.22   Partitioning of 16-QAM constellation.

The figure depicts the constellation itself, and the 2, 4, and 8 subsets resulting from three successive applications of the rule. These subsets share the common property that the minimum Euclidean distances between their individual points follow an increasing pattern: $d_0 < d_1 < d_2 \ldots$.

Figure 8.22 illustrates the partitioning of a rectangular constellation corresponding to 16-QAM. Here again we see that the subsets have increasing within-subset Euclidean distances: $d_0 < d_1 < d_2 \ldots$.

Based on the subsets resulting from successive partitioning of a two-dimensional constellation, we may devise relatively simple and yet highly effective coding schemes. Figure 8.23 shows the general structure of encoder/modulator combination for trellis-coded modulation. Specifically, to send n bits/symbol with *quadrature modulation* (i.e., one that has in-phase and quadrature components), we start with a two-dimensional constellation of 2^{n+1} signal points appropriate for the modulation format of interest; a circular grid is used for M-ary PSK, and a rectangular one for M-ary QAM. In any event, the constellation is partitioned into 4 or 8 subsets. One or two incoming bits per symbol enter a rate-1/2 or rate-2/3 binary convolutional encoder, respectively; the resulting two or three coded bits per symbol determine the selection of a particular subset. The remaining uncoded data bits determine which particular point from the selected subset is to be signaled. This class of trellis codes is known as *Ungerboeck codes*.

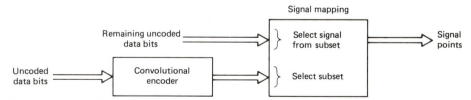

Figure 8.23   Block diagram of encoder-modulator for trellis-coded modulation.

Since the modulator has memory, we may use the Viterbi algorithm to perform maximum-likelihood sequence detection at the receiver. Each branch in the trellis of the Ungerboeck code corresponds to a subset rather than an individual signal point. The first step in the detection is to determine the signal point within each subset that is closest to the received signal point in the Euclidean sense. The signal point so determined and its metric (i.e., the squared Euclidean distance between it and the received point) may be used thereafter for the branch in question, and the Viterbi algorithm may then proceed in the usual manner.

(1) Ungerboeck Codes for 8-PSK

The scheme of Fig. 8.24a depicts the simplest Ungerboeck 8-PSK code for the transmission of 2 bits/symbol. The scheme uses a rate-1/2 convolutional encoder; the corresponding trellis of the code is shown in Fig. 8.24b, which has

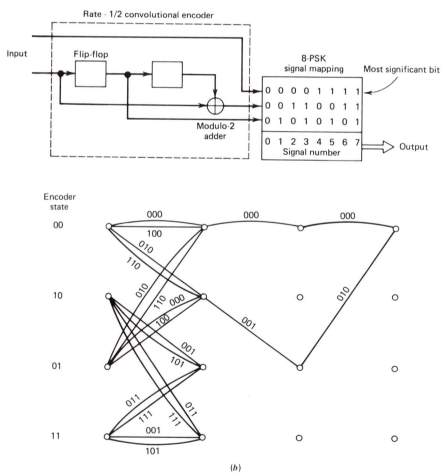

Figure 8.24 (a) Four-state Ungerboeck code for 8-PSK. (b) Trellis.

four states. Note that the most significant bit of the incoming binary word is left uncoded. Therefore, each branch of the trellis may correspond to two different output values of the 8-PSK modulator or, equivalently, to one of the four 2-point subsets shown in Fig. 8.21. The trellis of Fig. 8.24*b* also includes the minimum distance path.

The scheme of Fig. 8.25*a* depicts another Ungerboeck 8-PSK code for transmitting 2 bits/sample; it is next in the level of complexity. This second scheme uses a rate-2/3 convolutional encoder. Therefore, the corresponding trellis of the code has eight states, as shown in Fig. 8.25*b*. In this case, both bits of the incoming binary word are encoded. Hence, each branch of the trellis corresponds to a specific output value of the 8-PSK modulator. The trellis of Fig. 8.25*b* also includes the minimum distance path.

Figures 8.24*b* and 8.25*b* also include the encoder states. In Fig. 8.24, the state of the encoder is defined by the contents of the two-stage shift register; see Section 8.5(3). In Fig. 8.25, it is defined by the content of the single-stage (top) shift register followed by that of the two-stage (bottom) shift register.

(2)   Asymptotic Coding Gain

Following the discussion in Section 8.7, we define the *asymptotic coding gain* of Ungerboeck codes as

$$G_a = 10 \log_{10}\left(\frac{d_{\text{free}}^2}{d_{\text{ref}}^2}\right) \tag{8.98}$$

where d_{free} is the *free Euclidean distance* of the code, and d_{ref} is the minimum Euclidean distance of an uncoded modulation scheme operating with the same energy per bit. For example, by using the Ungerboeck 8-PSK code of Fig. 8.24, the signal constellation has 8 message points, and we send 2 bits per message point. Hence, uncoded transmission requires a signal constellation with 4 message points. We may therefore regard uncoded 4-PSK as the reference for the Ungerboeck 8-PSK code of Fig. 8.24.

The Ungerboeck 8-PSK code of Fig. 8.24*a* achieves an asymptotic coding gain of 3 dB, calculated as follows (see Fig. 8.26):

1. Each branch of the trellis in Fig. 8.24*b* corresponds to a subset of two antipodal signal points. Hence, the free Euclidean distance d_{free} of the code can be no larger than the Euclidean distance d_2 between the antipodal signal points of such a subset. We therefore write

    $$d_{\text{free}} = d_2 = 2$$

    where the distance d_2 is shown defined in Fig. 8.26*a*; see also Fig. 8.21.

2. The minimum Euclidean distance of an uncoded QPSK, viewed as a reference operating with the same energy per bit, equals (see Fig. 8.26*b*)

    $$d_{\text{ref}} = \sqrt{2}$$

    Hence, the use of Eq. 8.98 yields an asymptotic coding gain of $10 \log_{10} 2 = 3$ dB, as stated.

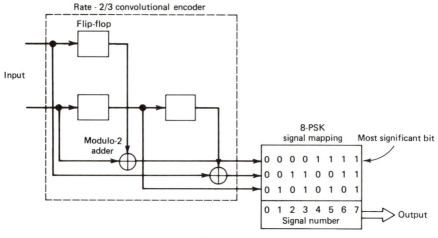

(a)

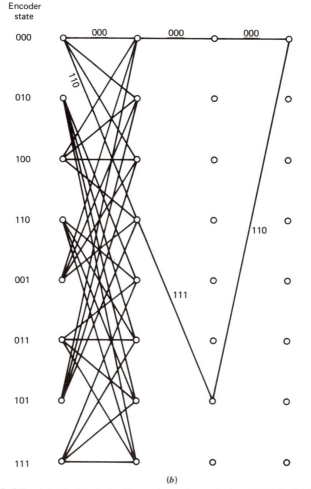

(b)

Figure 8.25 (a) Eight-state Ungerboeck code for 8-PSK. (b) Trellis.

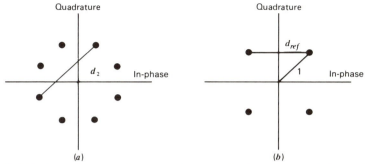

Figure 8.26   Signal-space diagrams for calculation of asymptotic coding gain.

The asymptotic coding gain achievable with Ungerboeck codes increases with the number of states in the convolutional encoder. Table 8.9 presents the asymptotic coding gain (in dB) for Ungerboeck 8-PSK codes for increasing number of states, expressed with respect to uncoded 4-PSK. Note that improvements in the order of 6 dB require codes with a very large number of states.

8.10   APPLICATIONS

In this section of the chapter, we discuss four applications of error-control coding techniques. We first consider the use of coding for white Gaussian noise channels that are representative of deep-space communication and satellite channels. We then consider the use of coding for compound-error channels that exhibit a combination of independent and burst errors. The third application involves the use of block codes for error control in data storage systems. For the last application, we describe the use of trellis-coded signaling for the efficient utilization of bandwidth and power.

(1)   Coding for White Gaussian Noise Channels

The use of forward-error correcting codes is well suited for channels that can be modeled as white Gaussian noise channels. Typical applications include line-of-sight radio links such as *satellite* and *deep-space communication links*. In this class of applications, the motivation is to achieve a *coding gain*, which may be translated either into reduced transmitted power or reduced antenna size, or into an increased bit rate for the same error performance. *The coding gain is defined as the difference in the bit energy-to-noise power density ratio, E_b/N_0, required for coded and uncoded systems to provide a specified bit error rate when operating on an ideal additive white Gaussian noise* (AWGN) *chan-*

Table 8.9[a]   Asymptotic Coding Gain of Ungerboeck 8-PSK codes, with respect to uncoded 4-PSK

Number of States	4	8	16	32	64	128	256	512
Coding Gain (dB)	3	3.6	4.1	4.6	4.8	5	5.4	5.7

[a] This table is adapted from Ungerboeck (1982).

nel. We stress that E_b is the energy per message bit. Thus, for a rate-1/2 convolutional code, for example, the transmitted symbol energy E is 3 dB less than the E_b (see Eq. 8.84).

For *moderate* values of coding gain, we may use *short constraint-length convolutional codes with soft-decision Viterbi decoding.* This assumes that the incoming message sequence is long. In Table 8.10, we present the coding gain for this coding–decoding approach for bit error rate BER = 10^{-3}, 10^{-5}, and 10^{-7}. All coding gains are relative to the value of E_b/N_0 shown in the left-most column required for uncoded binary PSK to achieve the specified BER. Note that the coding gain increases with increasing E_b/N_0. The values included in the bottom row of the table represent the asymptotic (upper bound) coding gain, calculated in accordance with Eq. 8.92 for the binary input AWGN channel. The short constraint-length convolutional codes with soft-decision Viterbi decoding have been applied in a large number of communication systems with varying data rates. Typically, they have outperformed block coding approaches for the same level of complexity. They have two natural advantages: (1) synchronization is simple, and (2) demodulator soft decisions are easily utilized.

For *high* values of coding gain (i.e., 6.5–7.5 dB at BER = 10^{-5}), we may use *concatenated coding* or *very long constraint-length convolutional codes with sequential decoding.*[*] The latter technique was discussed in Section 8.8. Concatenated coding[†] is described next.

In Fig. 8.27, we show a scheme for the *concatenation* of two error-control codes. We assume that the discrete memoryless channel is a binary input channel. The incoming data stream is first encoded with an *outer code* that consists of a nonbinary (n,k) block code. It is nonbinary in that it utilizes a J-bit byte for each symbol. The code has a block length of n bytes, the first k bytes of which are message symbols (i.e., the code is systematic). A second stage of

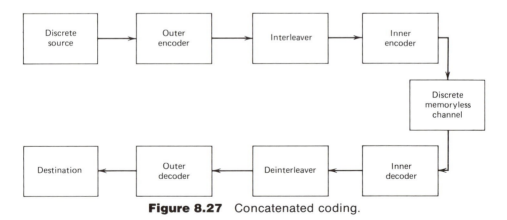

Figure 8.27 Concatenated coding.

* Clark and Cain (1981, pp. 342–343).

† The idea of concatenated coding was introduced by Forney (1966). For a discussion of different concatenated coding schemes, see Clark and Cain (1981, pp. 331–341), Michelson and Levesque (1985, pp. 375–384), and Berlekamp, Pelle and Pope (1987).

Table 8.10[a] Coding Gain for Convolutional Codes With Soft-decision Viterbi Decoding

E_b/N_0 (Uncoded) dB	BER	Coding Gain (dB)									
		Code Rate r									
		$\frac{1}{3}$		$\frac{1}{2}$			$\frac{2}{3}$		$\frac{3}{4}$		
	Constraint Length K	7	8	5	6	7	6	8	6	9	
6.8	10^{-3}	4.2	4.4	3.3	3.5	3.8	2.9	3.1	2.6	2.6	
9.6	10^{-5}	5.7	5.9	4.3	4.6	5.1	4.2	4.6	3.6	4.2	
11.3	10^{-7}	6.2	6.5	4.9	5.3	5.8	4.7	5.2	3.9	4.8	
—	Upper bound	7.0	7.3	5.4	6.0	7.0	5.2	6.7	4.8	5.7	

[a] This table is adapted from Jacobs (1974).

encoding is applied by the *inner encoder* that uses a short constraint-length convolutional code. At the receiving end, the *inner decoder* performs hard-decision Viterbi decoding of the convolutional code, and the *outer decoder* performs decoding of the block code. There are two factors that require attention:

1. At the transmitter, the J-bit symbols from the outer (block) code have to be converted to a serial form for convolutional encoding. Correspondingly, at the receiver, the Viterbi decoder output has to be converted to a parallel form for outer (block) decoding.
2. The errors at the Viterbi decoder output tend to occur in *bursts*.

To cater for these two factors, the concatenated coding scheme of Fig. 8.27 includes a symbol scrambling buffer, called the *interleaver*, after the outer block encoder, and a symbol unscrambling buffer, called the *deinterleaver*, before the outer block decoder. With this approach, error bursts out of the Viterbi decoding operation are spread among several consecutive code blocks rather than being contained within a single block, and the block code is thereby made more effective.

The only block codes that have been used as outer codes in the concatenated coding scheme of Fig. 8.27 are the nonbinary Reed–Solomon (RS) codes, which were briefly discussed in Section 8.4. The resulting scheme, called the *Reed–Solomon/Viterbi concatenated coding*, offers high reliability and modest complexity. These two desirable features make it a prime candidate for deep-space communication and other demanding satellite applications.* A deep-space communication system differs from other radio systems in that it involves radio propagation over very large distances. The resulting channel is noisy, requiring the use of coding for reliable communication.

Typical inner and outer code parameters used in the concatenated coding scheme of Fig. 8.27 for deep-space communication are as follows,† respectively:

1. *Convolutional code*

   $$\text{Code rate} \qquad r = 1/2$$
   $$\text{Constraint length} \quad K = 7$$

2. *RS block code*

   $$\text{Bits per byte} \qquad J = 8 \text{ bits}$$
   $$\text{Block length} \qquad n = 2^J - 1 = 255 \text{ bytes}$$

---

* For an overview paper on deep-space communications, including historical notes, see Posner and Stevens (1984).

† The Reed–Solomon/Viterbi concatenated coding scheme described herein is the NASA standard for deep-space communications. It achieves a bit-error rate of 10^{-6} at an energy per bit-to-noise power spectral density ratio of 2.53 dB. For details, see Yuen (1983, pp. 248–255). This particular code is employed in the image communication system of NASA's *Voyager* mission. At the Voyager flyby of Uranus 1.875 billion miles away in January 1986, source and channel coded imaging observations were sent back to Earth for decoding. This remarkable accomplishment was truly a triumph of coding over propagation distance.

Message portion $\quad k = 223$ bytes
Parity portion $\quad n - k = 32$ bytes
Correctable errors $\quad t = \frac{1}{2}(n - k) = 16$ bytes

3. *Interleaving*

Depth $\quad I = 4$

Note that, with this particular RS code, the overhead associated with the parity symbols is only about 15 percent.

A variety of interleaving/deinterleaving approaches are possible. Conceptually, the simplest approach is that of block *interleaving*, in which the interleaver memory is viewed as a rectangular array of storage locations. This is illustrated in Fig. 8.28 that shows the array representation of *I* RS code words, where *I* denotes the number of interleaving levels. The interleaver/deinterleaver may then operate in one of two ways:

1. The *J*-bit bytes are written in memory by columns; when the memory array is full, they are read out by rows. The deinterleaver performs the inverse operation by writing the received *J*-bit bytes into the rows of a similar memory array and reading them out by columns after the array is full.
2. The *J*-bit bytes are written in and read out in the opposite orders to those in Method 1.

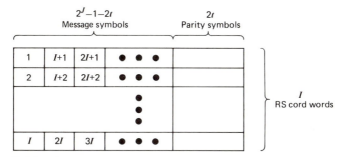

Figure 8.28 Array representation of Reed–Solomon (RS) code words.

(2) Coding for Compound-error Channels

Many real communication channels exhibit a mixture of *independent* and *burst error statistics*. We refer to such channels as *compound-error channels*. In *telephone channels*, for example, bursts of errors result from *impulse noise* on circuits due to lightning, and transients in central office switching equipment. In *radio channels*, bursts of errors are produced by atmospherics, multipath fading, and interferences from other users of the frequency band.

An effective method for error protection over compound-error channels is based on *automatic request for retransmission* (ARQ), which is by far the oldest and most widely used scheme for error control in data communication systems.* In Fig. 8.29, we show the block diagram of a basic ARQ system. The

---

* For a survey of various ARQ schemes, see Lin, Costello, and Miller (1984).

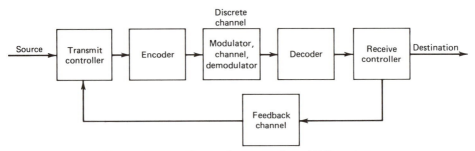

Figure 8.29 Block diagram of an ARQ system.

system requires the use of a *feedback channel* for requesting message retransmission. As usual, encoded bits are applied to a modulator for transmission over the channel, and the received bits are demodulated and decoded. The system includes *transmit and receive controllers* that exchange information via the feedback channel. The simplest ARQ system employs a *start and stop strategy*, which proceeds as follows:

1. A block of data is encoded into a code word for transmission over the communication channel.
2. The transmitter stops and waits until it receives (via the feedback channel) acknowledgment of correct reception of the code word or request for retransmission.

Clearly, this strategy requires only error detection (rather than both error detection and correction) in the receiver. Virtually error-free data transmission can be attained by the proper choice of a code for error detection. (Cyclic redundancy check codes are well-suited for this application). Moreover, the feedback channel has to be practically noiseless. This is not a severe restriction. Since the need for information transfer over the feedback channel is at a low rate, a substantial amount of redundancy can be used to ensure reliable feedback transmission.

Another straightforward and effective method to apply coding on a burst-error channel is to use *interleaving*. With this method, the channel is effectively transformed into an independent-error channel for which many forward-error correction coding techniques are applicable. The block diagram of such a system is shown in Fig. 8.30. The transmitter includes an encoder followed by an *interleaver* that scrambles the encoded data stream in a deterministic manner. Specifically, successive bits (or symbols) transmitted over the channel are separated as widely as possible. In the receiver, a *deinterleaver* is used to perform the inverse operation. That is, the received data are unscrambled, so that the decoding operation may proceed properly. Whereas the original data goes through interleaving in the transmitter and then deinterleaving in the receiver, error bursts are processed by the deinterleaver only. Accordingly, after deinterleaving, error bursts that occur on the channel are spread out in the data sequence to be decoded, thereby spanning many code words. The combination of interleaving and forward-error correction thus provides an effective means of combating the effect of error bursts.

Figure 8.30 Block diagram of interleaving method for burst-error channel.

A variation of the interleaving method described in Fig. 8.30 is commercially exploited in the *compact disc digital audio system.** In this system, optical recording is used to produce a master disc, from which compact discs are manufactured by galvanic processing. Production of the master disc and its replication into compact discs may be viewed as the "channel" of the digital audio system. An important characteristic of the channel is that it exhibits a burst-like error behavior. There are several factors responsible for this behavior. First, small unwanted particles or air bubbles trapped in the plastic material cause errors to occur when the recorded information is optically read out. Second, fingerprints or scratches on the disc, and surface roughness cause additional error bursts. For protection against channel errors, the recorded digital information includes parity bytes derived in two Reed–Solomon encoders. Moreover, interleaving is used to spread the errors out. In particular, the data streams entering the first encoder, between the two encoders, and those leaving the second encoder are scrambled by means of sets of delay lines. Accordingly, burst errors will, after deinterleaving on playback, be spread over a longer time interval, so that they can be corrected more easily. The two Reed–Solomon (n,k) codes have the values (32,28) and (28,24), so that each uses four parity bytes. Hence, each code can correct up to four error/erasure bytes.

(3) Block Codes for Error Control in Data Storage
Computer memory, like that of a human being, is inclined to make errors occasionally. For example, semiconductor memories suffer from errors due to their high density per silicon chip and exposure to radiation. When the error is organized such that each bit in a word visits a distinct set of logic (i.e., bit-oriented memory,), the errors tend to occur in a *random* manner. A single fault in any logic set affects a single bit per word, and does not induce more than a single error in any one word. Accordingly, the reliability of bit-oriented semiconductor memories can be improved significantly by using *single-error correcting and double-error detecting (SEC-SED) codes*; an extended Hamming code is an example of such a code (see Problem 8.3.9). Indeed, the use of error-correcting block codes for improving the reliability of semiconductor memories has become a standard design feature.†

Block codes are also widely used to provide error control for magnetic tapes, mass storage systems, magnetic disks, and other data storage systems.

---

* For a description of the various signal-processing operations used in compact disc digital audio systems, see Peek (1985).

† For a detailed discussion of the use of block codes for error control in data storage systems, see Lin and Costello (1983).

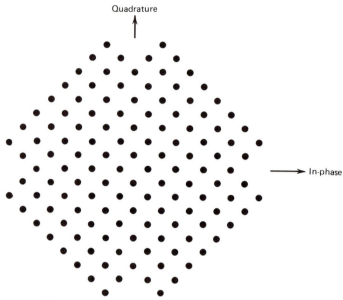

Figure 8.31 128-point signal constellation.

(4)   Trellis-coded Modulation for Efficient Utilization of Bandwidth and Power

Traditional forward-error-correcting codes, with hard-decision decoding, achieve their coding gains at the expense of increased channel bandwidth. However, bandwidth is a very precious commodity. Accordingly, the use of such codes has found little favor in band-limited applications (e.g., data communication via modems over voice-grade telephone channels).

We may obtain the advantages of error-correction coding without incurring the penalty of increased bandwidth by using trellis-coded modulation. As described in Section 8.9, this technique combines convolutional coding and modulation into a single function.

An important application of trellis-coded modulation is in the new generation of modems being developed for the telephone channel. In this context, we may mention a rate-2/3, 8-state trellis-coded 128-point QAM modem. For every 6-bit sequence of input data, two bits are convolutionally encoded in the rate-2/3 encoder, thereby resulting in a total of seven bits. These seven bits map a QAM constellation of $2^7 = 128$ signal points, as depicted in Fig. 8.31. A modulation (signaling) rate of 2400 bauds or, equivalently, a data rate defined by $2.4 \times 6 = 14.4$ kb/s has been achieved with the modem over good quality leased telephone channels.*

8.11   SUMMARY AND DISCUSSION

In this chapter, we covered *error-control coding* techniques that provide for reliable communication over noisy channels. We considered two major types of error control, namely, *forward error correction (FEC)* and *automatic request*

---

* See Payton and Qureshi (1985).

for retransmission (ARQ). In both types, the effect of errors occurring during transmission is reduced by adding *redundancy* to the data prior to transmission. Nevertheless they use different strategies for exploiting the added redundancy. In FEC, the redundancy is used to enable a decoder in the receiver to correct errors. There is no need for a return path with this approach. In ARQ, on the other hand, the redundancy is used to enable the receiver to detect errors, whereupon a retransmission is requested. A return path is required by ARQ.

In FEC codes, decoding is usually more complex than encoding. These codes may be divided into block codes and convolutional codes. In *block codes*, the encoder splits up the incoming data stream into blocks and processes each block individually by adding redundancy in accordance with a prescribed algorithm. Likewise, the decoder processes each block individually; it corrects errors by exploiting the redundancy.

Many of the important block codes are *cyclic codes*. The encoders for these codes can be implemented by using linear shift registers with feedback. Moreover, because of the inherent algebraic structure of cyclic codes, the decoders for them are conceptually quite simple. Examples of cyclic codes are *Hamming codes* and *Reed–Solomon codes*. In particular, Reed–Solomon codes are nonbinary codes, capable of combating combinations of both random and burst errors. Also, virtually all codes used for error detection are cyclic codes; such codes are called *cyclic redundancy check codes*.

In *convolutional codes*, the encoder processes the incoming data stream continuously. For decoding, the *Viterbi algorithm* is most frequently employed. The Viterbi algorithm provides a recursive procedure for maximum likelihood sequence estimation. Complexity of the Viterbi algorithm confines its application to convolutional codes with a constraint length K not exceeding 11. At conventional bit rates, the value $K = 7$ is considered appropriate.

To achieve a combination of large coding gain and low bit error rates (less than 10^{-5}) with convolutional codes, a large constraint length is required. In this situation, sequential decoding is more suitable than Viterbi decoding. The performance of *sequential decoding* is relatively independent of constraint length. Hence, although sequential decoding is inferior to Viterbi decoding in performance, it makes up for this defect by allowing the use of a large constraint length.

In the traditional approach to forward-error correction, coding is performed separately from modulation and bandwidth efficiency is reduced by the addition of redundancy. In *trellis-coded modulation*, on the other hand, the operations of coding and modulation are combined so as to permit significant coding gains over conventional uncoded multilevel modulation without sacrificing bandwidth efficiency. In an important class of trellis codes known as *Ungerboeck codes*, multilevel (amplitude and/or phase) modulation is combined with convolutional coding. At the receiver, the Viterbi algorithm is used to perform a maximum-likelihood sequence detection. Thus, coding gains of 3 to 6 dB are attained at bandwidth efficiencies equal to or larger than 2 bits per second per hertz. These are the values for which operation on many band-limited channels is desired.

Another type of combined coding-modulation technique currently receiving wide attention is that of lattice codes. A *lattice code* consists of a finite set of vectors derived from a lattice, to which binary information is mapped in a one-to-one fashion. The components of the lattice vector specify the amplitudes of a block of pulses in a pulse-amplitude modulation system. Thus in terms of coding, *encoding* is the process of mapping binary information into a lattice vector, and *decoding* consists of finding the closest lattice vector to the received vector and performing the inverse mapping to obtain the original binary message. Many of us are familiar with crystal lattice structures from chemistry, such as the body-centered cubic and face-centered cubic lattices. Also, the face-centered cubic lattice can be seen in such day-to-day encounters as the packing of fruit at the local market. This lattice packing is important because it is the densest known arrangement of spheres in three dimensions and as such makes the most efficient use of available space. This is an important concept for bandwidth-efficient communications, because we are generally confined to a finite signal space by a given average power or average energy constraint, and thus want to make the most efficient use of that space. The problem of packing spheres in three-dimensional space can be generalized to packings of n-dimensional spheres. The n-dimensional sphere may be described by a set of algebraic equations, which are of the same form as that of the sphere in three-dimensions. The relationship of sphere-packing to coding is quite simple. The number of spheres touching a central sphere, known as the *kissing number* in sphere packing, defines the number of nearest neighbors at the free distance of the code (equivalent to the number of paths at the free distance in trellis coding). The *free-distance* is then twice the radius of the spheres in the packing or, in other words, the distance from the center of one sphere to the center of a nearest neighbor sphere. Lattice packings of spheres are very useful for coding because they are very regular (i.e., all points in the lattice have the same number of nearest neighbors), easy to construct, and easy to decode.* Along with providing efficiency through their dense sphere-packings, lattices also have a greater distance between nearest neighbor signal vectors for a given energy constraint than do conventional bandwidth efficient modulations such as QAM, resulting in substantial coding gains over these modulation schemes.

In conclusion, the designer of a data communication system has at his disposal several powerful error-control coding schemes. For a particular coding scheme, the recommended procedure is to analyze it under one or more simple types of noise, and present the results in comparison to those obtained with an uncoded version of the system.† The analysis may also include other coding schemes, which are shown to be inferior and therefore presented as "strawmen." Thus, the designer will have identified a tangible scheme, and its appropriateness is questioned in the light of the designer's knowledge of the channel.

* For a highly readable description of sphere packings and lattices, see Sloane (1984); see also Thompson (1983). The second reference also presents a historical account of the origins of coding theory.
† The procedure summarized herein is credited to Jacobs and Viterbi by Berlekamp, Pelle, and Pope (1987). This paper presents some particular examples of data communications where error control has been applied.

PROBLEMS

P8.2 DISCRETE MEMORYLESS CHANNELS

Problem 8.2.1 Consider a binary input Q-ary output discrete memoryless channel. The channel is said to be symmetric if the channel transition probability $p(j|i)$ satisfies the condition:

$$p(j|0) = p(Q - 1 - j|1) \qquad j = 0, 1, \ldots, Q - 1$$

Suppose that the channel input symbols, 0 and 1, are equally likely. Show that the channel output symbols are also equally likely; that is,

$$p(j) = \frac{1}{Q} \qquad j = 0, 1, \ldots, Q - 1$$

Problem 8.2.2 Consider the quantized demodulator for binary PSK signals shown in Fig. 8.3a. The quantizer is a four-level quantizer, normalized as in Fig. P8.1. Evaluate the transition probabilities of the binary input-quaternary output discrete memoryless channel so characterized. Hence, show that it is a symmetric channel. Assume that the transmitted signal energy per bit is E_b, and the additive white Gaussian noise has zero mean and power spectral density $N_0/2$.

Problem 8.2.3 Consider a binary input AWGN channel, in which the binary symbols 1 and 0 are equally likely. The binary symbols are transmitted over the channel by means of phase-shift keying. The code symbol energy is E, and the AWGN has zero mean and power spectral density $N_0/2$.

(a) Show that the channel transition probability is given by

$$p(y|0) = \frac{1}{\sqrt{2\pi}} \exp\left[-\frac{1}{2}\left(y + \sqrt{\frac{2E}{N_0}}\right)^2\right] \qquad -\infty < y < \infty$$

(b) In asymptotic coding gain calculations, we use a parameter Z defined by

$$Z = \int_{-\infty}^{\infty} \sqrt{p(y|0)\, p(y|1)}\; dy$$

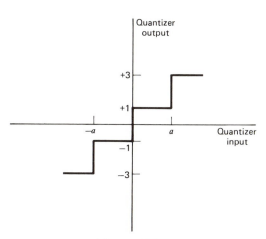

Figure P8.1

Show that for this channel,

$$Z = \exp\left(-\frac{E}{N_0}\right)$$

P8.3 LINEAR BLOCK CODES

Problem 8.3.1 In a *single-parity-check code*, a single parity bit is appended to a block of k message bits $(m_1, m_2, \ldots, m_k)$. The single parity bit, b_1, is chosen so that the code word satisfies the *even parity rule*:

$$m_1 + m_2 + \ldots + m_q + b_1 = 0 \qquad \text{mod-2}$$

For $k = 3$, set up the 2^k possible code words in the code defined by this rule.

Problem 8.3.2 Compare the parity-check matrix of the (7,4) Hamming code considered in Example 2 with that of a (4,1) repetition code.

Problem 8.3.3 Consider the (7,4) Hamming code considered in Example 2. The generator matrix **G** and the parity-check matrix **H** of the code are described in Eqs. 8.31 and 8.32, respectively. Show that these two matrices satisfy the condition (see Eq. 8.18)

$$\mathbf{HG}^T = \mathbf{0}$$

Problem 8.3.4 Taking the transpose of both sides of Eq. 8.18, we may also write

$$\mathbf{GH}^T = \mathbf{0}$$

This version of the equation suggests that every (n, k) linear block code with generator matrix **G** and parity-check matrix **H** has a *dual* code, with parameters $(n, n - k)$, generator matrix **H**, and parity-check matrix **G**.

(a) For the (7,4) Hamming code described in Example 2, construct the eight code words in the dual code.
(b) Find the minimum distance of the dual code determined in (a).

Problem 8.3.5 The *design distance* of an (n, k) linear block code is defined as being equal to $n - k + 1$. Show that the minimum distance of the code can never exceed its design distance.

Problem 8.3.6 Consider the (5,1) repetition code of Example 1. Evaluate the syndrome **s** for the following error patterns:

(a) All five possible single-error patterns.
(b) All 10 possible double-error patterns.

Problem 8.3.7 The *standard array* for an (n, k) linear block code consists of 2^k columns that are disjoint. The rows of the standard array are *cosets* of the code. The first entry in each row is called a *coset leader*. The coset leaders are the *correctable error patterns*.
Set up the standard array for the (7,4) Hamming code described in Example 2.

Problem 8.3.8 Show that the decoder for a Hamming code fails if there are two or more transmission errors in the received sequence.

Problem 8.3.9 The *extended $(n + 1, k)$ Hamming code* is obtained from the original (n, k) Hamming code by appending an overall parity bit. Let **H** denote the parity-check matrix of the original Hamming code. Then, the parity-check matrix of the extended Hamming matrix is defined by

$$\mathbf{H}_1 = \left[\begin{array}{c|ccccc} 1 & 1 & 1 & \cdots & 1 \\ \hline 0 & & & & \\ 0 & & \mathbf{H} & & \\ \vdots & & & & \\ 0 & & & & \end{array}\right]$$

Note that the first row of $\mathbf{H}_1$ consists of all 1s.

Consider the (7,4) Hamming code, whose parity-check matrix is given by

$$\mathbf{H} = \begin{bmatrix} 1 & 0 & 0 & 1 & 0 & 1 & 1 \\ 0 & 1 & 0 & 1 & 1 & 1 & 0 \\ 0 & 0 & 1 & 0 & 1 & 1 & 1 \end{bmatrix}$$

(a) Let b_{extra} denote the extra parity bit appended to the left of a code word in the original Hamming code C in order to obtain the corresponding code word in the extended (8,4) Hamming code. Show that

$$b_{\text{extra}} = \begin{cases} 1 & \text{if the code word in } C \text{ has odd weight} \\ 0 & \text{if the code word in } C \text{ has even weight} \end{cases}$$

(b) Show that the minimum distance of the extended Hamming code is 4.

(c) Determine the syndrome $\mathbf{s}$ of the extended Hamming code for (i) single-error patterns, and (ii) double-error patterns. Hence, demonstrate that single errors are correctable and double errors are at least detectable.

P8.4   CYCLIC CODES

Problem 8.4.1   Show that the extended (8,4) Hamming code considered in Problem 8.3.9 is a *noncyclic* code.

Problem 8.4.2   For an application that requires error detection *only*, we may use a *nonsystematic* code. In this problem, we explore the generation of such a cyclic code.

Let $g(D)$ denote the generator polynomial, and $m(D)$ denote the message polynomial. We define the code word polynomial $x(D)$ simply as

$$x(D) = m(D)g(D)$$

Hence, for a given generator polynomial, we may readily determine the code words in the code.

To illustrate this procedure, consider the generator polynomial for a (7,4) Hamming code:

$$g(D) = 1 + D + D^3$$

Determine the 16 code words in the code, and confirm the nonsystematic nature of the code.

Problem 8.4.3   The polynomial $1 + D^7$ has two primitive factors, namely, $1 + D + D^3$ and $1 + D^2 + D^3$. In Example 3, we used $1 + D + D^3$ as the generator polynomial for a (7,4) Hamming code. In this problem, we consider the adoption of $1 + D^2 + D^3$ as the generator polynomial. This should lead to a (7,4) Hamming code that is different from the code analyzed in Example 3.

Develop the encoder and syndrome calculator for the generator polynomial

$$g(D) = 1 + D^2 + D^3$$

Compare your results with those in Examples 4 and 5.

Problem 8.4.4 The generator polynomial of a (15,11) Hamming code is defined by

$$g(D) = 1 + D + D^4$$

Develop the encoder and syndrome calculator for this code, using a systematic form for the code.

Problem 8.4.5 The *expurgated (n, k − 1) Hamming code* is obtained from the original (n, k) Hamming code by discarding some of the code words. Let $g(D)$ denote the generator polynomial of the original Hamming code. The most common expurgated Hamming code is the one generated by

$$g_1(D) = (1 + D) g(D)$$

where $(1 + D)$ is a factor of $1 + D^n$.
Consider the (7,4) Hamming code generated by

$$g(D) = 1 + D^2 + D^3$$

(a) Construct the eight code words in the expurgated (7,3) Hamming code, assuming a systematic format. Hence, show that the minimum distance of the code is 4.
(b) Determine the generator matrix **G** and the parity-check matrix **H** of the expurgated Hamming code.
(c) Devise the encoder and syndrome calculator for the expurgated Hamming code. Hence, determine the syndrome for the received word 0111110.

Problem 8.4.6 The *extended (24,12) Golay code* is obtained by appending an overall parity bit to the original (23,12) Golay code. The extended Golay code is attractive in many applications because the code rate $r = k/n$ is exactly one half.
The *code word distribution* for the (23,12) Golay code is symmetric, as follows:

Hamming weight	0	7	8	11	12	15	16	23
Number of code words in the code	1	253	506	1288	1288	506	253	1

(a) Tabulate the code word distribution for the extended Golay code.
(b) What is the minimum distance of the extended Golay code?

Problem 8.4.7 Tabulate the generator polynomial for each of the binary BCH codes listed in Table 8.6.

Problem 8.4.8 Consider the (31,15) Reed–Solomon code.

(a) How many bits are there in a symbol of the code?
(b) What is the block length in bits?
(c) What is the minimum distance of the code?
(d) How many symbols in error can the code correct?
(e) What is the length of an in-phase burst that the code can correct?

P8.5 CONVOLUTIONAL CODES

Problem 8.5.1 Consider the rate $r = 1/2$, constraint length $K = 2$ convolutional encoder of Fig. P8.2. The code is systematic. Find the encoder output produced by the message sequence 10111 . . . .

Problem 8.5.2 A convolutional encoder has a single shift register with two stages, (i.e., constraint length $K = 3$), three modulo-2 adders, and an output multiplexer. The

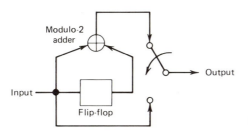

Figure P8.2

generator sequences of the encoder are as follows:

$$g^{(1)} = (1, 0, 1)$$

$$g^{(2)} = (1, 1, 0)$$

$$g^{(3)} = (1, 1, 1)$$

Draw the block diagram of the encoder.

Problem 8.5.3 Figure P8.3 shows the encoder for a rate $r = 1/2$, constraint length $K = 4$ convolutional code. Determine the encoder output produced by the message sequence 10111 . . . . , using the following two approaches:

(a) Time-domain approach, based on convolution.
(b) Transform-domain approach.

Problem 8.5.4 Consider the encoder of Fig. 8.10b for a rate $r = 2/3$, constraint length $K = 2$ convolutional code. Using the transform-domain approach, determine the code sequence produced by the message sequence 10111 . . . .

Problem 8.5.5 The code tree for the encoder of Fig. 8.10a, assuming that the incoming message sequence has length $L = 2$, is shown in Fig. P8.4. Validate this tree.

Problem 8.5.6 Construct the code tree for the convolutional encoder of Fig. P8.2. Trace the path through the tree that corresponds to the message sequence 10111 . . . , and compare the encoder output with that determined in Problem 8.5.1.

Problem 8.5.7 Construct the code tree for the encoder of Fig. P8.3. Trace the path

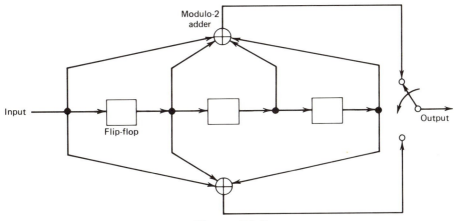

Figure P8.3

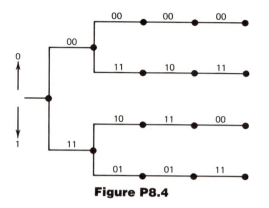

Figure P8.4

through the tree that corresponds to the message sequence 10111 . . . . Compare the resulting encoder output with that found in Problem 8.5.3.

Problem 8.5.8 Construct the trellis diagram for the encoder of Fig. P8.3, assuming a message sequence of length 5. Trace the path through the trellis, corresponding to the message sequence 10111 . . . . Compare the resulting encoder output with that found in Problem 8.5.3.

Problem 8.5.9 By viewing the minimum shift keying (MSK) scheme as a finite-state machine, construct the trellis diagram for MSK. (A description of MSK is presented in Section 7.3.)

Problem 8.5.10 Construct the state diagram for the encoder of Fig. P8.3. Starting with the all-zero state, trace the path that corresponds to the message sequence 10111 . . . , and compare the resulting code sequence with that determined in Problem 8.5.3.

Problem 8.5.11 Consider the encoder of Fig. 8.10b.

(a) Construct the state diagram for this encoder.
(b) Starting from the all-zero state, trace the path that corresponds to the message sequence 10111 . . . . Compare the resulting code sequence with that determined in Problem 8.5.4.

P8.6 MAXIMUM-LIKELIHOOD DECODING OF CONVOLUTIONAL CODES

Problem 8.6.1 The trellis diagram of a rate-1/2, constraint length-3 convolutional code is shown in Fig. P8.5. The all-zero sequence is transmitted, and the received sequence is 100010000 . . . . Using the Viterbi algorithm, compute the decoded sequence.

Problem 8.6.2 In Problem 8.6.1, the received sequence contains two transmission errors. Repeat the computation for the received sequence 11010000 . . . that contains three transmission errors.

P8.7 DISTANCE PROPERTIES OF CONVOLUTIONAL CODES

Problem 8.7.1 Reformulate the modified state diagram of Fig. 8.17, retaining information only on the Hamming weight of each path. Hence, show that the corresponding generating function of the code has the value

$$T(D) = \frac{D^5}{1 - 2D}$$

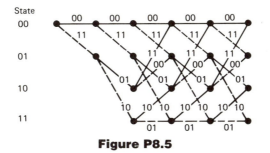

State

Figure P8.5

What does $T(D)$ tell us about the free distance of the code?

Problem 8.7.2   In this problem, we consider the path enumeration for the encoder of Fig. P8.3, emphasizing only the Hamming weight of each path.

(a)   Show that the generating function for this encoder is given by

$$T(D) = \frac{D^6 + D^7 - D^8}{1 - 2D - D^8}$$

(b)   What is the free distance of the code? How many errors can this code correct?
(c)   Enumerate the nonzero paths in the trellis diagram of the encoder.

Problem 8.7.3   Consider a rate-1/2, constraint length-7 convolutional code with free distance $d_{\text{free}} = 10$. Calculate the asymptotic coding gain of the channel for the following two channels:

(a)   Binary symmetric channel.
(b)   Binary input AWGN channel.

Problem 8.7.4   Consider a binary input Q-ary discrete memoryless channel that is symmetric. The parameter Z for such a channel is defined by

$$Z = \sum_{j=0}^{Q-1} \sqrt{p(j|0)\, p(j|1)}$$

(a)   Evaluate the asymptotic coding gain for a rate-r, free distance-d_{free} convolutional code used on such a channel, assuming that the number of quantizing levels $Q = 8$.
(b)   Show that the asymptotic coding gain calculated in (a) is within a small fraction of dB of the optimum performance achievable by a binary input AWGN channel (with no output quantization).

Assume that binary PSK (with coherent detection) is used for transmitting the code symbols, and that $E/N_0 = 4$.

P8.8   SEQUENTIAL DECODING OF CONVOLUTIONAL CODES

Problem 8.8.1   Consider a binary input-Q-ary output discrete memoryless channel. The channel is symmetric, and the input binary symbols are equally likely. Let $p(y_i|c_i)$ denote the probability of receiving y_i, given that bit c_i of a convolutional code is transmitted over the channel. Show that the Fano metric for a path of l branches may be expressed as follows

$$\Gamma(l) = \sum_{i=1}^{nl} \log_2 p(y_i|c_i) + nl\,(\log_2 Q - r)$$

where n is the number of bits per branch in the code tree, and r is the code rate. Explain the reasons for the fact that the *bias* represented by the second term is always positive. What is the significance of the bias increasing linearly with the path length?

Problem 8.8.2 The path metrics included in the code tree of Fig. 8.20 were calculated by using the code tree of an all-zero transmitted sequence, the received sequence (10,00,01,00) with two transmission errors, and the following values for the branch metric (see Example 11):

$$\gamma_j = \begin{cases} 1 & \text{for agreements on both branch bits} \\ -4 & \text{for agreement on one branch bit} \\ -9 & \text{for no agreement at all} \end{cases}$$

Starting from the code tree of Fig. P8.4, validate all the path metrics in Fig. 8.20.

Problem 8.8.3 In the code tree of Fig. 8.20, there are two nodes in the tree, A and C, where the decoder looks forward and sees two nodes at the same path metric. In Example 12, the choice was made in favor of the lower branch in both cases. Reformulate the path taken by the decoder through the code tree of Fig. 8.20, assuming that in the case of a tie in path metrics, the decoder chooses the upper branch.

Problem 8.8.4 Given the code tree of Fig. P8.4, and the branch metrics specified in Problem 8.8.2, recalculate the path metrics at the nodes of the tree for an all-zero transmitted sequence, and the received sequence (10,00,01,10) with three transmission errors. Then, apply the Fano algorithm to decode the received sequence, under the assumptions

(a) The decoder chooses the lower branch in the case of a tie in path metrics.
(b) The decoder chooses the upper branch in the case of a tie in path metrics.

P8.9   TRELLIS CODES

Problem 8.9.1 Figure P8.6 depicts the signal constellation of a hybrid amplitude/phase modulation scheme. Partition this constellation into eight subsets. At each stage of the partitioning, indicate the within-subset (shortest) Euclidean distance.

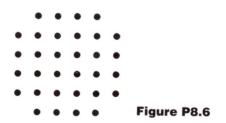

Figure P8.6

Problem 8.9.2 Construct the trellis, with all branches labeled, for an Ungerboeck 4-state code.

Problem 8.9.3 Table 8.9 gives the asymptotic coding gain for various Ungerboeck codes using M-ary PSK. Confirm the entries made in this table.

CHAPTER NINE

SPREAD-SPECTRUM MODULATION

A major issue of concern in the study of digital communications as considered in previous chapters of the book has been that of providing for the efficient utilization of bandwidth and power. The justification for preoccupation with this issue is simply the fact that bandwidth and power are the two primary communication resources, and it is therefore essential that they are used with care in the design of most communication systems. Nevertheless, there are situations where it is necessary to sacrifice the efficient utilization of these two resources in order to meet certain other design objectives. For example, the system may be required to provide a form of *secure* communication in a *hostile* environment such that the transmitted signal is not easily detected or recognized by unwanted listeners. This requirement is catered to by a class of signaling techniques known collectively as *spread-spectrum modulation*.

The primary advantage of a spread-spectrum communication system is its ability to reject *interference* whether it be the *unintentional* interference of another user simultaneously attempting to transmit through the channel, or the *intentional* interference of a hostile transmitter attempting to jam the transmission.

The definition of spread spectrum may be stated in two parts:*

1. *Spread spectrum is a means of transmission in which the data of interest occupies a bandwidth in excess of the minimum bandwidth necessary to send the data.*
2. *The spectrum spreading is accomplished before transmission through the*

---

* The definition given here is adapted from Pickholtz, Schilling, and Milstein (1982). This paper presents a tutorial review of the theory of spread-spectrum communications. For introductory papers on the subject, see Viterbi (1979) and Cook and Marsh (1983). For books on the subject, see Dixon (1984), Holmes (1982), Ziemer and Peterson (1985, pp. 327–649), Cooper and McGillem (1986, pp. 269–411), and Simon, Omura, Scholtz, and Levitt (1985, Volumes I, II and III). The three-volume book by Simon et al. is the most exhaustive treatment of spread-spectrum communications available in the open literature. The development of spread-spectrum communications dates back to about the mid-1950s. For a historical account of these techniques, see Scholtz (1982). This latter paper traces the origins of spread-spectrum communications back to the 1920s. Much of the historical material presented in this paper is reproduced in Chapter 2, Volume I, of the book by Simon et al.

use of a code that is independent of the data sequence. The same code is used in the receiver (operating in synchronism with the transmitter) to despread the received signal so that the original data may be recovered.

Although standard modulation techniques such as frequency modulation and pulse-code modulation do satisfy Part 1 of this definition, they are not spread-spectrum techniques because they do not satisfy Part 2 of the definition.

Spread-spectrum modulation was originally developed for military applications where resistance to jamming (interference) is of major concern. However, there are civilian applications that also benefit from the unique characteristics of spread-spectrum modulation. For example, it can be used to provide *multipath rejection* in a ground-based mobile radio environment. Yet another application is in *multiple-access* communication in which a number of independent users are required to share a common channel without an external synchronizing mechanism; here, for example, we may mention a ground-based mobile radio environment involving mobile vehicles that must communicate with a central station. Both of these applications are considered later in the chapter.

In this chapter, we discuss principles of spread-spectrum modulation, with emphasis on direct-sequence and frequency-hopping techniques. In a *direct-sequence spread-spectrum* technique, two stages of modulation are used. First, the incoming data sequence is used to modulate a wideband code. This code transforms the narrowband data sequence into a noise-like wideband signal. The resulting wideband signal undergoes a second modulation using a phase-shift keying technique. In a *frequency-hop spread spectrum* technique, on the other hand, the spectrum of a data-modulated carrier is widened by changing the carrier frequency in a pseudo-random manner. For their operation, both of these techniques rely on the availability of a noise-like spreading code called a *pseudo-random* or *pseudo-noise sequence*. Since such a sequence is basic to the operation of spread-spectrum modulation, it is logical that we begin our study by describing the generation and properties of pseudo-noise sequences.

9.1 PSEUDO-NOISE SEQUENCES

A *pseudo-noise* (PN) *sequence* is defined as a coded sequence of 1s and 0s with certain autocorrelation properties. The class of sequences used in spread-spectrum communications is usually *periodic* in that a sequence of 1s and 0s repeats itself exactly with a known period.* The *maximum-length sequence*, a type of cyclic code considered in Chapter 8, represents a commonly used periodic PN sequence. Such sequences have long periods and require simple instrumentation in the form of a *linear feedback shift register*. Indeed, they possess the longest possible period for this method of generation. A shift register of length m consists of m flip-flops (two-state memory stages) regulated by a single timing clock. At each pulse of the clock, the state of each flip-flop is shifted to the next one down the line. In order to prevent the shift register from emptying

---

* PN sequences may also be aperiodic. Such sequences are known as *Barker sequences*. Unfortunately, the longest Barker sequence found is 13 units. Indeed, it has been hypothesized that Barker sequences of length greater than 13 do not exist. Accordingly, Barker sequences are too short to be of practical use for spectrum spreading.

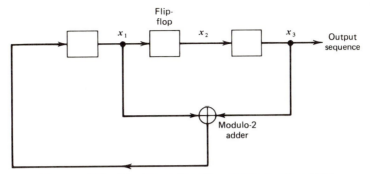

Figure 9.1 Maximum-length sequence generator.

by the end of m clock pulses, we use a logical (i.e., Boolean) function of the states of the m flip-flops to compute a feedback term, and apply it to the input of the first flip-flop. In a feedback shift register of the linear type, the feedback function is obtained using modulo-2 addition of the outputs of the various flip-flops. This operation is illustrated in Fig. 9.1 for the case of $m = 3$. Representing the states of the three flip-flops as x_1, x_2, and x_3, we see that in Fig. 9.1 the feedback function is equal to the modulo-2 sum of x_1 and x_3. A maximum-length sequence so generated is always periodic with a period of

$$N = 2^m - 1 \tag{9.1}$$

where m is the length of the shift register (equivalent to the degree of the generator polynomial). To get a feeling for parameter values of a PN sequence, we have listed a range of sequence (code) lengths in Table 9.1.

EXAMPLE 1
Consider the three-stage feedback shift register shown in Fig. 9.1. It is assumed that the initial state of the shift register is 100 (reading the contents of the three flip-flops from left to right). Then, the succession of states will be as follows:

$$100, 110, 111, 011, 101, 010, 001, 100, \ldots$$

Table 9.1 Range of PN Sequence Lengths

Length of Shift Register, m	PN Sequence Length, N
7	127
8	255
9	511
10	1023
11	2047
12	4095
13	8191
17	131071
19	524287

The output sequence (the last position of each state of the shift register) is therefore

$$0011101$$

which repeats itself with period 7.

Note that the choice of 100 as the initial state is an arbitrary one. Any of the other six states could serve equally well as an initial state. The resulting output sequence would then be some cyclic shift of the sequence given. Also, it should be noted that 000 is *not a* state of the shift register sequence since this results in a *catastrophic cyclic code*, (i.e., once the 000 state is entered, the shift register sequence cannot leave this state).

(1) Properties of Maximum-length Sequences

Maximum-length sequences* have many of the properties possessed by a truly *random binary sequence*. A random binary sequence is a sequence in which the presence of a binary symbol 1 or 0 is equally probable. Some properties of maximum-length sequences are listed below:

PROPERTY 1

In each period of a maximum-length sequence, the number of 1s is always one more than the number of 0s. This property is called the *balance property*.

PROPERTY 2

Among the runs of 1s and of 0s in each period of a maximum-length sequence, one-half the runs of each kind are of length one, one-fourth are of length two, one-eighth are of length three, and so on as long as these fractions represent meaningful numbers of runs. This property is called the *run property*.

By a "run" we mean a subsequence of identical symbols (1s or 0s) within one period of the sequence. The length of this subsequence is the length of the run. For a maximum-length sequence generated by a feedback shift register of length m, the total number of runs is $(m + 1)/2$.

PROPERTY 3

The autocorrelation function of a maximum-length sequence is periodic and binary-valued. This property is called the *correlation property*.

Let binary symbols 0 and 1 be represented by -1 volt and $+1$ volt, respectively. By definition, the *autocorrelation sequence* of a binary sequence $\{c_n\}$, so represented, equals

$$R_c(k) = \frac{1}{N} \sum_{n=1}^{N} c_n c_{n-k} \qquad (9.2)$$

* For further details on maximum-length sequences, see the following references: Golomb (1964, pp. 1–32), Ristenbatt (1965), Sawarte and Pursley (1980), Simon, Omura, Scholtz, and Levitt (1985, pp. 283–295), and Peterson and Weldon (1972). The last reference includes an extensive list of polynomials for generating maximum-length sequences.

where N is the *length* or *period* of the sequence and k is the *lag* of the autocorrelation sequence. For a maximum-length sequence of length N, the autocorrelation sequence is periodic with period N and two-valued, as shown by

$$R_c(k) = \begin{cases} 1 & k = lN \\ -\dfrac{1}{N} & k \neq lN \end{cases} \tag{9.3}$$

where l is any integer. When the length N is infinitely large, the autocorrelation sequence $R_c(k)$ approaches that of a completely random binary sequence.

EXAMPLE 2

Consider again the maximum-length sequence generated by the feedback shift register of Fig. 9.1. The output sequence (represented in terms of binary symbols 0 and 1) is

$$\{c_n\} = \underbrace{0011101}_{N=7} \ldots \tag{9.4}$$

In terms of the levels -1 and $+1$, the output sequence is

$$\{c_n\} = \underbrace{-1,-1,+1,+1,+1,-1,+1}_{N=7}, \ldots \tag{9.5}$$

We see that there are three 0s (or -1's) and four 1s (or $+1$'s) in one period of the sequence, which satisfies Property 1.

With $N = 7$, there are a total of four runs in one period of the sequence. Reading them from left to right in Eq. 9.4, the four runs are 00, 111, 0, and 1. Two of the runs (a half of the total) are of length one, and one run (a quarter of the total) is of length two, which satisfies Property 2.

Figure 9.2a shows two full periods of the maximum-length sequence. Figure 9.2b shows the corresponding autocorrelation function $R_c(\tau)$ plotted as a function of the time lag τ. In this figure, the parameter T_c denotes the duration of binary symbol 1 or 0 in the sequence, and N is the length of one period of the sequence. The periodic and two-valued correlation property of the sequence is clearly seen in Fig. 9.2b.

9.2   A NOTION OF SPREAD SPECTRUM

An important attribute of spread-spectrum modulation is that it can provide protection against externally generated interfering (jamming) signals with finite power. The jamming signal may consist of a fairly powerful broadband noise or multitone waveform that is directed at the receiver for the purpose of disrupting communications. (Jamming waveforms are discussed in Section 9.8.) Protection against jamming waveforms is provided by purposely making the information-bearing signal occupy a bandwidth far in excess of the minimum bandwidth necessary to transmit it. This has the effect of making the transmitted signal assume a noise-like appearance so as to blend into the background. The transmitted signal is thus enabled to propagate through the channel undetected by anyone who may be listening. We may therefore think of spread spectrum as a method of "camouflaging" the information-bearing signal.

One method of widening the bandwidth of an information-bearing (data)

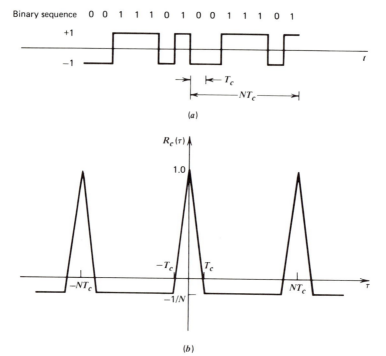

(a)

(b)

Figure 9.2 (a) Waveform of maximum-length sequence. (b) Autocorrelation of maximum-length sequence. Both parts refer to the output of the feedback shift register of Fig. 9.1.

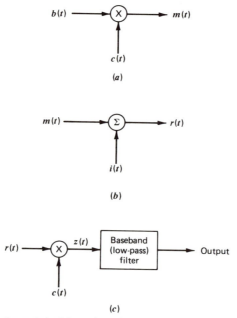

(a)

(b)

(c)

Figure 9.3 Idealized model of baseband spread-spectrum system. (a) Transmitter. (b) Channel. (c) Receiver.

sequence involves the use of *modulation*. Specifically, a data sequence $b(t)$ is used to modulate a wide-band *pseudo-noise* (*PN*) *sequence* $c(t)$ by applying these two sequences to a product modulator or multiplier, as in Fig. 9.3a. For this operation to work, both sequences are represented in their *polar* forms, that is, in terms of two levels equal in amplitude and opposite in polarity (e.g., -1 and $+1$). We know from Fourier transform theory that multiplication of two unrelated signals produces a signal whose spectrum equals the convolution of the spectra of the two component signals. Thus, if the data sequence $b(t)$ is narrowband and the PN sequence $c(t)$ is wideband, the product signal $m(t)$ will have a spectrum that is nearly the same as the PN sequence. In other words, in the context of our present application, the PN sequence performs the role of a *spreading code*.

By multiplying the information-bearing signal $b(t)$ by the spreading code $c(t)$, each information bit is "chopped" up into a number of small time increments, as illustrated in the waveforms of Fig. 9.4. These small time increments are commonly referred to as *chips*.

For *baseband* transmission, the product signal $m(t)$ represents the *transmitted signal*. We may thus express the transmitted signal as

$$m(t) = c(t)b(t) \qquad (9.6)$$

The received signal $r(t)$ consists of the transmitted signal $m(t)$ plus an additive *interference* denoted by $i(t)$, as shown in the channel model of Fig. 9.3b. Hence

$$\begin{aligned} r(t) &= m(t) + i(t) \\ &= c(t)b(t) + i(t) \end{aligned} \qquad (9.7)$$

To recover the original data sequence $b(t)$, the received signal $r(t)$ is applied to a *demodulator* that consists of a multiplier followed by a low-pass filter, as in

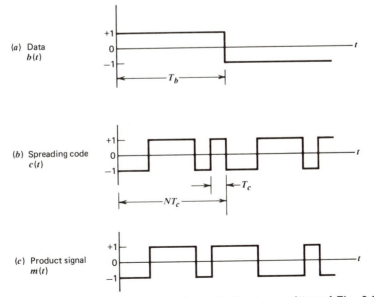

Figure 9.4   Illustrating the waveforms in the transmitter of Fig. 9.3a.

Fig. 9.3*c*. The multiplier is supplied with a locally generated PN sequence that is an exact *replica* of that used in the transmitter. Moreover, we assume that the receiver operates in perfect *synchronism* with the transmitter, which means that the PN sequence in the receiver is lined up exactly with that in the transmitter. The resulting demodulated signal is therefore given by

$$z(t) = c(t)r(t)$$
$$= c^2(t)b(t) + c(t)i(t) \tag{9.8}$$

Equation 9.8 shows that the desired signal $b(t)$ is multiplied *twice* by the spreading code $c(t)$, whereas the unwanted signal $i(t)$ is multiplied only *once*. The spreading code $c(t)$ alternates between the levels -1 and $+1$, and the alternation is destroyed when it is squared; hence

$$c^2(t) = 1 \qquad \text{for all } t \tag{9.9}$$

Accordingly, we may simplify Eq. 9.8 as

$$z(t) = b(t) + c(t)i(t) \tag{9.10}$$

We thus see from Eq. 9.10 that the data sequence $b(t)$ is reproduced at the multiplier output in the receiver, except for the effect of the interference represented by the additive term $c(t)i(t)$. Multiplication of the interference $i(t)$ by the locally generated PN sequence $c(t)$ means that the spreading code will affect the interference just as it did the original signal at the transmitter. We now observe that the data component $b(t)$ is narrowband, whereas the spurious component $c(t)i(t)$ is wideband. Hence, by applying the multiplier output to a baseband (low-pass) filter with a bandwidth just large enough to accommodate the recovery of the data signal $b(t)$, the spurious component $c(t)i(t)$ is made narrowband, thereby removing most of its power. The effect of the interference $i(t)$ is thus significantly reduced at the receiver output.

   In summary, the use of a spreading code (with pseudo-random properties) in the transmitter produces a wideband transmitted signal that appears *noise-like* to a receiver that has *no* knowledge of the spreading code. From the discussion presented in Section 9.1, we note that (for a prescribed data rate) the longer we make the period of the spreading code, the closer will the transmitted signal be to a truly random binary wave, and the harder it is to detect. Naturally, the price we have to pay for the improved protection against interference is increased transmission bandwidth, system complexity, and processing delay. However, when our primary concern is the security of transmission, these are not unreasonable costs to pay.

9.3 DIRECT-SEQUENCE SPREAD COHERENT BINARY PHASE-SHIFT KEYING

The spread-spectrum technique described in the previous section is referred to as *direct-sequence spread spectrum*. The discussion presented there was in the context of baseband transmission. To provide for the use of this technique over a band-pass channel (e.g., satellite channel), we may incorporate *coherent binary phase-shift keying* (PSK) into the transmitter and receiver, as shown in

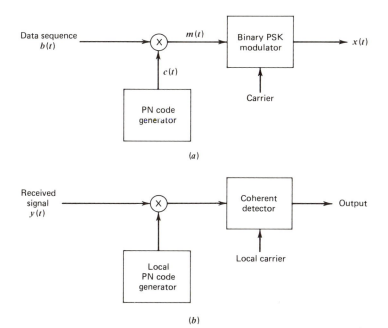

Figure 9.5   Direct-sequence spread coherent phase-shift keying. (a) Transmitter. (b) Receiver.

Fig. 9.5. The transmitter of Fig. 9.5a involves two stages of modulation. The first stage consists of a product modulator or multiplier with the data sequence and PN sequence as inputs. The second stage consists of a binary PSK modulator. The transmitted signal $x(t)$ is thus a *direct-sequence spread binary phase-shift-keyed* (DS/BPSK) *signal*. The phase modulation $\theta(t)$ of $x(t)$ has one of two values, 0 and π, depending on the polarities of the data sequence $b(t)$ and PN sequence $c(t)$ at time t in accordance with the truth table of Table 9.2. The receiver, shown in Fig. 9.5b, consists of two stages of demodulation. The received signal $y(t)$ and a locally generated replica of the PN sequence are applied to a multiplier. This multiplication represents the first stage of demodulation in the receiver. The second stage of demodulation consists of a coherent detector, the output of which provides an estimate of the original data sequence.

Figure 9.4 illustrates the waveforms for the first stage of modulation. Part of the modulated waveform shown in Fig. 9.4c is reproduced in Fig. 9.6a; the waveform shown here corresponds to one period of the PN sequence. Figure 9.6b shows the waveform of a sinusoidal carrier, and Fig. 9.6c shows the DS/BPSK waveform that results from the second stage of modulation.

Table 9.2   Truth Table for Phase Modulation $\theta(t)$, Radians

		Polarity of Data Sequence $b(t)$ at Time t	
		$+$	$-$
Polarity of PN sequence	$+$	0	π
$c(t)$ at time t	$-$	π	0

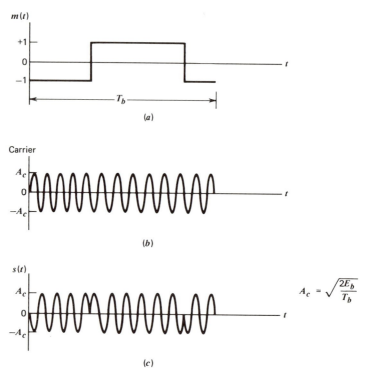

Figure 9.6   (a) Product signal m(t) = c(t)b(t). (b) Sinusoidal carrier. (c) DS/BPSK signal.

(1)   Model for Analysis

In the normal form of the transmitter, shown in Fig. 9.5a, the spectrum spreading is performed prior to phase modulation. For the purpose of analysis, however, we find it more convenient to interchange the order of these two operations, as in the model of Fig. 9.7. We are permitted to do this because the spectrum spreading and the binary phase-shift keying are both linear operations. The model of Fig. 9.7 also includes representations of the channel and the receiver. In this model, it is assumed that the interference $j(t)$ limits performance, so that the effect of channel noise may be ignored. Accordingly, the channel output is given by*

$$y(t) = x(t) + j(t)$$
$$= c(t)s(t) + j(t) \qquad (9.11)$$

where $s(t)$ is the binary PSK signal, and $c(t)$ is the PN sequence. In the receiver, the received signal $y(t)$ is first multiplied by the PN sequence $c(t)$ yielding an output that equals the coherent detector input $u(t)$. Thus,

---

* In the channel model included in Fig. 9.7, the interfering signal is denoted by $j(t)$. This notation is chosen purposely to be different from that used for the interference in Fig. 9.3b. The channel model in Fig. 9.7 is band-pass in spectral content, whereas that in Fig. 9.3b is in baseband form.

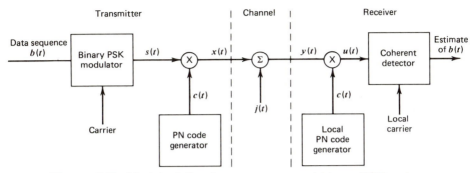

Figure 9.7 Model of direct-sequence spread binary PSK system.

$$u(t) = c(t)\, y(t)$$
$$= c^2(t)\, s(t) + c(t)\, j(t)$$
$$= s(t) + c(t)\, j(t), \tag{9.12}$$

In the last line of Eq. 9.12, we have noted that, by design, the PN sequence $c(t)$ satisfies the property described in Eq. 9.9, reproduced here for convenience:

$$c^2(t) = 1 \qquad \text{for all } t$$

Equation 9.12 shows that the coherent detector input $u(t)$ consists of a binary PSK signal $s(t)$ imbedded in additive code-modulated interference denoted by $c(t)j(t)$. The modulated nature of the latter component forces the interference signal (jammer) to spread its spectrum, such that the detection of information bits at the receiver output is afforded increased reliability.

9.4 SIGNAL-SPACE DIMENSIONALITY AND PROCESSING GAIN

Having developed a conceptual understanding of spread spectrum and a method for its implementation, we are ready to undertake a detailed mathematical analysis of the technique. The approach we have in mind is based on the signal-space theoretic ideas of Chapter 3. In particular, we develop signal-space representations of the transmitted signal and the interfering signal (jammer).

In this context, consider the set of orthonormal basis functions:

$$\phi_k(t) = \begin{cases} \sqrt{\dfrac{2}{T_c}} \cos(2\pi f_c t) & kT_c \le t \le (k+1)T_c \\ 0 & \text{otherwise} \end{cases} \tag{9.13}$$

$$\bar{\phi}_k(t) = \begin{cases} \sqrt{\dfrac{2}{T_c}} \sin(2\pi f_c t) & kT_c \le t \le (k+1)T_c \\ 0 & \text{otherwise} \end{cases} \tag{9.14}$$

$$k = 0, 1, \ldots, N-1$$

where T_c is the *chip duration*, and N is the number of chips per bit. Accordingly, we may describe the transmitted signal $x(t)$ for the interval of an information bit as follows:

$$x(t) = c(t)s(t)$$

$$= \pm \sqrt{\frac{2E_b}{T_b}} c(t) \cos(2\pi f_c t)$$

$$= \pm \sqrt{\frac{E_b}{N}} \sum_{k=0}^{N-1} c_k \phi_k(t) \qquad 0 \le t \le T_b \qquad (9.15)$$

where E_b is the signal energy per bit; the plus sign corresponds to information bit 1, and the minus sign corresponds to information bit 0. The code sequence $\{c_0, c_1, \ldots, c_{N-1}\}$ denotes the PN sequence, with $c_k = \pm 1$. The transmitted signal $x(t)$ is therefore N-dimensional in that it requires a minimum of N orthonormal functions for its representation.

Consider next the representation of the interfering signal (jammer), $j(t)$. Ideally, the jammer likes to place all of its available energy in exactly the same N-dimensional signal space as the transmitted signal $x(t)$; otherwise, part of its energy goes to waste. However, the best that the jammer can hope to know is the transmitted signal bandwidth. Moreover, there is no way that the jammer can have knowledge of the signal phase. Accordingly, we may represent the jammer by the general form

$$j(t) = \sum_{k=0}^{N-1} j_k \phi_k(t) + \sum_{k=0}^{N-1} \tilde{j}_k \tilde{\phi}_k(t) \qquad 0 \le t \le T_b \qquad (9.16)$$

where

$$j_k = \int_0^{T_b} j(t) \phi_k(t) \, dt \qquad k = 0, 1, \ldots, N-1 \qquad (9.17)$$

and

$$\tilde{j}_k = \int_0^{T_b} j(t) \tilde{\phi}_k(t) \, dt \qquad k = 0, 1, \ldots, N-1 \qquad (9.18)$$

Thus, the interference $j(t)$ is $2N$-dimensional; that is, it has twice the number of dimensions required for representing the transmitted DS/BPSK signal $x(t)$. In terms of the representation given in Eq. 9.16, we may express the average power of the interference $j(t)$ as follows:

$$J = \frac{1}{T_b} \int_0^{T_b} j^2(t) \, dt$$

$$= \frac{1}{T_b} \sum_{k=0}^{N-1} j_k^2 + \frac{1}{T_b} \sum_{k=0}^{N-1} \tilde{j}_k^2 \qquad (9.19)$$

Moreover, due to lack of knowledge of signal phase, the jammer can only place equal energy in the cosine and sine coordinates defined in Eqs. 9.17 and 9.18;

hence

$$\sum_{k=0}^{N-1} j_k^2 = \sum_{k=0}^{N-1} \tilde{j}_k^2 \tag{9.20}$$

Correspondingly, we may rewrite Eq. 9.19 as

$$J = \frac{2}{T_b} \sum_{k=0}^{N-1} j_k^2 \tag{9.21}$$

Our aim is to tie these results together by finding signal-to-noise ratios measured at the input and output of the DS/BPSK receiver in Fig. 9.7. To do this, we use Eq. 9.12 to express the coherent detector output as

$$v = \sqrt{\frac{2}{T_b}} \int_0^{T_b} u(t) \cos(2\pi f_c t)\, dt$$

$$= v_s + v_{cj} \tag{9.22}$$

where the components v_s and v_{cj} are due to the despread binary PSK signal, $s(t)$, and the spread interference, $c(t)j(t)$, respectively. These two components are defined as follows:

$$v_s = \sqrt{\frac{2}{T_b}} \int_0^{T_b} s(t) \cos(2\pi f_c t)\, dt \tag{9.23}$$

and

$$v_{cj} = \sqrt{\frac{2}{T_b}} \int_0^{T_b} c(t)j(t) \cos(2\pi f_c t)\, dt \tag{9.24}$$

Consider first the component v_s due to signal. The despread binary PSK signal $s(t)$ equals

$$s(t) = \pm \sqrt{\frac{2E_b}{T_b}} \cos(2\pi f_c t) \qquad 0 \le t \le T_b \tag{9.25}$$

where the plus sign corresponds to information bit 1, and the minus sign corresponds to information bit 0. Hence, assuming that the carrier frequency f_c is an integer multiple of $1/T_b$, we have

$$v_s = \pm \sqrt{E_b} \tag{9.26}$$

Consider next the component v_{cj} due to interference. Expressing the PN signal $c(t)$ in the explicit form of a sequence, $\{c_0, c_1, \ldots, c_{N-1}\}$, we may rewrite Eq. 9.24 in the corresponding form

$$v_{cj} = \sqrt{\frac{2}{T_b}} \sum_{k=0}^{N-1} c_k \int_{kT_c}^{(k+1)T_c} j(t) \cos(2\pi f_c t)\, dt \tag{9.27}$$

Using Eq. 9.13 for $\phi_k(t)$, and then Eq. 9.17 for the coefficient j_k, we may redefine the equivalent component v_{cj} as

$$v_{cj} = \sqrt{\frac{T_c}{T_b}} \sum_{k=0}^{N-1} c_k \int_0^{T_b} j(t)\phi_k(t)\, dt$$

$$= \sqrt{\frac{T_c}{T_b}} \sum_{k=0}^{N-1} c_k j_k \tag{9.28}$$

We next approximate the PN sequence as an *independent identically distributed binary sequence*. We emphasize the implication of this approximation by recasting Eq. 9.28 in the form

$$V_{cj} = \sqrt{\frac{T_c}{T_b}} \sum_{k=0}^{N-1} C_k j_k \tag{9.29}$$

where V_{cj} and C_k are random variables with sample values v_{cj} and c_k, respectively. In Eq. 9.29, the jammer is assumed to be fixed. With the C_k treated as independent identically distributed random variables, we find that the probability of the event $C_k = \pm 1$ equals

$$P(C_k = 1) = P(C_k = -1) = \frac{1}{2} \tag{9.30}$$

Accordingly, the mean value of the random variable V_{cj} is zero since, for fixed k, we have

$$E[C_k j_k | j_k] = j_k P(C_k = 1) - j_k P(C_k = -1)$$

$$= \frac{1}{2} j_k - \frac{1}{2} j_k$$

$$= 0 \tag{9.31}$$

For a fixed vector $\mathbf{j}$, representing the set of coefficients $j_0, j_1, \ldots, j_{N-1}$, the variance of V_{cj} is given by

$$\mathrm{Var}[V_{cj}|\mathbf{j}] = \frac{1}{N} \sum_{k=0}^{N-1} j_k^2 \tag{9.32}$$

Since the *spread factor* $N = T_b/T_c$, we may use Eq. 9.21 to express this variance in terms of the average interference power J as

$$\mathrm{Var}[V_{cj}|\mathbf{j}] = \frac{JT_c}{2} \tag{9.33}$$

Thus, the random variable V_{cj} has zero mean and variance $JT_c/2$.

From Eq. 9.26, we note that the signal component at the coherent detector output (during each bit interval) equals $\pm\sqrt{E_b}$, where E_b is the signal energy per bit. Hence, the peak instantaneous power of the signal component is E_b. Accordingly, we may define the *output signal-to-noise ratio* as the instantaneous peak power E_b divided by the variance of the equivalent noise component in Eq. 9.33. We thus write

$$(\mathrm{SNR})_O = \frac{2E_b}{JT_c} \tag{9.34}$$

The average signal power at the receiver input equals E_b/T_b. We thus define an *input signal-to-noise* ratio as

$$(\text{SNR})_I = \frac{E_b/T_b}{J} \tag{9.35}$$

Hence, eliminating E_b/J between Eqs. 9.34 and 9.35, we may express the output signal-to-noise ratio in terms of the input signal-to-noise ratio as

$$(\text{SNR})_O = \frac{2T_b}{T_c} (\text{SNR})_I \tag{9.36}$$

It is customary practice to express signal-to-noise ratios in decibels. We may thus write Eq. 9.36 in the equivalent form

$$10 \log_{10}(\text{SNR})_O = 10 \log_{10}(\text{SNR})_I + 3 + 10 \log_{10}(PG), \text{ dB} \tag{9.37}$$

where

$$PG = \frac{T_b}{T_c} \tag{9.38}$$

The 3-dB term on the right side of Eq. 9.37 accounts for the gain in SNR that is obtained through the use of coherent detection (which presumes exact knowledge of the signal phase by the receiver). This gain in SNR has nothing to do with the use of spread spectrum. Rather, it is the last term, $10 \log_{10}(PG)$, that accounts for the *gain in SNR obtained by the use of spread spectrum*. The ratio PG, defined in Eq. 9.38, is therefore referred to as the *processing gain*. Specifically, it represents the gain achieved by processing a spread-spectrum signal over an unspread signal. Note that both the processing gain PG and the spread factor N (i.e., PN sequence length) equal the ratio T_b/T_c. Thus the longer we make the PN sequence (or, correspondingly, the smaller the chip time T_c is), the larger will the processing gain be.

We may define the processing gain in another way by making two observations:

1. The *bit rate* of the binary data entering the transmitter input is given by

$$R_b = \frac{1}{T_b} \tag{9.39}$$

2. The bandwidth of the PN sequence $c(t)$, defined in terms of the main lobe of its spectrum, is given by

$$W_c = \frac{1}{T_c} \tag{9.40}$$

Note that both R_b and W_c are baseband parameters. Hence, we may reformulate the processing gain of Eq. 9.38 as

$$PG = \frac{W_c}{R_b} \tag{9.41}$$

9.5 PROBABILITY OF ERROR

Let the coherent detector output v in the direct-sequence spread BPSK system of Fig. 9.7 denote the sample value of a random variable V. Let the equivalent noise component v_{cj} produced by external interference denote the sample value of a random variable V_{cj}. Then, from Eqs. 9.22 and 9.26 we deduce that

$$V = \pm \sqrt{E_b} + V_{cj} \qquad (9.42)$$

where E_b is the transmitted signal energy per bit. The plus sign refers to sending symbol (information bit) 1, and the minus sign refers to sending symbol 0. The decision rule used by the coherent detector of Fig. 9.7 is to declare that the received bit in an interval $(0, T_b)$ is 1 if the detector output exceeds a threshold of zero volts; otherwise, a decision is made in favor of bit 0. With both information bits assumed equally likely, we find that (because of the symmetric nature of the problem) the average probability of error P_e is the same as the conditional probability of (say) the receiver making a decision in favor of symbol 1, given that symbol 0 was sent. That is,

$$
\begin{aligned}
P_e &= P(V > 0 | \text{symbol 0 was sent}) \\
&= P(V_{cj} > \sqrt{E_b}) \qquad (9.43)
\end{aligned}
$$

Naturally, the probability of error P_e depends on the random variable V_{cj} defined by Eq. 9.29. According to this definition, V_{cj} is the sum of N identically distributed random variables. Hence, from the *central limit theorem*, we deduce that for large N, the random variable V_{cj} assumes a Gaussian distribution. Indeed, the spread factor or PN sequence length N is typically large in the direct-sequence spread-spectrum systems encountered in practice; see Table 9.1. Earlier we evaluated the mean value and variance of V_{cj}; see Eqs. 9.31 and 9.33. We may therefore state that the equivalent noise component V_{cj} contained in the coherent detector output may be approximated as a Gaussian random variable with zero mean and variance $JT_c/2$, where J is the average interference power and T_c is the chip duration. With this approximation at hand, we may then proceed to calculate the probability of the event $V_{cj} > \sqrt{E_b}$, and thus express the average probability of error in accordance with Eq. 9.43 as

$$P_e \simeq \frac{1}{2} \operatorname{erfc}\left(\sqrt{\frac{E_b}{JT_c}}\right) \qquad (9.44)$$

This simple formula, invoking the Gaussian assumption, is appropriate for DS/BPSK binary systems with large spread factor N.

(1) Antijam Characteristics

It is informative to compare Eq. 9.44 with the formula for the average probability of error for a coherent binary PSK system reproduced here for convenience (see Eq. 7.13):

$$P_e = \frac{1}{2} \operatorname{erfc}\left(\sqrt{\frac{E_b}{N_0}}\right) \qquad (9.45)$$

Based on this comparison, we see that insofar as the calculation of bit error rate in a direct-sequence spread binary PSK system is concerned, the interference may be treated as wideband noise of power spectral density $N_0/2$, defined by

$$\frac{N_0}{2} = \frac{JT_c}{2} \tag{9.46}$$

This relation is simply a restatement of an earlier result given in Eq. 9.33.

Since the bit energy $E_b = PT_b$, where P is the average signal power and T_b is the bit duration, we may express the bit energy-to-noise density ratio as

$$\frac{E_b}{N_0} = \left(\frac{T_b}{T_c}\right)\left(\frac{P}{J}\right) \tag{9.47}$$

Using the definition of Eq. 9.38 for the processing gain PG, we may reformulate this result as

$$\frac{J}{P} = \frac{PG}{E_b/N_0} \tag{9.48}$$

The ratio J/P is termed the *jamming margin*. Accordingly, the jamming margin and the processing gain, both expressed in decibels, are related by

$$(\text{Jamming margin})_{dB} = (\text{Processing gain})_{dB} - 10 \log_{10}\left(\frac{E_b}{N_0}\right)_{\min} \tag{9.49}$$

where $(E_b/N_0)_{\min}$ is the minimum bit energy-to-noise density ratio needed to support a prescribed average probability of error.

EXAMPLE 3

A spread-spectrum communication system has the following parameters:

Information bit duration, $T_b = 4.095$ ms

PN chip duration, $T_c = 1$ μs

Hence, using Eq. 9.38, we find that the processing gain is

$$PG = 4095$$

Correspondingly, the required PN sequence is $N = 4095$, and the feedback shift length is $m = 12$.

For a satisfactory reception, we may assume that the average probability of error is not to exceed 10^{-5}. From the formula for a coherent binary PSK receiver, we find that $E_b/N_0 = 10$ yields an average probability of error equal to 0.387×10^{-5}. Hence, using this value for E_b/N_0, and the value calculated for the processing gain, we find from Eq. 9.49 that the jamming margin is

$$(\text{Jamming margin})_{dB} = 10 \log_{10}4095 - 10 \log_{10}(10)$$
$$= 36.1 - 10$$
$$= 26.1 \text{ dB}$$

That is, information bits at the receiver output can be detected reliably even when the noise or interference at the receiver input is up to 409.5 times the received signal power. Clearly, this is a powerful advantage against interference (jamming), which is realized through the clever use of spread spectrum.

9.6 FREQUENCY-HOP SPREAD SPECTRUM

In the type of spread-spectrum systems discussed previously, the use of a PN sequence to modulate a phase-shift-keyed signal achieves *instantaneous* spreading of the transmission bandwidth. The ability of such a system to combat the effects of jammers is determined by the processing gain of the system, which is a function of the PN sequence length. The processing gain can be made larger by employing a PN sequence with narrow chip duration, which, in turn, permits a greater transmission bandwidth and more chips per bit. However, the capabilities of physical devices used to generate the PN spread-spectrum signals impose a practical limit on the attainable processing gain. Indeed, it may turn out that the processing gain so attained is still not large enough to overcome the effects of some jammers of concern, in which case we have to resort to other methods. One such alternative method is to force the jammer to cover a wider spectrum by *randomly hopping* the data-modulated carrier from one frequency to the next. In effect, the spectrum of the transmitted signal is spread *sequentially* rather than instantaneously; the term "sequentially" refers to the pseudo-random-ordered sequence of frequency hops.

The type of spread spectrum in which the carrier hops randomly from one frequency to another is called *frequency-hop* (*FH*) *spread spectrum*. A common modulation format for FH systems is that of *M-ary frequency-shift keying* (MFSK). The combination is referred to simply as FH/MFSK. (A description of M-ary FSK was presented in Section 7.6).

Since frequency hopping does not cover the entire spread spectrum instantaneously, we are led to consider the rate at which the hops occur. In this context, we may identify two basic (technology-independent) characterizations of frequency hopping:

1. *Slow-frequency hopping*, in which the *symbol rate R_s* of the MFSK signal is an integer multiple of the *hop rate R_h*. That is, several symbols are transmitted on each frequency hop.
2. *Fast-frequency hopping*, in which the hop rate R_h is an integer multiple of the MFSK symbol rate R_s. That is, the carrier frequency will change or hop several times during the transmission of one symbol.

Obviously, slow-frequency hopping and fast-frequency hopping are the converse of one another. In the sequel, these two characterizations of frequency hopping are considered in turn.

(1) Slow-frequency Hopping

Figure 9.8*a* shows the block diagram of an FH/MFSK transmitter, which involves *frequency modulation* followed by *mixing*. First, the incoming binary data are applied to an M-ary FSK modulator. The resulting modulated wave and the output from a digital *frequency synthesizer* are then applied to a mixer that consists of a multiplier followed by a filter. The filter is designed to select the sum frequency component resulting from the multiplication process as the transmitted signal. In particular, successive (not necessarily disjoint) *k*-bit

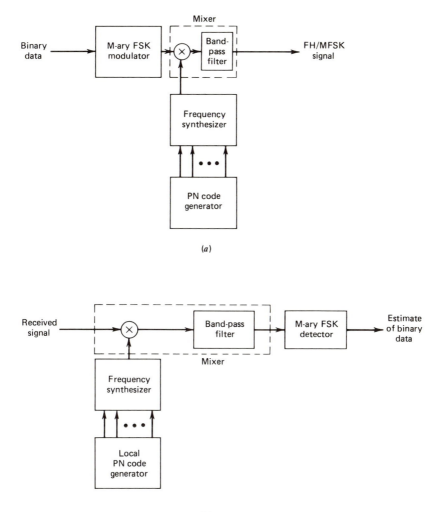

Figure 9.8 Frequency hop spread M-ary frequency-shift keying. (a) Transmitter. (b) Receiver.

segments of a PN sequence drive the frequency synthesizer, which enables the carrier frequency hop over 2^k distinct values. On a single hop, the bandwidth of the transmitted signal is the same as that resulting from the use of a conventional M-ary frequency-shift-keying (MFSK) format with an alphabet of $M = 2^K$ orthogonal signals. However, for a complete range of 2^k-frequency hops, the transmitted FH/MFSK signal occupies a much larger bandwidth. Indeed, with present-day technology, FH bandwidths of the order of several GHz are attainable, which is an order of magnitude larger than that achievable with direct-sequence spread spectra. An implication of these large FH bandwidths is that coherent detection is possible only within each hop, because frequency synthesizers are unable to maintain phase coherence over successive hops. Accordingly, most frequency-hop spread-spectrum communication systems use non-coherent M-ary modulation schemes.

In the receiver depicted in Fig. 9.8*b*, the frequency hopping is first removed by *mixing* (down-converting) the received signal with the output of a local frequency synthesizer that is synchronously controlled in the same manner as that in the transmitter. The resulting output is then band-pass filtered, and subsequently processed by a *noncoherent* M-ary FSK detector. To implement this M-ary detector, we may use a bank of *M* noncoherent matched filters, each of which is matched to one of the MFSK tones. (Noncoherent matched filters were described in Section 3.9). An estimate of the original symbol transmitted is obtained by selecting the largest filter output.

An individual FH/MFSK tone of shortest duration is referred to as a *chip*; this terminology should not be confused with that used in Section 9.3, describing DS/BPSK. The *chip rate*, R_c, for an FH/MFSK system is defined by

$$R_c = \max(R_h, R_s) \tag{9.50}$$

where R_h is the *hope rate*, and R_s is the *symbol rate*.

A slow FH/MFSK signal is characterized by having multiple symbols transmitted per hop. Hence, each symbol of a slow FH/MFSK signal is a chip. Correspondingly, in a slow FH/MFSK system, the bit rate R_b of the incoming binary data, the symbol rate R_s of the MFSK symbol, the chip rate R_c, and the hop rate R_h are related by

$$R_c = R_s = \frac{R_b}{K} \geqslant R_h \tag{9.51}$$

where $K = \log_2 M$.

EXAMPLE 4

Figure 9.9*a* illustrates the variation of the frequency of a slow FH/MFSK signal with time for one complete period of the PN sequence. The period of the PN sequence is $2^4 - 1 = 15$. The FH/MFSK signal has the following parameters:

Number of bits per MFSK symbol	$K = 2$
Number of MFSK tones	$M = 2^K = 4$
Length of PN segment per hop	$k = 3$
Total number of frequency hops	$2^k = 8$

In this example, the carrier is hopped to a new frequency after transmitting two symbols or equivalently, four information bits. Figure 9.9*a* also includes the input binary data, and the PN sequence controlling the selection of FH carrier frequency. It is noteworthy that although there are eight distinct frequencies available for hopping, only three of them are utilized by the PN sequence.

Figure 9.9*b* shows the variation of the dehopped frequency with time. This variation is recognized to be the same as that of a conventional MFSK signal produced by the given input data.

At each hop, the MFSK tones are separated in frequency by an integer multiple of the chip rate $R_c = R_s$, ensuring their orthogonality. The implication

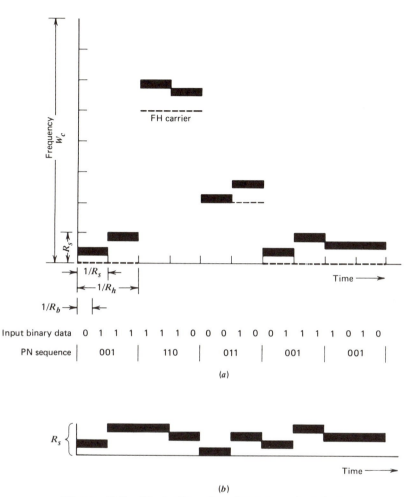

Figure 9.9 Illustrating slow-frequency hopping.

of this condition is that any transmitted symbol will not produce any crosstalk in the other $M - 1$ noncoherent matched filters constituting the MFSK detector of the receiver in Fig. 9.8b; By "crosstalk" we mean the spillover from one filter output into an adjacent one. The resulting performance of the slow FH/MFSK system is the same as that for the noncoherent detection of conventional (unhopped) MFSK signals in additive white Gaussian noise (AWGN). Thus the interfering (jamming) signal has an effect on the FH/MFSK receiver, in terms of average probability of symbol error, equivalent to that of additive white Gaussian noise on a conventional noncoherent M-ary FSK receiver experiencing no interference.

Assuming that the jammer decides to spread its average power J over the entire frequency-hopped spectrum, the jammer's effect is equivalent to an AWGN with power spectral density $N_0/2$, where $N_0 = J/W_c$ and W_c is the FH bandwidth. The spread-spectrum system is thus characterized by the *symbol energy-to-noise density ratio*:

$$\frac{E}{N_0} = \frac{P/J}{W_c/R_s} \tag{9.52}$$

where the ratio P/J is the reciprocal of the jamming margin. The other ratio is the processing gain of the slow FH/MFSK system, defined by

$$PG = \frac{W_c}{R_s}$$

$$= 2^k \tag{9.53}$$

That is, the processing gain (expressed in decibels) is equal to $10 \log_{10} 2^k = 3k$, where k is the length of the PN segment employed to select a frequency hop.

This result assumes that the jammer spreads its power over the entire FH spectrum. However, if the jammer decides to concentrate on just a few of the hopped frequencies, then the processing gain realized by the receiver would be less than $3k$ decibels.

(2) Fast-frequency Hopping

A fast FH/MFSK system differs from a slow FH/MFSK system in that there are multiple hops per M-ary symbol. Hence, in a fast FH/MFSK system, each hop is a chip. In general, fast-frequency hopping is used to defeat a smart jammer's tactic that involves two functions: measurement of the spectral content of the transmitted signal, and retuning of the interfering signal to that portion of the frequency band. Clearly, to overcome the jammer, the transmitted signal must be hopped to a new carrier frequency *before* the jammer is able to complete the processing of these two functions.

EXAMPLE 5

Figure 9.10*a* illustrates the variation of the transmitted frequency of a fast FH/MFSK signal with time. The signal has the following parameters:

Number of bits per MFSK symbol	$K = 2$
Number of MFSK tones	$M = 2^K = 4$
Length of PN segment per hop	$k = 3$
Total number of frequency hops	$2^k = 8$

In this example, each MFSK symbol has the same number of bits and chips; that is, the chip rate R_c is the same as the bit rate R_b. After each chip, the carrier frequency of the transmitted MFSK signal is hopped to a different value, except for few occasions when the k-chip segment of the PN sequence repeats itself.

Figure 9.10*b* depicts the time variation of the frequency of the dehopped MFSK signal, which is the same as that in Example 4.

For data recovery at the receiver, noncoherent detection is used. However, the detection procedure is quite different from that used in a slow FH/MFSK receiver. In particular, two procedures may be considered:

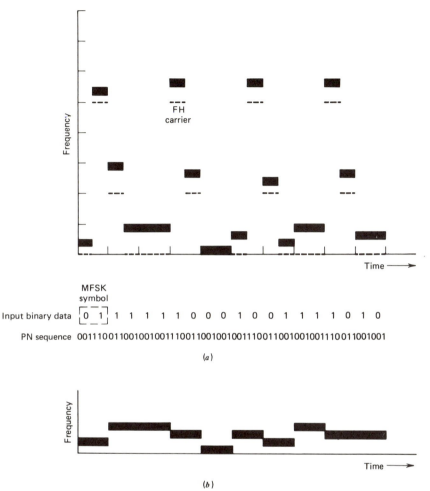

Input binary data: 0 1 1 1 1 1 1 0 0 0 1 0 0 1 1 1 1 0 1 0

PN sequence: 0011100110010010011100110010010011100110010010011100110010010011100110010010011100110010010011

(a)

(b)

Figure 9.10 Illustrating fast-frequency hopping.

1. For each FH/MFSK symbol, separate decisions are made on the K frequency-hop chips received, and a simple rule based on *majority vote* is used to make an estimate of the dehopped MFSK symbol.
2. For each FH/MFSK symbol, likelihood functions are computed as functions of the total signal received over K chips, and the larger one is selected.

A receiver based on the second procedure is optimum in the sense that it minimizes the average probability of symbol error for a given E_b/N_0.

9.7 APPLICATIONS
Probably the single most important application of spread-spectrum techniques is that of protection against jammers. This issue has been discussed at length in previous sections. In this section of the chapter, we briefly consider two other

applications of spread-spectrum techniques, namely, *code-division multiple access,* and *multipath suppression.*

(1) Code-division Multiple Access

As mentioned previously in Section 7.13(3), the two most common multiple access techniques for satellite communications are frequency-division multiple access (FDMA) and time-division multiple access (TDMA). In FDMA, all users access the satellite channel by transmitting simultaneously but using disjoint frequency bands. In TDMA, all users occupy the same RF bandwidth of the satellite channel, but they transmit sequentially in time. When, however, all users are permitted to transmit simultaneously and also occupy the same RF bandwidth of the satellite channel, then some other method must be provided for separating the individual signals at the receiver. *Code-division multiple access* (CDMA) is the method that makes it possible to perform this separation.

To accomplish CDMA, spread spectrum is always used.* In particular, each user is assigned a code of its own, which performs the direct-sequence or frequency-hop spread-spectrum modulation. The design of the codes has to cater for two provisions:

1. Each code is approximately *orthogonal* (i.e., has low cross-correlation) with all the other codes.
2. The CDMA system operates *asynchronously*, which means that the transition times of a user's data symbols do not have to coincide with those of the other users.

The second requirement complicates the design of good codes for CDMA.†

The use of CDMA offers three attractive features over TDMA:

1. CDMA does not require an external synchronization network, which is an essential feature of TDMA.
2. CDMA offers a gradual degradation in performance as the number of users is increased. It is therefore relatively easy to add new users to the system.
3. CDMA offers an external interference rejection capability (e.g., multipath rejection or resistance to deliberate jamming).

(2) Multipath Suppression

In many radio channels, the transmitted signal reaches the receiver input via more than one path. For example, in a *mobile communication* environment, the transmitted signal is reflected off a variety of *scatterers* such as buildings, trees, and moving vehicles. Thus, in addition to the *direct path* from the transmitter to the receiver, there are several other *indirect paths* (arising from the presence of the scatterers) that contribute to the composition of the received signal.

---

* For analytic considerations, the evaluation of energy and bandwidth in CDMA, selective calling and identification requirements, see Cooper and McGillen (1986, pp. 378–395).

† The optimum design of good codes for CDMA systems is covered in Sawarte and Pursley (1980).

Naturally, the contributions from these indirect paths exhibit different signal attenuations and time delays relative to that from the direct path. Indeed, they may interfere with the contribution from the direct path either constructively or destructively at the receiver input. The interference caused by these indirect paths is called *multipath interference* or simply *multipath*. The variation in received signal amplitude due to this interference is called *fading,* as the signal amplitude tends to fade away when destructive interference occurs between the contributions from the direct and indirect paths. The description of multipath fading is also complicated by whether the mobile receiving unit and nearby scatterers are all standing still, whether the mobile receiving unit is standing still but some of the scatterers are moving, or whether the mobile receiving unit is moving as well as some (or all) of the scatterers.*

In a slow-fading channel, we may combat the effects of multipath by applying spread spectrum.† Specifically, in a direct-sequence spread-spectrum system, we find that if the reflected signals at the receiver input are delayed (compared with the direct-path signal) by more than one chip duration of the PN code, then the reflected signals are treated by the matched filter or correlator of the receiver in the same way as any other uncorrelated input signal. Indeed, the higher the chip rate of the PN code, the smaller will the degradation due to multipath be.

In a frequency-hop spread-spectrum system, improvement in system performance in the presence of multipath is again possible, but through a mechanism different from that in a direct-sequence spread-spectrum system. In particular, the effect of multipath is diminished, provided that the carrier frequency of the transmitted signal hops fast enough relative to the differential time delay between the desired signal from the direct path and the undesired signals from the indirect paths. Under this condition, all (or most) of the multipath energy will (on the average) fall in frequency slots that are orthogonal to the slot occupied currently by the desired signal, and degradation due to multipath is thereby minimized.

9.8 SUMMARY AND DISCUSSION

Direct-sequence M-ary phase shift keying (DS/MPSK) and *frequency-hop M-ary frequency shift-keying* (FH/MFSK) represent two principal categories of spread-spectrum communications. Both of them rely on the use of a pseudo-noise (PN) sequence, which is applied differently in the two categories.

In a DS/MPSK system, the PN sequence makes the transmitted signal assume a noise-like appearance by spreading its spectrum over a broad range of frequencies simultaneously. For the phase-shift keying, we may use binary

---

* For a description of multipath fading in a mobile communication environment, see Lee (1982, pp. 25–33).

† For detailed discussion of spread-spectrum communication systems operating over slow-fading multipath channels, see Simon et al. (1985, Vol. II, pp. 40–51), Turin (1980), and Kavehrad and McLane (1987).

PSK (i.e., $M = 2$) with a single carrier, as discussed in Section 9.3. Alternatively, we may use QPSK (i.e., $M = 4$), in which case the data are transmitted using a pair of carriers in phase quadrature. The motivation here is to provide for improved bandwidth efficiency; see Section 7.3. In a spread-spectrum system, bandwidth efficiency is usually not of prime concern. Nevertheless, the use of QPSK is important, since it is less sensitive to some types of interference (jamming).

In an FH/MFSK system, the PN sequence makes the carrier hop over a number of frequencies in a pseudo-random manner, with the result that the spectrum of the transmitted signal is spread in a sequential manner.

Naturally, the direct-sequence and frequency-hop spectrum-spreading techniques may be employed in a single system. The resulting system is referred to as a *hybrid* DS/FH *spread-spectrum system*. The reason for seeking a hybrid approach is that advantages of both the direct-sequence and frequency-hop spectrum-spreading techniques are realized in the same system. Indeed, the hybrid approach is the only practical way of realizing extremely wide-spectrum spreading.

For its proper operation, spread-spectrum communication requires that the locally generated PN sequence used in the receiver to despread the received signal be *synchronized* to the PN sequence used to spread the transmitted signal in the transmitter. A solution to the synchronization problem consists of two parts: *acquisition* and *tracking*. In acquisition, or *coarse* synchronization, the two PN codes are aligned to within a fraction of a chip in as short a time as possible. Once the incoming PN code has been acquired, tracking, or *fine* synchronization, takes place. Typically, PN acquisition proceeds in two steps. First, the received signal is multiplied by a locally generated PN code to produce a measure of *correlation* between it and the PN code used in the transmitter. Next, an appropriate *decision-rule and search strategy* is used to process the measure of correlation so obtained to determine whether the two codes are in synchronism and what to do if they are not. As for tracking, it is accomplished using phase-lock techniques very similar to those used for the local generation of coherent carrier references. The principal difference between them lies in the way in which phase discrimination is implemented.*

A discussion of spread-spectrum communications would be incomplete without some reference to jammer waveforms. The jammers encountered in practice include the following types:

1. *The barrage noise jammer*, which consists of band-limited white Gaussian noise of high average power. The barrage noise jammer is a brute-force jammer that does not exploit any knowledge of the antijam communication system except for its spread bandwidth.

2. *The partial-band noise jammer*, which consists of noise whose total power is evenly spread over some frequency band that is a subset of the total

---

* For detailed discussion of the synchronization problem in spread-spectrum communications, see Ziemer and Peterson (1985, Chapters 9 and 10) and Simon et al. (1985, Volume III).

spread bandwidth. Owing to the smaller bandwidth, the partial-band noise jammer is easier to generate than the barrage noise jammer.

3. *The pulsed noise jammer*, which involves transmitting wideband noise of power

$$J_{peak} = \frac{J}{p}$$

for a fraction p of the time, and nothing for the remaining fraction $1 - p$ of the time. The average noise power equals J.

4. *The single-tone jammer*, which consists of a sinusoidal wave whose frequency lies inside the spread bandwidth. As such, it is the easiest of all jamming signals to generate.

5. *The multitone jammer*, which is the tone equivalent of the partial-band noise jammer.

In addition to these five, there are many other kinds of jamming waveforms that do occur in practice. In any event, there is no single jamming waveform that is worst for all spread-spectrum systems, and there is no single spread-spectrum system that is best against all possible jamming waveforms.

The pulsed-noise jammer is particularly effective against DS/BPSK systems. Correspondingly, the partial-band noise and multitone jammers are most effective against FH/MFSK systems. These jamming strategies are effective because they are able to concentrate jamming resources on some fraction of the transmitted symbols, thereby resulting in *bursts of errors* at the receiver output. To deal with this problem, a spread-spectrum system will rely on very powerful *error-correcting codes combined with interleaving*. Specifically, in the transmitter, the incoming data are first encoded, interleaved, and then applied to a spread-spectrum modulator. In the receiver, the received signal is despread, demodulated, deinterleaved, and then detected. Indeed, when error-correcting codes and interleaving are combined with hybrid DS/FH spread spectrum (i.e., pseudo-random chipping and pseudo-random frequency hopping), the result is a digital communication system that can provide very significant protection against external noise, unintentional interference, and intentional jamming.

PROBLEMS

P9.1 PSEUDO-NOISE SEQUENCES

Problem 9.1.1 A pseudo-noise (PN) sequence is generated using a feedback shift register of length $m = 4$. The chip rate is 10^7 chips per second. Find the following parameters

(a) PN sequence length.
(b) Chip duration of the PN sequence.
(c) PN sequence period.

Problem 9.1.2 Figure P9.i shows a four-stage feedback shift register. The initial state of the register is 1000. Find the output sequence of the shift register.

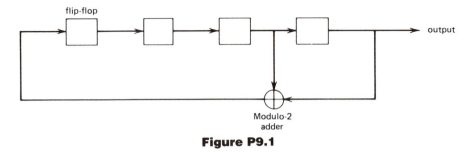

Figure P9.1

Problem 9.1.3 For the feedback shift register given in Problem 9.1.2, demonstrate the balance property and the run property of a PN sequence. Also, calculate and plot the autocorrelation function of the PN sequence produced by this shift register.

Problem 9.1.4 Show that the power spectral density of a PN sequence (alternating between -1 and $+1$ volt) is defined by

$$S_c(f) = \frac{1}{N^2} \delta(f) + \left(\frac{1 + N}{N^2}\right) \sum_{\substack{n=-\infty \\ n \neq 0}}^{\infty} \text{sinc}^2\left(\frac{n}{N}\right) \delta\left(f - \frac{n}{NT_c}\right)$$

where N is the sequence length, T_c is the chip duration, and $\delta(f)$ is a delta function. Plot $S_c(f)$ for the sequence length $N = 7$.

What is the limiting form of this power spectral density as N approaches infinity?

P9.3   DIRECT-SEQUENCE SPREAD COHERENT BINARY PHASE-SHIFT KEYING

Problem 9.3.1 Show that the truth table given in Table 9.2 can be constructed by combining the following two steps:

1. The data sequence $b(t)$ and PN sequence $c(t)$ are added modulo-2.
2. Symbols 0 and 1 at the modulo-2 adder output are represented by phase shifts of $0°$ and $180°$, respectively.

P9.4   SIGNAL-SPACE DIMENSIONALITY AND PROCESSING GAIN

Problem 9.4.1 A single-tone jammer

$$j(t) = \sqrt{2J} \cos(2\pi f_c t + \theta)$$

is applied to a DS/BPSK system. The N-dimensional transmitted signal $x(t)$ is described by Eq. 9.15. Find the $2N$ coordinates of the jammer $j(t)$.

Problem 9.4.2 The processing gain of a spread-spectrum system may be expressed as the ratio of the spread bandwidth of the transmitted signal to the despread bandwidth of the received signal. Justify this statement for the DS/BPSK system.

Problem 9.4.3 A direct-sequence spread binary phase-shift keying system uses a feedback shift register of length 19 for the generation of the PN sequence. Calculate the processing gain of the system.

P9.5 PROBABILITY OF ERROR

Problem 9.5.1 In a DS/BPSK system, the feedback shift register used to generate the PN sequence has length $m = 19$. The system is required to have a probability of error due to externally generated interfering signals that does not exceed 10^{-5}. Calculate the following system parameters in decibels:

(a) Processing gain.
(b) Antijam margin.

P9.6 FREQUENCY-HOP SPREAD SPECTRUM

Problem 9.6.1 A slow FH/MFSK system has the following parameters:

The number of bits per MFSK symbol = 4

The number of MFSK symbols per hop = 5

Calculate the processing gain of the system.

Problem 9.6.2 A fast FH/MFSK system has the following parameters:

The number of bits per MFSK symbol = 4

The number of hops per MFSK symbol = 4

Calculate the processing gain of the system.

CHAPTER TEN

DATA NETWORKS

Much of the material on coding and modulation techniques presented in previous chapters of the book has dealt with the *transfer* of data from one point to another. The data may consist of digitized voice, digitized video, computer-generated data, and so on. In this final chapter of the book, we consider the use of a data (communication) network designed to *transmit* data between various users (devices). Basically, a *data network* is a structure that enables a data user at one location to have access to some data processing function or service available at another location. Data networks range from a small network interconnecting computers (processors) and terminals inside a building or campus-like complex to a geographically distributed network interconnecting hundreds of computers of various sizes and tens of thousands of terminals.

Data networks find application in a variety of fields. They are used to provide access to a remote computer system for computational tasks. They are used for accessing remote data bases such as financial services. They are used for the remote updating of data bases in addition to accessing the data, as in airline reservation systems. Yet another popular application of a data network is in electronic mail. The list of applications involving data networks, in one form or another, goes on.

In this chapter, we present an introductory treatment of data networks.* In particular, we discuss topics of fundamental importance concerning topologies and architectures of data networks. We also consider the issue of standards, which play an important role in this field. We begin our treatment of data networks by identifying the need for a *communication network,* which provides for the sharing of transmission facilities among a number of devices such as computers, terminals, and telephones.

10.1 COMMUNICATION NETWORKS

In data transmission terminology, a *station* refers to a computer, terminal, telephone, or some other communication device. The *channel* or *link* connecting a pair of stations may be simplex, half-duplex, or full-duplex. In the *simplex* operation, signals are transmitted in one direction only, with one station acting as the transmitter and the other station acting as the receiver. When both

---

* Data networks are also referred to as *computer networks* in the literature. For detailed treatment of the subject, see Bertsekas and Gallager (1987), Schwartz (1987), Stallings (1985), Hayes (1984), Chou (1983, 1985), Green (1982), and Tanenbaum (1981).

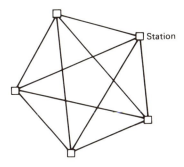

Figure 10.1   Fully connected mesh topology.

stations transmit, but only one at a time, the operation is said to be *half-duplex*. In the *full-duplex* operation, both stations transmit simultaneously.

Two stations (e.g., computers) are said to be *interconnected* if they are capable of exchanging information with each other. For the transmission medium, we may use copper wires, optical fibers, radio links, or satellites. Consider then the problem of providing communication among a number of stations, each of which may require a communication path to many of the other stations at various times. We may solve this problem by using a *fully connected* or *mesh topology,* shown illustrated in Fig. 10.1 for the case of five stations. In general, such an approach requires $(N - 1)$ *input-output* (I/O) *ports* and a total of $N(N - 1)/2$ *full-duplex links* where N is the number of stations. The cost of providing for links between any pair of stations therefore grows as N^2, which makes the system impractical for large N.

We may resolve this practical difficulty by employing a *communication network* or *subnet* as illustrated in Fig. 10.2. The subnet consists of an interconnection of a number of *nodes* made up of intelligent processors (e.g., microcomputers). The primary purpose of these nodes* is to route data through the

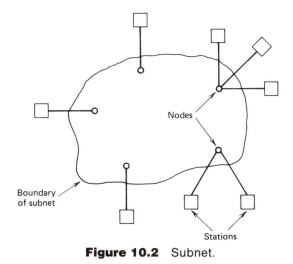

Figure 10.2   Subnet.

* In the literature, nodes are also referred to as *intermediate message processors* (IMPs) or *switches*.

subnet. Each node has one or more stations attached to it. The subnet is designed to serve as a *shared resource for the sole purpose of moving data exchanged between stations in an efficient manner*. Also, the use of a subnet reduces the requirement on I/O ports in that each station requires a single I/O port rather than $(N - 1)$ ports.

If the attached stations are made up of computers (processors) and terminals, then the subnet plus the attached stations is referred to as a *data network*.

A data network is usually required to be *reliable*, even in the face of unreliable nodes and links. To achieve this requirement, the network must be *redundant*. A sufficiently redundant network may lose a small number of components and yet continue to function properly, albeit with reduced performance.

Consider, for example, the subnet shown in Fig. 10.3. The transmission of data, originating from station *a* and intended for station *e*, say, may follow several paths. The data may enter node 1, and from there it may be routed via nodes 5 and 6. Alternatively, the data may be routed to its destination via nodes 2, 3, 4, and 6. Indeed, there are a total of six paths along which the data may be routed from station *a* to station *e*. The presence of alternative paths gives a subnet the redundancy it needs for it to become reliable.

The structural makeup of a subnet is characterized as follows:

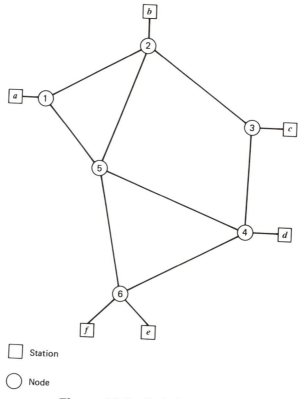

☐ Station

○ Node

Figure 10.3 Switched network.

1. The network may contain two kinds of nodes: *internal nodes* and *boundary nodes*. An internal node is connected only to other nodes. The sole function of an internal node is therefore to switch data. On the other hand, a boundary node may have one or more stations attached to it. Hence, a boundary node accepts data from and delivers data to the stations attached to it, in addition to its own switching function.
2. The connections between stations and nodes are provided by *point-to-point links*.
3. The network may provide more than one possible path between any pair of stations, thereby enhancing reliability of service. This is achieved by using a fully connected or partially connected topology; the latter one is the preferred method, particularly for large data networks.

When two users of a data network wish to send messages to each other, they first set up a *session*, much like a telephone call. To facilitate the transmission of data for the various sessions, the subnet may use circuit switching or store-and-forward switching. In the next two sections, we will describe these two general approaches to the design of a subnet.

10.2   CIRCUIT SWITCHING

A distinguishing feature of *circuit switching* is that it provides a dedicated communication path or *circuit* between two stations. The circuit consists of a connected sequence of links from source to destination. The links may consist of time slots in a time-division multiplexed (TDM) system or frequency slots in a frequency-division multiplexed (FDM) system.* The circuit, once in place, remains uninterrupted for the entire duration of transmission.

Circuit switching is usually controlled by a centralized hierarchical control mechanism with knowledge of the network's organization. To establish a circuit-switched connection, an available path through the network is seized and then dedicated to the exclusive use of the two stations wishing to communicate with each other. In particular, a call-request signal must propagate all the way to the destination, and be acknowledged, before data transmission can begin. Once the circuit is in place, the network is effectively transparent to the users. This means that during the connection time, the bandwidth and resources allocated to the circuit are essentially "owned" by the two stations, until the circuit is disconnected. The circuit thus represents an efficient use of resources only to the extent that the allocated bandwidth is properly utilized.

Circuit switching is used almost universally for telephone networks. Although the telephone network is used to transmit data (see Section 7.13), nevertheless, voice constitutes the bulk of the network's traffic. Indeed, circuit switching is well suited to the transmission of voice signals, since voice conversations tend to be of long duration (about 2 minutes on the average) compared

---

* In TDM, a number of independent message signals are permitted to share a common communication path (channel) by separating them in time; TDM was described in Section 4.7. In FDM, on the other hand, the sharing of a common communication path is accomplished by using a bank of bandpass filters to separate the message signals in frequency.

to the time required for setting up the circuit (about 0.1–0.5 seconds). Moreover, in most voice conversations, there is information flow for a relatively large percentage of the connection time, which makes circuit switching all the more suitable for voice conversations.

Circuit switching is also used in a digital *private branch exchange* (PBX) installed in an organization's building, principally for the purpose of switching telephone calls between parties located within and outside the facility.* However, as data traffic in the office continues to rise, PBXs are being required increasingly to switch data as well as voice.

10.3 STORE-AND-FORWARD SWITCHING

In circuit switching, a communication link is shared between the different sessions using that link on a *fixed* allocation basis. In *store-and-forward switching*, on the other hand, the sharing is done on a *demand* basis. Accordingly, store-and-forward switching has an advantage over circuit switching in that when a link has traffic to send, it may be more fully utilized.

There are two basic types of store-and-forward switching, namely, message switching and packet switching. In the sequel, we consider the two of them in that order.

In a *message-switched network*, the message is routed in its entirety from source to destination by working its way through the subnet, link by link, being queued at each node along its path. Thus, at each node, the entire message is received and then *stored in a buffer* until the link to the next node becomes available for forward transmission. Message switching is one of the earlier forms of data communication service.

With message switching, it is not necessary that a dedicated communication path be established between two stations, nor does it require a call setup. Nevertheless, message switching has two major shortcomings:

1. The intermediate processor at each node of the subnet must be equipped with a buffer large enough to store (possibly) long messages, because there is no limit on the duration of an incoming message.
2. It is possible for a single message, if it is too long, to tie up a link for a very long time.

The second shortcoming is particularly serious from an operational viewpoint, as it may render message switching useless for real-time or interactive traffic. To overcome the shortcomings of message switching, the customary practice is to use packet switching.

In a *packet-switched network*, any message larger than a specified size is subdivided into segments not exceeding the specified size, prior to transmission. The segments are commonly referred to as *packets*. The original message is reassembled at the destination, on a packet-by-packet basis.

We may view a packet-switched network as a distributed pool of productive

---

* For discussion of PBX, see Kasson and Kagan (1985, Chapter 5).

resources (i.e., channels, buffers, and switching processors) whose capacity is *shared dynamically* by a community of competing users (stations) wishing to communicate with each other. In contrast, in a circuit-switched network, resources are dedicated to a pair of stations for the entire period they are in session. Accordingly, packet switching is far better suited to a computer-communication environment in which bursts of data are exchanged between stations on an occasional basis. The use of packet switching, however, requires that careful control be exercised on user demands; otherwise, the network may be seriously abused.

10.4 LAYERED ARCHITECTURE

The design of a data network may proceed in an orderly way by looking at the network in terms of a *layered architecture*, regarded as a hierarchy of nested layers. A *layer* refers to a process or device inside a computer system, which is designed to perform a specific function. Naturally, the designers of a layer will be intimately familiar with its internal details and operation. At the system level, however, a user views the layer merely as a "black box" that is described in terms of the inputs, the outputs, and the functional relation between outputs and inputs. In a layered architecture, each layer regards the next lower layer as one or more black boxes with some given functional specification to be used by the given higher layer. Thus, the highly complex communication problem in data networks is resolved as a manageable set of well-defined interlocking functions.

It is this line of reasoning that has led to the development of the *open systems interconnection* (OSI) *reference model.** The term "open" refers to the ability of any two systems conforming to the reference model and its associated standards to interconnect.

In the OSI reference model, the communications and related-connections functions are organized as a series of *layers* or *levels* with well-defined *interfaces*, and with each layer built on its predecessor. In particular, each layer performs a related subset of primitive functions, and it relies on the next lower layer to perform additional primitive functions. Moreover, each layer offers certain services to the next higher layer, and shields the latter from the implementation details of those services. Between each pair of layers, there is an *interface*. It is the interface that defines the services offered by the lower layer to the upper layer.

The OSI model is composed of seven layers, as illustrated in Fig. 10.4. Layer k on system A, say, communicates with layer k on some other system B in accordance with a set of rules and conventions, collectively constituting the layer k *protocol,* where $k = 1, 2, \ldots, 7$. The entities that comprise the corresponding layers on different systems are referred to as *peer processes*. In other words, communication is achieved by having the peer processes in two

---

* The OSI reference model was developed by a subcommittee of the International Organization for Standardization (ISO) in 1977.

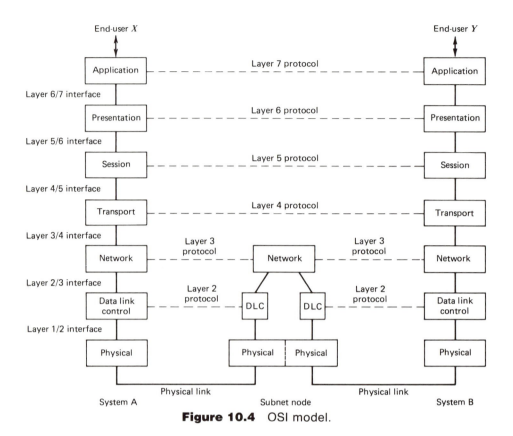

Figure 10.4 OSI model.

different systems communicate via a protocol, with the protocol itself being defined by a set of rules of procedure.*

Communication between applications is the ultimate goal in the two systems. Suppose end-user X wishes to send a message to end-user Y. The procedure for achieving communication starts with end-user X involving layer 7, the application layer, on system A. Layer 7 of system A establishes a peer relationship with layer 7 of the target system B. This is done by using a layer 7 protocol, which requires services from layer 6, the presentation layer. Then, layer 6 of system A establishes a peer relationship with layer 6 of the target system B, using a layer 6 protocol. This procedure is repeated downward, layer by layer, until layer 1 is reached. Finally, layer 1, the physical layer, passes the bits through a physical link, and communication is thereby established.

Note that physical communication between peer processes exists only at layer 1. On the other hand, layers 2 through 7 are in *virtual communication* with their distant peers. However, each of these six layers can exchange data and

---

* The term "protocol" has been borrowed from common usage, describing conventional social behavior between human beings.

control information with its neighboring layers (below and above) through layer-to-layer interfaces. In Fig. 10.4 physical communication is shown by solid lines and virtual communication by dashed lines.

The difference between a protocol and an interface should also be carefully noted. A protocol is a logical concept only, referring to a set of rules between *similar* processes. An interface, on the other hand, may be used in a logical context referring to a set of rules between *dissimilar* processes, or in a physical context referring to a connection between two separate devices.

As mentioned previously, the OSI reference model has seven layers. The major principles involved in arriving at this number are as follows:*

1. Each layer performs well-defined functions.
2. A boundary is created at a point where the description of services offered is small, and the number of interactions across the boundary is the minimum possible.
3. A layer is created from easily localized functions, so that the architecture of the model may permit modifications to the layer protocol to reflect changes in technology without affecting the other layers.
4. A boundary is created at some point with an eye toward standardization of the associated interface.
5. A layer is created only when a different level of abstraction is needed to handle the data.
6. The number of layers employed should be large enough to assign distinct functions to different layers, and yet small enough to maintain a manageable architecture for the model.

In the sequel, we describe the seven layers of the OSI model, one by one, starting with the bottom layer.

(1) Physical Layer

The function of the *physical layer,* layer 1, is to provide a virtual link for the transmission of raw bits of data over a physical communication channel joining a station (computer site) and a node or a pair of nodes. Such a link is referred to as a *virtual bit pipe.*

The physical layer performs its function by means of a *physical interface module* placed on each side of the physical link. At the transmitting end, the physical interface module maps the incoming stream of bits from the next higher layer, the data link control layer, into a signal suitable for transmission over the link. At the receiving end, the other physical interface module maps the received signal back into bits. In the case of an analog communication channel used for the link, the physical interface module takes on the form of a modem, which was discussed in Section 7.13.

---

* For a listing of the principles involved in arriving at the seven layers of the OSI model, and a description of the layers themselves, see Zimmerman (1982, Chapter 2).

(2)   Data Link Control Layer

The second layer in the OSI model is the *data link control (DLC) layer*. The main function of this second layer is *error control* over a single data link (from one node to the next). Thus, an unreliable bit pipe offered by layer 1, the physical layer, is transformed into a link that appears free of transmission errors to the next higher layer (i.e., the network layer).

(3)   Network Layer

The third layer in the OSI model is the *network layer*. The network layer has two major functions, namely, *routing* of packets through the network and *flow control*.

Routing refers to the ability to establish cost-effective paths across the subnet. Specifically, at a node of the subnet, the network layer process performs routing by deciding where to send all packets arriving at that node. If the packets are destined for some other node or station, they are sent on to the appropriate link. At a station (computer site), all packets arriving there are passed by the network layer process up to the higher level, the transport layer.

The goal of *flow control* is to guarantee good performance over a communication path found by the routing algorithm. As such, flow control is used to mean both *congestion control* and the control exercised by a receiving node to prevent its *buffer* (allocated for the storage of incoming packets) from overflowing. Typically, there is a queue of packets at each node of the subnet for an outgong channel; hence, the need for a buffer of some maximum length. When the rate at which packets arrive at a node of the subnet is in excess of the rate at which packets leave the node, then the queue size at that node naturally keeps on growing and, in a corresponding way, the *delay* experienced by the transmission of packets through that node becomes increasingly larger. The function of the network layer is to prevent such an undesirable situation from happening; it does so by matching the rate of transmission by a sender to the rate of acceptance by a receiver. Thus the network layer determines when to accept packets from the higher layer, the transport layer, and when to transmit packets to other subnet nodes or stations.

In conceptual terms, the network layer is the most sophisticated of the layered architecture, since all the peer processes of this layer must work together for the data network as a whole to function properly.

(4)   Transport Layer

The purpose of the *transport layer,* layer 4, is to provide a reliable and transparent transport service in association with the underlying services provided by the lower layers. The service is *reliable* in the sense that data units are delivered (virtually) error-free, in sequence, and with no losses or duplications. Indeed, the transport layer provides a truly *end-to-end* (i.e., source-to-destination) error recovery and flow control.

(5) Session Layer

The *session layer,* layer 5, assists in the support of interactions between two cooperating users (technically speaking, two presentation layer processes). Once a session has been established between the two users, the session layer manages the dialogue between them in an orderly manner.

(6) Presentation Layer

The function of the *presentation layer,* layer 6, is to transform the input data so as to provide the set of services that may be selected by the application layer. One reason for the transformation is to resolve differences in format and data representation (syntax), thereby providing independence to the application layer from such differences. Another transformation that the presentation layer may also perform is *encryption* to provide security.*

(7) Application Layer

The *application layer* is the highest layer in the OSI architecture, and the other six layers exist to support it. Specifically, the application layer provides a means for an end-user to access the OSI environment. The content of this layer is entirely up to the individual.

Figure 10.5 illustrates operational features of the OSI model. The exchange of information begins with the top layer, the application layer, receiving (from an end-user) a block of data referred to as the *application data unit* in Fig. 10.5. As this data unit is transferred downward, layer by layer, control information is added to the data for the purpose of implementing protocol functions; this function is referred to as *encapsulation.* The first encapsulation is performed when the data unit is transferred from the application layer to the layer below, the presentation layer. The encapsulation is performed by adding a *header* to the data, which contains control information pertaining to the presentation layer. The original application data, plus the presentation header, is then passed as a unit to the session layer. The whole unit is treated by the session layer as data. Accordingly, a second encapsulation is performed with the session layer appending a header of its own. This procedure continues on, all the way down to layer 2, the data link control layer. The latter layer usually adds both a *header* and a *trailer;* the header contains control information, while the trailer contains information required for error control. The layer 2 unit, called a *frame,* is then passed by the physical layer on to the communication channel. At the receiving end, the procedure is performed in reverse. In particular, as the data move up, each layer strips off the outermost header (and trailer in the

---

* For discussions of a hardware-implementable algorithm, namely, the *data encryption standard* (DES), for enciphering data to provide a high level of cryptographic protection, see Davis (1978) and Morris (1978). These two overview papers appear in a special issue of the 1978 *Communications Magazine* devoted to communications privacy. The issue of computer and communications security is also discussed in Walker (1985, Chapter 8).

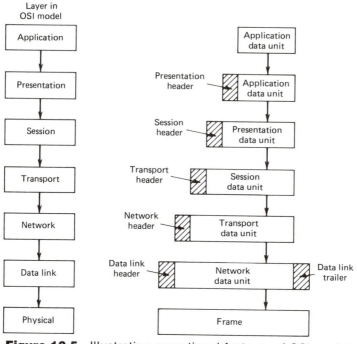

Figure 10.5 Illustrating operational features of OSI model.

case of layer 2), acts on the control information contained in the associated header, and passes the remainder up to the next layer for action. This goes on until the message is delivered to its destination.

It is noteworthy that at each stage of the process described, the data unit received by a particular layer (from the layer, one level higher) may be fragmented into several parts. This is done in order to accommodate the requirements of that layer. The data units resulting from any such fragmentation are reassembled by the corresponding peer before being passed up.

10.5 PACKET NETWORKS

A basic and yet controversial issue involved in the design of a packet-switched (point-to-point) network concerns the kind of service the network layer should offer to the transport layer. Conceptually, there are two alternatives, namely, *virtual circuit* service and *datagram* service. With virtual circuit service, the network layer attempts to provide the transport layer with a "perfect" channel by exercising strict error control and delivering all packets to the destination in their proper sequence. (The description of the network layer for the OSI model in Section 10.4 follows this prescription.) With datagram service, on the other hand, the network layer simply accepts packets from the transport layer and treats each packet independently of all previous and succeeding ones. Typically, networks offering datagram service make a reasonable attempt to provide reliable transfer of information from source to destination but do not guarantee error-free communication, nor do they guarantee that all packets arrive at their destination in sequence. Indeed, datagram is the more primitive of the two services.

There are three distinct phases to virtual circuit service; they are the setup procedure, followed by data transfer, and last the disconnect procedure. Virtual circuit service offers enhanced functionality in that, once in place, the two users involved in the data transfer are provided with the illusion of a dedicated channel with end-to-end sequencing, error control, and flow control, even though the operations performed by the network to make these provisions are undoubtedly very complicated. Thus, virtual circuit service is analogous to the service offered by a public telephone network (ignoring its relatively high error rate) when placing a telephone call.

Datagram service may allow for the efficient use of a packet-switched network. We say this for two reasons. First, there is no need for a call setup or disconnect procedure. Second, there is no provision for error control. This second feature makes the use of datagrams desirable for real-time applications. In this case, the functions of sequencing and error control may be provided by the transport layer protocol (i.e., the station).

In Table 10.1, we present a summary of major differences between virtual circuit service and datagram service. It should, however, be noted that the relative merits of these two services have been the subject of considerable debate in the literature.*

The characterization of a packet-switched network in terms of virtual circuits and datagrams presented thus far has been in the context of the interface between the network layer and the transport layer; that is, the interface between the subnet and the attached stations. There is another dimension to this characterization in that the same two models, virtual circuit and datagram, may also be used to describe the internal structure of the subnet itself. Accordingly,

Table 10.1   Comparison of Virtual Circuit and Datagram

Function	Virtual Circuit Service	Datagram Service
Error control	Provided by the network layer protocol	Explicitly provided by the station
Flow control (end-to-end)	Provided by the network layer protocol	Not provided by the network layer protocol
Packet sequencing	Packets are always delivered to the station in the same sequence in which they were sent	Packets are delivered to the station in the sequence at which they arrive at the attached node
Initial setup and disconnect	Required	Not required

* For a discussion of the various issues involved in virtual circuit and datagram services, see Roberts (1978).

there are four possible combinations, depending on the internal and external design decisions made:

1. *Datagram service implemented by using datagrams inside the subnet,* which is the simplest combination. In this case, each packet is treated independently from both the user's and subnet's point of view.
2. *Virtual circuit service implemented by using virtual circuits inside the subnet.* In this case, all packets follow the same dedicated route through the subnet, which is set up when the user requests a virtual circuit.
3. *Virtual circuit service implemented by using datagrams inside the subnet,* which is a mixture of virtual circuit and datagram models. In this hybrid combination, each packet is handled separately by the subnet, with the result that packets intended for the same virtual circuit take different routes inside the subnet. Nevertheless, the packets are delivered to the destination in their proper sequence.
4. *Datagram service implemented by using virtual circuits inside the subnet,* which provides another mixture of virtual circuit and datagram models. This second hybrid option makes no practical sense, however, because it combines a costly implementation of the subnet and yet provides no beneficial service.

As representative examples of data networks, we may mention ARPANET, TYMNET, and Datapac.* ARPANET may fit under combination 1 or 3 in that it provides both virtual circuit and datagram services, and functions as a datagram subnet. TYMNET is an example of combination 2 in that it uses virtual circuits both internally and externally. The Canadian Datapac is an example of combination 3 in that it uses virtual circuit externally and datagram service internally.

(1) Routing

As noted earlier, a major function of the network layer protocol is to handle the *routing* of packets as they move across the network. The packets may be required to follow a virtual circuit or move in a datagram mode. Regardless of which service is used, a path or route must be selected through the subnet so as to enable the data network to accept packets from a source and deliver them to a destination. It is the function of a *routing algorithm* to do this. Typical features of the routing algorithm include the following:

1. A rapid delivery of packets, which presumes an ability to detect low-delay paths.

---

* ARPANET is a packet-switched network originally developed by the Advanced Research Projects Agency (ARPA) of the U.S. Department of National Defense. TYMNET is a packet-switched network originally developed by Tymshare to provide cost-effective connections of terminals to central time-sharing computers. The Canadian Datapac was the first public data network in the world; it began operations in 1976. For a description of these networks and others, see Schwartz (1987, pp. 283–313).

2. An adaptive capability to find alternative paths when the current path becomes unusable due to link or node failures.

3. An adaptive capability to detect an alternate path when the current path becomes congested.

Four classes of routing algorithms can be identified, depending on where in the virtual circuit subnet, for example, the routing computation is performed and the type of network status that is required. The four cases are as follows:*

1. *Isolated routing algorithms,* in which each node uses local information to perform the routing function, independently of other nodes. Moreover, no exchange of information on network status or routing between the nodes is provided.

2. *Distributed routing algorithms,* in which all the nodes perform the routing computation in parallel and in a cooperative manner; the computation is based on partial status information exchanged between the nodes.

3. *Centralized routing algorithms,* in which a network control center assembles information on the global state of the subnet, attempts to find *minimum-delay* routes through the subnet, and distributes routing commands to all the nodes.

4. *Mixed routing algorithms,* which combine features derived from two or all three of the classes described before.

(2) Flow Control

The key advantages of a packet-switched network are the speed and flexibility with which data are transmitted across the network, and the efficient use to which the network resources are put. These advantages, however, are attainable only if careful control is exercised on user demands. Otherwise, the network may be abused by its users, with the result that the network becomes congested and the delay and efficiency advantages of packet switching are neutralized.

The type of congestion that occurs in an overloaded packet-switched network is somewhat analogous to that observed in a highway network.† During peak hours, the traffic may exceed capacity of the highway, thereby giving rise to large backlogs. Moreover, the interference between transit traffic on the highway and on-ramp and off-ramp traffic may further reduce the effective throughput of the highway, causing the backlog to increase even more rapidly. The highway thus experiences a form of *positive feedback.* Now, if this situation is allowed to persist, traffic on the highway may come to a standstill. However, by properly *monitoring and controlling traffic* through the use of access ramp traffic lights, interference between transit traffic on the highway

---

* For more details on the four classes of routing algorithms and the discussion of flow control procedures, see the overview paper by Gerla (1984).

† The analogy described herein is taken from Gerla (1984).

and incoming traffic is maintained within tolerable limits, and incoming traffic rate is prevented from exceeding the highway capacity.

The key point made here is the need for exercising *flow (congestion) control*. In the context of a packet-switched network, the overall purpose of flow control is that of efficient dynamic control of network resources (i.e., channels, buffers, and switching processors). Specifically, functions of flow (congestion) control are threefold:

1. The prevention of throughput degradation and loss of efficiency due to overload.
2. The prevention of deadlocks.
3. A fair allocation of the pool of resources among competing users of the network.

The problems of throughput degradation and deadlocks arise when the traffic accepted into the network exceeds the nominal capacity of the network, that is, when network resources are overallocated. By "throughput" we mean the total rate of data (in bits per second) transmitted between stations. The flow control algorithm overcomes these problems by including a set of *constraints* that effectively limit traffic access to selected parts of the subnet. The constraints may be fixed or dynamically adjusted in response to changes in traffic conditions. Whatever the form of constraints, their implementation translates into an overhead that tends to reduce throughput for some offered loads; the *offered load* is the total rate of data presented to the subnet. A typical trade-off between the gain in throughput due to the imposition of flow (congestion) control and the loss in throughput due to the associated overhead is illustrated in Fig. 10.6, showing the throughput plotted as a function of the offered load. The curve labeled "uncontrolled" results from the lack of flow control, and the one labeled "controlled" represents a typical performance obtained with an actual flow (congestion) control algorithm.

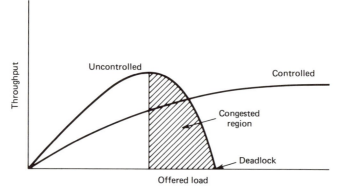

Figure 10.6 Comparison of controlled and uncontrolled flow.

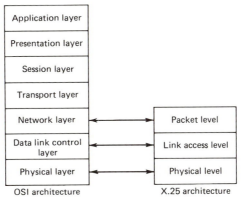

Figure 10.7   Illustrating the relation between OSI and X.25 architectures.

10.6   PACKET-SWITCHED NETWORK-ACCESS PROTOCOL: THE X.25 PROTOCOL

The *X.25 protocol** is a packet-mode interface specification that spells out the detailed protocols required to provide access to a packet-switched network. It is used in virtually all public data networks.

X.25 is an example of a structured protocol that has three layers, corresponding to the lowest three layers of the OSI model. Figure 10.7 depicts the relation between the X.25 and OSI architectures. The three layers of X.25, using its own terminology, are called the physical level, link-access level, and packet level. The *physical level,* level 1, specifies how 1s and 0s are represented, how contact with the subnet is established, timing aspects, and so forth. That is, level 1 specifies the electrical, physical, and procedural interface between a station and a node.† The functions of the *link-access level,* level 2, include synchronous operation of the receiver with respect to the transmitter, and error control. That is, level 2 manages the link between a station and a node. The *packet level,* level 3, specifies virtual circuit service through the subnet.

The link-access procedure (LAP)‡ at level 2 of X.25 is a *bit-oriented protocol* that permits sending data with an arbitrary number of bits. Figure 10.8 shows

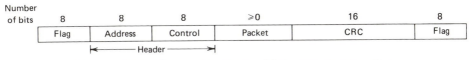

Figure 10.8   Frame structure of link-access procedure.

---

* X.25 was originally approved by the CCITT plenary session in October 1976. The original specifications included virtual circuit as well as datagram service. In a revision made in 1984, datagram service was dropped due to lack of interest.

† The "station" and "node" in our terminology refer to Data Terminal Equipment (DTE) and Data Circuit-termination Equipment (DCE) in CCITT terminology, respectively.

‡ To be more precise, the level 2 protocol is referred to as LAP-B (*link access procedure-balanced*). It is almost identical with the *high-level data link control* (HDLC), developed by the *International Organization for Standardization* (ISO).

the structure of a *frame*. Each frame begins and ends with the unique 8-bit pattern 01111110, called a *flag*. The flag is used to synchronize the receiver with the incoming frame. Also, a string of seven continuous 1s may be sent to indicate the existence of a problem on the link. Moreover, 15 or more continuous 1s keep the channel in *idle* state.

To avoid confusion between spurious appearances of the flag within a frame and the actual flag indicating end of the frame, a technique known as bit stuffing is employed. The *bit stuffing and flag/abort checking procedure* is as follows:*

1. At the transmitter, the incoming sequence is stuffed by inserting a 0 after each successive string of five 1s. This ensures that the flag will not appear anywhere in a frame except at its two ends. The integrity of frame-level synchronizing information is thereby preserved.
2. At the receiver, the incoming sequence of bits is continuously monitored. After receiving a 0 followed by a continuous string of five 1s, the receiver examines the next (i.e., seventh) bit. Depending on the identity of this bit, it takes one of two actions:
   (i) If the seventh bit is a 0, the received sequence is *destuffed* by deleting that bit.
   (ii) If, on the other hand, the seventh bit is a 1, the receiver examines the next (i.e., eighth) bit. If the eighth bit is a 0, it recognizes that a flag has been received, and the frame is therefore over. But if the eighth bit is a 1, it knows that an *abort* or *idle* signal has been received. More specifically, if the eighth bit is a 1 followed by at least 7 but fewer than 15 continuous 1s, the receiver knows that the signal is an "abort" signal. If, however, the eighth bit is a 1 followed by 15 or more continuous 1s, the receiver knows that the signal is an "idle channel" signal. Whichever of these two signals is received, the receiver takes appropriate action.

The *address* field normally consists of one byte (8 bits). It is used to identify the secondary station that transmitted the frame or the secondary station that is to receive the frame. Clearly, for point-to-point links, the address field is not needed; nevertheless, it is always included for the purpose of uniformity.

The *control* field, also consisting of 8 bits, is used to define three types of frames known as *information, supervisory,* and *unnumbered frames*. Information frames carry the data to be transmitted from one station to another. Note that these "data" may contain network-level information in addition to user data. Supervisory frames provide the ARQ mechanism for error control; this form of error control was discussed in Section 8.10. Unnumbered frames provide supplemental link control functions. Naturally, the three frames have different control field formats.

The *packet field* contains data to be transferred across the link. The data may be of any length. Frequently, the packet length is a multiple of 8 bits.

---

* Black (1987, p. 93).

All frames include a 16-bit *CRC* (*cyclic redundancy check*) *code,* prior to the closing flag sequence. This inclusion is made for the purpose of *error detection.* The code uses the CRC-CCITT generator polynomial described by (see Section 8.4):

$$g(D) = D^{16} + D^{12} + D^5 + 1$$

The CRC code is designed to check the entire frame (not including the flags).

The total amount of *overhead* included in a frame is 48 bits.

We conclude discussion of X.25 by expanding on the function of the third or packet layer. As mentioned previously, the packet layer provides for the routing and management of a virtual circuit. In particular, when one station wishes to communicate with another station, a virtual circuit must be first set up between the two stations. The sequence of events for doing this proceeds as follows:

1. The originating station devises a CALL REQUEST packet and passes it to its associated node in the subnet. The packet includes a number for identifying the virtual circuit as well as source and destination addresses.
2. The subnet, in turn, sends a CALL REQUEST packet to the destination. If the destination station is able to accept the call, it sends a CALL ACCEPTED packet back.
3. When the originating station receives the CALL ACCEPTED packet, the virtual circuit is established.
4. The two stations may then use the full-duplex virtual circuit to exchange data packets.
5. When the two stations have conversed long enough, one of them sends a CLEAR REQUEST packet to the subnet node. The latter sends back a CLEAR CONFIRMATION packet as an acknowledgment.

10.7 MULTIPLE-ACCESS COMMUNICATION

In a packet-switched network, as described in Section 10.5, the subnet consists of nodes that are connected together by means of point-to-point communication links (e.g., twisted wire pairs, coaxial cables, optical fibers, microwave radio links). Accordingly, we find that on each link, the received signal (except for additive noise) depends only on the transmitted signal. In a *multiple-access* or *multi-access communication network,** on the other hand, the various nodes of the network share a *multi-access medium* (e.g., satellite, radio broadcast, multitap bus). In this case, we find that the received signal at any node depends on the transmitted signals at two or more nodes.

In traditional multiple-access techniques, namely, *frequency-division multiple access* (FDMA) and *time-division multiple access* (TDMA), the channel is allocated to various users by partitioning the time-bandwidth space into preas-

---

* For detailed treatment of multiple-access communication networks, see the following references: Bertsekas and Gallager (1987, Chapter 4), Tobagi (1982, Chapter 6), and Lam (1983, Chapter 4).

signed (fixed) slots. In FDMA, each user is assigned a fraction of the channel bandwidth. In TDMA, on the other hand, each user is assigned a fixed time slot on the channel. FDMA and TDMA represent *fixed assignment techniques.* In another fixed assignment technique, known as *code-division multiple access* (CDMA), user transmissions are permitted to overlap in the frequency and time coordinates, but the individual users are now identified by assigning distinct signaling codes to them. The use of CDMA provides resistance to external interference and requires no synchronizing mechanism. TDMA was described at some length in Section 7.13, and CDMA was described briefly in Section 9.7.

Fixed assignment techniques are suitable for voice traffic and some data traffic. In a computer communication environment, however, the data traffic has a bursty nature; that is, the peak-to-average ratio in the data transmission rate is inherently high. If fixed assignment techniques are used to handle this kind of traffic, then we must provide a channel capacity high enough to meet the peak transmission rate requirement of each user. Consequently, the use of fixed assignment techniques for bursty traffic yields a low channel utilization. In the case of computer communications, a more efficient approach is to use other multiple access techniques, namely, random access and polling. *Random-access techniques** range from the simplest "free-for-all" approach in which a user (station) transmits whenever it has packets to send, to techniques in which a user is "constrained" to transmit only in certain time intervals, and on to more complicated techniques in which a user "listens" before transmitting and transmits only when it "senses" the multi-access medium to be idle. *Polling techniques,* on the other hand, rely on the use of a "perfectly scheduled" approach in which the nodes (users) of the network are permitted to access the medium during reserved time intervals according to some order. In the sequel, we discuss the use of these techniques for packet satellite, packet radio, and local area networks.

However, before we begin this discussion, it should be stressed that the layered architecture described in Section 10.4 on the OSI model is not quite appropriate for these networks. In particular, between the physical layer and the data link control layer, we need an extra sub-layer called the *medium access control* (MAC) *layer.* The function of this extra layer is to allocate the multi-access medium among the various nodes of the network.

(1) Packet Satellite Networks

A *packet satellite network* involves a satellite normally in geostationary orbit, and numerous fixed or mobile nodes (stations). Figure 10.9 depicts a simplified representation of the network. The key ingredients of the network are twofold: the stations are uncoordinated, and the only way in which they can communicate with each other is via the satellite channel. The fact that the channel has to

---

* Random-access protocols are also referred to as *contention protocols.* The alternative terminology emphasizes the fact that these protocols regulate the sharing of a multi-access medium by a group of users *contending* for time on the channel.

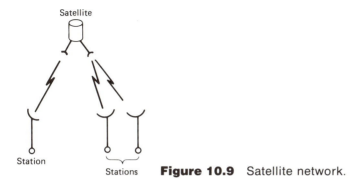

Figure 10.9 Satellite network.

be shared among a large number of stations, with each station contending for time on the network, raises the basic question as to how the issue of conflicts resulting from contention can be resolved. Clearly, to resolve this issue, we need a *channel-access protocol* encompassing a set of rules that governs the individual packet transmissions. Otherwise, chaos will result, for if every station in a satellite network transmits without regard to the other stations in the network, there may well be no communication at all.*

We emphasize the practical importance of this issue by making the following observations. If two stations in a packet satellite network transmit simultaneously, their packets overlap in time at the satellite. In such a case, a *collision* occurs,† and the satellite rebroadcasts a "destroyed" packet that carries no useful information about the contents or sources of the transmitted packets.

A packet satellite network exhibits the perfect *feedback* property of packet broadcasting in that every station in the network is able to listen to its own packet, one *round-trip time* (the time taken for the packet to propagate to the satellite and back) after sending it. With a parity-check code built into the packet, the station is capable of determining whether or not a collision has occurred. Indeed, this important feature of packet satellite networks is exploited in a type of random-access protocol known as ALOHA.‡

In ALOHA, or *pure* ALOHA as it is sometimes called, the stations are not synchronized in that each station is permitted to transmit a data packet whenever it is ready. In the event that two or more packets collide (i.e., overlap in

---

* For material on the issues involved in the design of packet satellite and packet radio networks, and discussion of the choices available, see the overview papers by Leiner, Nielson, and Tobagi (1987) and Tobagi, Binder, and Leiner (1984).

† The assumption of a collision occurring ignores the possibility of using a *time capture* technique, which provides a station the ability to successfully demodulate a packet assigned with a known code in spite of the presence of other overlapping packets with the same code (Pursley, 1987). In the discussion presented in Section 10.7, this issue is ignored.

‡ The ALOHA system was originally developed in 1970 by Abramson at the University of Hawaii (Abramson, 1970). Although this system originally used a packet-switched ground-based radio network, nevertheless, the basic idea of ALOHA is applicable to any distributed-processing system in which uncoordinated stations compete for the use of a common channel, as in a packet satellite network.

time) at the satellite, each station detects the occurrence of such an event one-round trip time after sending its own packet. Then, the station just waits a random amount of time, and transmits the packet again. The waiting time has to be chosen randomly, or else the same packets will continue to collide, over and over again.

However, due to conflicts and idle channel time, the maximum channel throughput attainable using pure ALOHA may be *rather low*. We demonstrate this by invoking the so-called *infinite-user population model,* which is described as follows:

1. All transmitted messages consist of single packets with a common *packet time* equal to the packet length (in bits) divided by the bit rate.
2. The *channel traffic,* made up of new packet transmissions and retransmissions of packets that previously suffered collisions, is described as a *Poisson process* with a rate of G packets per packet time. In effect, G is the *offered load,* defined as the total rate of data (in bits per second) presented to the network for transmission.
3. Statistical equilibrium is assumed to exist in the network.
4. The channel is noise-free.

To proceed with the analysis, we note from Fig. 10.10 that (due to partial overlap of packets) each transmitted packet has a *vulnerable period* equal to two packet times. There will be no collision if, and only if, no other packet is transmitted within the vulnerable period. With the channel traffic described by a Poisson distribution, the probability that a transmitted packet is successful (i.e., the probability of no other packet being transmitted during the entire vulnerable period) is equal to

$$P_0 = \exp(-2G) \tag{10.1}$$

Now, a key measure of performance of a multiple-access protocol is the *channel throughput,* defined as the total rate of data (in bits per second) transmitted between stations. It is natural that, due to collisions, the throughput S, say, is less than the offered load G. Figure 10.11 illustrates, in conceptual terms, the

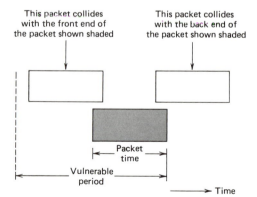

Figure 10.10 ALOHA vulnerable period of a transmitted packet.

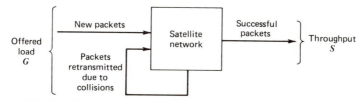

Figure 10.11   Conceptual model for calculating channel throughput.

interrelation between the offered load G and throughput S. With P_0 denoting the probability of a successful packet transmission, we find that the throughput S is given in terms of the offered load G as follows:

$$S = P_0 G$$
$$= G \exp(-2G) \qquad (10.2)$$

This result is shown plotted as "pure ALOHA" in Fig. 10.12. In this figure, we see that the maximum throughput occurs at the offered load $G = 0.5$, with

$$S_{\text{pure,max}} = \frac{1}{2e} \simeq 0.184 \qquad (10.3)$$

In other words, the maximum channel utilization attainable with pure ALOHA is limited to around 18 percent, assuming an infinite user population model. Note that in Fig. 10.12, both the offered load G and channel throughput S are normalized with respect to the channel capacity.

Under the assumption of an infinite user population model, the maximum throughput of an ALOHA system may be doubled by requiring that the channel

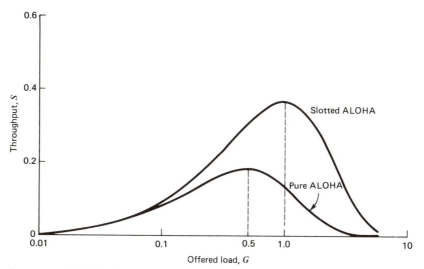

Figure 10.12   Performance curves for pure ALOHA and slotted ALOHA.

be "slotted" in time. The resulting system is thus known as *slotted* ALOHA. In slotted ALOHA, time is divided up into discrete intervals or slots, with each slot equal to the packet time; the network is thus turned into a discrete-time system. The competing stations now access the satellite channel in a *synchronized* manner by having the satellite broadcast a *pip* at the start of each interval, just like a clock. Although these pips reach the earth stations 270 ms later, the time taken for one round trip, each station in the network receives the pips at about the same time, thereby helping define the start times of its packet transmissions. The synchronization requirement has the effect of avoiding packet collisions due to partial overlaps, and therefore reducing the vulnerable period of a transmitted packet to just one packet time. Consequently, under the same assumptions made for pure ALOHA, we find that the throughput of slotted ALOHA is defined by

$$S = G \exp(-G) \tag{10.4}$$

This result is shown plotted as "slotted ALOHA" in Fig. 10.12. We see that the throughput S is maximized for $G = 1$, and the maximum value of S is given by

$$S_{\text{slotted, max}} = \frac{1}{e} \approx 0.368 \tag{10.5}$$

Comparing Eqs. 10.3 and 10.5, we see that the maximum channel throughput of slotted ALOHA is twice that of pure ALOHA, assuming an infinite user-population model.

The validity of the theoretical results given above for pure ALOHA and slotted ALOHA has been investigated using computer simulations. The results of these simulations show that Eqs. 10.3 and 10.5, defining the maximum channel throughput for pure ALOHA and slotted ALOHA, are quite robust.* In particular, the use of these equations yields fairly accurate results for a number of stations $n \geq 10$, and a mean randomized delay $\bar{k}$ for retransmission, limited as $\bar{k} \geq 5$.

(2) Packet Radio Networks

A desirable feature of radio is the ease of *broadcasting,* which makes its integration with packet-switching technology highly attractive. In a *packet radio network,*† there is an aggregation of radios that operate in the mutual support of each other to transport data from a source to a destination. The network nodes may be fixed or mobile, and the network may employ directive or omnidirectional antennas. Moreover, the network nodes are intelligent in that they can store and selectively forward data to other nodes in the network.

A packet radio network offers topological flexibility, as illustrated in Fig. 10.13 for a network that has six nodes. Fig. 10.13a depicts a star radio network, assuming the use of directive antennas and frequency assignments (up to two

---

* See Lam (1983, Chapter 4).

† The material presented here is based largely on Nielson (1985, Chapter 12).

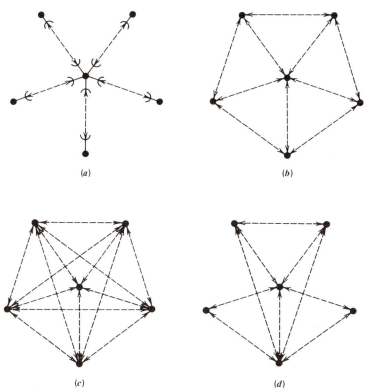

(a)   (b)

(c)   (d)

Figure 10.13   Various topologies for radio network.

per outer node). The outer nodes communicate only with the central node. Hence, node-to-node communication is indirect, regulated by the central node. This topology represents the basis of a *centralized* packet radio network that may provide user access to a central time-sharing computer (resource).

By replacing the directional antennas with ominidirectional ones, we may exploit the broadcast property of radio. The resulting architecture of the packet radio network takes on a *distributed* form, as illustrated by the topologies shown in Figs. 10.13b, 10.13c, and 10.13d. In the topology of Fig. 10.13b, the radios are assumed to be of *limited range*. On the other hand, in the fully connected topology of Fig. 10.13c, the radios are assumed to have *full range,* so that any node may communicate with all the other nodes in the network. Lastly, in Fig. 10.13d, the topology is *power-dependent* in that the radios are individually adjusted to create the result shown there. Thus, by exercising power control in a packet radio network, particularly if it is carried out dynamically, we are able to create network architectures that are difficult to duplicate with wire-based networks. Moreover, if one or more nodes in the network are mobile, packet radio offers a unique network capability.

However, packet radio has constraints of its own in that the greatest impediment to its use comes from the allocation of adequate spectrum. Specifically, the radio network may not be able to provide bandwidths characteristic of cables or optical fibers because of the carrier frequencies employed and the competition for spectral resource from other users.

Given the bursty nature of computer traffic in a packet radio network, the use of *random-access protocols* for sharing a radio channel leads to a better use of the channel than fixed assignments; in this context, packet radio and packet satellite networks behave alike. Two widely used types of random-access protocols are ALOHA and carrier-sense protocol. ALOHA is the least controlled of all multiple-access protocols, as described previously. A carrier-sense protocol, on the other hand, acquires knowledge of transmission activity and then acts accordingly to avoid collisions.

ALOHA is well-suited for a centralized packet radio network* having the star technology shown in Fig. 10.13a. In pure ALOHA, when a station has a packet to send, it just goes ahead and sends it. When the central computer receives the packet successfully, it transmits a positive acknowledgment back to the station. If the station does not receive the acknowledgment within a preset time, the station assumes that a collision occurred and retransmits the packet. To avoid the occurrence of continuously repeated conflicts, the retransmission delay is randomized. For example, the retransmission delay may be selected from a uniform distribution between the limits of 200 and 1500 ms.

Even though pure ALOHA is simple, it is of limited value to distributed packet radio networks, since it would not be logical for any radio station capable of both transmitting and receiving to interrupt a packet being received just to transmit at an arbitrary time. If, on the other hand, each station has to wait until the channel is free, the protocol becomes more controlled and access randomization is thereby reduced.

In slotted ALOHA, transmission is permitted only during well-defined time intervals. As explained previously, this modest control in the protocol decreases the chance of packets overlapping and therefore increases the maximum throughput, both by a factor of two (assuming an infinite user population model). The use of slotted ALOHA, however, needs synchronization of network nodes, which is required to be accurate to within a small fraction of packet time.

Packet radio does not suffer from the long propagation delay inherent in packet networks using geostationary satellites. Consequently, a packet radio network can provide *faster* feedback to the stations about the state of the network, thereby permitting the channel throughput to be pushed above the $1/e$ limit imposed by slotted ALOHA (for the infinite user population model). This is achieved by using *carrier-sense multiple access* (CSMA), which involves listening before transmitting. Specifically, a radio station wishing to transmit first listens to the network to determine whether or not the channel is busy. The action taken by the station, based on the information derived about the state of the network, determines the particular form of the CSMA. In a simple version

---

* The pioneering work done by Abramson on the ALOHA system was indeed performed on a single-hop terminal-access network with a star topology (Abramson, 1970).

of CSMA, known as *nonpersistent* CSMA, the protocol proceeds as follows:

1. If the channel is idle, transmit the packet.
2. If the channel is busy, wait a random amount of time (with retransmission delay of the packet drawn from some probability distribution), and repeat step 1.

The use of nonpersistent CSMA offers substantial improvements in performance over pure ALOHA. For example, in a single-hop fully connected packet radio network when the ratio of propagation time to transmission time of a packet is about 1 percent (which corresponds to a packet radio network with 20-mile radius, 100 kb/s data rate, and 1000 bits per packet), the use of nonpersistent CSMA has been shown to achieve a maximum channel throughput equal to 85 percent, assuming an infinite user population model.* However, this improvement is attainable only if all network nodes are in line-of-sight and within range of each other. Thus, all network nodes can hear each other's transmissions indiscriminantly. Otherwise, some degradation in performance is inevitable. This phenomenon is known as the *hidden terminal problem*.

Long channel idle time may be experienced with nonpersistent CSMA. This disadvantage is eliminated by using another carrier sense protocol called *1-persistent* CSMA.† This second protocol proceeds as follows:

1. If the channel is idle, transmit the packet.
2. If the channel is busy, continue to listen until the channel is sensed to be idle; then transmit the packet immediately.
3. If a collision occurs, wait a random amount of time, and repeat steps 1 and 2.

This second protocol is referred to as 1-persistent, because whenever the station finds the channel to be idle, it transmits with a probability of 1. It may therefore be viewed as more "greedy" than nonpersistent CSMA.

(3) Local-area Networks

A *local-area network* is a resource-sharing data-communication network with three general characteristics:‡

1. It is limited in geographic scope to the range of 0.1–10 km.

---

* See Tobagi, Binder, and Leiner (1984).

† Yet another protocol, called *p-persistent* CSMA, represents a compromise protocol that attempts to reduce collisions, as in nonpersistent CSMA, and also reduce channel idle time, as in 1-persistent CSMA. In the *p*-persistent CSMA protocol, the transmission by a station that finds the channel busy occurs immediately following the current one with probability *p*. The 1-persistent CSMA is a special case of the *p*-persistent version with $p = 1$. For more details, see Tanenbaum (1981, pp. 291–292).

‡ For an overview paper on local area networks, see Tsao (1984). For additional discussion of the subject, see also the books by Bertsekas and Gallager (1987, pp. 254–274), Schwartz (1987, Chapter 9), and Stallings (1987).

2. It provides high data rates (in excess of 1 Mb/s) over relatively inexpensive transmission media.
3. It is usually privately owned.

The most obvious application of a local area network is in a multiple computer environment. Specifically, it is used to interconnect computers, terminals, workstations, and other intelligent processors within a single building or several buildings that constitute a campus-like complex. Another important application of local-area networks is in office automation.

A local-area network is usually characterized in terms of its *topology*. The three topologies known as bus, ring, and star are the most common. The *bus* topology, shown in Fig. 10.14*a*, uses a multitap transmission medium, with its taps connected individually to the nodes. In the *ring* topology, shown in Fig. 10.14*b*, the various nodes (stations) are connected together in the form of a closed *loop* through *ring interface units*. In this topology, data travels around the ring in a unidirectional manner. Finally, the *star* topology, shown in Fig. 10.14*c*, employs a *central* node to establish a path between any pair of stations wishing to communicate with each other.* A disadvantage of this third network topology is dependency on the central node; failure of this node affects the entire network.

The type of protocol used to provide medium access to a local-area network (LAN) depends on the particular topology of the network. The popular medium access control protocols for LANs include *carrier-sense multiple-access with collision detection* (CSMA/CD), *token bus*, and *token ring*.†

The medium access control operates as a sublayer of the data link control layer, as depicted in Fig. 10.15. The remaining sublayer, called the *logical link control*, is used for communicating with higher layers of the OSI reference model.

The *CSMA/CD* protocol is implemented on a *multitapped bus*. Each node of a local-area network so implemented can communicate with every other node; but, as with a satellite channel, if multiple nodes transmit at the same time, the received signal becomes garbled. CSMA/CD represents a refinement on CSMA in that it involves *talking while listening*. Specifically, it has the following features:

1. A sender continuously listens to the network so as to detect collisions.
2. It transmits only when there is no activity detected in the medium.

---

* The star topology is used principally in conjunction with a digital *private branch exchange* (PBX) that supports circuit-switched communications for voice and data.

† The *Institute of Electrical and Electronic Engineers* (IEEE) has developed a set of six standards, 802.1 to 802.6, for local area networks. They are categorized as follows: 802.1 is a standard dealing with interfacing LAN protocols to higher layers. 802.2 is a *logical link control* standard. 802.3 to 802.5 are medium access control standards; they refer to CSMA/CD, token bus, and token ring, respectively. 802.6 is a *metropolitan-area network* standard, referring to networks that fall between a local- and a wide-area network. The IEEE standards conform to the OSI reference model. For description of the IEEE 802 local area network standards, see Stallings (1987, Chapter 5), Schwartz (1987, pp. 467–480), Bertsekas and Gallager (1987, pp. 257–267), and Black (1987, pp. 133–152).

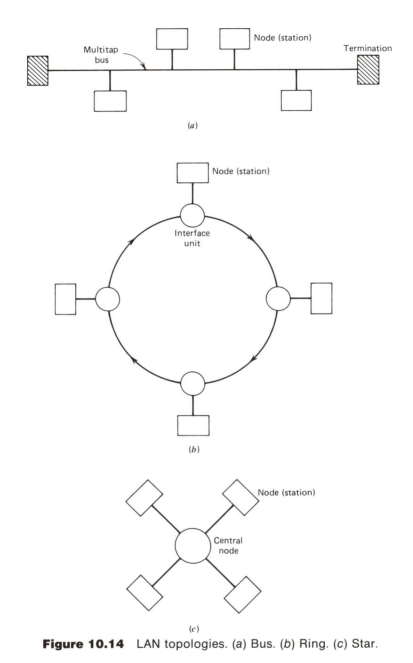

Figure 10.14   LAN topologies. (a) Bus. (b) Ring. (c) Star.

3. If a collision is detected, the sender aborts transmission and then reschedules it.

By following this procedure, the wastage of bandwidth due to collisions (especially, the collision of packets that are long compared with the propagation time) is reduced; whenever two packets collide, the channel remains unusable

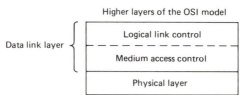

Figure 10.15 Relation of LAN protocol to OSI reference model.

for the transmission duration of the damaged packets. CSMA/CD has many versions as does the original CSMA (e.g., nonpersistent, 1-persistent). The most common choice is the 1-persistent verion. This version of the CSMA/CD is used in *Ethernet*.*

The *token bus* technique is a form of *polling*. It may be implemented on a multitapped bus as in CSMA/CD. However, token bus is more complicated than CSMA/CD. In a local-area network using token bus, the nodes (stations) form a *logical* ring by assigning them logical positions in some ordered sequence, with the first element of the sequence following the last. The right of a node to transmit data is determined by means of an *explicit token packet*. When a node receives the token, it is granted *permission to transmit* data. When the node completes the transmission of its data packets, or its allotted time has expired, it passes the token on to the next node in logical sequence. This node is now permitted to transmit data of its own. If, however, a node has nothing to transmit, it simply passes the token on to the next node.†

Finally, the *token ring*‡ technique is based on the use of a token that circulates around the nodes of a local-area network arranged in the form of a *topological* ring. By so doing, the token indicates that the ring is available for packet transmission by any node. A node is permitted to transmit a data packet only when it receives an *idle* token. The station must then change the token to a busy state. The *busy* state is represented by another unique bit pattern, which differs from the token's pattern in only one bit; hence, conversion from one state to the other is readily accomplished by inverting that bit. At the end of data packet transmission, the node issues another idle token, and so it goes on.

The apparent simplicity of token passing access protocols is rather deceiving. In reality, they have to guard against loss of the token, its duplicates, and other single points of failure, all of which contribute to additional complexity of the protocol.

---

* *Ethernet* is a local-area network, pioneered by workers at Xerox (Metcalf and Boggs, 1976). Subsequently, it was developed into a detailed specification jointly by Digital Equipment Corporation, Intel Corporation, and Xerox Corporation. The IEEE 802.3 standard for CSMA/CD is almost identical to the Ethernet specification. For discussion of Ethernet, see Schwartz (1987, pp. 469–474).

† The version of bus token described herein is less sophisticated than the IEEE Standard 820.4. For details of 820.4 standard, see Stallings (1987, pp. 142–147) and Black (1987, pp. 149–151).

‡ Token ring is also referred to as the *Newhall ring*; it is probably the oldest ring control technique (Farmer and Newhall, 1969). The IEEE 802.5 standard specification for token ring is a refinement of the procedure described herein; for details, see Stallings (1987, pp. 157–159).

10.8 SUMMARY AND DISCUSSION

A data network relies on the availability of a *subnet* that consists of an interconnection of *nodes*, each one of which is made up of an intelligent processor. One or more *stations* (computers, terminals, etc.) are attached to each boundary node of the subnet. It is helpful to think of the subnet nodes as being logically distinct from the stations outside the subnet.

There are two general approaches for the design of a subnet, namely, *circuit switching* and *store-and-forward switching*. There are two versions of the latter, namely, *message switching* and *packet switching*. Of these techniques, circuit switching is suitable for voice traffic, while packet switching is suitable for computer-generated traffic that is *bursty* in nature. A source of information is said to be "bursty" if it has nothing to send most of the time; that is, the ratio of peak to average data rate is high. In a packet-switched network, there is no dedicated circuit. Instead, an incoming packet is stored and then forwarded at each intermediate switching node of the network, until the packet reaches its destination.

Data networks are designed using a highly structured or layered architecture that permits a network of open systems to be viewed as a logical composition of a series of layers (or levels), with each layer encompassing the lower layers and isolating them from the higher layers. In this context, the *open systems interconnected (OSI) reference model* is important, since it provides not only a framework for developing standards but also the terms of reference for discussing the design of data networks. The OSI model has seven layers, each one of which performs a set of well-defined functions that enhance those performed by the lower layers. The three layers of X.25 protocol are examples of layers 1 through 3 of the OSI model.

A key issue in the design of a packet-switched network is the nature of the service offered by the network to its users. The issue of interest, stated in another way, concerns the division of labor between the subnet nodes and the stations. One alternative is to have the subnet take on responsibility for error control, routing, and flow control, in which case the subnet attempts to provide a "perfect" channel to its users. We refer to this kind of service as *virtual-circuit* service. Alternatively, the subnet offers *datagram* service, in which case the stations (users) must undertake to do most of the work. The use of virtual circuits offers enhanced functionality. On the other hand, datagram allows for a more efficient use of the subnet.

Packet satellite networks differ from terrestrial store-and-forward packet networks in that they have a single channel that must be shared among competing users. With bursty traffic, we are led to the use of ALOHA as a random-access protocol for sharing the channel. In *pure* ALOHA, the users transmit at any time they desire. Then, each user listens to detect the presence or absence of packet collisions, and retransmits packets, if necessary. The retransmission delay is randomized across the transmitting stations, so as to avoid continuously repeated conflicts. Assuming an infinite user population model, the maximum *throughput* (of about 18 percent) attainable with ALOHA may be doubled by using a slotted version of it, referred to as *slotted* ALOHA. But this improvement in channel utilization is attained at the expense of network synchronization, which adds complexity to the network.

The throughput is not the only parameter of interest to the user of a packet network. Another parameter of interest is the *mean packet delay*, defined as the average time elapsed between sending a packet and delivering it to the destination. With all packet networks, the requirements of high throughput and low packet delay are in conflict with each other. Specifically, improvement in throughput can be achieved only at the expense of a degradation in packet delay, or vice versa.

Packet radio networks differ from their satellite cousins in two important respects:

1. The propagation delay in packet radio is small compared to the packet time (i.e., the time taken to transmit a single packet). Accordingly, collisions may be avoided in a packet radio network by listening to the carrier of another user's transmission before transmitting, and backing off from transmission if the channel is sensed busy. This feature gives rise to a random-access protocol known as *carrier-sense multiple access* (CSMA). Many variants and modifications of CSMA are possible; they offer different improvements of their own, compared to ALOHA.

2. The usable range between stations in a packet radio network is limited, which necessitates the use of repeaters. Accordingly, the issue of routing assumes an important role in packet radio networks, just as it does in packet-switched point-to-point networks.

The performance of packet radio (and satellite) networks may be enhanced by exploiting properties of *spread spectrum*, where the transmission bandwidth used is in excess of that needed to transmit the data packets. Specifically, spread spectrum may be used to achieve the following characteristics:*

1. *Antijamming capability*, which enables a station to receive packets in the presence of narrow-band interference (jamming).

2. *Antimultipath capability*, which permits the transmission of packets over a link that exhibits multiple transmission paths.

3. *Time capture*, which refers to the ability of a station to successfully demodulate a packet assigned with a known code in spite of the presence of other overlapping packets with the same code; capture is achieved by distinguishing between the packets on the basis of their power levels or arrival times.

4. *Code-division multiple-access capability*, which refers to the fact that packet transmissions from multiple stations (assigned with orthogonal spreading codes) may overlap in time, with little or no effect on each other.

Packet satellites and packet radio networks are examples of *multiple-access* communication networks. Another important class of multiple-access net-

---

* For discussions of the role of spread spectrum in packet radio networks, see Pursley (1987) and Leiner, Nielson, and Tobagi (1987).

works is local-area networks. A *local-area network* (LAN) is defined as a data communication network that provides interconnections of various computing and related devices confined to a small area of approximately 0.1–10 km. Accordingly, the complicated routing and control algorithms for long-haul networks are not needed for LANs. We may classify local-area networks by topology or media access protocols. The *bus*, *ring*, and *star* represent three important LAN topologies. For medium access control, we may use *carrier-sense multiple access with collision detection* (CSMA/CD), which involves talking while listening, or *token passing*, which is based on the use of a token consisting of some unique bit pattern. The CSMA/CD protocol is used with buses, and token passing is used with buses and rings.

A discussion of data networks would be incomplete without some mention of *integrated services digital networks* (ISDNs).* The main feature of an ISDN is the support of a wide range of voice, video, facsimile, and computer data applications in the same network. Key elements of service integration for an ISDN are the implementation of a limited set of connection types and user-network interface arrangements that support a multiplicity of bearer services and other enhanced telecommunication services. Access to an ISDN will be determined with a layered functional set of protocols. The impact of ISDN will be to transform our society from a labor-intensive to a knowledge-intensive society, with the many ramifications thereof.

PROBLEMS

P10.1 COMMUNICATION NETWORKS

Problem 10.1.1 Consider the communication network shown in Fig. 10.2.
(a) Enumerate all the possible paths for communication between stations a and d.
(b) Suppose the processor at node 5 breaks down. What is the effect of this failure on the communication paths between stations a and d?

P10.3 STORE-AND-FORWARD SWITCHING

Problem 10.3.1 Tabulate the advantages and disadvantages of circuit switching, message switching, and packet switching.

P10.4 LAYERED ARCHITECTURE

Problem 10.4.1 Tabulate the advantages and disadvantages of the OSI reference model, based on the layered approach to protocols.

Problem 10.4.2 In Section 10.4, we outlined the principles used to develop the OSI model. Use these principles to justify the seven-layer composition of the OSI model.

Problem 10.4.3 Two computers A and B are directly coupled with each other. Both computers are described by the OSI model. An application *APX* on computer A sends a message to application *APY* on computer B. Illustrate the operations involved in the transport of this message.

---

* For an overview paper on integrated services digital networks, see Kostas (1984).

P10.5   PACKET NETWORKS

Problem 10.5.1   Are there any possible situations when a virtual circuit may deliver packets out of sequence? Explain your answer.

Problem 10.5.2   Justify the analogy between datagram service offered by a packet-switched network and the postal service.

Problem 10.5.3   Virtual circuit service offered by a packet-switched network requires the use of a destination address on a packet only during the setup procedure, whereas datagram service needs it on every packet. Justify the validity of these two statements.

Problem 10.5.4   Figures P10.1a and P10.1b illustrate the two services, virtual circuit and datagram, offered by a packet-switched network. Identify which is which, and justify your answer.

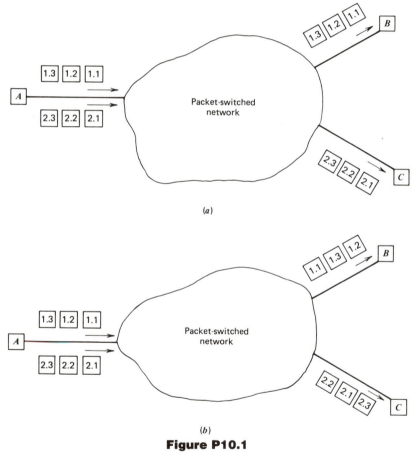

(a)

(b)

Figure P10.1

P10.6   PACKET-SWITCHED NETWORK-ACCESS PROTOCOL: THE X.25 PROTOCOL

Problem 10.6.1   The incoming data sequence at a node of a packet-switched network using X.25 protocol is as follows:

$$0111111001011111111111011111100$$

How is this sequence modified by the bit-stuffing operation performed before the data are transmitted? Perform destuffing on the stuffed sequence.

P10.7   MULTIPLE-ACCESS COMMUNICATION

Problem 10.7.1   Consider a packet satellite network using the pure ALOHA protocol for routing. Let G denote the offered load, and S denote the throughput. Assume an infinite user population model for traffic flow in the network. Using qualitative arguments, show the following:

(a)   For a small load, $G \simeq S$.
(b)   For a large load, $G > S$.

Problem 10.7.2   Calculate an approximate value for the throughput of a packet satellite network, using pure ALOHA, for the following values of offered load:

(a) $G = 0.5$
(b) $G = 1.0$
(c) $G = 1.5$
(d) $G = 2$

Assume an infinite user population model for the network.

Problem 10.7.3   Repeat the calculations in Problem 10.7.2 for a packet satellite network using slotted ALOHA.

Problem 10.7.4   Consider a packet satellite network that has the following parameters:

$$\text{Transmitted data rate} = 9.6 \text{ kb/s}$$

$$\text{Packet size} = 768 \text{ bits}$$

$$\text{Offered load} = 0.5$$

Calculate an approximate value for the load on the system in packets per second, assuming an infinite user population model.

Problem 10.7.5   Explain how a packet satellite network using ALOHA fits into the layered OSI model.

Problem 10.7.6   Consider the four radio network topologies shown in Fig. 10.13. Explain how the operation of each network is influenced by the breakdown of the processor at the central node.

Problem 10.7.7   Explain the reasons for the fact that token-passing-access protocols exhibit a more predictable delay than CSMA/CD protocol.

APPENDIX A

DISCRETE FOURIER TRANSFORM

In the *discrete Fourier transform* (DFT), both the input and the output consist of *sequences of numbers* defined at uniformly spaced points in time and frequency, respectively. Accordingly, the DFT lends itself directly to numerical evaluation on a digital computer. Moreover, the computation can be implemented most efficiently using a class of algorithms, called *fast Fourier transform (FFT) algorithms*. These fast algorithms are playing an increasingly important role in signal processing applications, not only in communications but also in many other fields.

In this appendix, we discuss the discrete Fourier transform, its properties, computation, and use in linear filtering and spectral analysis.

A.1 DEFINITIONS

Consider a finite *data sequence* $\{g_0, g_1, \ldots, g_{N-1}\}$. For brevity, we refer to the sequence as g_n, in which the subscript is the *time index* $n = 0, 1, \ldots, N - 1$. Such a sequence may represent the result of sampling an *analog signal* $g(t)$ at times $t = 0, T_s, \ldots, (N - 1)T_s$, where T_s is the sampling interval. The ordering of the data sequence defines the sample time in that $g_0, g_1, \ldots, g_{N-1}$ denote samples of $g(t)$ taken at times $0, T_s, \ldots, (N - 1)T_s$, respectively. Thus we have $g_n = g(nT_s)$.

We define the discrete Fourier transform (DFT) of g_n as

$$G_k = \sum_{n=0}^{N-1} g_n \exp\left(-\frac{j2\pi}{N} kn\right) \qquad k = 0, 1, \ldots, N - 1 \qquad \text{(A.1)}$$

The sequence $G_0, G_1 \ldots, G_{N-1}$ is called the *transform sequence*. For brevity, we refer to this sequence as G_k, in which the subscript is *frequency index* $k = 0, 1, \ldots, N - 1$. Correspondingly, we define the *inverse discrete Fourier transform* (IDFT) of G_k as

$$g_n = \frac{1}{N} \sum_{k=0}^{N-1} G_k \exp\left(\frac{j2\pi}{N} kn\right) \qquad n = 0, 1, \ldots, N - 1 \qquad \text{(A.2)}$$

The DFT and the IDFT form a transform pair. Specifically, given a data sequence g_n, we may use the DFT to compute the transform sequence G_k, and given G_k, we may use the IDFT to recover the original sequence g_n.

A distinctive feature of the DFT is that for the finite summations defined in Eqs. A.1 and A.2, there is no question of convergence.

When discussing the DFT (and algorithms for its computation), the words "sample" and "point" are used interchangeably to refer to a sequence value. Also, it is common practice to refer to a sequence of length N as an N-point sequence, and the DFT of a data sequence of length N as an N-point DFT.

EXAMPLE 1

Consider the exponential sequence

$$g_n = a^n \qquad n = 0, 1, \ldots, N - 1 \tag{A.3}$$

where a is a constant. Applying Eq. A.1, we obtain the DFT of g_n as

$$G_k = \sum_{n=0}^{N-1} a^n \exp\left(-\frac{j2\pi}{N} kn\right)$$

$$= \sum_{n=0}^{N-1} \left[a \exp\left(-\frac{j2\pi}{N} k\right)\right]^n \qquad k = 0, 1, \ldots, N - 1$$

This represents a *geometric series* with the following characteristics: a first term of unity, a geometric ratio of $a \exp(-j2\pi k/N)$, and a number of terms equal to N. Hence, using the formula for the sum of a geometric series, we get

$$G_k = \frac{1 - a^N \exp(-j2\pi k)}{1 - a \exp(-j2\pi k/N)} \qquad k = 0, 1, \ldots, N - 1$$

Since $\exp(-j2\pi k) = 1$ for all k, we may simplify this result as

$$G_k = \frac{1 - a^N}{1 - a \exp(-j2\pi k/N)} \qquad k = 0, 1, \ldots, N - 1 \tag{A.4}$$

Figure A.1 shows plots of the time sequence $\{g_n\}$, and the amplitude and phase of $\{G_k\}$ for $a = 0.9$, $N = 16$.

A.2 INTERPRETATION OF THE DFT AND THE IDFT

We may visualize the DFT process, described in Eq. A.1, as a collection of N *complex heterodyning* and *averaging* operations, as shown in Fig. A.2a. We say that the heterodyning is complex in that samples of the data sequence are multiplied by *complex exponential sequences*. There are a total of N complex exponential sequences to be considered, corresponding to the frequency index $k = 0, 1, \ldots, N - 1$. Their periods have been selected in such a way that each complex exponential sequence has precisely an integer number of cycles in the total interval 0 to $N - 1$. The zero-frequency response, corresponding to $k = 0$, is the only exception.

For the interpretation of the IDFT process, described in Eq. A.2, we may use the scheme shown in Fig. A.2b. Here we have a collection of N *complex signal generators*, each of which produces a complex exponential sequence:

$$\exp\left(\frac{j2\pi}{N} kn\right) = \cos\left(\frac{2\pi}{N} kn\right) + j \sin\left(\frac{2\pi}{N} kn\right)$$

$$= \left[\cos\left(\frac{2\pi}{N} kn\right), \sin\left(\frac{2\pi}{N} kn\right)\right] \tag{A.5}$$

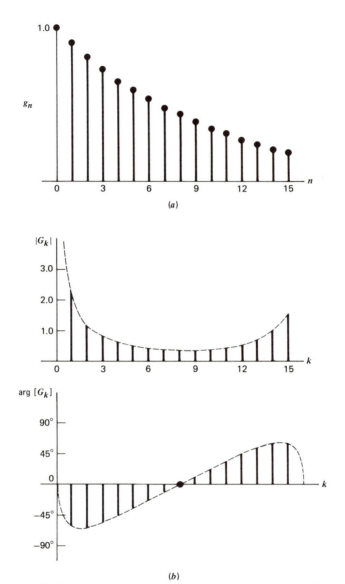

Figure A.1 (a) Data sequence g_n. (b) Amplitude response $|G_k|$ and phase response $\arg[G_k]$.

Thus, each complex signal generator, in reality, consists of a pair of generators that outputs a cosinusoidal and a sinusoidal sequence of k cycles per observation interval. The output of each complex signal generator is weighted by the complex Fourier coefficient G_k. At each time index n, an output is formed by summing the weighted complex generator outputs.

It is noteworthy that although the DFT and the IDFT are so similar in their mathematical formulations, as described in Eqs. A.1 and A.2, their interpretations, as depicted in Figs. A.2a and A.2b, are completely different.

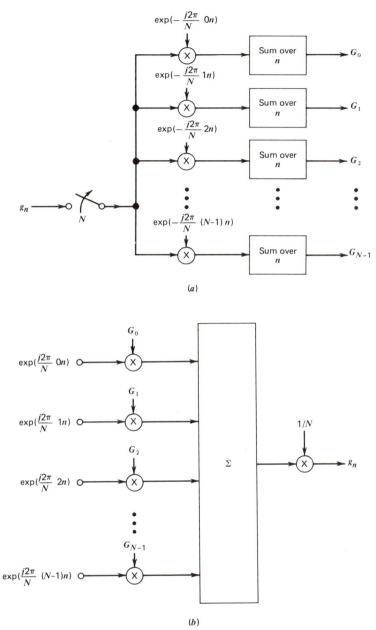

Figure A.2   Interpretation of (a) the DFT and (b) the IDFT.

Also, the addition of harmonically related periodic signals, as in Figs. A.2a and A.2b, suggests that their outputs, G_k and g_n, must be periodic. Moreover, the processors shown in Figs. A.2a and A.2b must be linear, suggesting that the DFT and IDFT must also be linear. These properties of the DFT are discussed formally in the next section.

A.3 PROPERTIES OF THE DFT

To develop a proper understanding of the DFT, we have to know its properties. Most of the properties have their counterpart in the continuous Fourier transform. In this section, we consider some of them.

PROPERTY 1 LINEARITY

Let g_n and h_n denote two sequences, each consisting of N samples. Let G_k and H_k denote their respective DFTs. Then, the DFT of $ag_n + bh_n$ is equal to $aG_k + bH_k$, where a and b are arbitrary constants.

PROPERTY 2 PERIODICITY

The data sequence g_n and transform sequence G_k exhibit *periodicity*, each with a period of N samples. That is,

$$g_{n+mN} = g_n \qquad m = \pm1, \pm2, \ldots \tag{A.6}$$

and

$$G_{k+lN} = G_k \qquad l = \pm1, \pm2, \ldots \tag{A.7}$$

The implication of this property is that although we defined g_n and G_k for $n, k = 0, 1, \ldots, N - 1$, nevertheless, we have periodic extensions of them. Indeed, the periodic properties of the DFT and IDFT are manifestations of uniform sampling in time as well as frequency.

PROPERTY 3 SYMMETRY

When g_n is real, its DFT exhibits *conjugate symmetry*, as shown by

$$G_{-k} = G_k^* \tag{A.8}$$

Moreover, the periodic property of the DFT requires that

$$G_{(N/2)-k} = G_{(N/2)+k}^* \tag{A.9}$$

Accordingly, G_k exhibits conjugate symmetry about $k = 0$ as well as $k = N/2$ when g_n is real.

PROPERTY 4 CIRCULAR SHIFT

Suppose a sequence g_n defined for the interval 0 to $N - 1$ is shifted to the right by n_0 time units, obtaining g_{n-n_0}. Owing to the inherent periodicity of g_n, we find that as a sample leaves the interval between 0 and $N - 1$, an identical sample enters the interval at the other end. This kind of behavior is known as the *wrap-around effect*. The resulting shift is referred to as a *circular shift*. It is illustrated in Fig. A.3 for $N = 6$ and $n_0 = 2$.

Let G_k denote the N-point DFT of g_n. Then, the N-point DFT of the shifted sequence g_{n-n_0} is equal to $G_k \exp(-j2\pi kn_0/N)$.

PROPERTY 5 CIRCULAR CONVOLUTION

Let x_n and h_n denote two periodic sequences, each of which has a period of N samples. The *circular convolution* of these two sequences is defined by

$$y_n = \sum_{l=0}^{N-1} x_l h_{n-l} \qquad n = 0, 1, \ldots, N - 1 \tag{A.10}$$

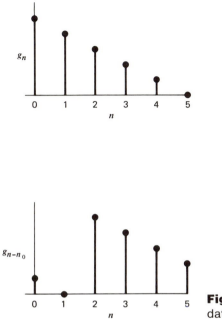

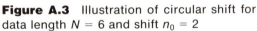

Figure A.3   Illustration of circular shift for data length $N = 6$ and shift $n_0 = 2$

The convolution is circular in the sense that y_n exhibits the same periodicity as x_n or h_n. The DFT of y_n is defined by

$$Y_k = \sum_{n=0}^{N-1} y_n \exp\left(-\frac{j2\pi}{N} kn\right)$$

$$= \sum_{n=0}^{N-1} \sum_{l=0}^{N-1} x_l \, h_{n-l} \exp\left(-\frac{j2\pi}{N} kn\right)$$

Interchanging the orders of the two summations, and recognizing that

$$\exp\left(-\frac{j2\pi}{N} kn\right) = \exp\left(-\frac{j2\pi}{N} kl\right) \exp\left(-\frac{j2\pi}{N} k(n - l)\right)$$

we may rewrite the expression for Y_k as

$$Y_k = \sum_{l=0}^{N-1} x_l \exp\left(-\frac{j2\pi}{N} kl\right) \sum_{n=0}^{N-1} h_{n-l} \exp\left(-\frac{j2\pi}{N} k(n - l)\right) \qquad (A.11)$$

Next, we recognize that

$$\sum_{n=0}^{N-1} h_{n-l} \exp\left(-\frac{j2\pi}{N} k(n - l)\right) = \sum_{n=l}^{N-l-1} h_n \exp\left(-\frac{j2\pi}{N} kn\right)$$

$$= H_k$$

where H_k is the DFT of h_n. Here we used the periodicity of h_n, which implies that the summation from l to $N - l - 1$ is the same as from 0 to $N - 1$. We also recognize that

$$\sum_{l=0}^{N-1} x_l \exp\left(-\frac{j2\pi}{N} kl\right) = X_k$$

where X_k is the DFT of x_n. Accordingly, we may simplify Eq. A.11 as

$$Y_k = H_k X_k \qquad (A.12)$$

We may therefore state that the *circular convolution of two periodic sequences is transformed into the product of their respective DFTs.*

A.4 FAST FOURIER TRANSFORM ALGORITHMS

The *fast Fourier transform** (FFT) refers to a class of *efficient algorithms* for computing the DFT. The algorithms are efficient in that they use a greatly reduced number of arithmetic operations as compared to the brute force computation of the DFT. Basically, an FFT algorithm attains its computational efficiency by following a "divide and conquer" strategy, whereby the original DFT computation is decomposed successively into smaller DFT computations. In this section, we describe one version of a popular FFT algorithm, the development of which is based on such a strategy.

To proceed, we first rewrite Eq. A.1, defining the DFT of g_n, in the convenient form

$$G_k = \sum_{n=0}^{N-1} g_n W^{nk} \qquad k = 0, 1, \ldots, N-1 \qquad (A.13)$$

where

$$W = \exp\left(-\frac{j2\pi}{N}\right) \qquad (A.14)$$

We readily see that

1. $W^N = 1$ (A.15)
2. $W^{N/2} = -1$ (A.16)
3. $W^{(k+lN)(n+mN)} = W^{kn}$ $m,l = 0, \pm1, \pm2, \ldots$ (A.17)

That is, W^{kn} is periodic with period N. The periodicity of W^{kn} is a key feature in the development of FFT algorithms.

Let N, the number of points in the data sequence, be an integer power of two, as shown by

$$N = 2^L$$

where L is an integer. Since N is an even integer, $N/2$ is an integer, and so we

---

* Fast Fourier transform (FFT) algorithms were brought into prominence by the publication of a paper by Cooley and Tukey (1965). For discussions of FFT algorithms, see Oppenheim and Schafer (1975, pp. 290–321), Rabiner and Gold (1975, pp. 357–381), and Elliott and Rao (1982, pp. 58–177). The last reference includes treatment of the newer FFT algorithms that have a reduced number of multiplications. It also includes a detailed mathematical analysis of the relation between the discrete Fourier transform and continuous Fourier transform.

may divide the data sequence into the first half and the last half of the points. Thus, we may rewrite Eq. A.13 as

$$G_k = \sum_{n=0}^{(N/2)-1} g_n W^{kn} + \sum_{n=N/2}^{N-1} g_n W^{kn}$$

$$= \sum_{n=0}^{(N/2)-1} g_n W^{kn} + \sum_{n=0}^{(N/2)-1} g_{n+N/2} W^{k(n+N/2)}$$

$$= \sum_{n=0}^{(N/2)-1} (g_n + g_{n+N/2} W^{kN/2}) W^{kn} \qquad (A.18)$$

Since $W^{N/2} = -1$, as in Eq. A.16, we have

$$W^{kN/2} = (-1)^k$$

Accordingly, the factor $W^{kN/2}$ in Eq. A.18 takes on only one of two possible values, $+1$ or -1, depending on whether the frequency index k is even or odd.

First, let k be *even*, so that $W^{kN/2} = 1$. Also let

$$k = 2l \qquad l = 0, 1, \ldots, \frac{N}{2} - 1$$

and

$$x_n = g_n + g_{n+N/2} \qquad (A.19)$$

Then, we may put Eq. A.18 into the form

$$G_{2l} = \sum_{n=0}^{(N/2)-1} x_n W^{2ln}$$

$$= \sum_{n=0}^{(N/2)-1} x_n (W^2)^{ln} \qquad l = 0, 1, \ldots, \frac{N}{2} - 1 \qquad (A.20)$$

From the definition of W given in Eq. A.14, we note that

$$W^2 = \exp\left(-\frac{j4\pi}{N}\right)$$

$$= \exp\left(-\frac{j2\pi}{N/2}\right)$$

Hence, we recognize the sum on the right side of Eq. A.20 as the $(N/2)$-point DFT of the sequence x_n.

Next, let k be *odd*, so that $W^{kN/2} = -1$. Also, let

$$k = 2l + 1 \qquad l = 0, 1, \ldots, \frac{N}{2} - 1$$

and

$$y_n = g_n - g_{n+N/2} \qquad (A.21)$$

Then, we may put Eq. A.18 into the corresponding form

$$G_{2l+1} = \sum_{n=0}^{(N/2)-1} y_n W^{(2l+1)n}$$

$$= \sum_{n=0}^{(N/2)-1} [y_n W^n](W^2)^{ln} \qquad l = 0, 1, \ldots, \frac{N}{2} - 1 \qquad (A.22)$$

We recognize the sum on the right side of Eq. A.22 as the $(N/2)$-point DFT of the sequence $y_n W^n$. The parameter W^n associated with y_n is called a *twiddle factor*.

Equations A.20 and A.22 show that the even- and odd-valued samples of the transform sequence G_k can be obtained from the $(N/2)$-point DFTs of the sequences x_n and $y_n W^n$, respectively. The sequences x_n and y_n are themselves related to the original data sequence g_n by Eqs. A.19 and A.21, respectively. Thus, the problem of computing an N-point DFT is reduced to that of computing two $(N/2)$-point DFTs. The procedure is repeated a second time, whereby an $(N/2)$-point is decomposed into two $(N/4)$-point DFTs. The decomposition procedure is continued in this fashion until (after $L = \log_2 N$ stages) we reach the trivial case of N single-point DFTs.

Figure A.4a illustrates the computations involved in applying the formulas of Eqs. A.20 and A.22 to an 8-point data sequence; that is, $N = 8$. In constructing the left-hand portion of the figure, we have used *signal-flow graph** notation. The computation of an 8-point DFT is reduced to that of two 4-point DFTs.

The procedure for the 8-point DFT may be mimicked to simplify the computation of the 4-point DFT. This is illustrated in Fig. A.4b, where the computation of a 4-point DFT is reduced to that of two 2-point DFTs. The computation of a 2-point DFT is shown in Fig. A.4c.

Combining the ideas described in Fig. A.4, we obtain the complete signal-flow graph of Fig. A.5 for the computation of the 8-point DFT. A repetitive structure, called a *butterfly*, can be discerned in the FFT algorithm of Fig. A.5; a butterfly has two inputs and two outputs. Examples of butterflies (for the three stages of the algorithm) are shown by the bold-faced lines in Fig. A.5.

For the general case of $N = 2^L$, the algorithm requires $L = \log_2 N$ stages of computation. Each stage requires $(N/2)$ butterflies. Each butterfly involves one complex multiplication and two complex additions (to be precise, one addition and one subtraction). Accordingly, the FFT structure described requires $(N/2)\log_2 N$ complex multiplications and $N\log_2 N$ complex additions. (Actually, the number of multiplications quoted is pessimistic, because we may omit all

---

* A *signal-flow graph* consists of an interconnection of *nodes* and *branches*. The *direction* of signal transmission along a branch is indicated by an arrow. A branch multiplies the variable at a node (to which it is connected) by the branch *transmittance*. A node sums the outputs of all incoming branches. The convention used for branch transmittances in Fig. A.4 is as follows. When no coefficient is indicated on a branch, the transmittance of that branch is assumed to be unity. For other branches, the transmittance of a branch is indicated by -1 or an integer power of W, placed alongside the arrow on the branch.

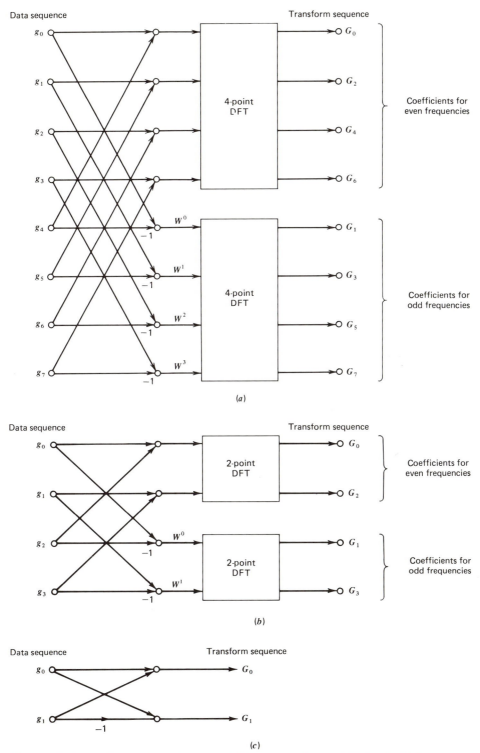

Figure A.4 (a) Reduction of 8-point DFT into two 4-point DFTs. (b) Reduction of 4-point DFT into two 2-point DFTS. (c) Trivial case of 2-point DFT.

Data sequence   Transform sequence

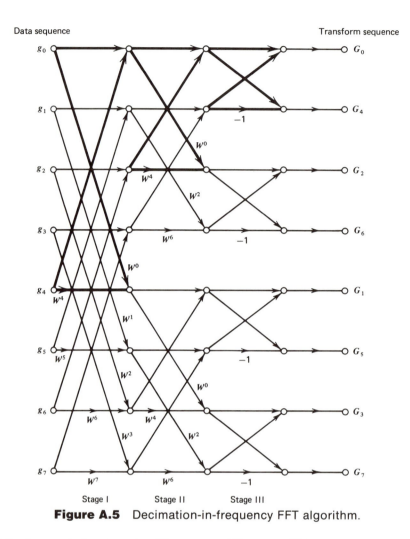

Stage I   Stage II   Stage III

Figure A.5 Decimation-in-frequency FFT algorithm.

twiddle factors $W^0 = 1$ and $W^{N/2} = -1$, $W^{N/4} = j$, $W^{3N/4} = -j$.) This computational complexity is significantly smaller than that of the N^2 complex multiplications and $N(N - 1)$ complex additions required for the *direct* computation of the DFT. The computational savings made possible by the FFT algorithm become more substantial as we increase the data length N.

We may establish two other important features of the FFT algorithm by carefully examining the signal-flow graph shown in Fig. A.5:

1. At each stage of the computation, the new set of N complex numbers resulting from the computation can be stored in the same memory locations used to store the previous set. This kind of computation is referred to as *in-place computation*.

2. The samples of the transform sequence X_k are stored in a *bit-reversed order*. To illustrate the meaning of this terminology, consider Table A.1

Table A.1   Illustrating Bit Reversal

Frequency Index, k	Binary Representation	Bit-reversed Binary Representation	Bit-reversed Index
0	000	000	0
1	001	100	4
2	010	010	2
3	011	110	6
4	100	001	1
5	101	101	5
6	110	011	3
7	111	111	7

constructed for the case of $N = 8$. At the left of the table, we show the eight possible values of the frequency index k (in their natural order) and their 3-bit binary representations. At the right of the table, we show the corresponding bit-reversed binary representations and indices. We observe that the bit-reversed indices in the right-most column of Table A.1 appear in the same order as the indices at the output of the FFT algorithm in Fig. A.5.

The FFT algorithm depicted in Fig. A.5 is referred to as a *decimation-in-frequency algorithm*, because the transform (frequency) sequence X_k is divided successively into smaller subsequences. In another popular FFT algorithm, called a *decimation-in-time* algorithm, the data (time) sequence x_n is divided successively into smaller subsequences. Indeed, the signal-flow graph for the decimation-in-time algorithm may be obtained by a *transposition* of the signal-flow graph for the decimation-in-frequency algorithm. Accordingly, both algorithms have the same computational complexity. They differ from each other in two respects. First, for decimation-in-frequency, the input is in natural order, whereas the output is in bit-reversed order. The reverse is true for decimation-in-time. Second, the butterfly for decimation-in-time is slightly different from that for decimation-in-frequency.

(1) Computation of the IDFT

The IDFT of the transform G_k is defined by Eq. A.2. We may rewrite this equation in terms of the parameter W as

$$g_n = \frac{1}{N} \sum_{k=0}^{N-1} G_k W^{-kn} \qquad n = 0, 1, \ldots, N - 1 \qquad (A.23)$$

Taking the complex conjugate of Eq. A.23, and multiplying by N, we get

$$N g_n^* = \sum_{k=0}^{N-1} G_k^* W^{kn} \qquad n = 0, 1, \ldots, N - 1 \qquad (A.24)$$

The right side of Eq. A.24 is recognized as the N-point DFT of the complex-conjugated sequence G_k^*. Accordingly, Eq. A.24 suggests that we may compute

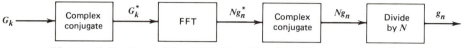

Figure A.6 Use of the FFT algorithm for computing the IDFT.

the desired sequence g_n using the scheme shown in Fig. A.6, based on an N-point FFT algorithm. Thus, an FFT algorithm can be used to compute the IDFT as well as the DFT.

A.5 LINEAR FILTERING

Consider a *linear filter*, whose *impulse response* is defined by a sequence h_n of finite length L; that is, the time index $n = 0, 1, \ldots, L - 1$. Let the *input signal* be defined by another sequence x_n of length M; that is $n = 0, 1, \ldots, M - 1$. The resulting output signal of the filter, denoted by y_n, is defined by the *linear convolution* of x_n and h_n, as shown by

$$y_n = \sum_{m=0}^{L-1} h_m x_{n-m} \tag{A.25}$$

Since h_m is zero outside the interval $0 \leq m \leq L - 1$, and x_{n-m} is zero outside the interval $0 \leq n - m \leq M - 1$, it follows that y_n is zero outside the interval defined by $0 \leq n \leq L + M - 2$. That is, the output signal of the filter consists of a sequence y_n of length $L + M - 1$.

From the circular convolution property of the DFT, we know that if we multiply the DFTs of two finite-duration sequences and then compute the IDFT of the product, the result is equivalent to the circular convolution of two periodic sequences created from the given finite-duration sequences. Moreover, we know that when two periodic sequences of a common period are convolved, the resulting sequence is also periodic, with the same period as the inputs. Accordingly, we may formulate an *indirect method* for computing the linear convolution of the sequences h_n and x_n as follows:

1. Add at least $M - 1$ zero-valued samples at the end of the sequence h_n of duration L, obtaining

$$\bar{h}_n = \begin{cases} h_n & 0 \leq n \leq L - 1 \\ 0 & L \leq n \leq N \end{cases} \tag{A.26}$$

Add at least $L - 1$ zero-valued samples at the end of the sequence $x(n)$ of duration M, obtaining

$$\tilde{x}_n = \begin{cases} x_n & 0 \leq n \leq M - 1 \\ 0 & M \leq n \leq N \end{cases} \tag{A.27}$$

The *augmented sequences* $\bar{h}_n$ and $\tilde{x}_n$ have the same duration, namely, $N \geq L + M - 1$.

2. Compute the N-point DFTs of the augmented sequences $\bar{h}_n$ and $\tilde{x}_n$. Let them be denoted by $\bar{H}_k$ and $\tilde{X}_k$, respectively.

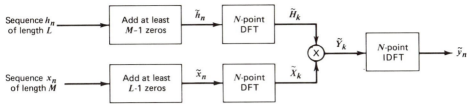

Figure A.7   Computation of linear convolution.

3.   Multiply $\tilde{H}_k$ by $\tilde{X}_k$, obtaining

$$\tilde{Y}_k = \tilde{H}_k \tilde{X}_k \tag{A.28}$$

4.   Compute the IDFT of the product $\tilde{Y}_k$, obtaining the N-point sequence $\bar{y}_n$. The first $L + M - 1$ samples of $\bar{y}_n$ constitute the desired output sequence y_n as shown by

$$\bar{y}_n = \begin{cases} y_n & 0 \leqslant n \leqslant L + M - 2 \\ 0 & L + M - 1 \leqslant n \leqslant N \end{cases} \tag{A.29}$$

Figure A.7 depicts a block diagram of the method. We refer to the method as indirect because the time function of interest, y_n, is obtained indirectly by moving out of the time domain into the frequency domain and then back into the time domain.

When the DFT and the IDFT operations indicated in Fig. A.7 are performed with the use of an N-point FFT algorithm, the method is called *fast convolution*. The transform length N is an integer power of two, equal to or greater than $L + M - 1$.

The fast convolution method permits the efficient computation of the output signal y_n of a linear filter whose impulse response h_n and input signal x_n consist of finite-length sequences. But in some applications, such as the linear filtering of a speech waveform, for example, the length M of the filter input x_n is essentially unlimited. In order to use the fast convolution method for such applications, we must modify it slightly. Basically, the modification entails sectioning the filter input x_n into *blocks* and then processing them sequentially as each block is formed. However, we must ensure that no artifacts are generated in the filter output y_n due to the artificial boundaries produced by sectioning the filter input x_n into blocks.*

A.6   DFT FILTER RESPONSE

An FFT algorithm provides an efficient technique for accomplishing narrow-band filtering. For a sinusoidal input, it produces an output in the form of a complex coefficient that describes the amplitude and phase of the input. In this context, we may refer to the DFT output as originating from a filter.

---

* For a discussion of "block" filtering techniques, see Oppenheim and Schafer (1975, pp. 110–115) and Rabiner and Gold (1975, pp. 63–67). For an overview of fast convolution and related issues, see Harris (1982).

To explore this issue, consider an input (analog) signal $g(t)$ whose sample values are denoted by g_n, $n = 0, 1, \ldots, N - 1$. For a sampling period T_s, the total *observation interval* is given by

$$T = NT_s \tag{A.30}$$

With g_n as input, the DFT output G_k is defined by Eq. A.1, reproduced here for convenience:

$$G_k = \sum_{n=0}^{N-1} g_n \exp\left(-\frac{j2\pi}{N} kn\right)$$

According to the sifting property of a delta function, we have

$$g_n \exp\left(-\frac{j2\pi}{N} kn\right) = \int_0^T g(t) \exp\left(-\frac{j2\pi kt}{T}\right) \delta(t - nT_s)dt \tag{A.31}$$

where we have used Eq. A.30 and the notation that $g_n = g(nT_s)$. We may therefore rewrite the DFT output as

$$G_k = \sum_{n=0}^{N-1} \int_0^T g(t) \exp\left(-\frac{j2\pi kt}{T}\right) \delta(t - nT_s)dt$$

$$= \int_0^T \sum_{n=0}^{N-1} \delta(t - nT_s) g(t) \exp\left(-\frac{j2\pi kt}{T}\right) dt \tag{A.32}$$

We may extend the limits of integration in Eq. A.32 from $-\infty$ to ∞ without affecting the value of G_k. Hence, we may write

$$G_k = \int_{-\infty}^{\infty} \gamma(t) g(t) \exp\left(-\frac{j2\pi kt}{T}\right) dt \tag{A.33}$$

where $\gamma(t)$ consists of N uniformly spaced delta functions, as shown by

$$\gamma(t) = \sum_{n=0}^{N-1} \delta(t - nT_s) \tag{A.34}$$

The integral in Eq. A.33 may be viewed as the Fourier transform of a time function equal to $\gamma(t)g(t)$ evaluated at the frequency variable $f = k/T$, as shown by

$$G_k = F[\gamma(t)g(t)] \qquad \text{evaluated at } f = k/T \tag{A.35}$$

We also know from the multiplication property of the Fourier transform that multiplication in the time domain is equivalent to convolution in the frequency domain. Let $G(f)$ denote the Fourier transform of the (analog) signal $g(t)$. Let $\Gamma(f)$ denote the Fourier transform of the time function $\gamma(t)$. We may therefore express G_k as

$$G_k = \Gamma(f) \star G(f) \qquad \text{evaluated at } f = k/T \tag{A.36}$$

where the star $\star$ denotes convolution. Using the definition of convolution in

Eq. A.36, we get

$$G_k = \int_{-\infty}^{\infty} \Gamma\left(\frac{k}{T} - \nu\right) G(\nu)d\nu \qquad k = 0, 1, \ldots, N - 1 \qquad (A.37)$$

The frequency function $\Gamma(f)$ is called the *DFT filter response*. Taking the Fourier transform of the corresponding time function $\gamma(t)$ defined in Eq. A.34, we obtain

$$\Gamma(f) = \sum_{n=0}^{N-1} \exp(-j2\pi nT_s f)$$

$$= \frac{1 - \exp(j2\pi NT_s f)}{1 - \exp(j2\pi T_s f)} \qquad (A.38)$$

Define the *normalized frequency variable*

$$\lambda = NT_s f$$
$$= Tf \qquad (A.39)$$

Then, we may redefine the DFT filter response in terms of λ as

$$\Gamma(\lambda) = \frac{1 - \exp(-j2\pi\lambda)}{1 - \exp(-j2\pi\lambda/N)}$$

$$= \frac{\sin(\pi\lambda)}{\sin(\pi\lambda/N)} \exp\left[-j\pi\lambda\left(1 - \frac{1}{N}\right)\right] \qquad (A.40)$$

The amplitude of $\Gamma(\lambda)$ is a *periodic* function of λ with period N. The phase of $\Gamma(\lambda)$ is a *linear* function of λ.

In Fig. A.8, the amplitude and the phase of the DFT filter response $\Gamma(\lambda)$ are shown plotted versus λ for $N = 16$. The main lobe of the response, centered at $\lambda = 0, \pm N, \ldots$, has a normalized width equal to 2. Equivalently, in terms of the frequency f, the main lobe has a width equal to $2/T$.

In terms of the normalized frequency variable λ, we may rewrite the DFT output G_k given in Eq. A.36 as

$$G_k = \Gamma(\lambda) \star G(\lambda) \qquad \text{evaluated at } \lambda = k \qquad (A.41)$$

Note that the variable λ is continuous, whereas the variable k is discrete. Also, recognizing that $k = 0, 1, \ldots, N - 1$, the DFT represents a *bank* of N narrow-band filters. The attractiveness of such a filter bank, implemented using the FFT algorithm, is the economy of computation.

A.7   SPECTRAL RESOLUTION AND LEAKAGE

In *DFT-based spectral analysis,* we are interested in the smallest separation between frequency components of the spectrum $G(f)$ that can be measured from the DFT output G_k. This frequency separation is called *spectral resolution,* Δf, measured in hertz. As a rule of thumb, the spectral resolution of a DFT is approximately the reciprocal of the observation interval, T, in seconds.

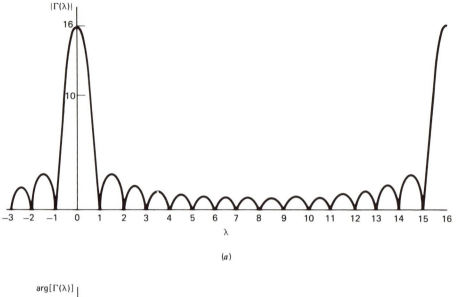

(a)

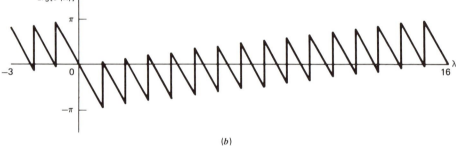

(b)

Figure A.8 (a) Amplitude response $|\Gamma(\lambda)|$. (b) Phase response $\arg[\Gamma(\lambda)]$.

That is,

$$\Delta f \simeq \frac{1}{T} \tag{A.42}$$

We may justify the validity of this rule by considering the convolution of the amplitude spectrum $|G(\lambda)|$ of an analog signal $g(t)$ and the amplitude $|\Gamma(\lambda)|$ of the DFT filter response as depicted in Fig. A.9a. The nonsymmetric nature of $|G(\lambda)|$ implies that $g(t)$ may be complex-valued. In Fig. A.9a, it is assumed that $g(t)$ is properly band-limited in that $|G(\lambda)|$ is essentially zero outside the interval $|\lambda| \le N/2$, where N is the number of samples used in the DFT computation. The amplitude $|\Gamma(\lambda)|$ is positioned such that its peak occurs at the frequency bin $\lambda = k$. If we were to ignore the phase terms associated with $G(\lambda)$ and $\Gamma(k - \lambda)$, the area under the curve of the product $|G(\lambda)||\Gamma(k - \lambda)|$ provides a measure of the amplitude of the DFT output G_k, as illustrated in Fig. A.9b. This area is dominated by the main lobe of $|\Gamma(k - \lambda)|$, which extends from $k - 1$ to $k + 1$. Conversely, the amplitude $|G_k|$ is a measure of the frequency content of the analog signal $g(t)$, centered at $\lambda = k$ and extending by one frequency bin on

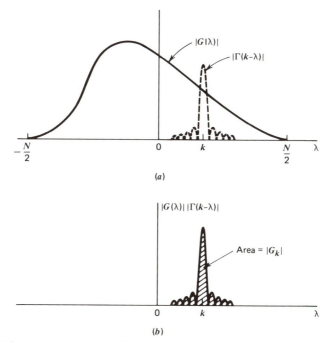

Figure A.9   Illustrating spectral resolution and leakage.

either side. This is equivalent to saying that the normalized spectral resolution $\Delta\lambda$ of the DFT is equal to one frequency bin, which is intuitively satisfying. From the relation between the frequency f and the normalized frequency λ, given in Eq. A.39, we may also state that the spectral resolution Δf, measured in hertz, is equal to $1/T$. This is precisely the rule described in Eq. A.42.

There is another important observation that we can make from Fig. A.9b. Specifically, the DFT output is smeared by a contribution (albeit minor) due to those frequencies of the input signal $g(t)$ that lie in the *sidelobes* of the DFT filter response $\Gamma(\lambda)$, on either side of the main lobe centered at $\lambda = k$. This effect is called *spectral leakage*. It is an inherent feature of the DFT, arising from the unavoidable time-domain truncation of the DFT input. The DFT filter response may be viewed as the result of applying a rectangular window to the input signal $g(t)$.

To reduce spectral leakage, it is necessary to weight the input signal $g(t)$ by means of a *window function* whose Fourier transform is characterized by sidelobes smaller than those of the DFT filter response. A reasonable window function is the *von Hann window,** defined by

$$w(t) = \begin{cases} 1 - \cos(2\pi t/T) & 0 \le t \le T \\ 0 & \text{otherwise} \end{cases} \quad\quad (A.43)$$

---

* The von Hann window is credited to the Austrian meteorologist Julius von Hann. It is also referred to in the literature as the *Hanning window* or *cosine squared weighting*. For a discussion of windows, see Harris (1978).

where T is the observation interval. The function $w(t)$ is shown plotted in Fig. A.10. The Fourier transform of $w(t)$ is given by

$$W(f) = \delta(f) - \frac{1}{2} \delta\left(f + \frac{1}{T}\right) - \frac{1}{2} \delta\left(f - \frac{1}{T}\right) \tag{A.44}$$

Correspondingly, in terms of the normalized frequency variable λ we may write

$$W(\lambda) = \delta(\lambda) - \frac{1}{2} \delta(\lambda + 1) - \frac{1}{2} \delta(\lambda - 1) \tag{A.45}$$

Since multiplication in the time domain is equivalent to convolution in the frequency domain, the DFT filter response as modified by the application of the von Hann window is given by

$$\Gamma_H(\lambda) = \Gamma(\lambda) \star W(\lambda)$$

$$= \Gamma(\lambda) - \frac{1}{2} \Gamma(\lambda + 1) - \frac{1}{2} \Gamma(\lambda - 1) \tag{A.46}$$

The first term, $\Gamma(\lambda)$ is defined by Eq. A.40. For the second term, we replace λ in Eq. A.40 by $\lambda + 1$, obtaining (except for a scaling factor)

$$\Gamma(\lambda + 1) = \frac{\sin[\pi(\lambda + 1)]}{\sin[\pi(\lambda + 1)/N]} \exp\left[-j\pi(\lambda + 1)\left(1 - \frac{1}{N}\right)\right]$$

For a large data length N, we may approximate the complex exponential term as follows:

$$\exp\left[-j\pi(\lambda + 1)\left(1 - \frac{1}{N}\right)\right] \simeq \exp(-j\pi) \exp\left[-j\pi\lambda\left(1 - \frac{1}{N}\right)\right]$$

$$= -\exp\left[-j\pi\lambda\left(1 - \frac{1}{N}\right)\right]$$

We may thus write

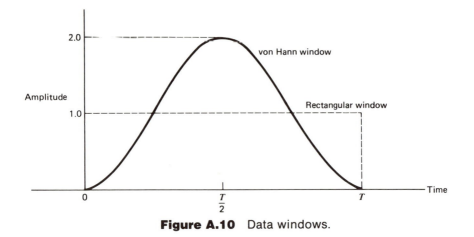

Figure A.10 Data windows.

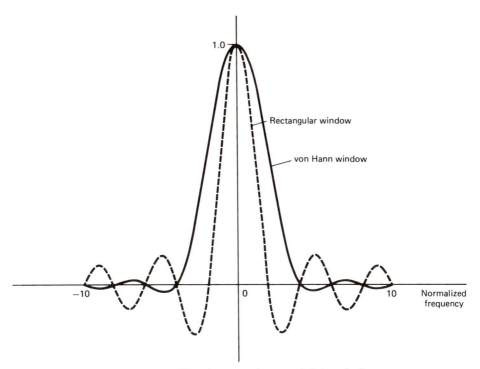

Figure A.11   Fourier transforms of data windows.

$$\Gamma(\lambda + 1) \simeq - \frac{\sin[\pi(\lambda + 1)]}{\sin[\pi(\lambda + 1)/N]} \exp\left[-j\pi\lambda \left(1 - \frac{1}{N}\right)\right] \qquad (A.47)$$

Similarly, except for a scaling factor, we may write for the third term in Eq. A.46:

$$\Gamma(\lambda - 1) \simeq - \frac{\sin[\pi(\lambda - 1)]}{\sin[\pi(\lambda - 1)/N]} \exp\left[-j\pi\lambda \left(1 - \frac{1}{N}\right)\right] \qquad (A.48)$$

Thus, using Eqs. A.40, A.47, and A.48 in Eq. A.46, we get

$$\Gamma_H(\lambda) \simeq \left\{ \frac{\sin(\pi\lambda)}{\sin(\pi\lambda/N)} + \frac{1}{2} \frac{\sin[\pi(\lambda + 1)]}{\sin[\pi(\lambda + 1)/N]} \right.$$
$$\left. + \frac{1}{2} \frac{\sin[\pi(\lambda - 1)]}{\sin[\pi(\lambda - 1)/N]} \right\} \exp\left[-j\pi\lambda \left(1 - \frac{1}{N}\right)\right] \qquad (A.49)$$

In Fig. A.11, we show the two DFT filter responses $\Gamma(\lambda)$ and $\Gamma_H(\lambda)$, for $N = 16$. The response $\Gamma(\lambda)$, shown as a dashed curve, corresponds to the use of a rectangular window. The second response $\Gamma_H(\lambda)$, shown as a continuous curve, corresponds to the use of a von Hann window. We find that the peak side lobe is reduced from -13 dB for $\Gamma(\lambda)$ to -30 dB for $\Gamma_H(\lambda)$, with respect to the main lobe. However, this significant reduction in side-lobe level is attained at the cost of a main lobe that is increased in width by a factor of two. That is, we have exchanged reduced spectral resolution for reduced spectral leakage.

APPENDIX B

PROPERTIES OF THE FOURIER TRANSFORM

Let the time function $g(t)$ denote a signal that may be real- or complex-valued. We assume that the signal has finite energy; that is

$$\int_{-\infty}^{\infty} |g(t)|^2 \, dt < \infty$$

The *Fourier transform* of the signal $g(t)$ is defined by

$$G(f) = \int_{-\infty}^{\infty} g(t) \exp(-j2\pi f t) dt \tag{B.1}$$

where the real variable f denotes frequency. Given $G(f)$, the original signal $g(t)$ is obtained by using the *inverse Fourier transform*:

$$g(t) = \int_{-\infty}^{\infty} G(f) \exp(j2\pi f t) df \tag{B.2}$$

The signal is thus completely defined by specifying its *time-domain description* $g(t)$ or *frequency-domain description* $G(f)$. The interrelationship between $g(t)$ and $G(f)$ is expressed by using the "shorthand" notation

$$g(t) \rightleftharpoons G(f)$$

Using the relations of Eqs. B.1 and B.2, we may readily establish the following properties of the Fourier transform*:

PROPERTY 1 LINEARITY (SUPERPOSITION)
Let $g_1(t) \rightleftharpoons G_1(f)$ and $g_2(t) \rightleftharpoons G_2(f)$. Then for all constants a and b, we have

$$ag_1(t) + bg_2(t) \rightleftharpoons aG_1(f) + bG_2(f)$$

PROPERTY 2 TIME SCALING
Let $g(t) \rightleftharpoons G(f)$. Then,

$$g(at) \rightleftharpoons \frac{1}{|a|} G\left(\frac{f}{a}\right).$$

where a is a scaling factor.

---

* The material presented in this appendix is summarized from Haykin (1989).

528

PROPERTY 3 DUALITY

If $g(t) \rightleftharpoons G(f)$, then

$$G(t) \rightleftharpoons g(-f).$$

where $G(t)$ is obtained by replacing f in $G(f)$ with t, and $g(-f)$ is obtained by replacing t in $g(t)$ with $-f$.

PROPERTY 4 TIME SHIFTING

If $g(t) \rightleftharpoons G(f)$, then

$$g(t - t_0) \rightleftharpoons G(f)\exp(-j2\pi f t_0)$$

where t_0 is a real constant.

PROPERTY 5 FREQUENCY SHIFTING

If $g(t) \rightleftharpoons G(f)$, then

$$\exp(j2\pi f_c t)g(t) \rightleftharpoons G(f - f_c)$$

PROPERTY 6 DIFFERENTIATION IN THE TIME DOMAIN

Let $g(t) \rightleftharpoons G(f)$, and assume that the first derivative of $g(t)$ is Fourier transformable. Then

$$\frac{d}{dt}\, g(t) \rightleftharpoons j2\pi f G(f)$$

That is, differentiation of a time function $g(t)$ has the effect of multiplying its Fourier transform $G(f)$ by the factor $j2\pi f$.

PROPERTY 7 INTEGRATION IN THE TIME DOMAIN

Let $g(t) \rightleftharpoons G(f)$. Then, provided $G(0) = 0$, we have

$$\int_{-\infty}^{t} g(\tau)d\tau \rightleftharpoons \frac{1}{j2\pi f}G(f)$$

That is, integration of a time function $g(t)$ has the effect of dividing its Fourier transform $G(f)$ by the factor $j2\pi f$, assuming that $G(0)$ is zero.

PROPERTY 8 CONJUGATE FUNCTIONS

If $g(t) \rightleftharpoons G(f)$, then for a complex-valued time function $g(t)$ we have

$$g^*(t) \rightleftharpoons G^*(-f)$$

where the asterisk denotes the complex conjugate operation.

PROPERTY 9 MULTIPLICATION IN THE TIME DOMAIN

Let $g_1(t) \rightleftharpoons G_1(f)$ and $g_2(t) \rightleftharpoons G_2(f)$. Then

$$g_1(t)g_2(t) \rightleftharpoons \int_{-\infty}^{\infty} G_1(\lambda)G_2(f - \lambda)d\lambda$$

That is, the multiplication of two signals in the time domain is transformed into the convolution of their individual Fourier transforms in the frequency domain.

In a discussion of convolution, the following shorthand notation is frequently used:

$$G_{12}(f) = G_1(f) \star G_2(f)$$

Note that convolution is commutative, that is

$$G_{12}(f) = G_{21}(f)$$

or

$$G_1(f) \star G_2(f) = G_2(f) \star G_1(f)$$

PROPERTY 10 CONVOLUTION IN THE TIME DOMAIN
Let $g_1(t) \rightleftharpoons G_1(f)$ and $g_2(t) \rightleftharpoons G_2(f)$. Then

$$\int_{-\infty}^{\infty} g_1(\tau)g_2(t - \tau)d\tau \rightleftharpoons G_1(f)G_2(f)$$

That is, the convolution of two signals in the time domain is transformed into the multiplication of their individual Fourier transforms in the frequency domain.

PROPERTY 11 RAYLEIGH'S ENERGY THEOREM
Let $g(t) \rightleftharpoons G(t)$. Then, the signal energy

$$\int_{-\infty}^{\infty} |g(t)|^2 \, dt = \int_{-\infty}^{\infty} |G(f)|^2 \, df$$

This relation may be viewed as a special case of Property 9 or 10. The squared amplitude response, $|G(f)|^2$, is called the *energy spectral density* of the signal $g(t)$.

Table of Fourier Transform Pairs
In Table B.1, we present Fourier-transform pairs of various time functions of interest. The *rectangular pulse*, *sinc pulse*, and *triangular pulse* are examples of energy signals. Hence, they are Fourier transformable.

The table also includes the *Dirac delta function* $\delta(t)$ that is defined by

$$\delta(t) = \begin{cases} 0 & t \neq 0 \\ \infty & t = 0 \end{cases}$$

and

$$\int_{-\infty}^{\infty} \delta(t)dt = 1$$

The delta function $\delta(t)$ may be viewed as the limiting form of a pulse of unit area as the duration of the pulse approaches zero. Hence, the Fourier transform of $\delta(t)$ may be defined in the limit, and thus shown to equal unity. This Fourier transform pair and other related ones are also included in Table B.1.

Table B.1 Fourier-transform Pairs

Time Function	$g(t)$	$G(f)$
Rectangular pulse	$\begin{cases} 1 & \|t\| < T/2 \\ 0 & \|t\| > T/2 \end{cases}$	$T \operatorname{sinc}(fT)$
Sinc pulse	$\operatorname{sinc}(2Wt)$	$\begin{cases} \dfrac{1}{2W} & \|f\| < W \\ 0 & \|f\| > W \end{cases}$
Triangular pulse	$\begin{cases} 1 - \dfrac{\|t\|}{T} & \|t\| < T \\ 0, & \|t\| \geq T \end{cases}$	$T \operatorname{sinc}^2(fT)$
Dirac delta function	$\delta(t)$	1
dc	1	$\delta(f)$
Delayed delta function	$\delta(t - t_0)$	$\exp(-j2\pi f t_0)$
Complex exponential	$\exp(j2\pi f_c t)$	$\delta(f - f_c)$
Cosine function	$\cos(2\pi f_c t)$	$\dfrac{1}{2}[\delta(f - f_c) + \delta(f + f_c)]$
Sine function	$\sin(2\pi f_c t)$	$\dfrac{1}{2j}[\delta(f - f_c) - \delta(f + f_c)]$
Sampling function	$\displaystyle\sum_{i=-\infty}^{\infty} \delta(t - iT_0)$	$\dfrac{1}{T_0}\displaystyle\sum_{n=-\infty}^{\infty} \delta\!\left(f - \dfrac{n}{T_0}\right)$

Definitions

1. sinc function

$$\operatorname{sinc}(x) = \frac{\sin(\pi x)}{\pi x}$$

2. Dirac delta function

$$\delta(t) = \begin{cases} \infty & t = 0 \\ 0 & t \neq 0 \end{cases}$$

$$\int_{-\infty}^{\infty} \delta(t)\, dt = 1$$

APPENDIX C

BAND-PASS SIGNALS AND SYSTEMS

C.1 COMPLEX REPRESENTATION OF BAND-PASS SIGNALS

We say that a signal $g(t)$ is a *band-pass signal* if its Fourier transform $G(f)$ is nonnegligible only in a band of frequencies of total extent $2W$, say, centered about some frequency $\pm f_c$. This is illustrated in Fig. C.1a. We refer to f_c as the *carrier frequency*. In the majority of communication signals, we find that the bandwidth $2W$ is small compared with f_c; more precisely, we have $W \ll f_c$. We refer to such a signal as a *narrow-band signal*. However, a precise statement about how small the bandwidth must be in order for the signal to be considered narrow-band is not necessary for our present discussion.

Let $\bar{g}(t)$ denote a complex-valued low-pass signal whose spectrum is limited to the band $-W \leq f \leq W$ and centered at the origin. Moreover, we assume that the spectrum of $\bar{g}(t)$, except for a scaling factor, is obtained by frequency-shifting the right-hand portion of the spectrum (Fourier transform) of the band-pass signal $g(t)$ to the left by an amount equal to f_c. The resulting amplitude spectrum of $\bar{g}(t)$ is depicted in Fig. C.1b, where the scaling factor (for convenience of subsequent development) is 2. The signal $\bar{g}(t)$ so defined is called the *complex envelope* of the band-pass signal.*

Using the frequency-shifting property of the Fourier transform, we may define the band-pass signal $g(t)$ in terms of the complex envelope $\bar{g}(t)$ as

$$g(t) = \text{Re}[\bar{g}(t)\exp(j2\pi f_c t)] \tag{C.1}$$

where $\text{Re}[\cdot]$ denotes the "real part of" the quantity contained inside the square brackets. Since, in general, $\bar{g}(t)$ is complex-valued, we may expand it as

$$\bar{g}(t) = g_I(t) + jg_Q(t) \tag{C.2}$$

where $g_I(t)$ is the real part of $\bar{g}(t)$, and $g_Q(t)$ is the imaginary part. Correspondingly, the exponential term $\exp(j2\pi f_c t)$ may be expanded as

$$\exp(j2\pi f_c t) = \cos(2\pi f_c t) + j\sin(2\pi f_c t) \tag{C.3}$$

Hence, substituting Eqs. C.2 and C.3 in Eq. C.1, we get the following canonical

---

* The material presented in this appendix is summarized from Haykin (1983).

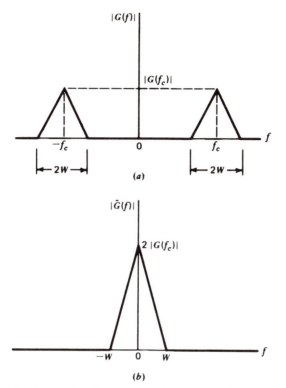

Figure C.1   (a) Amplitude spectrum of band-pass signal g(t). (b) Amplitude spectrum of complex envelope $\tilde{g}(t)$.

description for the band-pass signal:

$$g(t) = g_I(t)\cos(2\pi f_c t) - g_Q(t)\sin(2\pi f_c t) \qquad (C.4)$$

We refer to $g_I(t)$ as the *in-phase component* of the band-pass signal $g(t)$ and to $g_Q(t)$ as the *quadrature component* of the signal, both with respect to the *carrier* $\cos(2\pi f_c t)$. The complex envelope $\tilde{g}(t)$ may thus be pictured as a *time-varying phasor* at the origin of the $g_I g_Q$-plane. The end of the phasor moves about in the plane, and at the same time the plane rotates with an angular velocity equal to $2\pi f_c$ radians per second. The given signal $g(t)$ is the projection of this time-varying phasor on a fixed line.

Both $g_I(t)$ and $g_Q(t)$ are limited to the band $-W \leqslant f \leqslant W$. Hence, except for scaling factors, they may be derived from the band-pass signal $g(t)$ by using the scheme shown in Fig. C.2a, where both low-pass filters are identical, each having a bandwidth equal to W. To reconstruct $g(t)$ from its in-phase and quadrature components, we may use the scheme shown in Fig. C.2b. The two schemes shown in Fig. C.2 are basic to the study of all *linear modulation systems*.

Alternatively, we may express the complex envelope $\tilde{g}(t)$ in the form

$$\tilde{g}(t) = a(t)\exp[j\phi(t)] \qquad (C.5)$$

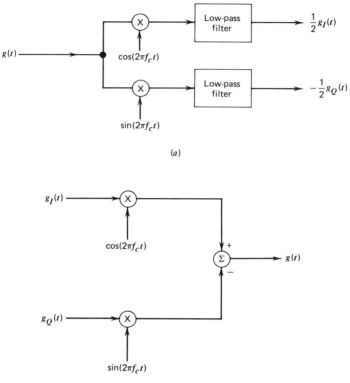

Figure C.2   (a) Scheme for deriving in-phase and quadrature components of a band-pass signal. (b) Scheme for reconstructing a band-pass signal from its in-phase and quadrature components.

where $a(t)$ and $\phi(t)$ are both real-valued low-pass functions. Based on this representation, the band-pass signal $g(t)$ is defined by

$$g(t) = a(t)\cos[2\pi f_c t + \phi(t)] \tag{C.6}$$

We refer to $a(t)$ as the *natural envelope* or the *envelope* of the band-pass signal $g(t)$ and to $\phi(t)$ as the *phase* of the signal.

It is therefore apparent that whether we represent the band-pass signal $g(t)$ in terms of its in-phase and quadrature components or in terms of its envelope and phase, the information content of the signal $g(t)$ is completely represented by the complex envelope $\tilde{g}(t)$. The particular virtue of using the complex envelope $\tilde{g}(t)$ to represent the band-pass signal is an analytical one, and will become evident in the next section.

C.2   COMPLEX REPRESENTATION OF BAND-PASS SYSTEMS

In this section we wish to develop a procedure for handling the analysis of band-pass systems. Specifically, we wish to show that the analysis of band-pass systems can be greatly simplified by establishing an analogy (or, more precisely, an isomorphism) between low-pass and band-pass systems.

Consider first a narrow-band signal $x(t)$, with its Fourier transform denoted by $X(f)$. We assume that the spectrum of the signal $x(t)$ is limited to frequencies within $\pm W$ Hz of the carrier frequency f_c, with $W < f_c$. Let this signal be represented as

$$x(t) = x_I(t)\cos(2\pi f_c t) - x_Q(t)\sin(2\pi f_c t)$$

where $x_I(t)$ is the in-phase component and $x_Q(t)$ is the quadrature component. Then, using $\tilde{x}(t)$ to denote the complex envelope of $x(t)$, we write

$$\tilde{x}(t) = x_I(t) + jx_Q(t)$$

The band-pass signal $x(t)$ thus takes on the complex representation:

$$\begin{aligned} x(t) &= \mathrm{Re}[\tilde{x}(t)\exp(j2\pi f_c t)] \\ &= \tfrac{1}{2}[\tilde{x}(t)\exp(j2\pi f_c t) + \tilde{x}^*(t)\exp(-j2\pi f_c t)] \end{aligned} \tag{C.9}$$

where $\tilde{x}^*(t)$ is the complex conjugate of $\tilde{x}(t)$. Let $X(f)$ denote the Fourier transform of $x(t)$. Hence, taking the Fourier transforms of both sides of Eq. C.9 and using the frequency-shifting and complex conjugation properties of the Fourier transform, we get

$$X(f) = \tfrac{1}{2}[\tilde{X}(f - f_c) + \tilde{X}^*(-f - f_c)] \tag{C.10}$$

where $\tilde{X}(f)$ is the Fourier transform of $\tilde{x}(t)$. Equation C.10 satisfies the requirement that $X^*(f) = X(-f)$ for a real-valued band-pass signal $x(t)$. Since $\tilde{X}(f)$ represents a low-pass transfer function limited to $|f| \leq W$, with $W < f_c$, we deduce from Eq. C.10 that

$$\tilde{X}(f - f_c) = \begin{cases} 2X(f) & f > 0 \\ 0 & f < 0 \end{cases} \tag{C.11}$$

Correspondingly, we have

$$\tilde{X}^*(-f - f_c) = \begin{cases} 0 & f > 0 \\ 2X^*(f) & f < 0 \end{cases} \tag{C.12}$$

Let the signal $x(t)$ be applied to a linear time-invariant band-pass system with impulse response $h(t)$ and transfer function $H(f)$. We assume that the frequency response of the system is limited to frequencies within $\pm B$ of the carrier frequency f_c. The system bandwidth $2B$ is usually narrower than or equal to the input signal bandwidth $2W$. We wish to represent the band-pass impulse response $h(t)$ in terms of two quadrature components, designated as $h_I(t)$ and $h_Q(t)$. Thus, by analogy to the representation of band-pass signals, we express $h(t)$ in the form

$$h(t) = 2h_I(t)\cos(2\pi f_c t) - 2h_Q(t)\sin(2\pi f_c t) \tag{C.13}$$

where the factor 2 has been introduced for convenience in subsequent analysis. Defining the *complex impulse response* of the band-pass system as

$$\tilde{h}(t) = h_I(t) + jh_Q(t), \tag{C.14}$$

we have the complex representation

$$h(t) = \text{Re}[2\tilde{h}(t)\exp(j2\pi f_c t)] \tag{C.15}$$

Note that $h_I(t)$, $h_Q(t)$, and $\tilde{h}(t)$ are all low-pass functions limited to the frequency band $-B \leq f \leq B$. Let $H(f)$ denote the *transfer function* of the band-pass system, defined as the Fourier transform of the impulse response $h(t)$. Then, following a procedure similar to that described for the complex representation of the band-pass signal $x(t)$, we may write

$$H(f) = \tilde{H}(f - f_c) + \tilde{H}^*(-f - f_c) \tag{C.16}$$

where

$$\tilde{H}(f - f_c) = \begin{cases} H(f) & f > 0 \\ 0 & f < 0 \end{cases} \tag{C.17}$$

and

$$\tilde{H}^*(-f - f_c) = \begin{cases} 0 & f > 0 \\ H(f) & f < 0 \end{cases} \tag{C.18}$$

The frequency function $\tilde{H}(f)$ is the Fourier transform of the complex impulse response $\tilde{h}(t)$.

The foregoing representations for band-pass signals and systems provide the basis of an efficient method for determining the output of a band-pass system driven by a band-pass signal. We assume that the spectrum of the input signal $x(t)$ and the transfer function $H(f)$ of the system are both centered around the same frequency f_c.* Let $y(t)$ denote the output signal of the system. It is clear that $y(t)$ is also a band-pass signal, so that we may represent it in terms of its low-pass complex envelope $\tilde{y}(t)$, as follows:

$$y(t) = \text{Re}[\tilde{y}(t)\exp(j2\pi f_c t)] \tag{C.19}$$

The output signal $y(t)$ is related to the input signal $x(t)$ and impulse response $h(t)$ of the system by the convolution integral

$$y(t) = \int_{-\infty}^{\infty} h(\tau)x(t - \tau)d\tau \tag{C.20}$$

Since convolution in the time domain is transformed into multiplication in the frequency domain, we find that Fourier transforming both sides of Eq. C.20 yields

---

* In practice, there is no need to consider a situation in which the carrier frequency of the input signal is not aligned with the mid-band frequency of the band-pass system, since we have considerable freedom in choosing the carrier or mid-band frequency. Thus, changing the carrier frequency of the input signal by an amount Δf_c, say, simply corresponds to absorbing (or removing) the factor $\exp(\pm j2\pi\Delta f_c t)$ in the complex envelope of the input signal or the complex impulse response of the band-pass system.

$$Y(f) = H(f)X(f) \tag{C.21}$$

where $Y(f)$ is the Fourier transform of the output signal $y(t)$. Substituting Eqs. C.10 and C.16 in Eq. C.21, and then simplifying, we get the result

$$Y(f) = \tfrac{1}{2}\tilde{H}(f - f_c)\tilde{X}(f - f_c) + \tfrac{1}{2}\tilde{H}^*(-f - f_c)\tilde{X}^*(-f - f_c) \tag{C.22}$$

The simplification results from the use of Eqs. C.11, C.12, C.17, and C.18. Specifically, in view of these four equations, both cross-product terms $\tilde{H}(f - f_c)\tilde{X}^*(-f - f_c)$ and $\tilde{H}^*(-f - f_c)\tilde{X}(f - f_c)$ are reduced to zero. Let $\tilde{Y}(f)$ denote the Fourier transform of the complex envelope $\tilde{y}(t)$ of the output signal. Then, we immediately deduce from Eq. C.22 that

$$\tilde{Y}(f - f_c) = \tilde{H}(f - f_c)\tilde{X}(f - f_c)$$

Equivalently, we may write

$$\tilde{Y}(f) = \tilde{H}(f)\tilde{X}(f) \tag{C.23}$$

Transforming this relation into the time domain, we get

$$\tilde{y}(t) = \int_{-\infty}^{\infty} \tilde{h}(\tau)\tilde{x}(t - \tau)d\tau, \tag{C.26}$$

or, using the shorthand notation for convolution,

$$\tilde{y}(t) = \tilde{h}(t) \,\star\, \tilde{x}(t)$$

In other words, the complex envelope $\tilde{y}(t)$ of the output signal of a band-pass system is obtained by convolving the complex impulse response $\tilde{h}(t)$ of the system with the complex envelope $\tilde{x}(t)$ of the input band-pass signal. Equation C.26 is the result of the isomorphism, for convolution, between band-pass functions and the corresponding low-pass functions. The scaling factor 2 was introduced in Eq. C.13 so that Eq. C.26 would have a familiar form.

The significance of this result is that in dealing with band-pass signals and systems, we need to deal only with the low-pass functions $\tilde{x}(t)$, $\tilde{y}(t)$, and $\tilde{h}(t)$, representing the excitation, the response, and the system, respectively. That is, the analysis of a band-pass system, which is complicated by the presence of the multiplying factor $\exp(j2\pi f_c t)$, is replaced by an equivalent but simpler low-pass analysis that completely retains the essence of the filtering process.

PROBABILITY THEORY AND RANDOM PROCESSES

The word "random" is used to describe erratic and apparently unpredictable variations of an observed signal. Indeed, random signals (in one form or another) are encountered in every practical communication system. Consider, for example, a radio communication system. The received signal in such a system is random in nature due to various causes. Ordinarily, the received signal consists of an information-bearing signal component, a random interference component, and receiver noise. The information-bearing signal component may represent, for example, a voice signal that, typically, consists of randomly spaced bursts of energy of random duration. The interference component represents the extraneous electromagnetic waves produced by other communication systems and atmospheric electricity. A major type of noise is thermal noise caused by the random motion of the electrons in conductors and devices at the front end of the receiver.

The important point is that, regardless of the underlying causes of randomness, we cannot predict the exact value of the received signal. Nevertheless, the received signal can be described in terms of its statistical properties such as its average power, or the spectral distribution of the average power. The mathematical discipline that deals with the statistical characterization of random signals is probability theory.*

D.1 A NOTION OF PROBABILITY
Probability theory is rooted in real-life situations that involve performing an experiment whose outcome is always subject to *chance*. Moreover, if the experiment is repeated, the outcome can be different due to the influence of an underlying random phenomenon or chance mechanism. Such an experiment is referred to as a *random experiment*. For example, the experiment may be the observation of the result of the tossing of a fair coin. In this experiment, the possible outcomes of trials are "heads" or "tails."

To be more precise in the description of a random experiment, we ask for three features:

---

* For a detailed treatment of probability theory and the related subject of random processes, see Blake (1979), Papoulis (1984), and Thomas (1986).

1. The experiment is repeatable under identical conditions.
2. On any trial of the experiment, the outcome is unpredictable.
3. For a large number of trials of the experiment, the outcomes exhibit *statistical regularity*. That is, a definite *average* pattern of outcomes is observed if the experiment is repeated a large number of times.

(1)   Relative-frequency Approach

Let *event A* denote one of the possible outcomes of a random experiment. For example, in the coin-tossing experiment, event A may represent "heads." In any case, suppose that in n trials of the experiment, event A occurs n_A times. We may then assign the ratio n_A/n to the event A. This ratio is called the *relative frequency* of the event A. Clearly, the relative frequency is a *nonnegative real number less than or equal to one*. That is,

$$0 \le \frac{n_A}{n} \le 1 \tag{D.1}$$

If event A occurs in none of the trials, $(n_A/n) = 0$. If, on the other hand, event A occurs in all of the n trials, $(n_A/n) = 1$.

We say that the experiment exhibits statistical regularity if for *any* sequence of n trials, the relative frequency n_A/n *converges to the same limit* as n becomes larger. Accordingly, it would seem natural for us to define the *probability* of event A as

$$P(A) = \lim_{n \to \infty} \left(\frac{n_A}{n}\right) \tag{D.2}$$

Thus, in the coin-tossing experiment, we may expect that out of a million tosses of a fair coin, about one half of them will show up heads.

The probability of an event is intended to represent the *likelihood* that a trial of the experiment will result in the occurrence of that event. For many engineering applications and games of chance, the use of Eq. D.2 to define the probability of an event is acceptable. However, for many other applications this definition is found to be inadequate. Consider, for example, the statistical analysis of the stock market: How are we to achieve repeatability of such an experiment? A more satisfying approach is to state the properties that any measure of probability is expected to have, postulating them as *axioms*, and then use relative-frequency interpretations to justify them.

(2)   Axioms of Probability

When we perform a random experiment, it is natural for us to be aware of the various outcomes that are likely to arise. In this context, it is convenient to think of an experiment and its possible outcomes as defining a space and its points. For each possible outcome of the experiment, we associate a point called the *sample point*, which we denote by s_k. The totality of sample points, corresponding to the aggregate of all possible outcomes of the experiment, is called the *sample space*, which we denote by $\mathcal{S}$. An event corresponds to either

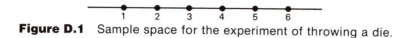

Figure D.1 Sample space for the experiment of throwing a die.

a single sample point or a set of sample points. In particular, the entire sample space $\mathcal{S}$ is called the *sure event*, the null set ϕ is called the *null* or *impossible event*, and a single sample point is called an *elementary event*.

Consider, for example, an experiment that involves the throw of a die. In this experiment there are six possible outcomes: the showing of one, two, three, four, five, or six dots on the upper face of the die. By assigning a sample point to each of these possible outcomes, we have a one-dimensional sample space that consists of six sample points, as shown in Fig. D.1. The elementary event describing the statement "a six shows" corresponds to the sample point {6}. On the other hand, the event describing the statement "an even number of dots shows" corresponds to the subset {2,4,6} of the sample space. Note that the term "event" is used interchangeably to describe the subset or the statement.

We are now ready to make a formal definition of probability. A *probability system* consists of the triple:

1. A *sample space $\mathcal{S}$ of elementary events (outcomes)*.
2. A *class $\mathcal{E}$ of events* that are subsets of $\mathcal{S}$.
3. A *probability measure $P(\cdot)$* assigned to each event A in the class $\mathcal{E}$ and having the following properties:

   (i) $P(\mathcal{S}) = 1$ (D.3)

   (ii) $0 \leqslant P(A) \leqslant 1$ (D.4)

   (iii) If $A + B$ is the *union of two mutually exclusive events* in the class $\mathcal{E}$, then

$$P(A + B) = P(A) + P(B) \tag{D.5}$$

Properties (i), (ii), and (iii) are known as the *axioms of probability*. Axiom (i) states that the probability of the sure event is unity. Axiom (ii) states that the probability of an event is a nonnegative real number that is less than or equal to unity. Axiom (iii) states that the probability of the union of two mutually exclusive events is the sum of the probabilities of the individual events.

Although the axiomatic approach to probability theory is abstract in nature, nevertheless, all three axioms have relative-frequency interpretations of their own. Axiom (ii) corresponds to Eq. D.1. Axiom (i) corresponds to the limiting case of Eq. D.1 when the event A occurs in all of the n trials. To interpret Axiom (iii), we note that if event A occurs n_A times in n trials and event B occurs n_B times, then the union event "A or B" occurs in $n_A + n_B$ trials (since A and B can never occur on the same trial). Hence, $n_{A+B} = n_A + n_B$, and so we have

$$\frac{n_{A+B}}{n} = \frac{n_A}{n} + \frac{n_B}{n}$$

which has a mathematical form similar to that of Axiom (iii).

(3)   Elementary Properties of Probability

Axioms (i), (ii), and (iii) constitute an implicit definition of probability. We may use these axioms to develop some other basic properties of probability:

PROPERTY 1

$$P(\bar{A}) = 1 - P(A) \tag{D.6}$$

where $\bar{A}$ is the complement of event A.

The use of this property helps us investigate the *nonoccurrence of an event*. To prove it, we express the sample space $\mathcal{S}$ as the union of two mutually exclusive events A and $\bar{A}$:

$$\mathcal{S} = A + \bar{A}$$

Then, the use of axioms (i) and (iii) yields

$$1 = P(A) + P(\bar{A})$$

from which Eq. D.6 follows directly.

PROPERTY 2

If M mutually exclusive events A_1, A_2, . . . , A_M have the exhaustive property

$$A_1 + A_2 + \cdots + A_M = \mathcal{S} \tag{D.7}$$

then

$$P(A_1) + P(A_2) + \cdots + P(A_M) = 1 \tag{D.8}$$

To prove this property, we generalize Axiom (iii) by writing

$$P(A_1 + A_2 + \cdots + A_M) = P(A_1) + P(A_2) + \cdots + P(A_M)$$

The use of Axiom (i) in Eq. D.7 yields

$$P(A_1 + A_2 + \cdots + A_M) = 1$$

Hence, the result of Eq. D.8 follows.

When the M events are *equally likely* (i.e., they have equal probabilities), then Eq. D.8 simplifies as

$$P(A_i) = \frac{1}{M} \qquad i = 1, 2, \ldots, M \tag{D.9}$$

PROPERTY 3

When events A and B are not mutually exclusive, then the probability of the union event "A or B" equals

$$P(A + B) = P(A) + P(B) - P(AB) \tag{D.10}$$

where P(AB) is the probability of the joint event "A and B."

The probability $P(AB)$ is called the *joint probability*. It has the following relative-frequency interpretation:

$$P(AB) = \lim_{n \to \infty} \left(\frac{n_{AB}}{n} \right)$$

where n_{AB} denotes the number of times the events A and B occur simultaneously in n trials of the experiment. Axiom (iii) is a special case of Eq. D.10; when A and B are mutually exclusive, $P(AB)$ is zero, and Eq. D.10 reduces to the same form as Eq. D.5.

(4) Conditional Probability

Suppose we perform an experiment that involves a pair of events A and B. Let $P(B|A)$ denote the probability of event B, given that event A has occurred. The probability $P(B|A)$ is called the *conditional probability of B given A*. Assuming that A has nonzero probability, the conditional probability $P(B|A)$ is defined by

$$P(B|A) = \frac{P(AB)}{P(A)} \tag{D.11}$$

where $P(AB)$ is the joint probability of A and B.

We justify the definition of conditional probability given in Eq. D.11 by presenting a relative-frequency interpretation. Suppose that we perform an experiment and examine the occurrence of a pair of events A and B. Let n_{AB} denote the number of times the joint event (AB) occurs in n trials. Suppose that in the same n trials the event A occurs n_A times. Since the joint event (AB) corresponds to both A and B occurring, it follows that n_A must include n_{AB}. In other words,

$$\frac{n_{AB}}{n_A} \leq 1$$

The ratio n_{AB}/n_A represents the relative frequency of B given that A has occurred. For large n, the ratio n_{AB}/n_A equals the conditional probability $P(B|A)$. That is,

$$P(B|A) = \lim_{n \to \infty} \left(\frac{n_{AB}}{n_A} \right)$$

or equivalently

$$P(B|A) = \lim_{n \to \infty} \left(\frac{n_{AB}/n}{n_A/n} \right)$$

Recognizing that

$$P(AB) = \lim_{n \to \infty} \left(\frac{n_{AB}}{n} \right)$$

and

$$P(A) = \lim_{n \to \infty} \left(\frac{n_A}{n} \right)$$

the result of Eq. D.11 follows.

We may rewrite Eq. D.11 as

$$P(AB) = P(B|A)P(A) \tag{D.12}$$

It is apparent that we may also write

$$P(AB) = P(A|B)P(B) \tag{D.13}$$

Equations D.12 *and* D.13 *state that the joint probability of two events may be expressed as the product of the conditional probability of one event, given the other, times the elementary probability of the other.* Note that the conditional probabilities $P(B|A)$ and $P(A|B)$ have essentially the same properties as the various probabilities previously defined.

Situations may exist where the conditional probability $P(A|B)$ and the probabilities $P(A)$ and $P(B)$ are easily determined directly, but the conditional probability $P(B|A)$ is desired. From Eqs. D.12 and D.13, it follows that, provided $P(A) \neq 0$, we may determine $P(B|A)$ by using the relation

$$P(B|A) = \frac{P(A|B)P(B)}{P(A)} \tag{D.14}$$

This relation is a special form of *Bayes' rule*.

Suppose that the conditional probability $P(B|A)$ is simply equal to the elementary probability of occurrence of event B, that is,

$$P(B|A) = P(B)$$

It then follows that the probability of occurrence of the joint event (AB) is equal to the product of the elementary probabilities of the events A and B:

$$P(AB) = P(A)P(B),$$

so that

$$P(A|B) = P(A)$$

That is, the conditional probability of the event A, assuming the occurrence of the event B, is simply equal to the elementary probability of the event A. We thus see that in this case a knowledge of the occurrence of one event tells us no more about the probability of occurrence of the other event that we knew without that knowledge. Events A and B that satisfy such relations are said to be *statistically independent*.

D.2 RANDOM VARIABLES

It is customary to think of an experiment as a variable whose value is determined by the outcome of the experiment. *A function whose domain is a sample space and whose range is some set of real numbers is called a random variable* *of the experiment.* * Thus when the outcome of the experiment is s, the random variable is denoted as $X(s)$ or simply X. For example, the sample space representing the outcomes of the throw of a die is a set of six sample points which may be taken to be the integers $1, 2, \ldots, 6$, as illustrated in Fig. D.1. Then if we identify the sample point k with the event that k dots show when the die is thrown, the function $X(k) = k$ is a random variable such that $X(k)$ equals the number of dots that show when the die is thrown. In this example, the random variable takes on only a discrete set of values. In such a case we say that we are dealing with a *discrete random variable*. More precisely, *the random variable X is a discrete random variable if X can take on only a finite number of values in any finite interval. If, however, the random variable X can take on any value in a finite interval, X is called a continuous random variable.* For example, the random variable that represents the amplitude of a noise voltage at a particular instant of time is a continuous random variable because, in theory, it may take on any value between plus and minus infinity.

To proceed further, we need a probabilistic description of random variables that works equally well for both discrete and continuous random variables. Let us consider the random variable X and the probability of the event $X \leq x$. We denote this probability by $P(X \leq x)$. It is apparent that this probability is a function of the *dummy variable x*. To simplify our notation, we write

$$F_X(x) = P(X \leq x) \tag{D.15}$$

The function $F_X(x)$ is called the *cumulative distribution function* or simply the *distribution function* of the random variable X. Note that $F_X(x)$ is a function of x, not of the random variable X. However, it depends on the assignment of the random variable X, which accounts for the use of X as subscript. For any point x, the distribution function $F_X(x)$ expresses a probability.

The distribution function $F_X(x)$ has the following properties, which follow directly from Eq. (D.15):

1. The distribution function $F_X(x)$ is bounded between zero and one.
2. The distribution function $F_X(x)$ is a monotone-nondecreasing function of x; that is,

$$F_X(x_1) \leq F_X(x_2) \qquad \text{if } x_1 < x_2 \tag{D.16}$$

An alternative description of the probability distribution of the random variable X is often useful. It is the first derivative of the distribution function, as shown by

$$f_X(x) = \frac{d}{dx} F_X(x) \tag{D.17}$$

---

* The term "random variable" is somewhat confusing: first, because the word "random" is not used in the sense of equal probability of occurrence, for which it should be reserved. Second, the word "variable" does not imply dependence (upon the experimental outcome), which is an essential part of the meaning. Nevertheless, the term is so deeply imbedded in the literature of probability that its usage has persisted.

which is called the *probability density function*. Note that the differentiation in Eq. D.17 is with respect to the dummy variable x. The name, density function, arises from the fact that the probability of the event $x_1 < X \leq x_2$ equals

$$P(x_1 < X \leq x_2) = P(X \leq x_2) - P(X \leq x_1)$$
$$= F_X(x_2) - F_X(x_1)$$
$$= \int_{x_1}^{x_2} f_X(x)\,dx \qquad \text{(D.18)}$$

Since $F_X(\infty) = 1$, corresponding to the probability of a certain event, and $F_X(-\infty) = 0$, corresponding to the probability of an impossible event, it follows immediately from Eq. D.18 that

$$\int_{-\infty}^{\infty} f_X(x)\,dx = 1 \qquad \text{(D.19)}$$

Also, as mentioned earlier, a distribution function must always be monotone nondecreasing. Hence, its derivative or the probability density function must always be nonnegative. *A probability density function must always be a nonnegative function, and with the total area under its curve equal to one.*

(1) Several Random Variables

Thus far we have focused attention on situations involving a single random variable. However, we find frequently that the outcome of an experiment requires several random variables for its description. We now consider situations involving two random variables. The probabilistic description developed in this way may be readily extended to any number of random variables.

Consider two random variables X and Y. *We define the joint distribution function $F_{X,Y}(x, y)$ as the probability that the random variable X is less than or equal to a specified value x and that the random variable Y is less than or equal to a specified value y.* The variables X and Y may be two separate one-dimensional random variables or the components of a single two-dimensional random variable. The joint distribution function $F_{X,Y}(x, y)$ is the probability that the outcome of an experiment will result in a sample point lying inside the quadrant $(-\infty < X \leq x, -\infty < Y \leq y)$ of the joint-sample space. That is

$$F_{X,Y}(x, y) = P(X \leq x, Y \leq y) \qquad \text{(D.20)}$$

Suppose that the joint distribution function $F_{X,Y}(x, y)$ is continuous everywhere, and that the partial derivative

$$f_{X,Y}(x, y) = \frac{\partial^2 F_{X,Y}(x, y)}{\partial x \partial y} \qquad \text{(D.21)}$$

exists and is continuous everywhere. We call the function $f_{X,Y}(x, y)$ the *joint probability density function* of the random variables X and Y. The joint distribution function $F_{X,Y}(x, y)$ is a monotone-nondecreasing function of both x and y. Therefore, from Eq. D.21 it follows that the joint probability density function

$f_{X,Y}(x, y)$ is always nonnegative. Also the total volume under the graph of a joint probability density function must be unity, as shown by

$$\int_{-\infty}^{\infty} \int_{-\infty}^{\infty} f_{X,Y}(\xi, \eta)d\xi d\eta = 1 \qquad (D.22)$$

The probability density function for a single random variable (X, say) can be obtained from its joint probability density function with a second random variable (Y, say) in the following way. We first note that

$$F_X(x) = \int_{-\infty}^{\infty} \int_{-\infty}^{x} f_{X,Y}(\xi, \eta)d\xi d\eta \qquad (D.23)$$

Therefore, differentiating both sides of Eq. D.23 with respect to x, we get the desired relation:

$$f_X(x) = \int_{-\infty}^{\infty} f_{X,Y}(x, \eta)d\eta \qquad (D.24)$$

Thus the probability density function $f_X(x)$ may be obtained from the joint probability density function $f_{X,Y}(x, y)$ by simply integrating it over all possible values of the undesired random variable, Y. The use of similar arguments in the other dimension yields $f_Y(y)$. The probability density functions $f_X(x)$ and $f_Y(y)$ are called *marginal densities*. Hence, the joint probability density function $f_{X,Y}(x, y)$ contains all the possible information about the joint random variables X and Y.

Suppose that X and Y are two continuous random variables with joint probability density function $f_{X,Y}(x, y)$. The *conditional probability density function* of Y given that $X = x$ is defined by

$$f_Y(y|X = x) = \frac{f_{X,Y}(x, y)}{f_X(x)} \qquad (D.25)$$

provided that $f_X(x) > 0$, where $f_X(x)$ is the marginal density of X. The function $f_Y(y|X = x)$ may be thought of as a function of the variable y, with the variable x arbitrary but fixed. Accordingly, it satisfies all the requirements of an ordinary probability density function, as shown by

$$f_Y(y|X = x) \geq 0$$

and

$$\int_{-\infty}^{\infty} f_Y(y|X = x)dy = 1$$

If the random variables X and Y are *statistically independent*, then knowledge of the outcome of X can in no way affect the distribution of Y. The result is that the conditional probability density function $f_Y(y|X = x)$ reduces to the marginal density $f_Y(y)$, as shown by

$$f_Y(y|X = x) = f_Y(y)$$

In such a case, we may express the joint probability density function of the random variables X and Y as the product of their respective marginal densities, as shown by

$$f_{X,Y}(x, y) = f_X(x)f_Y(y)$$

(2) Statistical Averages

Having discussed probability and some of its ramifications, we now seek ways for determining the *average* behavior of the outcomes arising in random experiments.

The *mean* or *expected value* of a random variable X is commonly defined by

$$m_X = E[X] = \int_{-\infty}^{\infty} x f_X(x)dx \tag{D.26}$$

where E denotes the *expectation operator*. That is, the mean m_X locates the center of gravity of the area under the probability density curve of the random variable X. Similarly, the expected value of a function of X, namely, $g(X)$, is defined by

$$E[g(X)] = \int_{-\infty}^{\infty} g(x) f_X(x)dx \tag{D.27}$$

For the special case of $g(X) = X^n$ we obtain the nth *moment* of the probability distribution of the random variable X; that is,

$$E[X^n] = \int_{-\infty}^{\infty} x^n f_X(x)dx \tag{D.28}$$

By far the most important moments of X are the first two moments. Thus putting $n = 1$ in Eq. D.28 gives the mean value of the random variable as discussed earlier, whereas putting $n = 2$ gives the *mean-square value* of X:

$$E[X^2] = \int_{-\infty}^{\infty} x^2 f_X(x)dx \tag{D.29}$$

We may also define *central moments*, which are simply the moments of the difference between a random variable X and its mean value m_X. Thus the nth central moment is

$$E[(X - m_X)^n] = \int_{-\infty}^{\infty} (x - m_X)^n f_X(x)dx \tag{D.30}$$

For $n = 1$, the central moment is, of course, zero, whereas for $n = 2$ the second central moment is referred to as the *variance* of the random variable:

$$\text{Var}[X] = E[(X - m_X)^2] = \int_{-\infty}^{\infty} (x - m_X)^2 f_X(x)dx \tag{D.31}$$

The variance of a random variable X is commonly denoted as σ_X^2. The square root of the variance, namely, σ_X, is called the *standard deviation* of the random variable X.

The variance σ_X^2 of a random variable X is a measure of the variable's dispersion. By specifying the variance σ_X^2, we essentially define the effective width of the probability density function $f_X(x)$ of the random variable X about the mean m_X. A precise statement of this definition is due to Chebyshev. The *Chebyshev inequality* states that for any positive number ε, we have

$$P(|X - m_X| \geqslant \varepsilon) \leqslant \frac{\sigma_X^2}{\varepsilon^2}$$

From this inequality we see that the mean and variance of a random variable give a partial description of its probability distribution.

The expectation operator is linear in that the expectation of the sum of two random variables is equal to the sum of the individual expectations of the two random variables. Hence, expanding $E[(X - m_X)^2]$ and using the linearity property of the expectation operator, we find that the variance σ_X^2 and mean-square value $E[X^2]$ are related by

$$\begin{aligned} \sigma_X^2 &= E[X^2 - 2m_X X + m_X^2] \\ &= E[X^2] - 2m_X E[X] + m_X^2 \\ &= E[X^2] - m_X^2 \end{aligned}$$

Therefore, if the mean m_X is zero, then the variance σ_X^2 and the mean-square value $E[X^2]$ of the random variable X are equal.

(3)   Joint Moments

Consider next a pair of random variables X and Y. A set of statistical averages of importance in this case are the *joint moments*, namely, the expected value of $X^j Y^k$, where j and k may assume any positive integer values. We may thus write

$$E[X^j Y^k] = \int_{-\infty}^{\infty} \int_{-\infty}^{\infty} x^j y^k f_{X,Y}(x, y)dx\, dy \tag{D.32}$$

A joint moment of particular importance is the *correlation* defined by $E[XY]$, which corresponds to $j = k = 1$ in Eq. D.32.

The correlation of the centered random variables $X - E[X]$ and $Y - E[Y]$, that is, the joint moment

$$\text{Cov}[XY] = E[(X - E[X])(Y - E[Y])] \tag{D.33}$$

is called the *covariance* of X and Y. Letting $m_X = E[X]$ and $m_Y = E[Y]$, we may expand Eq. D.33 to obtain

$$\text{Cov}[XY] = E[XY] - m_X m_Y \tag{D.34}$$

Let σ_X^2 and σ_Y^2 denote the variances of X and Y, respectively. Then the covariance of X and Y normalized with respect to $\sigma_X \sigma_Y$ is called the *correlation coefficient* of X and Y:

$$\rho = \frac{\text{Cov}[XY]}{\sigma_X \sigma_Y} \tag{D.35}$$

We say that the two random variables X and Y are uncorrelated if and only if their covariance is zero, that is, if and only if

$$\text{Cov}[XY] = 0$$

We say that they are orthogonal if and only if their correlation is zero, that is, if and only if

$$E[XY] = 0$$

From Eq. D.34 we observe that if one or both of the random variables X and Y have zero means, and if they are orthogonal random variables, then they are uncorrelated, and vice versa. Note also that if X and Y are statistically independent, then they are uncorrelated; however, the converse of this statement is not necessarily true.

(4) Transformation of Random Variables

Consider the problem of determining the probability density function of a random variable Y, which is obtained by a *one-to-one transformation* of a given random variable X. The simplest possible case is when the new random variable Y is a monotone-increasing differentiable function g of the random variable X (see Fig. D.2):

$$Y = g(X) \qquad (D.36)$$

In this case we have

$$\begin{aligned} F_Y(y) &= P(Y \leq y) \\ &= P(X \leq h(y)) \\ &= F_X(h(y)) \end{aligned}$$

where h is the *inverse transformation*

$$h(y) = g^{-1}(y) \qquad (D.37)$$

This inverse transformation exists for all y, because x and y are related one to one. Assuming that the given random variable X has a probability density function $f_X(x)$, we may write

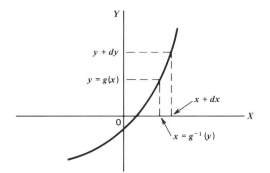

Figure D.2 A one-to-one transformation of a random variable X.

$$F_Y(y) = \int_{-\infty}^{h(y)} f_X(x)dx$$

Differentiating both sides of this relation, we get

$$f_Y(y) = f_X(h(y)) \frac{dh}{dy} \tag{D.38}$$

Consider next the case when g is a differentiable monotone-decreasing function with an inverse h. We may then write

$$F_Y(y) = \int_{h(y)}^{\infty} f_X(x)dx$$

which, on differentiation, yields

$$f_Y(y) = -f_X(h(y)) \frac{dh}{dy}$$

Since the derivative dh/dy is negative in this case, whereas it is positive in Eq. D.38, we may express both results by the single formula

$$f_Y(y) = f_X(h(y)) \left| \frac{dh}{dy} \right| \tag{D.39}$$

This is the desired formula for finding the probability density function of a one-to-one differentiable function of a given random variable.

(5) Jacobian

We may view the one-to-one transformation of random variables in another way. Consider Fig. D.2, where a differential interval dy and the corresponding interval dx are shown. Regardless of whether y is montonically increasing or decreasing, the probability that the random variable Y lies in the interval dy is the same as the probability that the random variable X lies in the corresponding interval dx. In other words, we may write

$$f_Y(y)|dy| = f_X(x)|dx| \tag{D.40}$$

In Eq. D.40, we have used absolute value signs so as to ensure that both sides of the equation are always nonnegative. Rearranging Eq. D.40, we have

$$f_Y(y) = f_X(x)|J(y)| \tag{D.41}$$

where

$$J(y) = \frac{dx}{dy} \tag{D.42}$$

The function $J(y)$ is called the *Jacobian* of the transformation described in Eq. D.36.

Consider next the functional transformation of a bivariate transformation. Let the transformation be defined by

$$U = g_1(X,Y) \tag{D.43}$$

and

$$V = g_2(X,Y) \tag{D.44}$$

We assume that this pair of equations defines a one-to-one continuously differentiable transformation, the application of which transforms the random variables X and Y into U and V. The inverse transformation is defined by

$$X = h_1(U,V) \tag{D.45}$$

and

$$Y = h_2(U,V) \tag{D.46}$$

Correspondingly, in terms of sample values of the random variables, we have

$$x = h_1(u,v) \tag{D.47}$$

and

$$y = h_2(u,v) \tag{D.48}$$

In this case, the Jacobian of the transformation is defined by the *determinant*

$$J(u,v) = \begin{vmatrix} \dfrac{\partial x}{\partial u} & \dfrac{\partial x}{\partial v} \\ \dfrac{\partial y}{\partial u} & \dfrac{\partial y}{\partial v} \end{vmatrix} \tag{D.49}$$

The joint probability density function of the new random variables U and V is related to that of the random variables X and Y by

$$f_{U,V}(u,v) = f_{X,Y}(x,y)|J(u,v)| \tag{D.50}$$

Thus, to determine $f_{U,V}(u,v)$, we proceed as follows:

1. Find $|J(u,v)|$ as a function of u and v.
2. Use the inverse transformation defined by Eqs. D.47 and D.48 to replace x and y in the given joint probability density function $f_{X,Y}(x,y)$ by u and v, respectively.
3. Multiply the result of step 2 by $|J(u,v)|$, thereby obtaining $f_{U,V}(u,v)$.

D.3   RANDOM PROCESSES

A basic concern in the statistical analysis of communication systems is the characterization of random signals such as voice signals, television signals, digital computer data, and electrical noise. These random signals have two properties: first, the signals are functions of time, defined on some observation interval; second, the signals are random in the sense that before conducting an experiment, it is not possible to describe exactly the waveforms that will be observed. Accordingly, in describing random signals we find that each sample point in our sample space is a function of time. For example, in studying the fluctuations in the output of a transistor, we may assume the simultaneous

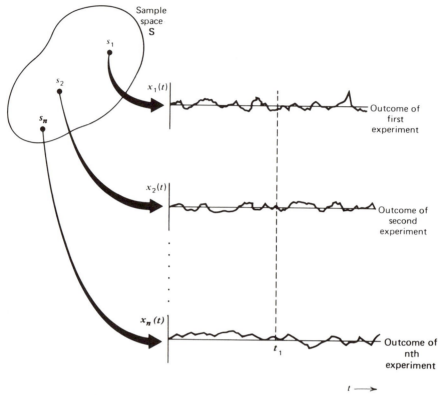

Figure D.3 An ensemble of sample functions.

testing of an indefinitely large number of identical transistors as a conceptual model of our problem. The output (measured as a function of time) of a particular transistor in the collection is then one sample point in our sample space. The sample space comprised of functions of time is called a *random* or *stochastic** *process*. As an integral part of this notion, we assume the existence of a probability distribution defined over an appropriate class of sets in the sample space, so that we may speak with confidence of the probability of various events. *We may thus define a random process as an ensemble of time functions together with a probability rule that assigns a probability to any meaningful event associated with an observation of one of these functions.*

Consider a random process $X(t)$ represented by the set of *sample functions* $\{x_j(t)\}, j = 1, 2, \ldots, n$, as illustrated in Fig. D.3. Sample function or waveform $x_1(t)$, with probability of occurrence $P(s_1)$, corresponds to *sample point* s_1 of the *sample space* S, and so on for the other sample functions $x_2(t), \ldots, x_n(t)$. Now suppose we observe the set of waveforms $\{x_j(t)\}, j = 1, 2, \ldots, n$, simultaneously at some time instant, $t = t_1$, as shown in the figure. Since each sample point s_j of the sample space S has associated with it a number $x_j(t_1)$ and a

---

* The word "stochastic" comes from Greek for "to aim (guess) at."

probability $P(s_j)$, we find that the resulting collection of numbers $\{x_j(t_1)\}, j = 1,$ $2, \ldots, n$, forms a *random variable*. We denote this random variable by $X(t_1)$. By observing the given set of waveforms simultaneously at a second time instant, say t_2, we obtain a different collection of numbers, hence a different random variable $X(t_2)$. Indeed, the set of waveforms $\{x_j(t)\}$ defines a different random variable for each choice of observation instant. The difference between a random variable and a random process is that for a random variable the outcome of an experiment is mapped into a number, whereas for a random process the outcome is mapped into a waveform that is a function of time.

(1) Random Vectors Obtained From Random Processes

By definition, a random process $X(t)$ implies the existence of an infinite number of random variables, one for each value of time t in the range $-\infty < t < \infty$. Thus we may speak of the distribution function $F_{X(t_1)}(x_1)$ of the random variable $X(t_1)$ obtained by observing the random process $X(t)$ at time t_1. More generally, for k time instants $t_1, t_2, \ldots, t_k$ we define the k random variables $X(t_1), X(t_2), \ldots, X(t_k)$, respectively, and express their *joint distribution function* as the probability of the joint event $X(t_1) \leq x_1, X(t_2) \leq x_2, \ldots, X(t_k) \leq x_k$ as shown by

$$F_{X(t_1),X(t_2),\ldots,X(t_k)}(x_1, x_2, \ldots, x_k) = P(X(t_1) \leq x_1, X(t_2) \leq x_2, \ldots, X(t_k) \leq x_k)$$

For convenience of notation, we write this joint distribution function simply as $F_{\mathbf{X(t)}}(\mathbf{x})$ where the *random vector* $\mathbf{X(t)}$ is defined by

$$\mathbf{X(t)} = \begin{bmatrix} X(t_1) \\ X(t_2) \\ \vdots \\ X(t_k) \end{bmatrix} \tag{D.51}$$

and the *dummy vector* $\mathbf{x}$ is defined by

$$\mathbf{x} = \begin{bmatrix} x_1 \\ x_2 \\ \vdots \\ x_k \end{bmatrix} \tag{D.52}$$

For a particular sample point s_j, the components of the random vector $\mathbf{X(t)}$ represent the values of the sample function $x_j(t)$ observed at times $t_1, t_2 \ldots, t_k$. Note also that the joint distribution function $F_{\mathbf{X(t)}}(\mathbf{x})$ depends on the random process $X(t)$ and the set of times $\{t_j\}, j = 1, 2, \ldots, k$.

The joint probability density function of the random vector $\mathbf{X(t)}$ equals

$$f_{\mathbf{X(t)}}(\mathbf{x}) = \frac{\partial^k}{\partial x_1 \, \partial x_2 \ldots \, \partial x_k} F_{\mathbf{X(t)}}(\mathbf{x}) \tag{D.53}$$

This function is always nonnegative, with a total volume underneath the curve in k-space equal to one.

D.4 STATIONARITY

Consider a set of times $t_1, t_2, \ldots, t_k$ in the interval in which a random process $X(t)$ is defined. A complete characterization of the random process $X(t)$ enables us to specify the joint probability density function $f_{\mathbf{X(t)}}(\mathbf{x})$. The random process $X(t)$ is said to be *strictly stationary* or *stationary in the strict sense* if this joint probability density function is invariant under shifts of the time origin, that is, if the equality

$$f_{\mathbf{X(t)}}(\mathbf{x}) = f_{\mathbf{X(t+}T)}(\mathbf{x}) \qquad (D.54)$$

holds for every finite set of time instants $\{t_j\}, j = 1, 2, \ldots, k$, and for every time shift T and dummy vector $\mathbf{x}$. The components of the random vector $\mathbf{X(t)}$ are obtained by observing the random process $X(t)$ at times $t_1, t_2, \ldots, t_k$. Correspondingly, the components of the random vector $\mathbf{X(t + }T)$ are obtained by observing the random process $X(t)$ at times $t_1 + T, t_2 + T, \ldots, t_k + T$, where T is the time shift.

Stationary processes are of great importance for at least two reasons:

1. They are frequently encountered in practice or approximated to a high degree of accuracy. It is not necessary that a random process be stationary for all time, but only for some observation interval that is long enough for the particular situation.
2. Many of the important properties of commonly encountered stationary processes are described by first and second moments. Consequently, it is relatively easy to develop a simple but useful theory to describe these processes.

Random processes that are not stationary are called *nonstationary*.

EXAMPLE 1

Suppose we have a random process $X(t)$ that is known to be strictly stationary. An implication of stationarity is that the probability that the set of sample functions of this process pass through the windows of Fig. D.4a is equal to the probability that the set of sample functions pass through the corresponding time-shifted windows of Fig. D.4b. Note, however, that it is not necessary that these two sets consist of the same sample functions.

D.5 MEAN, CORRELATION, AND COVARIANCE FUNCTIONS

In many practical situations we find that it is not possible to determine (by means of suitable measurements, say) the probability distribution of a random process. Then we must content ourselves with a *partial description* of the distribution of the process. Ordinarily, the mean, autocorrelation function, and autocovariance function of the random process are taken to give a crude but, nevertheless, useful description of the distribution.

Consider a real-valued random process $X(t)$. We define the *mean* of the process $X(t)$ as

$$m_X(t_k) = E[X(t_k)] \qquad (D.55)$$

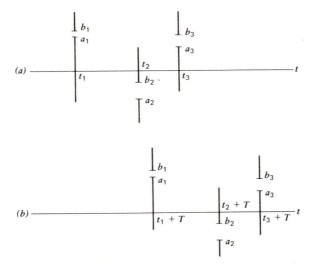

Figure D.4   Illustrating the concept of stationarity.

where E denotes the expectation operator, and $X(t_k)$ is the random variable obtained by observing the random process $X(t)$ at time t_k. Denoting the probability density function of this random variable by $f_{X(t_k)}(x)$, we may rewrite Eq. D.55 as

$$m_X(t_k) = \int_{-\infty}^{\infty} x f_{X(t_k)}(x)dx \qquad (D.56)$$

We define the *autocorrelation function* of the random process $X(t)$ to be a function of two time variables t_k and t_i, as shown by

$$R_X(t_k, t_i) = E[X(t_k)X(t_i)] \qquad (D.57)$$

where $X(t_k)$ and $X(t_i)$ are random variables obtained by observing the random process $X(t)$ at times t_k and t_i, respectively. Denoting the joint probability density function of these two random variables by $f_{X(t_k),X(t_i)}(x, y)$, we may rewrite Eq. D.57 as

$$R_X(t_k, t_i) = \int_{-\infty}^{\infty} \int_{-\infty}^{\infty} xy f_{X(t_k),X(t_i)}(x, y)dx\, dy \qquad (D.58)$$

The *autocovariance function* of the random process $X(t)$ is defined as

$$K_X(t_k, t_i) = E[(X(t_k) - m_X(t_k))(X(t_i) - m_X(t_i))] \qquad (D.59)$$

This may be expanded to yield a useful relation between the mean, autocorrelation and autocovariance functions, namely,

$$K_X(t_k, t_i) = R_X(t_k, t_i) - m_X(t_k)m_X(t_i) \qquad (D.60)$$

If the random process $X(t)$ has zero mean for all time t, we then find that $K_X(t_k, t_i) = R_X(t_k, t_i)$; otherwise, they are unequal.

For a strictly stationary process, all three quantities described above take on simpler forms. In particular, we find that the mean of the random process is a constant m_X (say), so that we may write

$$m_X(t_k) = m_X \qquad \text{for all } t_k \tag{D.61}$$

Also we find that both the autocorrelation and autocovariance functions depend only on the *time difference* $t_k - t_i$, as shown by

$$R_X(t_k, t_i) = R_X(t_k - t_i) \tag{D.62}$$

and

$$K_X(t_k, t_i) = K_X(t_k - t_i) \tag{D.63}$$

The conditions of Eqs. D.61, D.62, and D.63 are *not* sufficient to guarantee that the random process $X(t)$ is strictly stationary. However, a random process $X(t)$ which is not strictly stationary but for which these conditions hold is said to be *stationary in the wide sense*. Thus wide-sense stationarity represents a *weak* kind of stationarity in that all strictly stationary processes are also wide-sense stationary, but the converse is not necessarily true.

For convenience of notation, we define the autocorrelation function of a stationary process $X(t)$ as

$$R_X(\tau) = E[X(t + \tau)X(t)] \tag{D.64}$$

This autocorrelation function has several important properties:

1. The mean-square value of the process may be obtained from $R_X(\tau)$ simply by putting $\tau = 0$ in Eq. D.64 as, shown by

$$R_X(0) = E[X^2(t)] \tag{D.65}$$

2. The autocorrelation function $R_X(\tau)$ is an even function of τ, that is,

$$R_X(\tau) = R_X(-\tau) \tag{D.66}$$

3. The autocorrelation function $R_X(\tau)$ has its maximum magnitude at $\tau = 0$, that is,

$$|R_X(\tau)| \leq R_X(0) \tag{D.67}$$

The physical significance of the autocorrelation function $R_X(\tau)$ is that it provides a means of describing the interdependence of two random variables obtained by observing a random process $X(t)$ at times τ seconds apart. It is therefore apparent that the more rapidly the random process $X(t)$ changes with time, the more rapidly will the autocorrelation function $R_X(\tau)$ decrease from its maximum $R_X(0)$ as τ increases. This decrease may be characterized by a *decorrelation time* τ_0, such that for $\tau > \tau_0$, the magnitude of the autocorrelation function $R_X(\tau)$ remains below some prescribed value. We may thus define the decorrelation time τ_0 of a wide-sense stationary random process $X(t)$ of zero mean as the time taken for the magnitude of the autocorrelation function $R_X(\tau)$ to decrease to and remain below (say) 1 percent of its maximum value $R_X(0)$.

(1)   Cross-Correlation Functions

Consider two random processes $X(t)$ and $Y(t)$ with autocorrelation functions $R_X(t, u)$ and $R_Y(t, u)$, respectively. The two *cross-correlation functions* of $X(t)$ and $Y(t)$ may be defined by

$$R_{XY}(t, u) = E[X(t)Y(u)] \tag{D.68}$$

and

$$R_{YX}(t, u) = E[Y(t)X(u)] \tag{D.69}$$

where t and u denote values of time at which the processes are observed.

The cross-correlation function is not generally an even function of τ as was true for the autocorrelation function, nor does it have a maximum at the origin. However, it does obey a certain symmetry relationship as follows:

$$R_{XY}(\tau) = R_{YX}(-\tau) \tag{D.70}$$

D.6   WIENER–KHINTCHINE RELATIONS

The autocorrelation function $R_X(\tau)$ provides one parameter for describing the second-order statistics of a stationary process. Another important parameter is the power spectral density $S_X(f)$, defined as the Fourier transform of the auto-correlation function $R_X(\tau)$. We may thus define the power spectral density in terms of autocorrelation function as follows:

$$S_X(f) = \int_{-\infty}^{\infty} R_X(\tau)\exp(-j2\pi f\tau)d\tau \tag{D.71}$$

Conversely, we may define the autocorrelation function as the inverse Fourier transform of the power spectral density, as shown by

$$R_X(\tau) = \int_{-\infty}^{\infty} S_X(f)\exp(j2\pi f\tau)df \tag{D.72}$$

Equations D.71 and D.72 are basic relations in the theory of spectral analysis of random processes, and together they constitute what arc usually called the *Wiener–Khintchine relations*.

The Wiener–Khintchine relations show that if either the autocorrelation function or power spectral density of a random process is known, the other can be found exactly. However, these functions display different aspects of the correlation information about the process. It is commonly accepted that for practical purposes, however, the power spectral density is the more useful description.

Using the pair of relations D.71 and D.72, we may readily establish the following properties of the power spectral density of a wide-sense stationary process:

PROPERTY 1

The zero-frequency value of the power spectral density of a wide-sense stationary random process equals the total area under the graph of the autocorrelation function;

that is,

$$S_X(0) = \int_{-\infty}^{\infty} R_X(\tau)d\tau \tag{D.73}$$

PROPERTY 2
The mean-square value of a wide-sense stationary random process equals the total area under the graph of the power spectral density; that is,

$$E[X^2(t)] = \int_{-\infty}^{\infty} S_X(f)df \tag{D.74}$$

PROPERTY 3
The power spectral density of a wide-sense stationary random process is always non-negative; that is,

$$S_X(f) \geqslant 0 \qquad \text{for all } f \tag{D.75}$$

PROPERTY 4
The power spectral density of a real-valued random process is an even function of frequency; that is,

$$S_X(-f) = S_X(f) \tag{D.76}$$

We now consider two examples to illustrate application of the Wiener–Khintchine relations.

EXAMPLE 2   WHITE NOISE
A white noise process $W(t)$ is characterized by a *constant* power spectral density, as shown by

$$S_W(f) = \frac{N_0}{2} \tag{D.77}$$

where the factor $1/2$ has been included to indicate that half the power is associated with positive frequencies and half with negative frequencies, as illustrated in Fig. D.5a. The parameter N_0 is measured in watts per hertz. The autocorrelation function $R_W(\tau)$ is the inverse Fourier transform of $S_W(f)$; hence, we may write

$$R_W(\tau) = \frac{N_0}{2} \delta(\tau) \tag{D.78}$$

That is, the autocorrelation function of white noise consists of a delta function weighted by the factor $N_0/2$ and occurring at $\tau = 0$, as in Fig. D.5b. We note that $R_W(\tau) = 0$ for $\tau \neq 0$. Accordingly, any two different samples of white noise are uncorrelated.

EXAMPLE 3   RANDOM BINARY WAVE
Figure D.6 shows the sample function $x(t)$ of a random process $X(t)$ that consists of a random sequence of binary symbols 1 and 0. It is assumed that:

1. The symbols 1 and 0 are represented by rectangular pulses of amplitude $+a$ and $-a$ volts, respectively, and of duration T seconds.
2. The sample functions are not synchronized, so that the starting time of the first

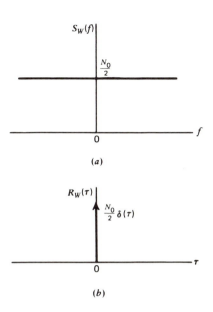

Figure D.5 Characteristics of white noise. (a) Power spectral density. (b) Autocorrelation function.

pulse, t_d, is equally likely to lie anywhere between zero and T seconds. That is, t_d is the sample value of a uniformly distributed random variable T_d, with its probability density function defined by

$$f_{T_d}(t_d) = \begin{cases} \dfrac{1}{T} & 0 \le t_d \le T \\ 0 & \text{elsewhere} \end{cases} \tag{D.79}$$

3. During any time interval $(n-1)T < t - t_d < nT$, where n is an integer, the presence of a 1 or a 0 is determined by tossing a fair coin; specifically, if the outcome is "heads," we have a 1 and if the outcome is "tails," we have a 0. These two symbols are thus equally likely, and the presence of a 1 or 0 in any one interval is independent of all other intervals.

For obvious reasons, the sample function $x(t)$ is referred to as a *random binary wave*.

Since the amplitude levels $-a$ and $+a$ occur with equal probability, it follows immediately that the mean of the random binary wave $X(t)$ is zero; that is, $E[X(t)] = 0$ for all t.

To find the autocorrelation function $R_X(t_k, t_i)$, we have to evaluate $E[X(t_k)X(t_i)]$,

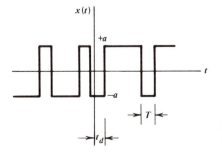

Figure D.6 Segment of a random binary wave.

where $X(t_k)$ and $X(t_i)$ are random variables obtained by observing the random binary wave $X(t)$ at times t_k and t_i, respectively.

Consider first the case when $|t_k - t_i| > T$. Then the random variables $X(t_k)$ and $X(t_i)$ occur in different pulse intervals and are therefore independent. We thus have

$$E[X(t_k)X(t_i)] = E[X(t_k)]E[X(t_i)] = 0 \qquad |t_k - t_i| > T$$

Consider next the case where $|t_k - t_i| < T$, with $t_k = 0$ and $t_i < t_k$. In such a situation we observe from Fig. D.6, that the random variables $X(t_k)$ and $X(t_i)$ occur in the same pulse interval if and only if the delay t_d satisfies the condition $t_d < T - |t_k - t_i|$. We thus obtain the *conditional expectation*:

$$E[X(t_k)X(t_i)|t_d] = \begin{cases} a^2 & t_d < T - |t_k - t_i| \\ 0 & \text{elsewhere} \end{cases}$$

Averaging this result over all possible values of t_d, we get

$$E[X(t_k)X(t_i)] = \int_0^{T-|t_k-t_i|} a^2 f_{T_d}(t_d)dt_d$$

$$= \int_0^{T-|t_k-t_i|} \frac{a^2}{T} dt_d$$

$$= a^2 \left(1 - \frac{|t_k - t_i|}{T}\right) \qquad |t_k - t_i| < T$$

By similar reasoning for any other value of t_k, we conclude that the autocorrelation function of a random binary wave, represented by the sample function shown in Fig. D.6, is only a function of the time difference $\tau = t_k - t_i$, as shown by

$$R_X(\tau) = \begin{cases} a^2 \left(1 - \frac{|\tau|}{T}\right) & |\tau| < T \\ 0 & |\tau| \geq T \end{cases} \tag{D.80}$$

This result is shown plotted in Fig. D.7a. The power spectral density of the process is therefore

$$S_X(f) = \int_{-T}^{T} a^2 \left(1 - \frac{|\tau|}{T}\right) \exp(-j2\pi f\tau)d\tau$$

Using the Fourier transform of a triangular function, we obtain

$$S_X(f) = a^2 T \operatorname{sinc}^2(fT) \tag{D.81}$$

which is shown plotted in Fig. D.7b.

The result of Eq. D.81 may be generalized as follows. We note that the energy spectral density of a rectangular pulse $g(t)$ of amplitude a and duration T is given by

$$\Psi_g(f) = a^2 T^2 \operatorname{sinc}^2(fT)$$

We may therefore rewrite Eq. D.81 in terms of $\Psi_g(f)$ as follows

$$S_X(f) = \frac{\Psi_g(f)}{T} \tag{D.82}$$

Equation D.82 states that, for a random binary wave in which binary symbols 1 and 0 are represented by pulses $g(t)$ and $-g(t)$, respectively, the power spectral density $S_X(f)$ is equal to the energy spectral density $\Psi_g(f)$ of the *symbol shaping pulse* $g(t)$ divided by the *symbol duration* T.

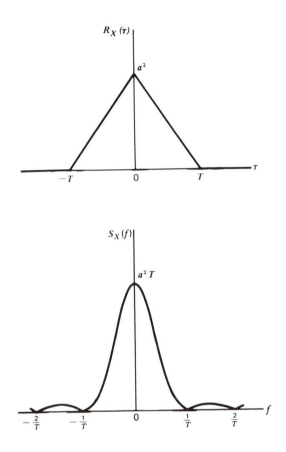

Figure D.7 Characteristics of a random binary wave. (*a*) Autocorrelation function. (*b*) Power spectral density.

D.7 GAUSSIAN PROCESS

Let us suppose that we observe a random process $X(t)$ for an interval that starts at time $t = 0$ and lasts until $t = T$. Suppose also that we weight the random process $X(t)$ by some function $g(t)$ and then integrate the product $g(t)X(t)$ over this observation interval, thereby obtaining a random variable Y defined by

$$Y = \int_0^T g(t)X(t)dt \tag{D.83}$$

We refer to Y as a *linear functional* of $X(t)$. The distinction between a function and a functional should be carefully noted. For example, Y defined as the sum $\sum_{i=1}^N a_i X_i$, where the a_i are constants and the X_i are random variables, is a linear *function* of the X_i; for each observed set of values for the random variables X_i, we have a corresponding value for the random variable Y. On the other hand, in Eq. D.83 the value of the random variable Y depends on the course of the *argument function* $g(t)X(t)$ over the observation interval 0 to T. Thus a functional is a quantity that depends on the entire course of one or more functions rather than on a number of discrete variables. In other words, the domain of a functional is a set or space of admissible functions rather than a region of a coordinate space.

If in Eq. D.83 the weighting function $g(t)$ is such that the mean-square value of the random variable Y is finite, and if the random variable Y is a *Gaussian-distributed* random variable for every $g(t)$ in this class of functions, then the process $X(t)$ is said to be a *Gaussian process*. In other words, the process $X(t)$ is a Gaussian process if every linear functional of $X(t)$ is a Gaussian random variable.

We say that the random variable Y has a Gaussian distribution if its probability density function has the form

$$f_Y(y) = \frac{1}{\sqrt{2\pi}\,\sigma_Y} \exp\left[-\frac{(y - m_Y)^2}{2\sigma_Y^2}\right] \tag{D.84}$$

where m_Y is the mean and σ_Y^2 is the variance of the random variable Y. We thus see that a Gaussian-distributed random variable is completely characterized by specifying its mean and variance. A plot of this probability density function is given in Fig. D.8 for the case when the mean m_Y is zero and the variance σ_Y^2 equals one.

A Gaussian process has two main virtues. First, the Gaussian process has many properties that make analytic results possible. Second, the random processes produced by physical phenomena are often such that a Gaussian model is appropriate. The *central limit theorem* provides the mathematical justification for using a Gaussian process as a model for a large number of different physical phenomena in which the observed random variable, at a particular instant of time, is the result of a large number of individual random events. Furthermore, the use of a Gaussian model to describe such physical phenomena is usually confirmed by experiments. Thus the widespread occurrence of physical phenomena for which a Gaussian model is appropriate, together with the ease with which a Gaussian process is handled mathematically, make the Gaussian process very important in the study of communication systems.

Some of the important properties of a Gaussian process are as follows:

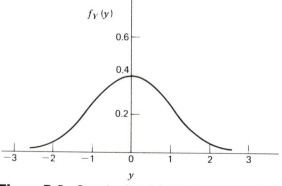

Figure D.8   Standardized Gaussian distribution.

PROPERTY 1

If a Gaussian process $X(t)$ is applied to a stable linear filter, then the random process $Y(t)$ developed at the output of the filter is also Gaussian.

PROPERTY 2

Consider the set of random variables or samples $X(t_1)$, $X(t_2)$, . . . , $X(t_n)$, obtained by observing a random process $X(t)$ at times t_1, t_2, . . . , t_n. If the process $X(t)$ is Gaussian, then this set of random variables are jointly Gaussian for any n, with their n-fold joint probability density function being completely determined by specifying the set of means

$$m_{X(t_i)} = E[X(t_i)] \qquad i = 1, 2, . . . , n$$

and the set of autocovariance functions

$$K_X(t_k, t_i) = E[(X(t_k) - m_{X(t_k)})(X(t_i) - m_{X(t_i)})] \qquad k, i = 1, 2, . . . , n$$

Property 2 is frequently used as the definition of a Gaussian process. However, this definition is more difficult to use than that based on Eq. D.83 for evaluating the effects of filtering on a Gaussian process.

PROPERTY 3

If a Gaussian process is wide-sense stationary, then the process is also stationary in the strict sense.

PROPERTY 4

If the set of random variables $X(t_1)$, $X(t_2)$, . . . , $X(t_n)$, obtained by sampling a Gaussian process $X(t)$ at times t_1, t_2, . . . , t_n, are uncorrelated, that is,

$$E[(X(t_k) - m_{X(t_k)})(X(t_i) - m_{X(t_i)})] = 0 \qquad i \neq k$$

then this set of random variables are statistically independent.

The implication of this property is that the joint probability density function of the set of random variables $X(t_1)$, $X(t_2)$, . . . , $X(t_n)$ can be expressed as the product of the probability density functions of the individual random variables in the set.

D.8   NARROW-BAND NOISE

Consider a *narrow-band noise* $n(t)$ representing the sample function of a Gaussian process $N(t)$ of zero mean. It is assumed that the power spectrum of the process $N(t)$ occupies a frequency band of width $2W$ centered around f_c, with the *half-bandwidth* $W < f_c$. The narrow-band noise $n(t)$ may be represented in the canonical form

$$n(t) = n_I(t) \cos(2\pi f_c t) - n_Q(t) \sin(2\pi f_c t) \qquad (D.85)$$

where $n_I(t)$ is the *in-phase noise component*, and $n_Q(t)$ is the *quadrature noise component*.

When the narrow-band noise $n(t)$ is drawn from a Gaussian wide-sense stationary process $N(t)$ with zero mean and variance σ_N^2, then the in-phase noise component $n_I(t)$ and quadrature noise component $n_Q(t)$ represent sample functions of independent wide-sense stationary Gaussian processes, each of which also has zero mean and variance σ_N^2.

D.9 RAYLEIGH DISTRIBUTION

The narrow-band noise $n(t)$ may also be represented in terms of its envelope and phase components as follows

$$n(t) = r(t)\cos[2\pi f_c t + \psi(t)] \tag{D.86}$$

where

$$r(t) = [n_I^2(t) + n_Q^2(t)]^{1/2} \tag{D.87}$$

and

$$\psi(t) = \tan^{-1}\left(\frac{n_Q(t)}{n_I(t)}\right) \tag{D.88}$$

The function $r(t)$ is called the *envelope* of $n(t)$, and the function $\psi(t)$ is called the *phase* of $n(t)$.

The sample functions $n_I(t)$ and $n_Q(t)$ refer to random processes $N_I(t)$ and $N_Q(t)$, respectively. Let X and Y denote random variables obtained by observing the random processes $N_I(t)$ and $N_Q(t)$ at some fixed time, respectively. These two random variables are represented by sample values x and y. We note that X and Y are statistically independent and Gaussian distributed with zero mean and variance σ_N^2. We may thus express their joint probability density function as

$$f_{X,Y}(x,y) = \frac{1}{2\pi\sigma_N^2} \exp\left(-\frac{x^2 + y^2}{2\sigma_N^2}\right) \tag{D.89}$$

Define the bivariate transformation

$$x = r\cos\psi \tag{D.90}$$

and

$$y = r\sin\psi \tag{D.91}$$

The Jacobian of the transformation is defined by (see Section D.2)

$$J(r,\psi) = \begin{vmatrix} \dfrac{\partial x}{\partial r} & \dfrac{\partial y}{\partial r} \\[2mm] \dfrac{\partial x}{\partial \psi} & \dfrac{\partial y}{\partial \psi} \end{vmatrix}$$

$$= \begin{vmatrix} \cos\psi & \sin\psi \\ -r\sin\psi & r\cos\psi \end{vmatrix}$$

$$= r\cos^2\psi + r\sin^2\psi$$

$$= r \tag{D.92}$$

Let R and Ψ denote the random variables resulting from the transformation of X and Y. The joint probability density function of R and Ψ is given by

$$f_{R,\Psi}(r,\psi) = f_{X,Y}(x,y)\, J(r,\psi)$$

$$= \frac{r}{2\pi\sigma_N^2} \exp\left(-\frac{r^2}{2\sigma_N^2}\right) \tag{D.93}$$

where we have made use of Eqs. D.89–D.91. This probability density function is not dependent on the angle ψ, which means that the random variables R and Ψ are statistically independent. We may thus express $f_{R,\Psi}(r, \psi)$ as the product of $f_R(r)$ and $f_\Psi(\psi)$. In particular, the random variable Ψ is uniformly distributed inside the range 0 to 2π, as shown by

$$f_\Psi(\psi) = \begin{cases} \dfrac{1}{2\pi} & 0 \leqslant \psi \leqslant 2\pi \\ 0 & \text{elsewhere} \end{cases} \tag{D.94}$$

This leaves the probability density function of R as

$$f_R(r) = \begin{cases} \dfrac{r}{\sigma_N^2} \exp\left(-\dfrac{r^2}{2\sigma_N^2}\right) & r \geqslant 0 \\ 0 & \text{elsewhere} \end{cases} \tag{D.95}$$

where σ_N^2 is the variance of the original narrow-band noise process $N(t)$. A random variable having the probability density function of Eq. D.95 is said to be *Rayleigh distributed*.

D.10   RICIAN DISTRIBUTION

Suppose next that we add the sinusoidal wave $a\cos(2\pi f_c t)$ to the narrow-band noise $n(t)$. It is assumed that a and f_c are both constants, and that the frequency f_c of the sinusoidal wave is the same as the nominal midband frequency for the noise. A sample function of the sinusoidal wave-plus-noise process is then expressed by

$$x(t) = a\cos(2\pi f_c t) + n(t) \tag{D.96}$$

Representing the narrow-band noise $n(t)$ in terms of its in-phase and quadrature components, we may write

$$x(t) = n_I'(t)\cos(2\pi f_c t) - n_Q(t)\sin(2\pi f_c t) \tag{D.97}$$

where

$$n_I'(t) = a + n_I(t) \tag{D.98}$$

We assume that $n(t)$ is drawn from a Gaussian process with zero mean and variance σ_N^2. Accordingly, we may state that:

1. The random processes $N_I'(t)$ and $N_Q(t)$, represented by the sample functions $n_I'(t)$ and $n_Q(t)$, are both Gaussian and statistically independent.
2. The mean of $N_I'(t)$ is a and that of $N_Q(t)$ is zero.
3. The variance of both $N_I'(t)$ and $N_Q(t)$ is σ_N^2.

Let $r(t)$ denote the envelope of $x(t)$ and $\psi(t)$ denote its phase. From Eq. D.98, we thus find that

$$r(t) = \{[n_I'(t)]^2 + n_Q^2(t)\}^{1/2} \tag{D.99}$$

and

$$\psi(t) = \tan^{-1}\left(\frac{n_Q(t)}{n_I'(t)}\right) \tag{D.100}$$

Following a procedure similar to that described in Section D.9 for the derivation of the Rayleigh distribution, we find that the joint probability density function of the random variables R and Ψ, obtained by sampling the envelope process $R(t)$ and phase process $\Psi(t)$ at some fixed time t, is given by

$$f_{R,\Psi}(r, \psi) = \frac{r}{2\pi\sigma_N^2} \exp\left(-\frac{r^2 + a^2 - 2ar\cos\psi}{2\sigma_N^2}\right) \tag{D.101}$$

We see that in this case, however, we cannot express the joint probability density function $f_{R,\Psi}(r, \psi)$ as a product $f_R(r)f_\Psi(\psi)$. This is because we now have a term involving the values of both random variables multiplied together, namely, $r\cos\psi$. Hence, R and Ψ are dependent random variables for nonzero values of the amplitude a of the sinusoidal wave component.

We are interested, in particular, in the probability density function of R. To determine this probability density function, we integrate Eq. D.101 over all possible values of ψ, obtaining the marginal density

$$f_R(r) = \int_0^{2\pi} f_{R,\Psi}(r, \psi)\, d\psi$$

$$= \frac{r}{2\pi\sigma_N^2} \exp\left(-\frac{r^2 + a^2}{2\sigma_N^2}\right) \int_0^{2\pi} \exp\left(\frac{ar}{\sigma_N^2}\cos\psi\right) d\psi \tag{D.102}$$

The integral in the right-hand side of Eq. D.102 can be identified in terms of the defining integral for the *modified Bessel function of the first-kind of zero order;* that is

$$I_0(x) = \frac{1}{2\pi} \int_0^{2\pi} \exp(x\cos\psi)\,d\psi \tag{D.103}$$

Thus, letting $x = ar/\sigma_N^2$, we may rewrite Eq. D.103 in the compact form:

$$f_R(r) = \frac{r}{\sigma_N^2} \exp\left(-\frac{r^2 + a^2}{2\sigma_N^2}\right) I_0\left(\frac{ar}{\sigma_N^2}\right) \tag{D.104}$$

A random variable R having the probability density function of Eq. D.104 is said to be *Rician distributed.*

The probability density function of the phase Ψ is given by

$$f_\Psi(\psi) = \int_0^\infty f_{R,\Psi}(r, \psi)\,dr$$

$$= \frac{1}{2\pi} \exp(-\gamma) + \sqrt{\frac{\gamma}{\pi}} \cos\psi \exp(-\gamma \sin^2\psi) \left[1 - \frac{1}{2} \operatorname{erfc}(\gamma \cos\psi)\right] \tag{D.105}$$

where

$$\gamma = \frac{a^2}{2\sigma_N^2} \tag{D.106}$$

The Rayleigh distribution is a special case of the Rician distribution. This follows from a property of the modified Bessel function, namely

$$I_0(0) = 1.0$$

Hence, putting the amplitude $a = 0$, which is equivalent to having the narrow-band noise $n(t)$ by itself, we find that Eq. D.104 reduces to the same form shown in Eq. D.95.

Also, for $a = \gamma = 0$, we find that Eq. D.105 reduces to the uniform distribution given in Eq. D.94.

APPENDIX E

ERROR FUNCTION

The *error function*, denoted by erf(u), is defined by

$$\text{erf}(u) = \frac{2}{\sqrt{\pi}} \int_0^u \exp(-z^2)dz \qquad (E.1)$$

The error function has the following properties:

1. $\text{erf}(-u) = -\text{erf}(u)$ (E.2)
   This is known as the *symmetry relation*.
2. As u approaches infinity, erf(u) approaches unity; that is

$$\frac{2}{\sqrt{\pi}} \int_0^\infty \exp(-z^2)dz = 1 \qquad (E.3)$$

3. Given that a random variable X is Gaussian distributed with mean m_X and variance σ_X^2, the probability that X lies in the interval $(m_X - a, m_X + a)$ is defined by

$$P(m_X - a < X \leq m_X + a) = \text{erf}\left(\frac{a}{\sqrt{2}\sigma_X}\right) \qquad (E.4)$$

The *complementary error function* is defined by

$$\text{erfc}(u) = \frac{2}{\sqrt{\pi}} \int_u^\infty \exp(-z^2)dz \qquad (E.5)$$

It is related to the error function as follows:

$$\text{erfc}(u) = 1 - \text{erf}(u) \qquad (E.6)$$

Table E.1 gives values of the error function erf(u) for u in the range 0 to 3.3.*

---

* The error function is tabulated extensively in Abramowitz and Stegun, (1965, pp. 297–316).

Table E.1   Error Function

u	erf(u)	u	erf(u)
0.00	0.00000	1.10	0.88021
0.05	0.05637	1.15	0.89612
0.10	0.11246	1.20	0.91031
0.15	0.16800	1.25	0.92290
0.20	0.22270	1.30	0.93401
0.25	0.27633	1.35	0.94376
0.30	0.32863	1.40	0.95229
0.35	0.37938	1.45	0.95970
0.40	0.42839	1.50	0.96611
0.45	0.47548	1.55	0.97162
0.50	0.52050	1.60	0.97635
0.55	0.56332	1.65	0.98038
0.60	0.60386	1.70	0.98379
0.65	0.64203	1.75	0.98667
0.70	0.67780	1.80	0.98909
0.75	0.71116	1.85	0.99111
0.80	0.74210	1.90	0.99279
0.85	0.77067	1.95	0.99418
0.90	0.79691	2.00	0.99532
0.95	0.82089	2.50	0.99959
1.00	0.84270	3.00	0.99998
1.05	0.86244	3.30	0.999998

E.1   BOUNDS ON THE COMPLEMENTARY ERROR FUNCTION

Substituting $u - x$ for z in Eq. E.5, we get

$$\text{erfc}(u) = \frac{2}{\sqrt{\pi}} \exp(-u^2) \int_{-\infty}^{0} \exp(2ux)\exp(-x^2)dx$$

For any real x, the value of $\exp(-x^2)$ lies between the successive partial sums of the power series

$$1 - \frac{x^2}{1!} + \frac{(x^2)^2}{2!} - \frac{(x^2)^3}{3!} + \cdots$$

Therefore, for $u > 0$, we find, on using $(n + 1)$ terms of this series, that $\text{erfc}(u)$ lies between the values taken by

$$\frac{2}{\sqrt{\pi}} \exp(-u^2) \int_{-\infty}^{0} \left(1 - x^2 + \frac{x^4}{2} - \cdots \pm \frac{x^{2n}}{n!}\right) \exp(2ux)dx$$

with the sign of the last term in the integrand depending on whether n is odd or even. Putting $2ux = -v$, and using the integral

$$\int_{0}^{\infty} v^n \exp(-v)dv = n!$$

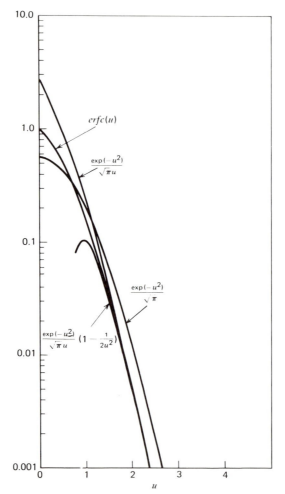

Figure E.1 The complementary error function and its bounds.

we obtain the following *asymptotic expansion* for erfc(u), assuming $u > 0$,

$$\text{erfc}(u) \simeq \frac{\exp(-u^2)}{\sqrt{\pi}u}\left[1 - \frac{1}{2u^2} + \frac{1 \cdot 3}{2^2 u^4} - \cdots \pm \frac{1 \cdot 3 \cdot 5 \cdots (2n - 1)}{2^n u^{2n}}\right] \quad (E.7)$$

For large positive values of u, the successive terms of the series on the right-hand side of Eq. E.7 decrease very rapidly. We thus deduce two simple bounds on erfc(u), as shown by*

$$\frac{\exp(-u^2)}{\sqrt{\pi}u}\left(1 - \frac{1}{2u^2}\right) < \text{erfc}(u) < \frac{\exp(-u^2)}{\sqrt{\pi}u} \quad (E.8)$$

For large positive u, a second bound on the complementary error function

* See Blachman (1966, pp. 5–6).

erfc(u) is obtained by omitting the multiplying factor $1/u$ in the upper bound of Eq. E.8:

$$\text{erfc}(u) < \frac{\exp(-u^2)}{\sqrt{\pi}} \tag{E.9}$$

In Fig. E.1 we have plotted erfc(u), the two bounds defined by Eq. E.8 and the upper bound of Eq. E.9. We see that for $u \geq 1.5$ the bounds on erfc(u), defined by Eq. E.8, become increasingly tight.

E.2 THE Q-FUNCTION

Consider a *standardized* Gaussian random variable X of zero mean and unit variance. The probability that an observed value of the random variable X will be greater than v is given by the *Q-function:*

$$Q(v) = \frac{1}{\sqrt{2\pi}} \int_v^\infty \exp\left(-\frac{x^2}{2}\right) \tag{E.10}$$

The Q-function defines the area under the standardized Gaussian tail. Inspection of Eqs. E.5 and E.10 reveals that the Q-function is related to the complementary error function as follows:

$$Q(v) = \frac{1}{2} \text{erfc}\left(\frac{v}{\sqrt{2}}\right) \tag{E.11}$$

Conversely, putting $u = v/\sqrt{2}$, we have

$$\text{erfc}(u) = 2Q(\sqrt{2}u) \tag{E.12}$$

APPENDIX F

KRAFT–McMILLAN INEQUALITY

The *Kraft–McMillan inequality** states that we can construct a *uniquely decodable* code with word-lengths $l_0, l_1, \ldots, l_{K-1}$ if and only if these lengths satisfy the condition

$$\sum_{k=0}^{K-1} r^{-l_k} \leq 1 \tag{F.1}$$

where r is the radix (number of symbols) of the alphabet of encoded symbols.

To prove the inequality F.1,† consider the quantity

$$\left(\sum_{k=0}^{K-1} r^{-l_k}\right)^n = (r^{-l_0} + r^{-l_1} + \cdots + r^{-l_{K-1}})^n \tag{F.2}$$

where n is a positive integer. When the right side of Eq. F.2 is expanded, we find that we have a sum of K^n terms, each of which has the form

$$r^{-l_{k_1} - l_{k_2} - \cdots - l_{k_n}} = r^{-i} \tag{F.3}$$

Here we have used the definition

$$l_{k_1} + l_{k_2} + \cdots + l_{k_n} = i \tag{F.4}$$

Let the smallest of the code word-lengths l_k be unity and the largest be l. Then, the integer i may take on a value extending from n to nl. Let N_i denote the number of terms of the form r^{-i} in Eq. F.2. We may then rewrite this equation as

$$\left(\sum_{k=0}^{K-1} r^{-l_k}\right)^n = \sum_{i=n}^{nl} N_i r^{-i} \tag{F.5}$$

We observe that N_i is also the number of code symbols of length i. Hence, if the

---

* The *Kraft inequality* (originally introduced in 1949) applies to *prefix codes*, which are a special case of *uniquely decodable codes* (Kraft, 1949). The fact that the same inequality is necessary for uniquely decodable codes was subsequently proved by McMillan (1956).

† The proof presented here follows Abramson (1963).

572

code is uniquely decodable, N_i cannot be greater than r^i, which is the number of distinct sequences of length i in an alphabet of radix r. We may thus write

$$\left(\sum_{k=0}^{K-1} r^{-l_k} \right)^n \leq \sum_{i=n}^{nl} r^i r^{-i} = nl - n + 1$$

Accordingly, we have

$$\left(\sum_{k=0}^{K-1} r^{-l_k} \right)^n \leq nl \qquad \text{for all } n \tag{F.6}$$

Taking the nth roots of both sides of this last inequality, we get

$$\sum_{k=0}^{K-1} r^{-l_k} \leq (nl)^{1/n} \qquad \text{for all } n \tag{F.7}$$

The inequality of F.6 provides us the proof we need. In particular, since we may choose n as large as we please, we may go to the limit and set $n \to \infty$. Then, recognizing that

$$\lim_{n \to \infty} (nl)^{1/n} = 1$$

we get the inequality given in F.1.

APPENDIX G

SCHWARZ'S INEQUALITY

Let $x(t)$ and $y(t)$ denote any pair of complex-valued signals. We assume that both $x(t)$ and $y(t)$ have finite energy, as shown by

$$\int_{-\infty}^{\infty} |x(t)|^2 \, dt < \infty$$

and

$$\int_{-\infty}^{\infty} |y(t)|^2 \, dt < \infty$$

Schwarz's inequality states that

$$\left| \int_{-\infty}^{\infty} x(t) y(t) dt \right|^2 \leq \int_{-\infty}^{\infty} |x(t)|^2 dt \int_{-\infty}^{\infty} |y(t)|^2 dt \qquad (G.1)$$

The equality holds if and only if

$$y(t) = kx^*(t) \qquad (G.2)$$

where k is some constant, and the asterisk denotes complex conjugation.

To justify the inequality, we first observe that the magnitude of the total area under a function is less than or equal to the total area under the magnitude of the function itself. That is,

$$\left| \int_{-\infty}^{\infty} x(t) y(t) dt \right| \leq \int_{-\infty}^{\infty} |x(t)| |y(t)| dt \qquad (G.3)$$

The equality holds only if $x(t)$ and $y(t)$ satisfy the condition of Eq. G.2, that is, the product $x(t)y(t)$ is real-valued.

For convenience of notation, let $|x(t)|$ and $|y(t)|$ be denoted by the real-valued functions $a(t)$ and $b(t)$, respectively, as shown by

$$a(t) = |x(t)| \qquad (G.4)$$

and

$$b(t) = |y(t)| \qquad (G.5)$$

Let $a(t)$ and $b(t)$ be defined in terms of a pair of orthonormal functions $\phi_1(t)$ and $\phi_2(t)$ as follows:

574

$$a(t) = a_1\phi_1(t) + a_2\phi_2(t) \qquad (G.6)$$

and

$$b(t) = b_1\phi_1(t) + b_2\phi_2(t) \qquad (G.7)$$

where

$$a_i = \int_{-\infty}^{\infty} a(t)\phi_i(t)dt \qquad i = 1, 2 \qquad (G.8)$$

and

$$b_i = \int_{-\infty}^{\infty} b(t)\phi_i(t)dt \qquad i = 1, 2 \qquad (G.9)$$

The orthonormal functions $\phi_1(t)$ and $\phi_2(t)$ are themselves related as follows:

$$\int_{-\infty}^{\infty} \phi_i(t)\phi_j(t)dt = \begin{cases} 1 & i = j \\ 0 & i \neq j \end{cases} \qquad (G.10)$$

The two time functions $a(t)$ and $b(t)$ may thus be represented by the vectors $\mathbf{a} = (a_1, a_2)$ and $\mathbf{b} = (b_1, b_2)$, respectively. The *cosine of the angle between the vectors $\mathbf{a}$ and $\mathbf{b}$* is defined by

$$\cos(\angle(\mathbf{a},\mathbf{b})) = \frac{(\mathbf{a},\mathbf{b})}{\|\mathbf{a}\| \, \|\mathbf{b}\|} \qquad (G.11)$$

where $(\mathbf{a},\mathbf{b})$ is the *inner product of the vectors* $\mathbf{a}$ and $\mathbf{b}$, and $\|\mathbf{a}\|$ and $\|\mathbf{b}\|$ are their *norms*. Since the cosine of an angle has a magnitude less than or equal to unity, it follows that

$$|(\mathbf{a},\mathbf{b})| \leq \|\mathbf{a}\| \, \|\mathbf{b}\| \qquad (G.12)$$

The equality holds if and only if

$$\mathbf{b} = k\mathbf{a} \qquad (G.13)$$

where k is some constant. Equation G.12 is a mathematical statement of Schwarz's inequality in vector form.

Using the definitions given in Eqs. G.6 to G.10, we may write

$$(\mathbf{a},\mathbf{b}) = \int_{-\infty}^{\infty} a(t)b(t)dt \qquad (G.14)$$

$$\|\mathbf{a}\| = \left[\int_{-\infty}^{\infty} a^2(t)dt \right]^{1/2} \qquad (G.15)$$

$$\|\mathbf{b}\| = \left[\int_{-\infty}^{\infty} b^2(t)dt \right]^{1/2} \qquad (G.16)$$

Accordingly, we may rewrite the inequality of G.12 as

$$\left[\int_{-\infty}^{\infty} a(t)b(t)dt \right]^2 \leq \int_{-\infty}^{\infty} a^2(t)dt \int_{-\infty}^{\infty} b^2(t)dt \qquad (G.17)$$

The equality holds if

$$b(t) = ka(t) \tag{G.18}$$

where k is some constant. Equation G.17 is a statement of Schwarz's inequality for real-valued functions. The more general version of Schwarz's inequality for complex-valued functions follows from the use of the inequality of G.3 and the definitions of Eqs. G.4 and G.5.

APPENDIX H

BINARY ARITHMETIC

In binary input–binary output coded systems, the alphabet consists only of symbols 0 and 1. A *field* is described simply as a set of elements in which we can perform addition, subtraction, multiplication, and division without leaving the set. The field in which we are interested is a *binary field* that contains only two elements: 0 and 1. In this section, we discuss arithmetic over the binary field.

In binary arithmetic we use *modulo-2* addition and multiplication. The rules for modulo-2 addition are as follows:

$$0 + 0 = 0$$

$$1 + 0 = 1$$

$$0 + 1 = 1$$

$$1 + 1 = 0$$

where the plus sign is intended to mean modulo-2 addition. There are no "carries" in modulo-2 addition. Hence, the sum of an odd number of 1s equals 1, and the sum of an even number of 1s equals 0. Note that since $1 + 1 = 0$, we may write $1 = -1$. Accordingly, in binary arithmetic, subtraction is the same as addition.

For modulo-2 multiplication, we have the following rules:

$$0 \times 0 = 0$$

$$1 \times 0 = 0$$

$$0 \times 1 = 0$$

$$1 \times 1 = 1$$

Division is trivial in that $1 \div 1 = 1$, $0 \div 1 = 0$, and division by 0 is not permitted.

It is of interest to note that modulo-2 addition is simply the EXCLUSIVE-OR operation in logic. Correspondingly, modulo-2 multiplication is the AND operation in logic.

ABBREVIATIONS

ac:	alternating current
ADM:	adaptive delta modulation
ADPCM:	adaptive differential pulse-code modulation
APB:	adaptive prediction with backward estimation
APF:	adaptive prediction with forward estimation
APK:	amplitude-phase keying
AQB:	adaptive quantization with backward estimation
AQF:	adaptive quantization with forward estimation
ARQ:	automatic request for retransmission
ASBC:	amplitude sub-band coding
ASK:	amplitude-shift keying
AWGN:	additive white Gaussian noise
BCH:	Bose–Chaudhuri–Hocquqenqhem
BER:	bit error rate
BPF:	band-pass filter
b/s:	bit per second
BSC:	binary symmetric channel
CCD:	charge-coupled device
CCITT:	Comité Consultatif International Téléphonique et Télégraphique
CPFSK:	continuous-phase frequency-shift keying
CRC:	cyclic redundancy check
CSMA:	carrier-sense multiple access
CSMA/CD:	carrier-sense multiple access with collision detection
CW:	continuous wave
dB:	decibel
dc:	direct current
DFT:	discrete Fourier transform
DM:	delta modulation
DMC:	discrete memoryless channel
DPCM:	differential pulse-code modulation
DPSK:	differential phase-shift keying
DS/BPSK:	direct sequence spread binary phase-shift keying
DS/QPSK:	direct sequence spread quadriphase-shift keying
exp:	exponential
FDM:	frequency division multiplexing
FDMA:	frequency division multiple-access
FEC:	forward error correction
FFT:	fast Fourier transform

FH:	frequency hop
FH/MFSK:	frequency hop spread M-ary frequency-shift keying
FSK:	frequency-shift keying
Hz:	hertz
IDFT:	inverse discrete Fourier transform
IF:	intermediate frequency
I/O:	input/output
ISI:	intersymbol interference
ISO:	international organization for standardization
LAN:	local-area network
LAP:	link-access procedure
LDM:	linear delta modulation
LMS:	least mean square
ln:	natural logarithm
log:	logarithm
LPF:	low-pass filter
MAP:	maximum a posteriori
μs:	microsecond
ML:	maximum likelihood
modem:	modulator–demodulator
MOS:	mean opinion score
MSK:	minimum shift keying
MUX:	multiplexer
NASA:	National Aerospace Administration
NRZ:	nonreturn-to-zero
OOK:	on–off keying
OPSK:	octal-phase-shift keying
OSI:	open systems interconnection
PAM:	pulse-amplitude modulation
PBX:	private branch exchange
PCM:	pulse-code modulation
PG:	processing gain
PLL:	phase-locked loop
PN:	pseudo-noise
PSK:	phase-shift kcying
QAM:	quadrature amplitude modulation
QPSK:	quadriphase shift-keying
RF:	radio frequency
rms:	root mean-square
RS:	Reed–Solomon
RZ:	return-to-zero
s:	second
SDR:	signal-to-distortion ratio
SNR:	signal-to-noise ratio
TCM:	trellis-coded modulation
TDM:	time-division multiplexing
TDMA:	time-division multiple access
VCO:	voltage-controlled oscillator
VLSI:	very large-scale integration
vocoder:	voice coder

REFERENCES AND BIBLIOGRAPHY

M. R. Aaron, "Digital Communications—The silent [R]evolution?", *IEEE Communications Magazine*, Vol. 17, pp. 16–26, January 1979.

J. E. Abate, "Linear and Adaptive Delta Modulation," *Proc. IEEE*, Vol. 55, pp. 298–308, March 1967.

M. Abramowitz and I. A. Stegun, *Handbook of Mathematical Functions* (Dover Publications, 1965).

N. Abramson, *Information Theory and Coding* (McGraw-Hill, 1963).

N. Abramson, "The ALOHA System—Another Alternative for Computer Communications," *AFIPS Conf. Proc.*, Vol. 37, pp. 281–285, 1970.

N. Abramson and F. Kuo (Editors), *Computer Networks* (Prentice-Hall, 1973).

F. Amoroso, "The Bandwidth of Digital Data Signals," *IEEE Communications Magazine*, Vol. 18, pp. 13–24, November 1980.

J. B. Anderson and R. de Buda, "Better Phase-modulation Error Performance Using Trellis Phase Codes," *Electronics Letters*, Vol. 12, pp. 587–588, October 1976.

J. B. Anderson and D. P. Taylor, "A Bandwidth-efficient Class of Signal-space Codes," *IEEE Trans. on Information Theory*, Vol. IT-24, pp. 703–712, November, 1978.

J. B. Anderson, T. Aulin, and C. E. Sundberg, *Digital Phase Modulation* (Plenum, 1986).

R. R. Anderson and J. Salz, "Spectra of Digital FM," *Bell System Tech. J.*, Vol. 44, pp. 1165–1189, July–August 1965.

E. Arthurs and H. Dym, "On the Optimum Detection of Digital Signals in the Presence of White Gaussian Noise—A Geometric Interpretation and a Study of Three Basic Data Transmission Systems," *IEEE Trans. on Communication Systems*, Vol. CS-10, pp. 336–372, December 1962.

V. Aschoff, "The Early History of the Binary Code," *IEEE Communications Magazine*, Vol. 21, pp. 4–10, January 1983.

R. B. Ash, *Information Theory* (Wiley, 1965).

P. Balaban, E. A. Sweedyk, and G. S. Axeling, "Angle Diversity with Two Antennas: Model and Experimental Results," *IEEE International Conference on Communications*, pp. 0846–0847, June 1987.

580

T. C. Bartee (Editor), *Data Communications, Networks, and Systems* (Howard W. Sams, 1985).

T. Bayes, "An Essay Towards Solving a Problem in the Doctrine of Chances," *Phil. Trans.,* Vol. 53, pp. 370–418, 1764.

Bell Telephone Laboratories, *Transmission Systems for Communications* (1970).

J. C. Bellamy, *Digital Telephony* (Wiley, 1982).

W. R. Bennett, "Spectra of Quantized Signals", *Bell System Tech. J.,* Vol. 27, pp. 446–472, July 1948.

W. R. Bennett and J. R. Davey, *Data Transmission* (McGraw-Hill, 1965).

W. R. Bennett, *Introduction to Signal Transmission* (McGraw-Hill, 1970).

S. Benedetto, E. Biglieri, and V. Castellani, *Digital Transmission Theory* (Prentice-Hall, 1987).

N. Benvenuto, et al., "The 32-kb/s ADPCM Coding Standard," *AT&T Technical Journal,* Vol. 65, pp. 12–22, September/October 1986.

T. Berger, *Rate Distortion Theory* (Prentice-Hall, 1971).

E. R. Berlekamp, *Algebraic Coding Theory* (McGraw-Hill, 1968).

E. R. Berlekamp, R. E. Pelle, and S. P. Pope, "The Application of Error Control to Communications," *IEEE Communications Magazine,* Vol. 25, pp. 44–57, April 1987.

D. Bertsekas and R. Gallager, *Data Networks* (Prentice-Hall, 1987).

V. K. Bhargava, D. Haccoun, R. Matyas, and P. Nuspl, *Digital Communications by Satellite* (Wiley, 1981).

V. K. Bhargava, "Forward Error Correction Schemes for Digital Communications," *IEEE Communications Magazine,* Vol. 21, pp. 11–19, January 1983.

J. G. Birdsall, "On Understanding the Matched Filter in the Frequency Domain," *IEEE Trans. on Education,* Vol. E-19, pp. 168–169, November 1976.

H. S. Black, *Modulation Theory* (Van Nostrand, 1953).

U. Black, *Computer Networks: Protocols, Standards, and Interfaces* (Prentice-Hall, 1987).

N. M. Blachman, *Noise and Its Effect in Communication* (McGraw-Hill, 1966).

I. F. Blake, *An Introduction to Applied Probability* (Wiley, 1979).

E. O. Brigham, *The Fast Fourier Transform* (Prentice-Hall, 1974).

L. Brillouin, *Science and Information Theory,* Second Edition (Academic Press, 1962).

J. L. Brown, Jr., "On the Error in Reconstructing a Non-bandlimited Function by Means of the Bandpass Sampling Theorem," *J. Mathematical Analysis and Applications,* Vol. 18, pp. 75–84, 1967.

A. R. Calderbank and J. E. Mazo, "A New Description of Trellis Codes," *IEEE Trans. on Information Theory,* Vol. IT-30, pp. 784–791, November 1984.

A. R. Calderbank and N. J. A. Sloane, "New Trellis Codes based on Lattices and Cosets," *IEEE Trans. Information Theory,* Vol. IT-33, pp. 177–195, March 1987.

A. B. Carlson, *Communication Systems,* Third Edition (McGraw-Hill, 1986).

K. W. Cattermole, *Principles of Pulse-code Modulation* (American Elsevier, 1969).

J. W. Chamberlin, et al., "Design and Field Test of a 256-QAM DIV Modem," *IEEE J. on Selected Areas in Communications,* Vol. SAC-5, pp. 349–356, April 1987.

W. Chou (Editor), *Computer Communications,* Vol. I: *Principles* (Prentice-Hall, 1983).

W. Chou (Editor), *Computer Communications,* Vol. II: *Systems and Applications* (Prentice-Hall, 1985).

G. C. Clark, Jr., and J. B. Cain, *Error-correction Coding for Digital Communications* (Plenum, 1981).

C. E. Cook and H. S. Marsh, "An Introduction to Spread Spectrum," *IEEE Communications Magazine,* Vol. 21, pp. 8–16, March 1983.

J. W. Cooley and J. W. Tukey, "An Algorithm for the Machine Calculation of Complex Fourier Series," *Mathematics of Computation,* Vol. 19, pp. 297–301, April 1965.

G. R. Cooper and C. D. McGillem, *Modern Communications and Spread Spectrum* (McGraw-Hill, 1986).

C. C. Cutler, *Differential Quantization of Communication Signals* (United States Patent, No. 2605361—1952).

W. R. Daumer, "Subjective Evaluation of Several Efficient Speech Coders," *IEEE Trans. on Communications,* Vol. COM-30, pp. 655–662, April 1982.

J. R. Davey, "Modems," *Proc. IEEE,* Vol. 60, pp. 1284–1292, November 1972.

R. M. Davis, "The Data Encryption Standard in Perspective," *IEEE Communications Magazine,* Vol. 16, pp. 5–9, November 1978.

R. de Buda, "Coherent Demodulation of Frequency-shift Keying with Low Deviation Ratio," *IEEE Trans. on Communications,* Vol. COM-20, pp. 429–435, June 1972.

R. de Buda and P. de Buda, "The Sampling Theorem and Aliasing Error," Communications Research Laboratory, McMaster University, Hamilton, Ontario, Internal Report No. 180, 1987.

F. E. De Jager, "Delta Modulation, A Method of PCM Transmission Using a 1-unit Code," *Philips Research Report,* pp. 442–466, December 1952.

R. C. Dixon, *Spread Spectrum Systems,* Second Edition (Wiley, 1984).

M. I. Doelz and E. H. Heald, *Minimum-shift Data Communication System* (U.S. Patent No. 2977417, March 28, 1961).

P. Elias, "Coding for Noisy Channels," *IRE Convention Record,* Part 4, pp. 37–46, March 1955.

D. F. Elliott and K. R. Rao, *Fast Transforms: Algorithms, Analyses, Applications* (Academic Press, 1982).

R. M. Fano, "A Heuristic Discussion of Probabilistic Coding," *IEEE Trans. on Information Theory,* Vol. IT-9, pp. 64–74, 1963.

W. D. Farmer and E. E. Newhall, "An Experimental Distributed Switching System to handle Bursty Computer Traffic," *Proceedings, ACM Symposium on Problems in the Optimization of Data Communications,* pp. 31–34, October 1969.

K. Feher, *Digital Communications: Microwave Applications* (Prentice-Hall, 1981).

R. A. Fisher, "Theory of Statistical Estimation," *Proc. Cambridge Philos. Soc.*, Vol. 22, p. 700, 1925.

J. L. Flanagan, *Speech Analysis, Synthesis and Perception* (Springer-Verlag, 1972).

J. L. Flanagan et al., "Speech Coding," *IEEE Trans. on Communications*, Vol. COM-27, pp. 710–737, April 1979.

W. Fleming, *Functions of Several Variables* (Addison-Wesley, 1965).

G. D. Forney, Jr., *Concatenated Codes* (MIT Press, 1966).

G. D. Forney, Jr., "The Viterbi Algorithm," *Proc. IEEE*, Vol. 61, pp. 268–278, March, 1973.

G. D. Forney, Jr., et al., "Efficient Modulation for Band-limited Channels," *IEEE J. on Selected Areas in Communications*, Vol. SAC-2, pp. 632–647, September 1984.

G. F. Franklin and J. D. Powell, *Digital Control of Dynamic Systems* (Addison-Wesley, 1980).

L. E. Franks, *Signal Theory* (Prentice-Hall, 1969).

S. L. Freeney, "TDM/FDM Translation as an Application of Digital Signal Processing," *IEEE Communications Magazine*, Vol. 18, pp. 5–15, January 1980.

K. E. Fultz and D. B. Penick, "TI Carrier System," *Bell System Tech. J.*, Vol. 44, pp. 1405–1451, September 1965.

R. M. Gagliardi, *Introduction to Communications Engineering* (Wiley, 1978).

R. G. Gallager, *Information Theory and Reliable Communication* (Wiley 1968).

M. Gerla, "Controlling Routes, Traffic Rates, and Buffer Allocation in Packet Networks," *IEEE Communications Magazine*, Vol. 22, pp. 11–23, November 1984.

A. Gersho, "Adaptive Equalization of Highly Dispersive Channels for Data Transmission," *Bell System Tech. J.*, Vol. 48, pp. 55–70, January 1969.

A. Gersho, "Quantization," *IEEE Communications Magazine*, Vol. 15, pp. 20–29, September 1977.

A. Gersho and V. Cuperman, "Vector Quantization: A Pattern-matching Technique for Speech Coding," *IEEE Communications Magazine*, Vol. 21, pp. 15–21, December 1983.

R. D. Gitlin and S. B. Weinstein, "Fractionally Spaced Equalization: An Improved Digital Transversal Equalizer," *Bell System Tech. J.*, Vol. 60, pp. 275–296, February 1981.

S. W. Golomb (Editor), *Digital Communications with Space Applications* (Prentice-Hall, 1964).

P. E. Green, Jr., *Computer Network Architectures and Protocols* (Plenum, 1982).

P. E. Green, Jr., "Computer Communications: Milestones and Prophecies," *IEEE Communications Magazine*, Vol. 22, pp. 49–63, May 1984.

W. D. Gregg, *Analog and Digital Communication* (Wiley, 1977).

S. A. Gronemeyer and A. L. McBride, "MSK and QPSK Modulation," *IEEE Trans. on Communications*, Vol. COM-24, pp. 809–820, August 1976.

D. W. Hagelbarger, "Recurrent Codes: Easily Mechanized, Burst-correcting, Binary Codes," *Bell System Tech. J.*, Vol. 38, pp. 969–984, 1959.

R. W. Hamming, "Error Detecting and Error Correcting Codes," *Bell System Tech. J.*, Vol. 29, pp. 147–160, April 1950.

R. W. Hamming, *Coding and Information Theory* (Prentice-Hall, 1980).

F. J. Harris, "On the Use of Windows for Harmonic Analysis with the Discrete Fourier Transform," *Proc. IEEE*, Vol. 66, pp. 51–83, January 1978.

F. J. Harris, "The Discrete Fourier Transform Applied to Time Domain Signal Processing," *IEEE Communications Magazine*, Vol. 20, May 1982.

R. V. L. Hartley, "Transmission of Information," *Bell System Tech. J.*, Vol. 7, pp. 535–563, 1928.

J. F. Hayes, "The Viterbi Algorithm Applied to Digital Data Transmission," *IEEE Communications Magazine*, Vol. 13, pp. 15–20, March 1975.

J. F. Hayes, *Modeling and Analysis of Computer Communications* (Plenum, 1984).

S. Haykin, *Communication Systems*, Second Edition (Wiley, 1983).

S. Haykin, *Adaptive Filter Theory* (Prentice-Hall, 1986).

S. Haykin, *An Introduction to Analog and Digital Communications* (Wiley, 1989).

F. G. Heath, "Origins of the Binary Code," *Scientific American*, Vol. 227, No. 2, pp. 76–83, August 1972.

C. W. Helstrom, *Statistical Theory of Signal Detection* (Pergamon Press, 1968).

H. H. Henning and J. W. Pan, "D2 Channel Bank: System Aspects," *Bell System Tech. J.*, Vol. 51, pp. 1641–1657, October 1972.

P. S. Henry, "Introduction to Lightwave Transmission," *IEEE Communications Magazine*, Vol. 23, pp. 12–16, May 1985.

J. R. Higgins, "Five Short Stories About the Cardinal Series," *Bulletin of the American Mathematical Society*, Vol. 12, pp. 45–89, 1985.

J. K. Holmes, *Coherent Spread Spectrum Systems* (Wiley, 1982).

H. Holzwarth, "PCM and Its Distortions by Logarithmic Quantization," (in German), *Archiv der elektrischen Übertragung*, pp. 277–285, January 1949.

D. A. Huffman, "A Method for the Construction of Minimum-redundancy Codes," *Proc. IRE*, Vol. 40, pp. 1098–1101, September 1952.

I. M. Jacobs, "Practical Applications of Coding," *IEEE Trans. on Information Theory*, Vol. IT-20, pp. 305–310, May 1974.

N. S. Jayant, "Adaptive Quantization with a One-word Memory," *Bell System Tech. J.*, Vol. 52, pp. 1119–1144, September 1973.

N. S. Jayant, "Digital Coding of Speech Waveforms: PCM, DPCM, and DM Quantizers," *Proc. IEEE*, Vol. 62, pp. 611–632, May 1974.

N. S. Jayant and P. Noll, *Digital Coding of Waveforms* (Prentice-Hall, 1984).

N. S. Jayant, "Coding Speech at Low Bit Rates," *IEEE Spectrum*, Vol. 23, pp. 58–63, August 1986.

A. J. Jerri, "The Shannon Sampling Theorem—Its Various Extensions and Applications: A Tutorial Review," *Proc. IEEE*, Vol. 65, pp. 1565–1596, November 1977.

A. E. Joel, Jr., "The Past 100 Years in Telecommunications Switching," *IEEE Communications Magazine*, Vol. 22, pp. 64–70 and p. 83, May 1984.

P. Kabal and S. Pasupathy, "Partial-response Signaling," *IEEE Trans. on Communications*, Vol. COM-23, pp. 921–934, September 1975.

T. Kailath, "A View of Three Decades of Linear Filtering Theory," *IEEE Trans. on Information Theory*, Vol. IT-20, pp. 146–181, March 1974.

M. Kanefsky, *Communication Techniques for Digital and Analog Signals* (Harper & Row, 1985).

H. Kaneko, "A Unified Formulation of Segment Companding Laws and Synthesis of Codecs and Digital Companders," *Bell System Tech. J.*, Vol. 49, pp. 1555–1588, September 1970.

C. K. Kao and G. A. Hockham, "Dielectric-fiber Surface Waveguides for Optical Frequencies," *Proc. IEE* (London), Vol. 113, pp. 1151–1158, 1966.

C. K. Kao, *Optical Fiber Systems* (McGraw-Hill, 1982).

J. M. Kasson and K. S. Kagan, "PBX Local Area Networks," In *Data Communications, Networks, and Systems*, Chap. 5, edited by T. C. Bartee (H. W. Sams and Co., 1985).

M. Kavehrad and P. J. McLane, "Spread Spectrum for Indoor Digital Radio," *IEEE Communications Magazine*, Vol. 25, pp. 32–40, June 1987.

D. B. Keck, "Fundamentals of Optical Waveguide Fibers," *IEEE Communications Magazine*, Vol. 23, pp. 17–22, May 1985.

D. J. Kostas, "Transition to ISDN—An Overview," *IEEE Communications Magazine*, Vol. 22, pp. 11–17, January 1984.

V. A. Kotel'nikov, *The Theory of Optimum Noise Immenity* (Dover Publications, 1960).

L. G. Kraft, "A Device for Quantizing, Grouping, and Coding Amplitude Modulated Pulses," M. S. Thesis, Electrical Engineering Department, Massachusetts Institute of Technology, March 1949.

E. R. Kretzmer, "Generalization of a Technique for Binary Data Communication," *IEEE Trans. on Communication Technology*, Vol. COM-14, pp. 67–68, February 1966.

S. S. Lam, "Multiple Access Protocols," In *Computer Communications*, Vol. I, Chap. 4, edited by W. Chou (Prentice-Hall, 1983).

B. P. Lathi, *Modern Digital and Analog Communication Systems* (Holt, Rinehart and Winston, 1983).

V. B. Lawrence, J. L. LoCicero, and L. B. Milstein (Editors), *Tutorials in Modern Communications* (Computer Science Press, 1983).

W. C. Y. Lee, *Mobile Communications Engineering* (McGraw-Hill, 1982).

B. M. Leiner, D. L. Nielson, and F. A. Tobagi, "Issues in Packet Radio Network Design," *Proc. IEEE*, Vol. 75, pp. 6–20, January 1987.

A. Lender, "The Duobinary Technique for High-speed Data Transmission," *IEEE Trans. on Communications and Electronics*, Vol. 82, pp. 214–218, May 1963.

A. Lender, "Correlative (Partial Response) Techniques and Applications to Digital Radio Systems," in *Digital Communications: Microwave Applications* edited by K. Feher (Prentice-Hall, 1981).

E. H. Lin, A. J. Giger, and G. D. Alley, "Angle Diversity in Line-of-sight Microwave Paths Using Dual-beam Dish Antennas," *IEEE International Conference on Communications*, pp. 0831–0841, June 1987.

S. Lin and D. J. Costello, Jr., *Error Control Coding: Fundamentals and Applications* (Prentice-Hall, 1983).

S. Lin, D. J. Costello, Jr., and M. J. Miller, "Automatic-Repeat-Request-Error Control Schemes," *IEEE Communications Magazine*, Vol. 22, pp. 5–17, December 1984.

D. A. Linden, "A Discussion of Sampling Theorems," *Proc. IRE*, Vol. 47, pp. 1219–1226, July 1959.

W. C. Lindsey, *Synchronization Systems in Communication and Control* (Prentice-Hall, 1972).

W. C. Lindsey and M. K. Simon, *Telecommunication Systems Engineering* (Prentice-Hall, 1973).

R. W. Lucky, "Automatic Equalization for Digital Communication," *Bell System Tech. J.*, Vol. 44, pp. 547–588, April 1965.

R. W. Lucky, J. Salz, and E. J. Weldon, Jr., *Principles of Data Communication* (McGraw-Hill, 1968).

F. J. MacWilliams and N. J. A. Sloane, *The Theory of Error-correcting Codes* (North-Holland, 1977).

J. Makhoul, S. Roucos, and H. Gish, "Vector Quantization in Speech Coding," *Proc. IEEE*, Vol. 73, pp. 1551–1588, November 1985.

J. D. Markel and A. H. Gray, Jr., *Linear Prediction of Speech* (Springer-Verlag, 1976).

J. L. Massey, "Variable-length Codes and the Fano Metric," *IEEE Trans. on Information Theory*, Vol. IT-18, pp. 196–198, 1972.

J. L. Massey, "Information Theory: The Copernican System of Communications," *IEEE Communications Magazine*, Vol. 22, pp. 26–28, December 1984.

R. J. McEliece, *The Theory of Information and Coding* (Addison Wesley, 1977).

B. McMillan, "Two Inequalities Implied by Unique Decipherability," *IRE Trans. Information Theory*, Vol. IT-2, pp. 115–116, December 1956.

R. M. Metcalf and D. R. Boggs, "Ethernet: Distributed Packet Switching for Local Computer Networks," *Communications of the ACM*, Vol. 19, pp. 395–404, July 1976.

A. M. Michelson and A. H. Levesque, *Error-control Techniques for Digital Communication* (Wiley, 1985).

R. Moreau, *The Computer Comes of Age* (MIT Press, 1984).

R. Morris, "The Data Encryption Standard—Retrospective and Prospects," *IEEE Communications Magazine*, Vol. 16, pp. 11–14, November 1978.

N. E. Nahi, *Estimation Theory and Applications* (Wiley, 1969).

J. Neyman and E. S. Pearson, "On the Problem of the Most Efficient Tests of Statistical Hypotheses," *Phil. Trans. Roy. Soc. London*, Vol. A231, p. 289, 1933.

D. Nielson, "Packet Radio: An Area-coverage Digital Radio Network," In *Computer Communications*, Vol. II, Chap. 12, edited by W. Chou (Prentice-Hall, 1985).

P. Noll, "A Comparative Study of Various Schemes for Speech Encoding," *Bell System Tech. J.*, Vol. 54, pp. 1597–1614, November 1975.

D. O. North, "An Analysis of the Factors which Determine Signal/Noise Discrimination in Pulsed-carrier Systems," *Proc. IEEE*, Vol. 51, pp. 1016–1027, July 1963.

H. Nyquist, "Certain Topics in Telegraph Transmission Theory," *Trans. AIEE*, Vol. 47, pp. 617–644, February 1928.

J. P. Odenwalder, "Error Control," in *Data Communications, Networks, and Systems*, Chap. 10, edited by T. C. Bartee (H. W. Sams and Co., 1985).

B. M. Oliver, J. R. Pierce, and C. E. Shannon, "The Philosophy of PCM," *Proc. IRE*, Vol. 36, pp. 1324–1331, November 1948.

J. K. Omura, "On the Viterbi Decoding Algorithm," *IEEE Trans. on Information Theory*, Vol. IT-15, pp. 177–179, January 1969.

A. V. Oppenheim and R. W. Schafer, *Digital Signal Processing* (Prentice-Hall, 1975).

A. Papoulis, *Probability, Random Variables, and Stochastic Processes* (second edition) (McGraw-Hill, 1984).

S. Pasupathy, "Correlative Coding: A Bandwidth-efficient Coding Scheme," *IEEE Communications Magazine*, Vol. 15, pp. 4–11, July 1977.

S. Pasupathy, "Minimum Shift Keying: A Spectrally Efficient Modulation," *IEEE Communications Magazine*, Vol. 17, pp. 14–22, July 1979.

J. Payton and S. Qureshi, "Trellis Encoding: What It Is and How It affects Data Transmission," *Data Communication*, Vol. 14, pp. 143–152, May 1985.

P. Z. Peebles, Jr., *Digital Communication Systems* (Prentice-Hall, 1987).

J. B. H. Peek, "Communications Aspects of the Compact Disc Digital Audio System," *IEEE Communications Magazine*, Vol. 23, pp. 7–15, February 1985.

S. D. Personick, *Fiber Optics* (Plenum, 1986).

W. W. Peterson and E. J. Weldon, Jr., *Error Correcting Codes*, Second Edition (MIT Press, 1972).

R. L. Pickholtz, D. L. Schilling, and L. B. Milstein, "Theory of Spread-Spectrum Communications—A Tutorial," *IEEE Trans. on Communications*, Vol. COM-30, pp. 855–884, May 1982.

R. L. Pikholtz, "Modems, Multiplexers, and Concentrators," in *Data Communications, Networks and Systems*, Chap. 3, edited by T. C. Bartee (H. W. Sams and Co., 1985).

J. R. Pierce, *Symbols, Signals and Noise* (Harper 1961).

E. C. Posner and R. Stevens, "Deep Space Communication—Past, Present, and Future," *IEEE Communications Magazine*, Vol. 22, pp. 8–21, May 1984.

V. K. Prabhu, "Error Probability Performance of M-ary CPSK Systems with Intersymbol Interference," *IEEE Trans. on Communications*, Vol. COM-21, pp. 97–109, February 1973.

V. K. Prabhu, "Spectral Occupancy of Digital Angle-modulated Signals," *Bell System Tech. J.*, Vol. 55, pp. 429–453, April 1976.

V. K. Prabhu and J. Salz, "On the Performance of Phase-shift Keying Systems," *Bell System Tech. J.*, Vol. 60, pp. 2307–2343, December 1981.

T. Pratt and C. W. Bostian, *Satellite Communications* (Wiley, 1986).

W. K. Pratt, *Digital Image Processing* (Wiley, 1978).

W. L. Pritchard, "The History and Future of Commercial Satellite Communications," *IEEE Communications Magazine*, Vol. 22, pp. 22–37, May 1984.

J. G. Proakis, "Advances in Equalization for Intersymbol Interference," *Advances in Communication Systems*, Vol. 4, pp. 123–198 (Academic Press, 1975).

J. G. Proakis, *Digital Communications* (McGraw-Hill, 1983).

M. P. Pursley, "The Role of Spread Spectrum in Packet Radio Networks," *Proc. IEEE*, Vol. 75, pp. 116–134, January 1987.

S. Qureshi, "Adaptive Equalization," *IEEE Communications Magazine*, Vol. 20, pp. 9–16, March 1982.

S. U. H. Qureshi, "Adaptive Equalization," *Proc. IEEE*, Vol. 73, pp. 1349–1387, September 1985.

L. R. Rabiner and B. Gold, *Theory and Application of Digital Signal Processing* (Prentice-Hall, 1975).

L. R. Rabiner and R. W. Schafer, *Digital Processing of Speech Signals* (Prentice-Hall, 1978).

S. S. Rappaport and D. M. Grieco, "Spread-spectrum Signal Acquisition: Methods and Technology," *IEEE Communications Magazine*, Vol. 22, pp. 6–21, June 1984.

I. S. Reed and G. Solomon, "Polynomial Codes over Certain Finite Fields," *J. SIAM*, Vol. 8, pp. 300–304, June 1960.

A. H. Reeves, "The Past, Present, and Future of PCM," *IEEE Spectrum*, Vol. 12, pp. 58–63, May 1975.

F. Reif, *Statistical Physics* (McGraw-Hill, 1965).

G. Retnadhas, "Satellite Multiple Access Protocols," *IEEE Communications Magazine*, Vol. 18, pp. 16–20, September 1980.

A. W. Rihaczek, *Principles of High-resolution Radar* (McGraw-Hill, 1969).

M. P. Ristenbatt, "Pseudo-random Binary Coded Waveforms," in *Modern Radar*, edited by R. S. Berkowitz, pp. 274–313 (Wiley, 1965).

L. G. Roberts, "The Evolution of Packet Switching," *Proc. IEEE*, Vol. 66, pp. 1307–1313, November 1978.

W. D. Rummler, R. P. Coutts, and M. Liniger, "Multipath Fading Channel Models for Microwave Digital Radio," *IEEE Communications Magazine*, Vol. 24, pp. 30–42, November 1986.

D. J. Sakrison, *Communication Theory: Transmission of Waveforms and Digital Information* (Wiley, 1968).

D. V. Sarwate and M. B. Pursley, "Cross-correlation Properties of Pseudorandom and Related Sequences," *Proc. IEEE*, Vol. 68, pp. 593–619, May 1980.

H. Scheuermann and G. Göklu, "A Comprehensive Survey of Digital Transmultiplexing Methods," *Proc. IEEE*, Vol. 69, pp. 1419–1450, November 1981.

R. A. Scholtz, "The Origins of Spread-spectrum Communications," *IEEE Trans. on Communications*, Vol. COM-30, pp. 822–854, May 1982.

J. S. Schouten, F. E. D. Jager, and J. A. Greefkes, "Delta Modulation, a Modulation System for Telecommunications," *Philips Technical Report*, pp. 237–245, March 1952.

M. Schwartz, W. R. Bennett, and S. Stein, *Communication Systems and Techniques* (McGraw-Hill, 1966).

M. Schwartz, *Information Transmission, Modulation and Noise*, Third Edition (McGraw-Hill, 1980).

M. Schwartz, *Telecommunication Networks* (Addison-Wesley, 1987).

M. I. Schwartz, "Optical Fiber Transmission—From Conception to Prominence in 20 Years," *IEEE Communications Magazine*, Vol. 22, pp. 38–48, May 1984.

M. Shafi, "Statistical Analysis/Simulation of a Three Ray Model for Multipath Fading with Applications in Outage Prediction," *IEEE J. on Selected Areas in Communication*, Vol. SAC-5, pp. 389–401, April 1987.

K. S. Shanmugam, *Digital and Analog Communication Systems* (Wiley, 1979).

C. E. Shannon, "A Mathematical Theory of Communication," *Bell System Tech. J.*, Vol. 27, pp. 379–423 and pp. 623–656, 1948.

C. E. Shannon, "Communication in the Presence of Noise," *Proc. IRE*, Vol. 37, pp. 10–21, January 1949.

C. E. Shannon and W. Weaver, *The Mathematical Theory of Communication* (University of Illinois Press, 1949).

O. Shimbo, R. J. Fang, and M. Celebiler, "Performance of M-ary PSK Systems in Gaussian Noise and Intersymbol Interference," *IEEE Trans. on Information Theory*, Vol. IT-19, pp. 44–58, January 1973.

M. K. Simon, J. K. Omura, R. A. Scholtz, and B. K. Levitt, *Spread Spectrum Communications*, Volumes I, II, and III (Computer Science Press, 1985).

B. Sklar, "A Structured Overview of Digital Communications—A Tutorial Review",

IEEE Communications Magazine, vol. 21, Part I, pp. 4–17, August 1983, and Part II, pp. 6–21, October 1983.

D. Slepian (Editor), *Key Papers in the Development of Information Theory* (IEEE Press, 1974).

N. J. A. Sloane, "The Packing of Spheres", *Scientific American,* vol. 250, pp. 116–125, January 1984.

B. Smith, "Instantaneous Companding of Quantized Signals," *Bell System Tech. J.,* Vol. 36, pp. 653–709, May 1957.

D. R. Smith, *Digital Transmission Systems* (Van Nostrand Reinhold, 1985).

J. J. Spilker, Jr., and D. T. Magill, "The Delay-code Discriminator—An Optimum Tracking Device," *Proc. IRE,* Vol. 49, pp. 1403–1416, September 1961.

J. J. Spilker, Jr., "Delay-code Tracking of Binary Signals," *IEEE Trans. on Space Electronics and Telemetry,* Vol. SET-9, pp. 1–8, March 1963.

J. J. Spilker, Jr., *Digital Communications by Satellite* (Prentice-Hall, 1977).

W. Stallings, *Data and Computer Communications* (Macmillan, 1985).

W. Stallings, *Local Networks,* Second Edition (Macmillan, 1987).

R. Steele, *Delta Modulation Systems* (Halsted Press, 1975).

S. Stein, "Unified Analysis of Certain Coherent and Noncoherent Binary Communication Systems," *IRE Trans, Information Theory,* Vol. IT-10, pp. 43–51, January 1964.

J. J. Stiffler, *Theory of Synchronous Communications* (Prentice-Hall, 1971).

F. L. H. M. Stumpers, "The History, Development, and Future of Telecommunications in Europe," *IEEE Communications Magazine,* Vol. 22, pp. 84–95, May 1984.

E. D. Sunde, "Ideal Binary Pulse Transmission by AM and FM," *Bell System Tech. J.,* Vol. 38, pp. 1357–1426, November 1959.

A. S. Tanenbaum, *Computer Networks* (Prentice-Hall, 1981).

D. P. Taylor and P. R. Hartman, "Telecommunications by Microwave Digital Radio," *IEEE Communications Magazine,* Vol. 24, pp. 11–16, August 1986.

J. B. Thomas, *An Introduction to Statistical Communication Theory* (Wiley, 1969).

J. B. Thomas, *Introduction to Probability* (Springer-Verlag, 1986).

T. M. Thompson, *From Error-correcting Codes Through Sphere Packings to Simple Groups* (The Mathematical Association of America, 1983).

F. A. Tobagi, "Multiaccess Protocols in Packet Communication Systems," *IEEE Trans. on Communications,* Vol. COM-28, pp. 468–488, April 1980.

F. A. Tobagi, "Multiaccess Link Control," *Computer Network Architectures and Protocols,* Chap. 6, edited by P. E. Green, Jr. (Plenum, 1982).

F. A. Tobagi, R. Binder, and B. Leiner, "Packet Radio and Satellite Networks," *IEEE Communications Magazine,* Vol. 22, pp. 24–40, November 1984.

C. D. Tsao, "A Local Area Network Architecture Overview," *IEEE Communications Magazine,* Vol. 22, pp. 7–11, August 1984.

G. L. Turin, "An Introduction to Matched Filters," *IRE Trans. on Information Theory*, Vol. IT-6, pp. 311–329, June 1960.

G. L. Turin, "An Introduction to Digital Matched Filters," *Proc. IEEE*, Vol. 64, pp. 1092–1112, July 1976.

G. L. Turin, "Introduction to Spread-spectrum Antimultipath Techniques and Their Application to Urban Digital Radio," *Proc. IEEE*, Vol. 68, pp. 328–353, March 1980.

G. Ungerboeck, "Channel Coding with Multilevel/phase Signals," *IEEE Trans. Information Theory*, IT-28, pp. 55–67, January 1982.

G. Ungerboeck, "Trellis-coded Modulation with Redundant Signal Sets, Part I: Introduction," *IEEE Communications Magazine*, Vol. 25, pp. 5–11, February 1987.

G. Ungerboeck, "Trellis-coded Modulation with Redundant Signal Sets, Part II: State of the Art," *IEEE Communications Magazine*, Vol. 25, pp. 12–21, February 1987.

H. L. Van Trees, *Detection, Estimation, and Modulation Theory*, Part I (Wiley, 1968).

H. L. Van Trees, *Detection, Estimation, and Modulation Theory*, Part III (Wiley, 1971).

J. H. Van Vleck and D. Middleton, "A Theoretical Comparison of the Visual, Aural and Meter Reception of Pulsed Signals in the Presence of Noise," *J. Appl. Phys.*, Vol. 17, pp. 940–971, November 1946.

A. J. Viterbi, *Principles of Coherent Communication* (McGraw-Hill, 1966).

A. J. Viterbi, "Error Bounds for Convolutional Codes and an Asymptotically Optimum Decoding Algorithm," *IEEE Trans. on Information Theory*, Vol. IT-13, pp. 260–269, 1967.

A. J. Viterbi and J. K. Omura, *Principles of Digital Communication and Coding* (McGraw-Hill, 1979).

A. J. Viterbi, "Spread Spectrum Communications—Myths and Realities," *IEEE Communications Magazine*, Vol. 17, pp. 11–18, May 1979.

S. T. Walker, "Computer and Communications Security," in *Data Communications, Networks, and Systems*, Chap. 8, edited by T. C. Bartee (H. W. Sams, 1985).

G. N. Watson, *A Treatise in the Theory of Bessel Functions*, Second Edition (Cambridge University Press, 1966).

P. Weiss, "An Estimate of the Error Arising from Misapplication of the Sampling Theorem," *Amer. Math. Soc. Notices*, Vol. 10, p. 351, 1963.

A. D. Whalen, *Detection of Signals in Noise* (Academic Press, 1971).

E. T. Whittaker, "On the Functions which are Represented by the Expansions of Interpolation Theory," *Proc. Royal Society*, Edinburgh, Vol. 35, pp. 181–194, 1915.

J. M. Whittaker, "On the Cardinal Function of Interpolation Theory," *Proc. Edinburgh Math. Soc.*, Vol. 1, pp. 41–46, 1929.

J. M. Whittaker, "On the 'Fourier Theory' of the Cardinal Function," *Proc. Edinburgh Math. Soc.*, Vol. 1, pp. 169–176, 1929.

B. Widrow, "Adaptive Filters", in *Aspects of Network and System Theory*, edited by R. F. Kalman and N. Declaris, pp. 563–587 (Holt, Rinehart and Winston, 1971).

B. Widrow and S. D. Stearns, *Adaptive Signal Processing* (Prentice-Hall, 1985).

N. Wiener, *The Extrapolation, Interpolation, and Smoothing of Stationary Time Series, with Engineering Applications* (Wiley, 1949).

R. A. Williams, *Communication Systems Analysis and Design: A Systems Approach* (Prentice-Hall, 1987).

P. M. Woodward, *Probability and Information Theory, with Applications to Radar,* Second Edition (Pergamon Press, 1964).

J. M. Wozencraft, "Sequential Decoding for Reliable Communication," *Tech. Rep.* 325, Research Laboratories for Electronics, MIT, 1957.

J. M. Wozencraft and I. M. Jacobs, *Principles of Communication Engineering* (Wiley, 1965).

A. D. Wyner, "Fundamental Limits in Information Theory," *Proc. IEEE*, Vol. 69, pp. 239–251, February 1981.

J. H. Yuen (Editor), *Deep Space Telecommunications Systems Engineering* (Plenum, 1983).

R. E. Ziemer and R. L. Peterson, *Digital Communications and Spread Spectrum Systems* (Macmillan, 1985).

R. E. Ziemer and W. H. Tranter, *Principles of Communications*, Second Edition, (Houghton Miffin, 1985).

H. Zimmermann, "A Standard Layer Model," in *Computer Network Architectures and Protocols,* Chap. 2, edited by P. E. Green, Jr. (Plenum, 1982).

INDEX